Advances in

VIRUS RESEARCH

VOLUME 60

ADVISORY BOARD

Advances in
VIRUS RESEARCH

Edited by

KARL MARAMOROSCH
Department of Entomology
Rutgers University
New Brunswick, New Jersey

FREDERICK A. MURPHY
School of Veterinary Medicine
University of California, Davis
Davis, California

AARON J. SHATKIN
Center for Advanced Biotechnology and Medicine
Piscataway, New Jersey

VOLUME 60
The Flaviviruses: Pathogenesis and Immunity

Edited by

THOMAS J. CHAMBERS
Department of Molecular Microbiology and Immunology
St. Louis University Health Sciences Center, School of Medicine
St. Louis, Missouri

THOMAS P. MONATH
Acambis, Inc. and Department of Microbiology and Immunology
Harvard School of Public Health
Cambridge, Massachusetts

ELSEVIER
ACADEMIC PRESS

AMSTERDAM • BOSTON • HEIDELBERG • LONDON
NEW YORK • OXFORD • PARIS • SAN DIEGO
SAN FRANCISCO • SINGAPORE • SYDNEY • TOKYO
Academic Press is an imprint of Elsevier

This book is printed on acid-free paper. ♾

Elsevier Academic Press
525 B Street, Suite 1900, San Diego, California 92101-4495, USA
84 Theobald's Road, London WC1X 8RR, UK
http://www.academicpress.com

International Standard Book Number: 0-12-039860-5

PRINTED IN THE UNITED STATES OF AMERICA
03 04 05 06 07 08 9 8 7 6 5 4 3 2 1

CONTENTS

Immune Modulation by Flaviviruses

Nicholas J. C. King, Bimmi Shrestha, and Alison M. Kesson

Mechanisms of Dengue Virus-Induced Cell Death

Marie-Pierre Courageot, Adeline Catteau, and Philippe Desprès

Dynamics of Flavivirus Infection in Mosquitoes

Laura D. Kramer and Gregory D. Ebel

Dynamics of Infection in Tick Vectors and at the Tick-Host Interface

P. A. Nuttall and M. Labuda

Pathogenesis of Flavivirus Encephalitis

Thomas J. Chambers and Michael S. Diamond

Pathogenesis and Pathophysiology of Yellow Fever

Thomas P. Monath and Alan D. T. Barrett

Immunology and Immunopathogenesis of Dengue Disease

Alan L. Rothman

Neutralization and Antibody-Dependent Enhancement of Dengue Viruses

Scott B. Halstead

PREFACE TO THE FLAVIVIRUSES

The Flavivirus family continues to provide great fascination for virologists, immunologists, entomologists, epidemiologists, and scientists in various other disciplines. Research over the past few decades has yielded considerable progress in many of these areas, but there remain a number of challenges surrounding our understanding of the behavior of flaviviruses in natural conditions and in the laboratory. At a time when continued global emergence of flaviviruses calls for the development and improvement of vaccines and antiviral agents, it is appropriate that a broad compendium of knowledge be made available that presents recent conceptual advances and reviews current information on the many different facets of these viruses. Certainly there have been some noteworthy scientific achievements. For instance, the molecular details of virus structure have been greatly advanced as a result of high-resolution analysis of the envelope protein and its organization at the level of the virion particle, which, together with functional studies, have revealed the uniqueness of this viral protein during replication and pathogenesis. The characterization of an increasing number of flavivirus strains at the sequence level has led to an improved taxonomic classification of the genus, enhanced our understanding of evolution, geographic variation, and epidemiology, and stimulated research on variation in viral virulence. Use of molecular clone technology has advanced from basic studies that have identified the functions and properties of viral proteins during RNA replication and virus assembly to the evaluation of candidate virulence determinants, engineering of live attenuated vaccines, and related applications.

Studies on the immunobiology of flaviviruses have led to the realization that these viruses interact with the host immune system in ways that differ from other small RNA viruses. The importance of neutralizing antibody responses for immunity continues to be an area of focus, and the basis for this protection at the epitope-specific level is gradually being dissected. However, there remain enigmatic aspects, such as the wide cross-reactivity elicited by these viruses and the phenomenon of antibody-dependent enhancement, both of which have important implications for pathogenesis and vaccine development, and require better molecular characterization. It is becoming clear that T-cell responses to flavivirus infections also have unusual properties that may contribute to pathogenesis through immunopathologic and/or immune-subverting events. Further characterization of these responses and their relationship to immune protection are avenues of research needed to optimize the use of the increasing range of vaccine modalities that are being pursued.

In conjunction with advances in flavivirus molecular virology and immunology, more and more attention is being directed to investigation of the pathogenesis of flavivirus diseases. Progress in this area has been heralded by the long-awaited identification of the molecular basis for genetic susceptibility of mice to flaviviruses. This will undoubtedly increase interest in the role of innate defenses in these infections and promote research into the genetic basis of flavivirus susceptibility in humans. Together with the use of modern techniques to identify critical target cells of infection, research in this area will expand our understanding of the cellular and molecular basis for flavivirus tropism. In this regard, the cell-surface molecules that interact with these viruses during entry have yet to be fully characterized, but progress continues to be made on this front. It remains somewhat frustrating that suitable animal models for some flavivirus diseases, particularly dengue hemorrhagic fever, are not available. However, data accumulated from human clinical studies are yielding insight into the pathogenesis of this disease, and similar studies with other pathogenic flaviviruses are anticipated in the future.

The interactions between flaviviruses and their arthropod hosts have been the subject of many classical studies that have now progressed to the molecular level as well. There are many secrets to these interactions that must be discovered to understand the process of virus persistence in molecular terms. These will be forthcoming with the use of modern technologies by creative investigators interested in vector biology. The improvement in molecular technologies has had concomitant impact on the ability to conduct molecular epidemiology at the "macro" and "micro" levels. In response to progressive emergence in recent years of dengue, Japanese encephalitis, West Nile, and tick-borne viruses, the application of such technologies for detection and surveillance in arthropod and vertebrate reservoirs has provided insight into the factors that support the global movements of flaviviruses. Yet, there is a tremendous amount of such data concerning virus evolution in the natural environment that is still needed to understand this process and possibly predict future trends. Additional molecular studies of these viruses as they are transmitted among vectors, reservoirs, and humans are needed to further our conceptual understanding of virus emergence.

The development of vaccines for flaviviruses has also benefited greatly from the availability of modern technologies, and new as well as next-generation vaccines for some viruses are on the horizon. As a better understanding of the immune responses to these viruses in the context of disease as well as vaccine-induced protection becomes available, the ability to control the growing worldwide burden of disease from these agents will likely be improved.

Clearly a comprehensive research approach in many scientific disciplines is needed to unravel the complexities of the virus-host interactions that these viruses have had the benefit of manipulating for centuries. In this three-volume edition on the flaviviruses, our goal has been to assemble a base of knowledge that encompasses these complexities, describes technologies that have contributed to this knowledge, and identifies the major problems faced in attempting to further understand the virus-host interactions that result in disease, and in using vaccine strategies for preventing them. We are grateful to the many contributors who have generously assisted in the preparation of this book series. We must also acknowledge that there are many other colleagues who are active in the field whose expertise has not been represented here.

Thomas J. Chambers,
St. Louis, Missouri, 2003

Thomas P. Monath
Cambridge, Massachusetts, 2003

PREFACE TO VOLUME 60
THE FLAVIVIRUSES: PATHOGENESIS AND IMMUNITY

Flaviviruses have a very broad host range, infecting a wide variety of arthropod and vertebrate hosts and exhibiting different tissue tropisms in these organisms. Interactions with human hosts commonly result in symptomatic outcomes that include either febrile illness or such syndromes as hemorrhagic fever and acute encephalitis, reflecting the high neurotropism and viscerotropism common to many members of the family. The molecular basis for these tropisms remains to be fully explained. The E protein is generally acknowledged as a principal factor in this process, although data are accumulating to implicate nonstructural and untranslated regions of the viral genome as well. Despite many studies identifying such determinants, little is known about how mutations in viral proteins and RNA structures affect virulence at the mechanistic level. Further genetic studies of the viruses and their interactions with receptor molecules and intracellular proteins in target cells of the CNS and the periphery in appropriate animal models are needed to understand the pathogenesis of these viruses.

The processes by which flaviviruses cause disease involve degenerative cytopathic effects and apoptosis, in some cases associated with immunopathologic responses. As for the alphaviruses, there is now some evidence to link flavivirus neurovirulence with apoptosis. However, neuronal responses to viral injury are very intricate, and effects of innate immune responses and adaptive responses recruited from the periphery add additional layers of complexity to the pathogenic process. Pathogenesis and immune control of flavivirus infections in the CNS may be similar to those of other neuronotropic viruses, but further research is needed to substantiate this. The propensity of flaviviruses to modulate immune responses through effects on immunoregulatory cell-surface molecules suggests that important differences may well exist. Mechanisms by which flaviviruses that cause hemorrhagic fever elicit cellular damage and death in lymphoid and visceral compartments in the periphery and physiological changes are still poorly understood, and research in this area is limited by lack of readily available animal models that are capable of reproducing the diseases.

Questions that remain for future research include: What is the basis for innate resistance to flaviviruses, and what role does it play in the susceptibility of human hosts to systemic infection? What are the mechanisms of neuroinvasion and what factors control differential susceptibility of neuronal populations to viral injury? What pathogenic cascades lead to flavivirus hemorrhagic fevers, and are they the same or different from those of other viruses capable of causing this syndrome? What role do T lymphocytes, particularly $CD8^+$ T cells, play in virus clearance vs. immunopathologic responses, and how does immunomodulation by these viruses influence the properties of the adaptive immune response? What is the significance of virus persistence for long-term consequences of flavivirus infections? Better understanding of these issues should facilitate the development of approaches to treat or prevent current and emergent flavivirus infections that result in severe encephalitic or hemorrhagic diseases in humans.

Thomas J. Chambers
Thomas P. Monath

ADVANCES IN VIRUS RESEARCH, VOL 60

MOLECULAR DETERMINANTS OF VIRULENCE: THE STRUCTURAL AND FUNCTIONAL BASIS FOR FLAVIVIRUS ATTENUATION

Robert J. Hurrelbrink and Peter C. McMinn

Department of Virology
Telethon Institute for Child Health Research
University of Western Australia
Perth WA 6008, Australia

I. Introduction

In recent years, the molecular and genetic basis for flavivirus attenuation has received increased attention. With the advent of molecular clone technology and reverse genetics, much work has focused on the impact of mutations in structural protein genes, such as the premembrane (prM) and envelope (E) proteins. For the most part, this work has been facilitated by the solution of the X-ray crystallographic structure of the E protein ectodomain from tick-borne encephalitis virus (TBE) (Rey *et al.*, 1995), a structure that is widely regarded as being representative of E proteins throughout the *Flavivirus* genus. Similar information is also emerging regarding the effect of mutations in nonstructural protein genes, such as the viral protease NS3 (for which an X-ray crystallographic structure is also available; Murthy *et al.*, 1999), the viral polymerase NS5, and the 5′- and 3′-untranslated regions (UTRs) of the viral genome.

Many flaviviruses, including dengue virus (DEN), yellow fever virus (YF), and members of the Japanese encephalitis (JE) and TBE

0065-3527/03 $35.00

serocomplexes, cause encephalitis in laboratory rodents after peripheral or intracerebral inoculation, providing an excellent model of arbovirus-mediated encephalitis in humans (Monath, 1986). The ability of a virus to replicate in peripheral tissues, induce viremia, and invade the central nervous system (CNS) of these animals is referred to as neuroinvasiveness, whereas the ability of a virus to infect cells of the CNS, initiate cytopathic infection, and cause encephalitis is referred to as neurovirulence. Mutations scattered throughout the flaviviral genome have the capacity to attenuate either neuroinvasiveness alone or both neuroinvasiveness and neurovirulence, and much of this attenuation appears to be due to a delay in virus spread from the site of infection to the CNS.

This review focuses on molecular determinants of virulence in flaviviruses, with an emphasis on the implications of these and other mutations for protein structure and function. Mutations affecting protein *structure*, such as the formation of biologically relevant salt bridges between amino acids in the three-dimensional structure of the E protein, are addressed. Similarly, mutations that affect protein *function*, such as those occuring in the catalytic His-Asp-Ser triad of the serine protease NS3, are reviewed. The structural and functional implications of mutations in the 3′ and 5′ UTRs are also discussed. In many cases, the effect of these mutations on virulence has not been directly assessed in an animal model; however, the observed reduction in viral replication kinetics seen *in vitro* may be predictive of a delay in virus spread from the site of inoculation *in vivo*. This is likely to attenuate neuroinvasiveness and/or neurovirulence by allowing the host immune system time to clear an infection before it can become established and/or reducing the immunopathological damage done to infected tissues during a productive infection.

II. Morphology and Genome Structure

Flavivirus virions are spherical in shape and have a diameter of between 40 and 50 nm. They contain a positive sense, single-stranded (ss) RNA genome that is surrounded by an electron-dense nucleocapsid (reviewed in Chambers *et al.*, 1990a). The nucleocapsid is ~30 nm in diameter and is predicted to have T = 3 icosahedral symmetry (Kuhn *et al.*, 2002; Stiasny *et al.*, 1996). It is composed of capsid (C) protein, which is further enclosed in a host cell-derived lipid bilayer containing two virally encoded type I membrane proteins: E and M (Brinton, 1991). Intracellular virions contain unprocessed prM protein, which is cleaved to the more mature M protein on release from the cell (reviewed in Lindenbach and Rice, 2001). The E protein is the major component of the viral envelope and forms head-to-tail dimers, which rearrange into trimers after fusion activation (Allison *et al.*, 1995). E dimers are slightly curved and are oriented parallel to the viral membrane, contrasting with most other viral fusion proteins, which assemble to form trimeric spikes (Allison *et al.*, 1999; Rey *et al.*, 1995).

The ssRNA genome of flaviviruses is approximately 11 kb in length and has a type I cap structure at its 5′ terminus (m^7GpppAmp; see Chambers *et al.*, 1990a). RNA from both mosquito-borne and tick-borne flaviviruses lacks a poly(A) tail, contains conserved dinucleotide sequences at the 5′ (AG) and 3′ (CU) termini, and possesses a single open reading frame (ORF) of greater than 10 kb. This ORF is flanked by a 5′ UTR of approximately 100 nucleotides and a 3′ UTR of between 100 and 700 nucleotides (reviewed in Lindenbach and Rice, 2001).

The flavivirus ORF encodes a single polyprotein that is cleaved posttranslationally by viral and cellular proteases into 10 separate proteins. These proteins have a common coding order and include the three structural proteins C, prM, and E, as well as the seven nonstructural proteins NS1, NS2A, NS2B, NS3, NS4A, NS4B, and NS5. The 3′-terminal 100 nucleotides of all flaviviruses, although variable in terms of their primary sequence, are predicted to form a conserved secondary structure that may be important in regulating transcription and/or translation (Blackwell and Brinton, 1997; Shi *et al.*, 1996; Zeng *et al.*, 1998).

III. Molecular Determinants of Virulence

Mutations scattered throughout the flaviviral genome, including mutations in the structural and nonstructural proteins and the 3′ UTR, have been implicated as molecular determinants of virulence. Early work in this field involved comparisons of the genome sequences of high and low virulence strains of a number of viruses. These included comparisons of YF strains Asibi and 17D-204 (Hahn *et al.*, 1987a), JE strains SA-14 and SA-14-14-2 (Nitayaphan *et al.*, 1990), DEN-2 strains 16681 and 16681-PDK53 (Blok *et al.*, 1992), and TBE strains Neudorfl and Hypr (Wallner *et al.*, 1996). Such studies identified a number of mutations, scattered throughout the respective genomes, which potentially contributed to observed differences in phenotype. For example, 67 nucleotide and 31 amino acid differences were identified between YF strains Asibi and 17D-204 (Hahn *et al.*, 1987a), whereas 45 nucleotide and 15 amino acid differences were identified between JE strains SA-14 and SA-14-14-2 (Nitayaphan *et al.*, 1990). Unfortunately, the number and distribution of these mutations made speculation about their individual effects on viral attenuation difficult. Subsequent work on different strains of these viruses has, however, narrowed down the list of mutations that may contribute to attenuation. For example, it appears that 13 of the observed 31 amino acid differences between YF-Asibi and YF-17D-204 are responsible for the differences in phenotype based on cross-comparison with other YF strains, including 17DD and 17D-213 (dos Santos *et al.*, 1995; Jennings *et al.*, 1993).

More specific information about virulence determinants has been obtained by the characterization of viruses derived by a variety of different methods. Such methods include plaque purification of uncloned virus populations, neutralization escape selection with monoclonal antibodies, limited passage in cell culture or mice, and γ-irradiation mutagenesis. More recently, site-directed mutagenesis of viral cDNA and construction of intratypic and heterotypic chimeras using infectious cDNA clones has enabled the identification of a great number of mutations and/or substitutions that affect the replicative ability of recombinant viruses. In many cases, however, the virulence of these viruses has not been assessed *in vivo*, hindering an assessment of the relative contribution of such mutations to virulence. Despite this, reasonable conclusions can be drawn as to the likely impact of these mutations on the attenuation of *in vivo* phenotype based on their structural and functional significance and their likely effect on delaying viral replication and subsequent spread during viral infection.

A. The Dynamics of Virulence: The Relationship between Attenuation and Disease Outcome

After subcutaneous inoculation, encephalitogenic flaviviruses are transported to, and replicate within, regional draining lymph nodes (Albrecht, 1962; Huang and Wong, 1963; Malkova and Frankova, 1959; McMinn *et al.*, 1996a). Studies of DEN virus infection in humans (Wu *et al.*, 2000) and MVE infection in mice (McMinn and Sammels, 1997) indicate that dendritic cells are the primary cellular targets of viral infection, both at the inoculation site and within regional draining lymph nodes. Furthermore, efficient replication in dendritic cells may be an important determinant of neuroinvasiveness. For example, a nonneuroinvasive neutralization escape variant of MVE

(BHv1), altered from serine to isoleucine at position 277 in the E protein (S277I; McMinn *et al.*, 1995) and defective in fusion function (McMinn *et al.*, 1996b), is unable to replicate within lymph node dendritic cells *in vivo*, in contrast to the neuroinvasive parental strain BH3479 (McMinn and Sammels, 1997). Defective replication of BHv1 in lymph node dendritic cells results in a 2-day delay and a 50-fold reduction in viremia, and a 3-day delay in invasion of the CNS, allowing mice to develop an immune response that clears virus from the CNS and leads to full recovery. The free movement of virus to the bloodstream via efferent lymphatics is also critical in determining the neuroinvasiveness phenotype. Surgical disruption of efferent lymphatic pathways causes a significant delay and reduction of TBE (Malkova and Frankova, 1959) and MVE (McMinn and Sammels, unpublished data) viremia and in attenuation of neuroinvasiveness in mice.

Virus entry into the brain typically occurs 4–5 days after peripheral inoculation (Huang and Wong, 1963; Malkova and Frankova, 1959; McMinn *et al.*, 1996a; Monath, 1986). Clinically apparent encephalitis then develops approximately 2 days later (~6–7 days postinoculation), when brain virus titers surpass $\sim 10^7$ plaque-forming units (pfu) per gram and CNS neurons contain large quantities of viral antigen (Albrecht, 1962) or RNA (McMinn *et al.*, 1996a). The mechanism by which encephalitogenic flaviviruses cross the blood–brain barrier remains uncertain. The direct relationship between the magnitude of viremia and the development of encephalitis (Malkova and Frankova, 1959; Reid and Doherty, 1971; Weiner *et al.*, 1970) and the simultaneous appearance of viral antigen at multiple sites within the CNS in some animal models (reviewed in Albrecht, 1962) has suggested that encephalitogenic flaviviruses enter the CNS by direct hematogenous spread. However, there is little direct experimental support for this hypothesis. Flavivirus antigen has rarely been observed in endothelial cells in experimentally infected animals (Albrecht, 1962) or in postmortem studies of fatal human cases (Johnson *et al.*, 1985). Furthermore, flaviviruses replicate poorly (WNV, YF) or not at all (MVE) in mouse brain endothelial cells *in vitro* (Dropulic and Masters, 1990), suggesting that flaviviruses do not cross the blood–brain barrier by replication in cerebral endothelial cells.

The olfactory nerve has long been recognized as a possible pathway for neurotropic virus entry into the CNS (reviewed in Johnson, 1982). Two studies have established the importance of the olfactory nerve in the entry of encephalitogenic flaviviruses into the CNS of laboratory rodents after peripheral inoculation. Monath *et al.* (1983) demonstrated the presence of SLE antigen in the olfactory neuroepithelium of

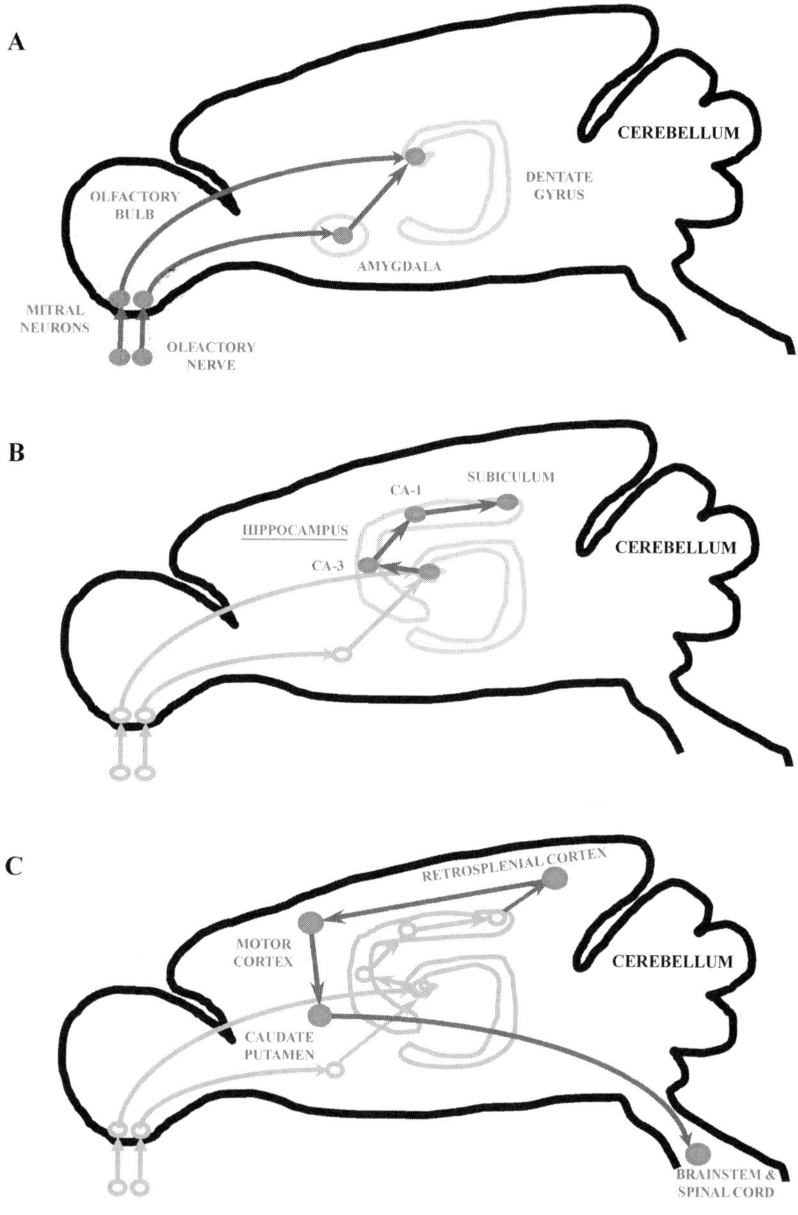

FIG 1. Diagrams of sagittal sections of mouse brain showing the mode of spread of Murray Valley encephalitis virus within the central nervous system of Swiss outbred mice after subcutaneous inoculation of 100 plaque-forming units of wild-type virus into the footpad. Specific structures within the mouse central nervous system were identified

peripherally inoculated hamsters and showed a clear progression of infection from extraneural tissues to sensory neurons. Virus spread via olfactory nerve axons to the mitral neurons of the olfactory bulb was also documented by electron microscopy. The authors argued that the olfactory neuroepithelium became infected during the viremic phase, allowing virus to infect olfactory neurons and thus enter the olfactory bulb of the CNS by retrograde axonal spread.

McMinn *et al.* (1996a) showed that both the virulent MVE strain BH3479 and its attenuated variant BHv1 entered the CNS of peripherally inoculated mice via olfactory bulb mitral neurons. BH3479 entered the CNS at 4 days pi and progressed by cell-to-cell spread along neuronal pathways linking the olfactory bulb, the hippocampal formation, and the major motor centers of the CNS to reach the spinal cord over a period of 4–5 days (McMinn *et al.*, 1996a; McMinn and Sammels, unpublished data). The spread of MVE within the hippocampal formation is of particular interest, as it appears to utilize a well-documented neuronal circuit comprising the dentate gyrus, the CA-3, and CA-1 regions of the hippocampus and the subiculum (Swanson *et al.*, 1987). In contrast, BHv1 entered the CNS at 7 days pi (3 days later than the virulent BH3479), infection was restricted to the olfactory bulb, and virus was cleared from the CNS during recovery (McMinn *et al.*, 1996a). A model for the entry and dissemination of virulent MVE throughout the murine CNS is presented in Fig. 1.

Accumulated evidence suggests that the attenuation of neuroinvasiveness in the BHv1 strain of MVE is caused by a delay in virus spread from the site of inoculation and that this delay is a direct result of the inability of the virus to replicate in dendritic cells. Given the observation that this virus is also defective in fusion, it appears that disruption of virus entry (endosomal fusion) is sufficient to cause a delay in viral replication kinetics. This directly affects the dynamics of neuroinvasiveness by delaying the onset of viremia and the subsequent invasion of the CNS, allowing the immune system sufficient time to clear the infection before the onset of encephalitis.

by reference to a stereotactic atlas of the mouse brain (Franklin and Paxinos, 1997). (A) Virus initially appears in the olfactory bulb at 4 days pi and spreads to the amygdaloid nucleus and the dentate gyrus at 5 days pi. (B) Virus spreads via neuronal connections within the hippocampus between days 5 and 6 pi, including the proximal CA-3 and distal CA-1 regions and the subiculum. (C) Virus spreads from the hippocampus to the retrosplenial cortex, the motor cortex, the caudate putamen, brainstem, and spinal cord between days 6 and 8 pi. (See Color Insert.)

More than likely, similar delays in the spread of virus from the site of inoculation to the CNS are responsible for the observed attenuation of neuroinvasiveness in other flaviviruses. Unfortunately, the neuroinvasiveness of many of these viruses has not been tested, with many researchers preferring to focus on neurovirulence as the sole indicator of pathogenicity. Our understanding of the dynamic nature of the virus–host relationship would undoubtedly be improved if both parameters were measured and would help to establish whether a loss of neurovirulence represents a higher degree of attenuation than a loss of neuroinvasiveness alone. Despite the need for increased work in the field, it is clear that perturbations in the structure and/or function of viral proteins can have serious implications for virulence and that an understanding of the factors contributing to both parameters is required. Evidence suggests that mutations in viral proteins and UTRs can cause host restriction, directly affecting the ability of some flaviviruses to replicate in particular cell types. This has obvious implications for virus spread, especially when the virus is acquired naturally, as is the case when passed from an arthropod vector to a vertebrate host.

B. The Structure and Function of Viral Envelope Proteins

The structural proteins of flaviviruses, including the C, prM, and E proteins, constitute approximately 20–25% of the viral polyprotein and are involved in various aspects of the virus life cycle, such as binding and entry, virion assembly, and viral egress. The C protein is the smallest of the three structural proteins and, in combination with genomic RNA, forms the core of the virus particle. In contrast, the prM and E proteins are integral viral membrane proteins, which together with host-derived lipids form the viral envelope. Due to their obvious importance in viral entry and their direct exposure to the host immune system, molecular determinants of virulence are found predominantly in the prM and E proteins.

The immature prM protein is a precursor of the more mature M protein, and the N-terminal "pr" portion of prM does not form part of the mature virion except in trace amounts. In contrast, the C-terminal "M" component of prM is predominantly hydrophobic and is an integral membrane protein present on the surface of mature extracellular virions. The cleavage of pr from M immediately precedes or is concomitant with the release of virions from infected cells (Murray *et al.*, 1993; Stadler *et al.*, 1997).

It is thought that prM, which forms a heterodimer with E during virus assembly (Wengler and Wengler, 1989), prevents the E protein from undergoing acid-catalyzed conformational change during transport through acidic intracellular compartments (Guirakhoo *et al.*, 1992; Heinz *et al.*, 1994). Shielding of the E protein from exposure to acidic pH prevents the premature conversion of E to a fusion active form, which is a functional requirement for the entry of flaviviruses into new host cells. Such a system is analogous to that seen in alphaviruses, where the p62 protein forms heterodimers with E1 and is subsequently cleaved by furin to form E2 and E3 just prior to viral egress (Lobigs and Garoff, 1990). In both flaviviruses and alphaviruses, inhibition of furin-mediated cleavage of pr from M (or E2 from E3) does not prevent the release of virions from the cell but does reduce their infectivity significantly (Guirakhoo *et al.*, 1991; Salminen *et al.*, 1992). This suggests that, in flaviviruses, prM is directly involved in preventing the premature dimer-to-trimer rearrangement of E required for fusion activation and that the cleavage of pr from M (on release from the cell) primes the E protein for reactivity upon exposure to the acidic conditions of the endosome.

In contrast to the relatively small and potentially unifunctional prM protein, the E protein has been well characterized and constitutes the major envelope protein of the flavivirus virion. It mediates a number of functions, including host–cell receptor binding and membrane fusion (reviewed in Lindenbach and Rice, 2001), and in addition to being the viral hemagglutinin is also the principal target for neutralizing antibodies (Heinz, 1986). Unlike prM, the three-dimensional structure of the E protein has been solved by X-ray crystallography (for TBE). This structure is considered to be representative of the tertiary structure of E for all flaviviruses due to the high amino acid sequence homology observed throughout the genus.

The E protein (illustrated in Fig. 2) consists of three domains. The central domain (domain I) is discontinuous and contains approximately 120 residues spanning 1–51, 137–189, and 285–302 (TBE numbering). It contains a single Asn-linked glycosylation site of the form Asn-Glu-Thr at residues 154–156 on the E_0F_0 loop (according to the nomenclature of Rey *et al.*, 1995). Most, but not all, flaviviruses are glycosylated at a site close to this region; however, the addition of an N-linked glycan does not appear to be required for the function of E (reviewed in Chambers *et al.*, 1990a).

Domain II forms a dimerization domain and has an extended finger-like structure (Rey *et al.*, 1995). Like domain I it is discontinuous and contains approximately 180 residues spanning 52–136 and 190–284. It contains the putative fusion peptide (amino acids 98–113 in the cd loop) and is the only domain containing flavivirus cross-reactive

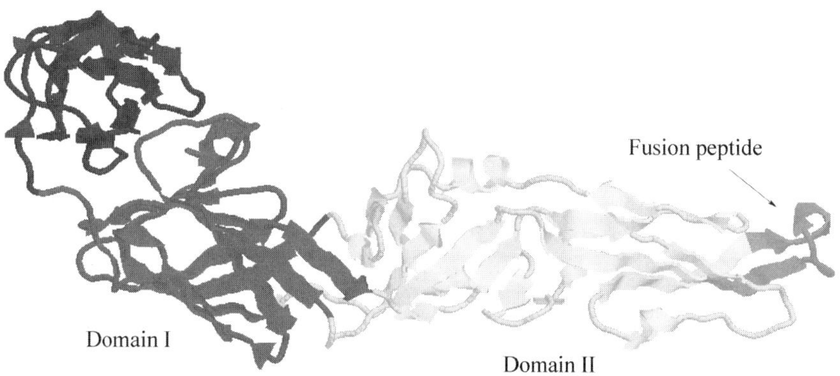

FIG 2. The three-dimensional structure of the TBE E protein ectodomain as determined by X-ray crystallography (Rey *et al.*, 1995). The protein consists of three domains, including the central domain (domain I in red), the dimerization domain (domain II in yellow), and the immunoglobulin-like domain (domain III in blue). The fusion peptide, conserved throughout the *Flavivirus* genus and located on the distal tip of domain II, is highlighted in green. (See Color Insert.)

epitopes. Type- and subtype-specific epitopes are found in domains I and III (Heinz, 1986; Mandl *et al.*, 1989a). Domain II appears to undergo the most extensive rearrangement of the three domains when exposed to acidic pH. The base of this domain has hinge-like characteristics, and according to the model of Rey *et al.* (1995), a hinge motion in this region would project the cd loop toward the target cell membrane for participation in membrane fusion. Other evidence for extensive rearrangement in domain II at acidic pH comes from studies on TBE, where the reactivities of monoclonal antibodies to domain II are reduced markedly by a pH-induced conformational change (Guirakhoo *et al.*, 1989).

Domain III forms the C-terminal end of the solubilized E protein and spans residues 303–395. It has an immunoglobulin-like conformation (Bork *et al.*, 1994) and contains regions thought to be important in receptor binding (Bhardwaj *et al.*, 2001; Chen *et al.*, 1997b; Crill and Roehrig, 2001; Mandl *et al.*, 2000). The region of E immediately following domain III, spanning residues 396–496, does not form part of the ectodomain of the protein. It contains regions important for membrane anchoring, interactions with prM, and pH-induced conformational change. Work by Allison *et al.* (1999) has shown that the stem-anchor region contains two predicted α-helical segments involved both in the trimerization of soluble E protein and in the stabilization of prM–E interactions. Furthermore, two transmembrane segments (TM1 and TM2) act as membrane anchors and/or signal sequences for the translocation of NS1 into the ER lumen (Mandl *et al.*, 1989b; reviewed in Lindenbach and Rice, 2001).

Whereas the three-dimensional structure of the E protein has been solved, little information exists as to the structure of prM or to its organization on the surface of the virion. Work by Ferlenghi *et al.* (2001), involving cryoelectron microscopy of recombinant subviral particles (RSPs) of TBE, has suggested that three copies of the mature M protein are centered on the three-fold axis of symmetry. These M proteins are themselves surrounded by three copies of the E protein, with each copy of the E protein representing one half of a dimer (Fig. 3). In contrast, a more recent reconstruction of the DEN virion by Kuhn *et al.* (2002) suggests that M proteins are located in areas of electron density centered on the holes between E protein dimers. Interestingly, the proposed T = 3 model for virion structure extrapolated by Ferlenghi *et al.* (2001) from the structure of RSPs and the model proposed by Kuhn *et al.* (2002) for virion structure after exposure to low pH are identical in terms of the location of E protein dimers on the particle surface. According to the transition model of Kuhn *et al.* (2002), however, the position of M would remain unchanged after exposure to low pH, contrasting with the position of M around the three-fold axis of symmetry as proposed by Ferlenghi *et al.* (2001). Regardless of the final position of M, however, the position of the protein relative to E provides the potential for it to interact with domains I and II of the E protein. The observation that the presence of prM alters the reactivity of certain mAbs with epitopes in domain II of E supports this hypothesis (Heinz *et al.*, 1994).

C. Molecular Determinants of Virulence in the prM Protein

Relatively few molecular determinants of virulence have been identified in the prM protein. While many studies have included sequence analyses of prM and other structural protein genes in an effort to identify attenuating mutations, the overwhelming majority of these mutations have been found in the E protein. As such, analysis of the prM protein has received little attention and has been limited to studies on the prM furin and glycosylation sites, as well as the prM/E signalase cleavage site.

Early work by Pletnev *et al.* (1993) used an infectious cDNA clone system to introduce mutations into the prM furin and glycosylation

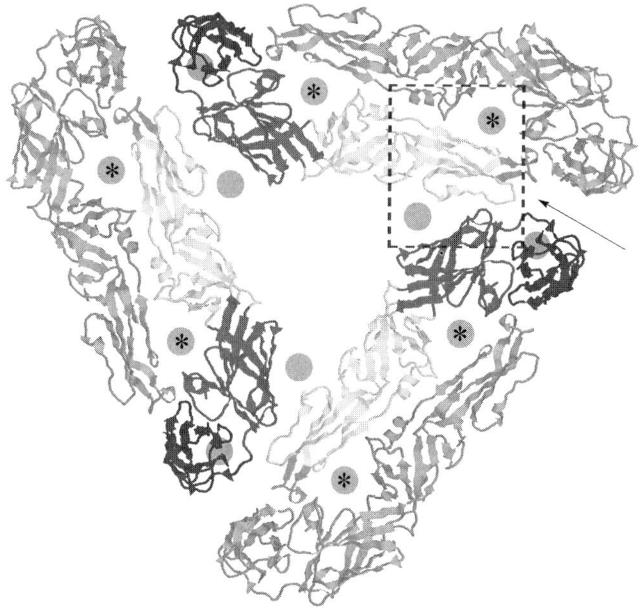

FIG 3. Proposed arrangement of structural proteins on the surface of recombinant subviral particles (RSPs) of TBE (adapted from Ferlenghi *et al.*, 2001, with permission). A similar arrangement has been proposed for DEN virions after exposure to low pH (Kuhn *et al.*, 2002). E proteins are believed to form head-to-tail dimers on the particle surface, which subsequently rearrange into trimers during acid-induced conformational change (fusion) in the endosome. Three E protein monomers (shown in color as in Fig. 2) participate in the formation of a single trimer and surround the threefold axis of symmetry. In RSPs, three copies of the prM protein are also centered on this axis (light blue circles); however, in virions, the prM protein is believed to sit inside E protein dimers rather than outside (light blue circles with asterisk). The boxed area, designated as the prM/E interface (see text), includes the stem-anchor region of E (pink circles), the prM protein, and domains I and II from adjacent E proteins. (See Color Insert.)

sites of a TBE/DEN-4 chimeric virus. This virus (vME) had previously been shown to have low neuroinvasiveness and high neurovirulence, contrasting with parental DEN-4, which had both low neuroinvasiveness and neurovirulence (Pletnev *et al.*, 1992). A Ser-Val substitution at the +1 position of the furin cleavage site of vME (S206V) resulted in a virus [vPreM (V_{206})] that exhibited a smaller plaque phenotype on both LLC-MK_2 and C6/36 cells and a concomitant 10-fold decrease in neurovirulence. Introduction of a second mutation at the −2 position of this cleavage site (A204V) abolished infectivity altogether. In contrast, abolition of the prM glycosylation site by the dual introduction of G143D and N144R mutations resulted in a virus [(vPreM (D_{143}, R_{144})] with increased neurovirulence, having a twentyfold lower LD_{50} than vME.

In a study with similar methodology, Pryor *et al.* (1998) introduced mutations into the −1, −3, and −4 positions of the prM/E signalase site using an infectious cDNA clone of DEN-2. Not surprisingly, virus could not be recovered from RNA transcribed from a clone with a Thr-Leu substitution (T166L) at the −1 position of this site, which strongly inhibited cleavage in a transient expression system. In contrast, introduction of a Ser-Ile substitution at the −3 position (S164I), which reduced but did not abolish cleavage, resulted in the production of virus (501prM-S164I) with a 5- to 10-fold lower titer when compared to clone-derived wild-type virus (MON501). Interestingly, 501prM-S164I was still highly neurovirulent for mice, even though the time to death was extended by up to 7 days at a dose of 10 pfu (Pryor *et al.*, 1998).

Mutations at or near the prM/E cleavage site of Langat (LGT) virus have also been shown to attenuate mouse neurovirulence. Holbrook *et al.* (2001) identified a number of variants that were unable to bind membrane receptor preparations (MRPs) at a range of pH values (pH 5.0, 7.0, and 7.6). In addition, a number of viruses were selected by neutralization escape using a monoclonal antibody that had been shown previously to bind epitopes on both M and E proteins (Iacono-Connors *et al.*, 1996). Specifically, escape variants with mutations at the −2 and −5 positions of the prM/E cleavage site (referred to as amino acid positions M-71 and M-74 and corresponding to V162A and Y165H substitutions, respectively) were shown to have reduced levels of neurovirulence in the range of 13- to 30-fold (Holbrook *et al.*, 2001).

The Holbrook *et al.* (2001) study is unique in that it is the only report of the attenuation of a flavivirus caused by what appears to be a single mutation in the prM protein. The isolation of a mouse MRP variant at pH 5.0, with an identical Y165H mutation to the two neutralization escape variants is of particular interest, as it suggests that the rearrangement of the E protein, which would normally be caused by exposure to low pH, is not sufficient to render these viruses noninfectious. The authors suggest that because these variants are able to bind and infect Vero cells and because variant and wild-type viruses had similar infectivity titers, the Y165H mutation may lie within a part of the M protein that is affected by the pH-induced conformational shift of the E protein (Holbrook *et al.*, 2001). In light of the observed differences in infectivity titers seen in the Pryor *et al.* (1998) study and given that LGT viruses with the Y165H mutation had greater than fivefold lower titres at 3 dpi when compared to wild-type (TP21) virus, it is also possible that the mutation affects cleavage at the prM/E cleavage site.

Other mutations identified in prM that may contribute to differences in virulence are limited to those identified during the comparison of different viral strains. In all cases, the relative contribution of these mutations to observed differences in virulence is difficult to ascertain, as they occur in combination with mutations in other regions of the genome and are outside known cleavage and glycosylation motifs. For example, a substitution at prM L124F (described as position M-35 in Wang *et al.*, 1995) has been implicated in virulence because it is present in eight vaccine strains of YF virus, all of which have been derived from different parental strains, including Asibi and French viscerotropic virus. Unfortunately, other mutations such as NS4B I95M have also been implicated in these viruses, making an assessment of the relative contribution of the prM mutation to virulence difficult. Similar difficulties arise in determining the relative contribution of prM (M38V) and E (D308A) mutations to the *in vivo* attenuation of a clone-derived TP21 strain of LGT (Campbell and Pletnev, 2000), as well as the relative contribution of two prM mutations in TBE strain 263 (Y73H and T149A) in the context of 10 other mutations scattered throughout the genome (Wallner *et al.*, 1996).

D. Molecular Determinants of Virulence in the E Protein

In contrast to the prM protein, many molecular determinants of virulence in the E protein have been identified (Table I). When analyzed in terms of their location on the three-dimensional structure of the protein, patterns begin to emerge that are suggestive of specific effects on structure and function. As shown in Figs. 4 and 5, superimposition of molecular determinants of virulence onto the TBE E protein model reveals five clusters of mutations: four located on the ectodomain of the protein (clusters A–D) and one located in the stem-anchor region (cluster E).

1. Cluster A: The Receptor Binding Region

Mutations in cluster A are found predominantly on the lateral face of domain III, a domain that has been implicated in receptor binding in various studies (Bhardwaj *et al.*, 2001; Chen *et al.*, 1997b; Crill and Roehrig, 2001; Mandl *et al.*, 2000). Interestingly, this region includes an Arg-Gly-Asp (RGD) motif in the FG loop of mosquito-borne

TABLE I

MOLECULAR DETERMINANTS OF NEUROINVASIVENESS AND NEUROVIRULENCE IN THE FLAVIVIRUS E PROTEIN [SEE COLOR INSERT.]

	Cluster/ location[a]	Virus[b]	Substitution	Relative position in TBE	Reference
Neuroinvasiveness	A	JE	G333D	337	Cecilia and Gould (1991)
		JE	S364F	367	Hasegawa *et al.* (1992)
		JE	N367I	368	Hasegawa *et al.* (1992)
		LGT	F333S	333	Pletnev and Men (1998)
		LGT	N389D	389	Campbell and Pletnev (2000)
		LGT	N389D	389	Pletnev and Men (1998)
		LGT	D308A	308	Campbell and Pletnev (2000)
		LIV[b]	D308N	308	Jiang *et al.* (1993)
		LIV[b]	D308N	308	Gao *et al.* (1994)
		LIV[b]	S310P	310	Jiang *et al.* (1993)
		MVE	Various at 390	387	Hurrelbrink *et al.* (2001)
		MVE	Various at 390	387	Lobigs *et al.* (1990)
		MVE	Various at 390	387	Lee and Lobigs (2000)
		TBE	Various at 308–311	308–311	Mandl *et al.* (2000)
		TBE	G368R	368	Holzmann *et al.* (1997)
		TBE	Y384H	384	Holzmann *et al.* (1990)
		WNV	K307E	309	Chambers *et al.* (1998)
		YF	Q303K	311	Jennings *et al.* (1994)
		YF	K326G	334	Chambers and Nickells (2001)
		YF	R380T	387	Chambers and Nickells (2001)
	B	LGT	G285S	285	Pletnev and Men (1998)
		JE	Q52R/K	52	Hasegawa *et al.* (1992)
		JE	I270S	272	Cecilia and Gould (1991)
		MVE	S277I	279	McMinn *et al.* (1995)
		MVE	Various at 277	279	Hurrelbrink *et al.* (2001)
		YF	R52G	52	Chambers and Nickells (2001)
	C	LGT	F119V	119	Campbell and Pletnev (2000)
		LGT	F119V	119	Pletnev and Men (1998)
		TBE	E84K	84	Labuda *et al.* (1994)
		TBE	A123K	123	Holzmann *et al.* (1997)
		WNV	L68P	68	Chambers *et al.* (1998)
	D	TBE	K171E	172	Mandl *et al.* (1989a)
		TBE	D181Y	181	Holzmann *et al.* (1997)
		YF	G155D	159	Jennings *et al.* (1994)
		YF	I173T	177	Chambers and Nickells (2001)
	E	LGT	H438Y	438	Campbell and Pletnev (2000)
		TBE	H496R	496	Gritsun *et al.* (2001)

(continues)

TABLE I (*continued*)

	Cluster/ location[a]	Virus[b]	Substitution	Relative position in TBE	Reference
Neurovirulence	A	DEN-1	V365I	363	dos Santos *et al.* (2000)
		DEN-2	Various at 383–385	387	Hiramatsu *et al.* (1996)
		DEN-2	H390N	391	Sanchez and Ruiz (1996)
		JE	G306E	309	Ni and Barrett (1998)
		YF	V305F	313	Schlesinger *et al.* (1996)
		YF	S305F	313	Ryman *et al.* (1998)
		YF	S325L	333	Ryman *et al.* (1998)
		YF	F380R	387	Schlesinger *et al.* (1996)
	B	DEN-1	M196V	197	dos Santos *et al.* (2000)
		DEN-2	K126E	128	Bray *et al.* (1998)
		DEN-2	K126E	128	Gualano *et al.* (1998)
		DEN-3	A54E	54	Lee *et al.* (1997)
		DEN-3	F277S	284	Lee *et al.* (1997)
		JE	E138K	140	Sumiyoshi *et al.* (1995)
		JE	E138K	140	Chen *et al.* (1996)
		JE	E138K	140	Chambers *et al.* (1999)
		JE	E138K	140	Ni *et al.* (1995)
		JE	E138K	140	Arroyo *et al.* (2001)
		JE	K279M	282	Chambers *et al.* (1999)
		JE	K279M	282	Arroyo *et al.* (2001)
		JE	K279M	282	Monath *et al.* (2002)
		YF	G52R	52	Schlesinger *et al.* (1996)
	X	JE	L107F	107	Chambers *et al.* (1999)
		JE	L107F	107	Arroyo *et al.* (2001)
	Y	TBE	N154L	154	Pletnev *et al.* (1993)
	C	DEN-2	D71E	71	Bray *et al.* (1998)
		YF	A240V	255	Ryman *et al.* (1997)
	D	DEN-3	A18S	18	Lee *et al.* (1997)
		DEN-4	T155I	157	Kawano *et al.* (1993)
		JE	I176V	172	Chambers *et al.* (1999)
		JE	I176V	172	Ni *et al.* (1995)
		YF	I173T	177	Ryman *et al.* (1997)
	E	DEN-1	T405I	406	dos Santos *et al.* (2000)
		DEN-3	E401K	406	Lee *et al.* (1997)
		DEN-3	T403I	408	Lee *et al.* (1997)
		DEN-3	K406E	409	Chen *et al.* (1995)
		DEN-4	F401L	403	Kawano *et al.* (1993)
		LGT	L416A	416	Holbrook *et al.* (2001)
		LGT	H438Y	438	Holbrook *et al.* (2001)
		LGT	V440A	440	Holbrook *et al.* (2001)
		LGT	N473K	473	Holbrook *et al.* (2001)

[a] The relative position of each mutation in the TBE E protein is shown using the same coloring scheme as that used in Fig. 4.

[b] DEN – dengue virus; JE – Japanese encephalitis virus; LGT – Langat virus; LIV – louping ill virus; MVE – Murray Valley encephalitis virus; TBE – tick-borne encephalitis virus; WNV – West Nile virus; YF – yellow fever virus.

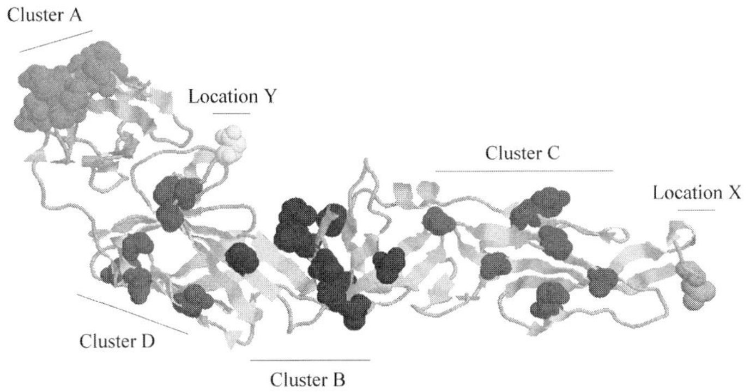

FIG 4. Superimposition of molecular determinants of virulence onto the three-dimensional structure of the TBE E protein ectodomain (Rey *et al.*, 1995). Four clusters of mutations can be seen: cluster A (green), cluster B (blue), cluster C (red), and cluster D (purple), as well as two isolated mutations located within the fusion peptide (location X in orange) and the glycosylation site (location Y in yellow), respectively. (See Color Insert.)

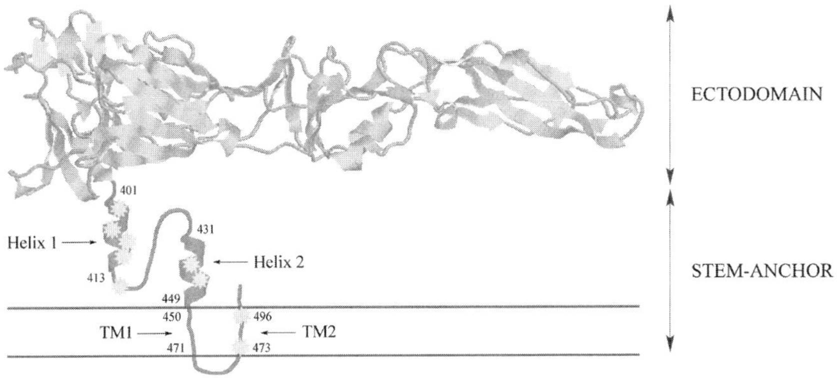

FIG 5. Schematic representation of the stem-anchor region of the TBE E protein (adapted from Allison *et al.*, 1999, with permission). Trypsin cleavage of the ectodomain (shown in gray) occurs at amino acid residue 395. Two α helices and two transmembrane domains (TM; predicted by Stiasny *et al.*, 1996) are indicated. The relative positions of mutations implicated in virulence (cluster E) are shown in blue. (See Color Insert.)

flaviviruses, and there was originally some suggestion that this motif may play a role in the binding of cell surface integrins. A study by Lobigs *et al.* (1990) first implicated this motif when it described the attenuation of MVE after passage in SW13 human adenocarcinoma cells.

This motif has since been the subject of various mutagenesis studies (Hurrelbrink and McMinn 2001; Lee and Lobigs, 2000; Lee and Lobigs, 2002; van der Most *et al.*, 1999), and while attenuation of neuroinvasiveness has been observed, the mechanism for this attenuation appears to relate to the affinity of the virus for glycosaminoglycans on the host–cell surface rather than integrin binding. Interestingly, mutations that increase the net positive charge of the E protein enhance virus binding to heparin, resulting in rapid removal of the virus from the bloodstream of infected mice and a failure of the virus to spread from extraneural sites of replication into the brain (Lee and Lobigs, 2002).

In contrast, the study by van der Most *et al.* (1999) suggested that the nature of attenuating mutations in the FG loop of flaviviruses is to cause incorrect folding of the protein at a posttranslational stage. Intracellular levels of the E protein were considerably lower in YF viruses with RGD motif mutations; however, this instability could be overcome by incubation at a suboptimal temperature of 30 °C. It is possible that in viruses with mutations in the RGD motif and/or the lateral face of domain III, misfolded E protein is retained in the ER and eventually degraded in the same way as that described for misfolded HA of influenza A virus (IAV) (Gething *et al.*, 1986). Such misfolding would cause a delay in virion assembly, possibly reducing viral titers and subsequently impacting the dynamics of neuroinvasiveness.

Even minor changes in the three-dimensional structure of the E protein may affect its ability to form heterodimers with prM or homodimers with E on the virion surface. More importantly, the ability of a receptor binding region to interact with its cognate receptor is likely to be severely disrupted by perturbations in the tertiary structure. Clear evidence for this has been described by Mandl *et al.* (2000), who showed that changes in and around residues involved in the formation of a salt bridge on the lateral face of domain III were sufficient to destabilize the virus (TBE) and cause a reduction in neuroinvasiveness. Interestingly, some of these viruses made compensatory mutations in adjacent residues after passage in baby mouse brain. For example, a D308K substitution, which was predicted to disrupt the formation of a salt bridge between residues 308 and 311, was compensated for by a K308E substitution. The importance of this salt bridge was confirmed by the construction of a double mutant designed to allow the formation of the bridge in an inverted orientation (Mandl *et al.*, 2000).

Whether the formation of a salt bridge (or other tertiary structure) is required for the stability of domain III in mosquito-borne viruses will require elucidation of the three-dimensional structure of the E protein from a mosquito-borne virus. It is interesting to note, however, that disruption of the Asp residue in the RGD motif of MVE by substitution of a polar residue (Tyr) or a nonpolar residue (Asn) causes a loss of neuroinvasiveness in mice, whereas the substitution of an alternate negatively charged residue such as Glu has no effect on virulence (Hurrelbrink and McMinn, 2001). It is possible that this and other residues in the RGD motif form structurally important salt bridges with adjacent residues on the lateral face of domain III or that charge interactions are important for the stability of receptor–ligand and/or E protein complexes.

2. *Cluster B: The Hinge Region*

Cluster B constitutes the second largest cluster of mutations in the E protein after cluster A and is centered on the polar interface linking domains I and II. Based on the three-dimensional structure of the TBE E protein, this region is believed to form part of a molecular hinge and is predicted to be involved in projection of the fusion peptide on the tip of domain II upward for contact with the endosomal membrane during fusion (Rey *et al.*, 1995). The importance of this region is evidenced in studies where mutations in the hinge appear to directly disrupt low pH-mediated fusion. McMinn *et al.* (1995) showed that a S277I substitution in a neutralization escape variant of MVE was directly responsible for a loss of neuroinvasiveness and the concomitant inability of the virus to hemagglutinate red blood cells. This virus was also found to be less efficient in fusion-from-within assays (McMinn *et al.*, 1996b). A similar defect in fusion was caused by an I270S substitution in JE virus (Cecilia and Gould, 1991) and by Ile, Val, Asn, and Pro substitutions at position 277 of MVE (Hurrelbrink and McMinn, 2001). In the latter study, two of the attenuating mutations caused an elevation of the pH threshold for fusion, reminiscent of that seen for mutants of IAV (Qiao *et al.*, 1998; Steinhauer *et al.*, 1996). Such a shift may be indicative of E protein instability and an overall reduction in the activation energy required to trigger the fusion reaction.

While the direct or indirect effect of mutations in the hinge region on membrane fusion has yet to be investigated, the kinetics of the fusion reaction itself have been well characterized for TBE (Corver *et al.*, 2000). Kinetic fusion studies, using well-characterized hinge mutant viruses, would undoubtedly provide valuable information as to the nature of attenuating mutations in this region of the protein. Information on putative defects in the formation and enlargement of fusion pores could also be obtained by examining the rate and extent of fusion of virions with RBC ghosts, which have been labeled with cytoplasmic fluorophores of varying size. Such an approach has been used successfully to analyze fusion pore development in other enveloped viruses such as IAV (Kozerski *et al.*, 2000).

Whether mutations in the hinge cause defects in fusion by perturbing the conformational change of the protein at low pH (the dimer-to-trimer transition) or by disrupting the receptor–ligand complex is unknown. The presence of a receptor has been shown to be dispensable for viral fusion in TBE (Corver *et al.*, 2000), suggesting that ongoing receptor binding in the endosome is not required for fusion to occur. Interestingly, MVE hinge mutants that have been bound to the cell surface at low temperature (4°C) appear to dissociate from their receptor at low pH (McMinn, unpublished data), suggesting that the fusion defect may be directly related to the affinity of the receptor–ligand complex. It may be that mutations in the hinge region cause dissociation of the receptor complex in the endosome, thereby preventing the correct insertion of the fusion peptide into the endosomal membrane. In addition, such mutations may prevent full motion of the hinge region, resulting in inefficient presentation of the fusion peptide at the membrane boundary. Regardless of the precise nature of the attenuation, it is clear that such mutations cause a delay in viral entry, having implications for the spread of the virus after peripheral inoculation and causing an attenuation of neuroinvasiveness.

3. *Locations X and Y: The Fusion Peptide and Glycosylation Site*

Locations X and Y are so named because they are not strictly clusters, containing only single mutations that are known to affect virulence. The mutation at location X is positioned within the fusion peptide at the tip of domain II (cd loop), which is conserved throughout the *Flavivirus* genus. While this hydrophobic region of the protein has long been considered to harbor residues involved in the initiation of fusion between viral and cellular membranes, direct evidence for its involvement was established only recently. Allison *et al.* (2001) confirmed the role of this region by constructing recombinant subviral particles (RSPs) of TBE with substitutions at position 107 of the E protein. Hydrophilic amino acid substitutions at this position strongly impaired or abolished fusion activity, as evidenced by their slower fusion kinetics and lack of interaction with liposomes in coflotation assays. Interestingly, RSPs with a Phe substitution at this position retained a significant degree of fusion activity (Allison *et al.*, 2001). Such an observation is consistent with chimeric JE/YF viruses containing an identical L107F substitution, which are viable but are significantly attenuated in mice (Arroyo *et al.*, 2001; Chambers *et al.*, 1999). It is likely that the mutations outlined in the Allison *et al.*

(2001) study would also cause an attenuation of virulence; however, these mutations are yet to be reverse engineered into an infectious virus.

Location Y represents the conserved glycosylation site of flaviviruses. Mutations at this site (N154L) have been shown to cause a reduction in the neurovirulence of TBE/DEN-4 chimeric viruses in mice (Pletnev *et al.*, 1993). Similarly, selection of DEN-2 fusion mutants by repeated exposure to acidic pH or ammonium chloride has been shown to induce mutations at the glycosylation site (N153Y), concomitant with an increase in the optimal pH for fusion (Guirakhoo *et al.*, 1993). This increase in optimal pH may be linked to the proposed interaction of the E_0F_0 loop (containing the Asn-linked glycosylation site) with the buried fusion peptide (cd loop) on an adjacent E protein monomer. It is possible that the N-linked glycan stabilizes the dimer, thereby preventing premature triggering of the dimer-to-trimer transition.

4. *Clusters C, D, and E: The prM/E Interface*

Mutations in clusters C, D, and E have been implicated as potential determinants of virulence in a number of flaviviruses, including DEN types 1 to 4 (Bray *et al.*, 1998; Chen *et al.*, 1995; dos Santos *et al.*, 2000; Kawano *et al.*, 1993; Lee *et al.*, 1997), YF (Chambers and Nickells, 2001; Jennings *et al.*, 1994; Ryman *et al.*, 1997), WNV (Chambers *et al.*, 1998), LGT (Campbell and Pletnev, 2000; Holbrook *et al.*, 2001; Pletnev and Men, 1998), and TBE (Gritsun *et al.*, 2001; Holzmann *et al.*, 1997; Labuda *et al.*, 1994; Mandl *et al.*, 1989b) (see Table I).

Mutations in cluster C occur toward the distal end of domain II in β strands b, d, and j, and in the bc loop, and are located relatively close to a region of the protein believed to be involved in trimer contacts with the proximal stem-anchor of a neighboring monomer (Ferlenghi *et al.*, 2001; see Fig. 3). In addition, their distal location on domain II places them close to the predicted fusion peptide at the tip of this domain. On examination of the relative positions of cluster C mutations on the three-dimensional structure of the protein, it appears that most are located on the exposed upper surface rather than on the basal or lateral faces where contacts with the stem-anchor are likely to occur. As such, it is likely that mutations in cluster C directly disrupt lateral interactions with prM and/or other E protein monomers. Alternatively, they may alter the conformation of the distal tip of domain II and perturb presentation of the fusion peptide during fusion.

The proposed packing model for flaviviral glycoproteins on the surface of RSPs and virions places prM at the junction of adjacent E protein monomers (the prM/E interface; see Fig. 3). In either arrangement, prM could potentially form contacts with domain I of one E protein monomer and domain II of another. Many of the mutations along the extended finger of domain II in cluster C are directly adjacent to the proposed position(s) for prM, as are some of the mutations in cluster D.

Interestingly, most of the mutations outside the ectodomain (in cluster E) are located in the predicted α-helical domains of the stem-anchor region (see Fig. 5), and these domains have been implicated in both the trimerization of soluble E protein and the stabilization of prM–E interactions (Allison *et al.*, 1999). While the stem-anchor region has been shown to be dispensible for the trimerization of E in the presence of cell membranes (Stiasny *et al.*, 2002), it is possible that mutations in clusters C–E directly impact on the stability of interactions at the prM/E interface (specifically in the ectodomain of E, the stem-anchor region of E, and prM) and that these mutations may disrupt the trimerization of E and the stability of the glycoprotein network on the virion surface.

E. The Structure and Function of Nonstructural Proteins

The nonstructural proteins of flaviviruses include NS1, NS2A, NS2B, NS3, NS4A, NS4B, and NS5 and together constitute approximately 75% of the viral ORF. These proteins perform a range of functions in the replication of the virus, including posttranslational processing of the polyprotein, formation of a replicase complex (RC) for the production of progeny RNA, and possibly assembly and/or release of infectious particles from host cells.

Translation of incoming viral RNA occurs in association with the rough endoplasmic reticulum (ER) and generally begins at the first AUG codon in the long ORF (Chambers *et al.*, 1990a; Crawford and Wright, 1987). This polyprotein precursor is rapidly co- and posttranslationally cleaved by host and viral proteases, both in the lumen of the ER and in the cytosolic compartment. The virally encoded protease, composed of a complex between NS2B and NS3, is required for cleavages at highly conserved dibasic sites at the N-terminal ends of NS2B, NS3, NS4A, and NS5 (Chambers *et al.*, 1991; Ryan *et al.*, 1998), as well as at the C-terminal ends of anchored-C and NS4A (Amberg *et al.*, 1994; Lin *et al.*, 1993; Nowak *et al.*, 1989). A conserved central domain of NS2B serves as an obligatory cofactor for the enzymatic activity of the NS3 protein (Falgout *et al.*, 1993; Yusof *et al.*, 2000); however, the protease function of the NS2B–NS3 complex is mediated by the N-terminal one third of NS3 (Chambers *et al.*, 1990b). This region contains a His-Asp-Ser catalytic triad typical of serine proteases

(Murthy *et al.*, 1999). The NS2B–NS3 complex also serves as a helicase and an RNA-dependent nucleoside triphosphatase (NTPase) during RNA replication, the motifs for which are located in the C-terminal two-thirds of the NS3 protein (Ryan *et al.*, 1998).

Based on numerous studies with KUN, the precursor RC is proposed to contain NS2A, NS3, and NS5, which together bind to a 3′ RNA stem–loop structure located in the terminal 100 nucleotides of the genome (Khromykh *et al.*, 1999). NS2A within this precursor RC is proposed to bind to membrane-bound NS4A, which is itself bound by hydrophilic extensions in the lumen (between transmembrane domains) to dimeric NS1 to form the complete RC (Khromykh *et al.*, 1999; Westaway *et al.*, 1997). The NS2A protein has also been implicated in the assembly and/or release of infectious particles from infected cells (Kümmerer and Rice, 2002). The NS5 protein, which contains RNA-dependent RNA polymerase (RDRP), *S*-adenosylmethionine methyltransferase (SAM), and importin-β binding motifs (Bartholomeusz and Thompson, 1999; Johansson *et al.*, 2001; Koonin, 1993), is believed to be responsible for cytoplasmic RNA replication by the RC.

F. Molecular Determinants of Virulence in Nonstructural Proteins

Numerous molecular determinants affecting the replicative ability of flaviviruses have been identified in nonstructural proteins, many of which are located either in the viral protease NS2B–NS3 or in other components of the RC. In particular, mutations in NS2B and NS3 have been shown to directly affect protease activity, as well as the physical interactions of these two proteins. Similarly, mutations in the RC have been shown to affect the helicase activity of NS3 and the RDRP and/or methyltransferase activities of NS5, as well as the physical interactions of NS1, NS3, NS4A, and NS5 with other components of the RC. Unfortunately, very few studies have analyzed the contribution of such mutations to attenuation of virulence *in vivo*. Nevertheless, the marked effect of such mutations on *in vitro* phenotype, such as a reduction in plaque size, increased temperature sensitivity, and delayed replication kinetics, will undoubtedly have implications for virulence, which may manifest as a reduction in neuroinvasiveness and/or neurovirulence *in vivo*.

1. The Viral Protease: NS2B–NS3

The N-terminal 180 amino acids of NS3 mediate the protease activity of the NS2B–NS3 protease. Four conserved regions (C1–C4; Fig. 6A) have been identified in this region of the protein based on the amino

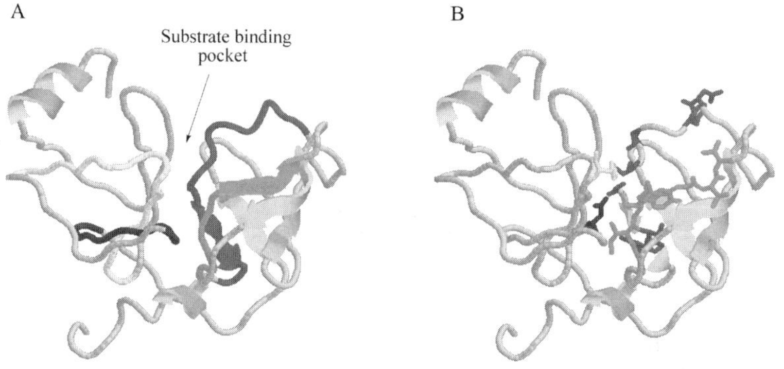

FIG 6. Three-dimensional structure of the NS3 protease domain from DEN-2 (Murthy *et al.*, 1999). Conserved regions C1–C4 (A), based on amino acid sequence alignment of NS3 with flaviviral and cellular proteases, are highlighted in yellow, blue, red, and green, respectively (Bazan and Fletterick, 1989; Murthy *et al.*, 1999). Amino acid substitutions in and around the substrate binding pocket affect replication and/or virulence (B), and are shown as stick representations using the same color scheme as A. (See Color Insert.)

acid sequence alignment of flaviviral and cellular proteases. These include residues flanking components of the His-Asp-Ser catalytic triad [His (C1), Asp (C2), and Ser (C3)], as well as others predicted to be involved in substrate binding specificity (C3 and C4) (Bazan and Fletterick, 1989; Gorbalenya *et al.*, 1989).

Mutations likely to affect virulence, based on their observed effects on *in vitro* phenotype, can be mapped onto the known three-dimensional structure of the protease component of the NS3 protein (Murthy *et al.*, 1999). As described for the E protein earlier, such modeling can provide insight into likely effects on function and the implications this may have for virulence. As one would expect, numerous mutations in and around the substrate binding pocket appear to influence protein function and viral replication (Fig. 6B). In particular, mutations in the His-Asp-Ser catalytic triad have been shown to severely inhibit the protease activity of NS3 and have inhibited subsequent viral replication *in vitro* when introduced into an infectious cDNA clone. In addition, mutations in the NS2B cofactor or in the NS2B–NS3 autocatalytic cleavage site also appear to reduce replicative ability *in vitro*.

Chambers *et al.* (1990b) showed that mutations in the His-Asp-Ser catalytic triad of NS3 from YF either completely (H53A, D77N, D77A, S138A) or severely (S138C) impaired the protease activity of the protein in a rabbit reticulocyte lysate system. Furthermore, introduction of the S138A and S138C mutations into viral RNA derived from an infectious cDNA clone rendered the transcripts noninfectious after transfection into BHK-21 cells (Chambers *et al.*, 1990b). A similar attenuation of protease activity has been observed after mutation of the His-Asp-Ser catalytic triad in WNV (H51A; Wengler *et al.*, 1991) and TBE (S138X; Pugachev *et al.*, 1993), as well as in a panel of 46 mutants of DEN-2, which included mutations in the C3 region similar to those outlined in the Chambers *et al.* (1990b) study (Valle and Falgout, 1998).

Interestingly, other mutations in the Valle and Falgout (1998) study, which were located outside the His-Asp-Ser catalytic triad but within the putative substrate binding pocket, had varying effects on protease activity. Mutations designed specifically to disrupt protease activity, based on known functional residues and motifs in other serine proteases, such as trypsin and chymotrypsin, completely abolished (G148A, L149A/R, Y150A/V/H, G151A, G153A/V), severely disrupted (D129K/R/L, F130A/S, G133A, G136A, N152A/Q), slightly disrupted (D129E/S/A, F130Y, I139L/A, I140A, G144P, L149I, Y150F, V154A, V155A), or had no effect (S131P/C, I140L, D141E/A, K142A/N, K143A/N, G144A, R184A) on the protease function of NS3.

Mutations outside the substrate binding pocket of the NS3 protease have also been shown to affect viral replication. Matusan *et al.* (2001a) used clustered charged-to-alanine mutagenesis to ascertain the effect of mutations in different regions of the NS3 protein. Five clusters of up to five amino acid residues (at least three of which were charged) were chosen for mutagenesis and subsequently tested in transient expression and/or infectious cDNA clone systems. Not all of the tested mutations caused a reduction in protease activity when expressed transiently in COS cells; however, all of the recombinant viruses produced had a smaller plaque phenotype and/or slower growth kinetics in cell culture (Matusan *et al.*, 2001a). Computer modeling of the NS2B–NS3 complex suggested that some of the mutations might have affected the stability of the complex, whereas others may have disrupted the formation of functionally relevant salt bridges and/or perturbed substrate binding specificity.

In addition to direct inhibition of the NS3 protease, the requirement for formation of an NS2B–NS3 complex in protease function has been clearly illustrated in YF. Chambers *et al.* (1993) showed that mutations in the conserved central domain of NS2B eliminate its ability to associate with NS3 as well as its *trans*-cleavage activity, whereas mutations in flanking hydrophobic domains have little effect. In contrast, all of the mutations in the Chambers *et al.* (1993) study were shown to be deleterious for viral replication, as evidenced by a complete lack of infectivity of transcripts in an infectious cDNA clone system. This suggests that mutations in the conserved central domain of NS2B directly affect its ability to form a stable complex with NS3. Conversely, mutations outside this domain may disrupt the ability of the NS2B–NS3 complex to oligomerize efficiently with other members of the replicase complex. It is possible that mutations in NS3, which are solvent exposed and are located outside known functional motifs, may also be involved in the attenuation of the SA14 vaccine strain of JE (A105G; Ni *et al.*, 1995), a temperature-sensitive vaccine strain of DEN-2 (Blaney *et al.*, 2001), and a naturally attenuated strain of JE isolated from a paddy-free islet in Taiwan (F109I, E122K; Chiou and Chen, 2001).

In addition to mutations within the NS2B and NS3 proteins, mutations in the autocatalytic cleavage site between the two proteins have also been shown to disrupt viral replication. Nestorowicz *et al.* (1994) and Chambers *et al.* (1995) introduced site-specific changes in the −4 through +1 positions of the NS2A–NS2B and NS2B–NS3 cleavage sites in YF and analyzed their effect on cell-free translation, cellular expression, and viral phenotype *in vitro*. Substitutions at the −3 and −4 positions had little to no effect on the cleavage efficiency of a NS2B–$NS3_{181}$ and NS2A-$NS3_{181}$ constructs when expressed in rabbit reticulocyte lysate or transfected SW-13 cells, whereas other conservative and nonconservative substitutions at the −2 and −1 positions either reduced or completely abolished cleavage activity. RNA transcripts containing mutations that abolished cleavage activity were also found to be noninfectious, whereas those containing mutations that reduced cleavage efficiency had lower specific infectivities and yielded viruses with smaller plaque phenotypes (Chambers *et al.*, 1995; Nestorowicz *et al.*, 1994). As such, mutations that directly or indirectly affect the protease function of the NS2B–NS3 complex, or the autocatalytic cleavage of these two proteins, can reduce replicative efficiency *in vitro*. Although yet to be tested, it is likely that such mutations will also have an impact on neuroinvasiveness and neurovirulence *in vivo*, and this is an area of research requiring attention.

2. *The Replicase Complex: NS1, NS2A, NS3, NS4A, and NS5*

The replicase complex, like most other cellular and viral replicases, is composed of a number of proteins that together perform all of the functions required for transcription. This includes stabilization and sequestration of the RC to intracellular membranes, as well as the unwinding of double-stranded templates, primer extension, and the addition of methylated caps to RNA transcripts. In flaviviruses, the helicase function is performed by the NS3 protein, whereas the polymerase and methyltransferase activities are performed by the NS5 protein. Both of these proteins form a complex on the cytosolic side of the ER, and their binding to the 3′ stem–loop of RNA is proposed to be mediated by the NS2A protein. NS2A also appears to serve as a linker, binding the NS3–NS5–RNA complex to ER membranes via its hydrophobic interaction with membrane-bound NS4A. The hydrophilic regions of NS4A also bind NS1 in the lumen of the ER to form a complete RC (see Khromykh *et al.*, 1999).

In a number of studies, mutations in the glycosylation sites of NS1 have been shown to affect dimerization and subsequently impact virulence. Using transient expression of COS cells, early studies by Pryor and Wright (1993, 1994) established the importance of the second glycosylation site in the stability of DEN-2 NS1 dimers *in vitro*. Subsequent work by this group showed that recombinant DEN-2 viruses with identical mutations had reduced neurovirulence in mice, even though the mutation had no effect on the dimerization of NS1 *in vivo* (Pryor *et al.*, 1998). Attenuation of neurovirulence has also been observed in recombinant chimeric TBE/DEN-4 (Pletnev *et al.*, 1993) and YF viruses (Muylaert *et al.*, 1996) with similar mutations. Viruses in the Muylaert *et al.* (1996) study also had a smaller plaque phenotype, a decreased virus yield, and were impaired in their ability to secrete NS1 into the culture medium.

Interestingly, work by Hall *et al.* (1999) showed that a P250L mutation in KUN NS1 is sufficient to prevent dimerization of the protein but does not affect the secretion of monomers into the culture medium. In addition, even though virus replication was delayed early in infection and the neuroinvasiveness of the virus was reduced 10-fold in weanling mice, viral titers *in vitro* were similar at 30–48 hpi. The finding that this virus was able to replicate efficiently, albeit more slowly in cell culture, while expressing NS1 in monomeric form, suggests that dimerization of the protein is not an absolute requirement for its function.

It is possible that mutations affecting the dimerization of NS1 cause instability in the RC, resulting in reduced replication kinetics and an attenuation of virulence. Such a requirement for efficient interaction of RC components is illustrated in work by Lindenbach and Rice (1999), who showed that the NS1 gene from DEN-2 was unable to functionally substitute for the corresponding gene in YF. Subsequent screening for replication-competent mutants yielded a virus with a N42Y substitution in the NS4A gene, which allowed functional substitution of the NS1 gene from DEN-2. This provides evidence for a direct interaction between NS1 and NS4A and illustrates the importance of RC stability in virus replication. A similar disruption of RC components has been suggested by Matusan *et al.* (2001b), who showed that viruses with mutations in the helicase region of NS3 (outside known functional motifs) were temperature sensitive and had a reduced plaque size. The authors suggested that these mutations (charged-to-alanine) may have affected protein–protein or RNA–protein interactions by destabilizing hydrogen-bonding and electrostatic interactions at the protein surface, thereby disrupting the stability of the RC (Matusan *et al.*, 2001b). Given that this region of NS3 is proposed to interact directly with NS5 (Johansson *et al.*, 2001), this would seem a reasonable conclusion. A L435S mutation, which appears to be located close to the RNA binding region of the NS3 helicase (based on homology with the known three-dimensional structure of the NS3 helicase from hepatitis C virus), may also play a role in the increased neurovirulence of a mouse brain-passaged strain of DEN-1 (dos Santos *et al.*, 2000). Furthermore, paired charged-to-alanine mutagenesis of NS5 has shown that mutations outside known functional motifs can reduce the replicative ability of DEN-4 *in vivo* after intracerebral inoculation (i.c.) inoculation into suckling mice (Hanley *et al.*, 2002).

Aside from the formation of a stable complex, mutations in RC components may directly affect the enzymatic activities of RC proteins and therefore have an impact on virulence. For example, mutations in conserved NTPase and helicase motifs, engineered into an enzymatically active but truncated form of DEN-2 NS3 produced in *Escherichia coli*, have been shown to reduce or abolish RNA helicase and/or NTPase activity *in vitro* (Matusan *et al.*, 2001b). Such mutations also reduce or abolish the infectivity of RNA transcripts produced from an infectious cDNA clone. Mutations in the methyltransferase and/or RNA polymerase motifs of NS5 have also been shown to affect the virulence of YF vaccines passaged in mouse brain or Vero cells (Holbrook *et al.*, 2000; Xie *et al.*, 1998) and to abolish the infectivity of clone-derived KUN virus (Khromykh *et al.*, 1998).

G. Untranslated Regions of the Genome

The 5′- and 3′-untranslated regions constitute between 4 and 7% of the flaviviral genome. Despite little homology at the nucleotide level, regions of conserved complementarity and predicted secondary structure are found in all flaviviruses and are believed to serve as *cis*-acting elements during RNA transcription. Mosquito-borne viruses in particular contain between one and three consensus sequences (CS) in the 3′ UTR, which are complementary to conserved regions in the 5′ UTR and have therefore been implicated in the cyclization of the genome (Hahn *et al.*, 1987b). Furthermore, putative stem–loop (SL) and/or pseudoknot structures have been predicted at the extreme 5′ and 3′ termini of all flaviviruses and are considered obligatory for the replication of positive- and negative-sense RNAs, respectively (Hahn *et al.*, 1987b; Proutski *et al.*, 1997; Rauscher *et al.*, 1997; Shi *et al.*, 1996). This function may require the involvement of cellular factors, such as elongation factor 1α, Mov34, TIA-1, and TIAR, which in addition to NS3 and NS5 (Chen *et al.*, 1997a) have been shown to interact with the 3′ SL of WNV (Blackwell and Brinton, 1997; Li *et al.*, 2002) and JE (Ta and Vrati, 2000).

The requirement for distal interactions between 5′ and 3′ termini during replication has been confirmed in studies with DEN-2, where self-primed RNA synthesis from the authentic 3′ end of template RNA was shown to be dependent on the presence of the 5′-terminal region, provided either in *cis* or in *trans* (You and Padmanabhan, 1999). UV cross-linking and site-directed mutagenesis experiments have further confirmed a direct interaction between cyclization sequences in the 5′ and 3′ termini of DEN-2 (You *et al.*, 2001) and KUN (Khromykh *et al.*, 2001), with the latter study showing that the base-pairing of conserved sequences, rather than the nucleotide sequence *per se*, was essential for RNA replication.

Early studies with DEN and TBE showed the effect of mutations in the 5′ and 3′ UTRs on replication. Eleven deletion mutants of DEN-4, reverse engineered to disrupt the formation of the predicted SL in the 5′ UTR, were shown to be either noninfectious or growth restricted *in vitro* (Cahour *et al.*, 1995). More specifically, mutations in the long stem region were found to be lethal, whereas those in the shorter stem or loop regions caused a reduction in replicative ability. Furthermore, deletion of nucleotides 82–87 resulted in a virus unable to replicate in mosquito cells, suggesting host restriction and the involvement of different cellular factors in diverse cell types (Cahour *et al.*, 1995).

Similarly, mutations in the 3′ UTR have also been shown to affect virulence directly. Deletion of the CS1 region, which is the conserved consensus region most proximal to the 3′ terminus, was found to be lethal for DEN-4 (Men *et al.*, 1996), whereas deletion of all or part of the CS2A and/or CS2B regions of DEN-2 (Durbin *et al.*, 2001) or DEN-4 (Men *et al.*, 1996) yielded viruses with smaller plaque phenotypes, delayed viral replication kinetics, and delayed viremia in monkeys. More recently, mutations introduced into the 3′ SL of DEN-1 have been shown to cause host-range restriction and an attenuation of neurovirulence in rhesus macaques (Markoff *et al.*, 2002). In TBE, a range of deletions extending into the conserved 340 nucleotide core element at the 3′ terminus had variable effects on virulence in an *in vivo* mouse model, with deletions having an increased biological effect as they extended further into the core region (Mandl *et al.*, 1998). Similar attenuation has been observed in a deletion mutant of LGT, which was 10^5-fold less neuroinvasive in mice than the TP21 strain from which it had been derived originally (Pletnev, 2001).

Chimeric flaviviruses with mutations or deletions in the 3′ UTR are also attenuated *in vitro*. Zeng *et al.* (1998) produced a range of chimeric DEN-2 viruses containing different regions of the WNV 3′ SL and showed that substitution of the terminal 96 nucleotides of WNV into the DEN-2 genome was sublethal. Because truncated DEN-2 3′ SL RNAs were able to compete efficiently for the same cellular factors as WNV 3′ SL RNAs in this study, the basis for the attenuation was likely to be disruption of the interaction of the SL with viral rather than cellular proteins. The observation in other studies that mutations in the 3′ SL affect host range and RNA cyclization does, however, suggest that attenuation can be caused by either the disruption of RC formation and/or the initiation of RNA synthesis.

IV. Conclusion

Much of the early work on virulence involved viruses derived by cell culture passage or neutralization escape selection, and it is for this reason that the majority of published works on molecular determinants of virulence have focused on mutations in the E protein. With the advent of infectious clone technology, more recent studies have begun to reverse engineer and assess the relative contributions of specific mutations in nonstructural regions and/or the UTRs. While relatively few of these studies assess virulence in an animal model, it is clear that reverse genetics has the potential to serve as a useful tool for investigations on virulence and pathogenesis.

At present there is a need for a more detailed analysis of the virus–host relationship, and it is likely that future studies will help explain this relationship in the context of observed effects on protein structure and function. One could reasonably predict that mutations in structural proteins would affect aspects of virus binding and entry, and/or the stability of the glycoprotein network on the surface of the virion. Such mutations may disrupt important aspects of the virus life cycle, such as the dimer-to-trimer transition associated with endosomal fusion or the ability of chaperones such as prM to shield the E protein from premature exposure to acidic pH during viral egress. In contrast, mutations in nonstructural proteins and/or 5′ and 3′ UTRs are likely to affect virulence by disrupting enzyme function, causing defects in transcription, translation, and assembly, and/or the interaction of viral and host factors with the replicative machinery.

It is clear from work on encephalitogenic viruses such as MVE, JE, and TBE that efficient viral replication in regional draining lymph nodes is required for the development of viremia and that the magnitude and duration of such viremia can impact directly on the ability of the virus to invade the CNS and cause disease (Huang and Wong, 1963; Malkova and Frankova, 1959; McMinn *et al.*, 1996a). Furthermore, dendritic cells appear to be the principal cell type responsible for viral replication *in vivo* early in infection (McMinn and Sammels, 1997; Wu *et al.*, 2000). In MVE, the temporal spread of virus from a peripheral site of infection to the CNS is affected directly by mutations that perturb viral protein function (McMinn *et al.*, 1996a). In particular, attenuating mutations causing defects in fusion appear to delay viral replication and spread in and from regional lymph nodes, causing a reduction in neuroinvasiveness. This delay is sufficient to allow the host immune system time to clear the infection before the onset of an acute inflammatory response associated with viral encephalitis. In contrast, the attenuated phenotype does not manifest when the virus is inoculated directly into the brain, presumably because circumvention of the peripheral immune system affords the virus sufficient time to overwhelm the host before a protective immune response can become effective.

Whether attenuation of neuroinvasiveness and/or neurovirulence caused by mutations in nonstructural and untranslated regions of the genome is caused by a similar delay in replication at peripheral sites will require a closer examination of pathogenicity *in vivo*. As evidenced by recent work in the field, reverse genetics is becoming increasingly popular as a tool for dissecting the relative contribution of such mutations to the *in vivo* phenotype of attenuated viruses. Furthermore, infectious cDNA clones provide a useful means of generating attenuated viruses for use as vaccines, and this approach is proving increasingly popular. For example, chimeric flaviviruses expressing the prM and E genes of JE, DEN-2, and LGT appear to be very effective immunogens for protection against homologous virus challenge (Caufour *et al.*, 2001; Chambers *et al.*, 1999; Guirakhoo *et al.*, 1999, 2000; Monath *et al.*, 1999; Pletnev *et al.*, 2001; van der Most *et al.*, 2000). Despite their promise, such vaccine constructs must be tested vigorously to ensure that they do not revert to virulence after cell culture or mouse brain passage. Viruses generated by *in vivo* passage often have increased replicative ability and can become more virulent as a result of adaptation to the host. For example, Schlesinger *et al.* (1996) were able to increase the neurovirulence of YF after serial passage of the virus in mouse brain. Chambers and Nickells (2001) subsequently showed that this virus had a number of mutations scattered across the genome and used reverse genetics to introduce these into an infectious cDNA clone to yield virus with a similar phenotype. It should now be possible to determine which of these mutations contribute to the increased virulence of neuroadapted YF, using a methodological approach similar to that currently being employed to ascertain the molecular basis for attenuation in chimeric JE/YF viruses (Arroyo *et al.*, 2001). Such studies will undoubtedly contribute to our understanding of pathogenesis and, in conjunction with more specific work, will help elucidate the structural and functional basis for flavivirus attenuation.

References

Albrecht, P. (1962). Pathogenesis of experimental infection with tick-borne encephalitis virus. *In* "Biology of Viruses of the Tick-Borne Encephalitis Complex" (H. Libikova, ed.), pp. 247–57. Academic Press, New York.

Allison, S. L., Schalich, J., Stiasny, K., Mandl, C. W., and Heinz, F. X. (2001). Mutational evidence for an internal fusion peptide in flavivirus envelope protein E. *J. Virol.* **75**(9):4268–4275.

Allison, S. L., Schalich, J., Stiasny, K., Mandl, C. W., Kunz, C., and Heinz, F. X. (1995). Oligomeric rearrangement of tick-borne encephalitis virus envelope proteins induced by an acidic pH. *J. Virol.* **69**(2):695–700.

Allison, S. L., Stiasny, K., Stadler, K., Mandl, C. W., and Heinz, F. X. (1999). Mapping of functional elements in the stem-anchor region of tick-borne encephalitis virus envelope protein E. *J. Virol.* **73**(7):5605–5612.

Amberg, S. M., Nestorowicz, A., McCourt, D. W., and Rice, C. M. (1994). NS2B-3 proteinase-mediated processing in the yellow fever virus structural region: In vitro and in vivo studies. *J. Virol.* **68**(6):3794–3802.

Arroyo, J., Guirakhoo, F., Fenner, S., Zhang, Z. X., Monath, T. P., and Chambers, T. J. (2001). Molecular basis for attenuation of neurovirulence of a yellow fever virus/Japanese encephalitis virus chimera vaccine (ChimeriVax-JE). *J. Virol.* **75**(2):934–942.

Bartholomeusz, A., and Thompson, P. (1999). *Flaviviridae* polymerase and RNA replication. *J. Viral Hepat.* **6**(4):261–270.

Bazan, J. F., and Fletterick, R. J. (1989). Detection of a trypsin-like serine protease domain in flaviviruses and pestiviruses. *Virology* **171**(2):637–639.

Bhardwaj, S., Holbrook, M., Shope, R. E., Barrett, A. D., and Watowich, S. J. (2001). Biophysical characterization and vector-specific antagonist activity of domain III of the tick-borne flavivirus envelope protein. *J. Virol.* **75**(8):4002–4007.

Blackwell, J. L., and Brinton, M. A. (1997). Translation elongation factor-1 alpha interacts with the 3′ stem-loop region of West Nile virus genomic RNA. *J. Virol.* **71**(9):6433–6444.

Blaney, J. E., Johnson, D. H., Firestone, C. Y., Hanson, C. T., Murphy, B. R., and Whitehead, S. S. (2001). Chemical mutagenesis of dengue virus type 4 yields mutant viruses which are temperature sensitive in Vero cells or human liver cells and attenuated in mice. *J. Virol.* **75**(20):9731–9740.

Blok, J., McWilliam, S. M., Butler, H. C., Gibbs, A. J., Weiller, G., Herring, B. L., Hemsley, A. C., Aaskov, J. G., Yoksan, S., and Bhamarapravati, N. (1992). Comparison of a dengue-2 virus and its candidate vaccine derivative: Sequence relationships with the flaviviruses and other viruses. *Virology* **187**(2):573–590.

Bork, P., Holm, L., and Sander, C. (1994). The immunoglobulin fold: Structural classification, sequence patterns and common core. *J. Mol. Biol.* **242**(4):309–320.

Bray, M., Men, R., Tokimatsu, I., and Lai, C. J. (1998). Genetic determinants responsible for acquisition of dengue type 2 virus mouse neurovirulence. *J. Virol.* **72**(2):1647–1651.

Brinton, M. (1991). Flaviviruses. *In* "Handbook of Neurovirology" (R. R. McKendall and W. G. Stroop, eds.), pp. 379–389. Dekker, New York.

Cahour, A., Pletnev, A., Vazielle-Falcoz, M., Rosen, L., and Lai, C. J. (1995). Growth-restricted dengue virus mutants containing deletions in the 5′ noncoding region of the RNA genome. *Virology* **207**(1):68–76.

Campbell, M. S., and Pletnev, A. G. (2000). Infectious cDNA clones of Langat tick-borne flavivirus that differ from their parent in peripheral neurovirulence. *Virology* **269**(1):225–237.

Caufour, P. S., Motta, M. C. A., Yamamura, A. M. Y., Vazquez, S., Ferreira, II, Jabor, A. V., Bonaldo, M. C., Freire, M. S., and Galler, R. (2001). Construction, characterization and immunogenicity of recombinant yellow fever 17D-dengue type 2 viruses. *Virus Res.* **79**(1–2):1–14.

Cecilia, D., and Gould, E. A. (1991). Nucleotide changes responsible for loss of neuroinvasiveness in Japanese encephalitis virus neutralization-resistant mutants. *Virology* **181**(1):70–77.

Chambers, T. J., Grakoui, A., and Rice, C. M. (1991). Processing of the yellow fever virus nonstructural polyprotein: A catalytically active NS3 proteinase domain and NS2B are required for cleavages at dibasic sites. *J. Virol.* **65**(11):6042–6050.

Chambers, T. J., Hahn, C. S., Galler, R., and Rice, C. M. (1990a). Flavivirus genome organization, expression, and replication. *Annu. Rev. Microbiol.* **44**(10):649–688.

Chambers, T. J., Halevy, M., Nestorowicz, A., Rice, C. M., and Lustig, S. (1998). West Nile virus envelope proteins: nucleotide sequence analysis of strains differing in mouse neuroinvasiveness. *J. Gen. Virol.* **79**(10):2375–2380.

Chambers, T. J., Nestorowicz, A., Amberg, S. M., and Rice, C. M. (1993). Mutagenesis of the yellow fever virus NS2B protein: Effects on proteolytic processing, NS2B-NS3 complex formation, and viral replication. *J. Virol.* **67**(11):6797–6807.

Chambers, T. J., Nestorowicz, A., Mason, P. W., and Rice, C. M. (1999). Yellow fever Japanese encephalitis chimeric viruses: Construction and biological properties. *J. Virol.* **73**(4):3095–3101.

Chambers, T. J., Nestorowicz, A., and Rice, C. M. (1995). Mutagenesis of the yellow fever virus NS2B/3 cleavage site: Determinants of cleavage site specificity and effects on polyprotein processing and viral replication. *J. Virol.* **69**(3):1600–1605.

Chambers, T. J., and Nickells, M. (2001). Neuroadapted yellow fever virus 17D: Genetic and biological characterization of a highly mouse-neurovirulent virus and its infectious molecular clone. *J. Virol.* **75**(22):10912–10922.

Chambers, T. J., Weir, R. C., Grakoui, A., McCourt, D. W., Bazan, J. F., Fletterick, R. J., and Rice, C. M. (1990b). Evidence that the N-terminal domain of nonstructural protein NS3 from yellow fever virus is a serine protease responsible for site-specific cleavages in the viral polyprotein. *Proc. Nat. Acad. Sci. USA* **87**(22):8898–8902.

Chen, C. J., Kuo, M. D., Chien, L. J., Hsu, S. L., Wang, Y. M., and Lin, J. H. (1997a). RNA-protein interactions: Involvement of NS3, NS5, and 3′ noncoding regions of Japanese encephalitis virus genomic RNA. *J. Virol.* **71**(5):3466–3473.

Chen, L. K., Lin, Y. L., Liao, C. L., Lin, C. G., Huang, Y. L., Yeh, C. T., Lai, S. C., Jan, J. T., and Chin, C. (1996). Generation and characterization of organ-tropism mutants of Japanese encephalitis virus in vivo and in vitro. *Virology* **223**(1):79–88.

Chen, W., Kawano, H., Men, R., Clark, D., and Lai, C. J. (1995). Construction of intertypic chimeric dengue viruses exhibiting type 3 antigenicity and neurovirulence for mice. *J. Virol.* **69**(8):5186–5190.

Chen, Y. P., Maguire, T., Hileman, R. E., Fromm, J. R., Esko, J. D., Linhardt, R. J., and Marks, R. M. (1997b). Dengue virus infectivity depends on envelope protein binding to target cell heparan sulfate. *Nature Med.* **3**(8):866–871.

Chiou, S. S., and Chen, W. J. (2001). Mutations in the NS3 gene and 3′-NCR of Japanese encephalitis virus isolated from an unconventional ecosystem and implications for natural attenuation of the virus. *Virology* **289**(1):129–136.

Corver, J., Ortiz, A., Allison, S. L., Schalich, J., Heinz, F. X., and Wilschut, J. (2000). Membrane fusion activity of tick-borne encephalitis virus and recombinant subviral particles in a liposomal model system. *Virology* **269**(1):37–46.

Crawford, G. R., and Wright, P. J. (1987). Characterization of novel viral polyproteins detected in cells infected by the flavivirus Kunjin and radiolabelled in the presence of the leucine analogue hydroxyleucine. *J. Gen. Virol.* **68**(2):365–376.

Crill, W. D., and Roehrig, J. T. (2001). Monoclonal antibodies that bind to domain III of dengue virus E glycoprotein are the most efficient blockers of virus adsorption to Vero cells. *J. Virol.* **75**(16):7769–7773.

dos Santos, C. N., Post, P. R., Carvalho, R., Ferreira, II, Rice, C. M., and Galler, R. (1995). Complete nucleotide sequence of yellow fever virus vaccine strains 17DD and 17D-213. *Virus Res.* **35**(1):35–41.

dos Santos, C. N. D., Frenkiel, M. P., Courageot, M. P., Rocha, C. F. S., Vazeille-Falcoz, M. C., Wien, M. W., Rey, F. A., Deubel, V., and Despres, P. (2000). Determinants in the envelope E protein and viral RNA helicase NS3 that influence the induction of apoptosis in response to infection with dengue type 1 virus. *Virology* **274**(2):292–308.

Dropulic, B., and Masters, C. L. (1990). Entry of neurotropic arboviruses into the central nervous system: An in vitro study using mouse brain endothelium. *J. Infect. Dis.* **161**(4):685–691.

Durbin, A. P., Karron, R. A., Sun, W., Vaughn, D. W., Reynolds, M. J., Perreault, J. R., Thumar, B., Men, R., Lai, C. J., Elkins, W. R., Chanock, R. M., Murphy, B. R., and Whitehead, S. S. (2001). Attenuation and immunogenicity in humans of a live dengue virus type-4 vaccine candidate with a 30 nucleotide deletion in its 3′-untranslated region. *Am. J. Trop. Med. Hyg.* **65**(5):405–413.

Falgout, B., Miller, R. H., and Lai, C. J. (1993). Deletion analysis of dengue virus type 4 nonstructural protein NS2B: Identification of a domain required for NS2B-NS3 protease activity. *J. Virol.* **67**(4):2034–2042.

Ferlenghi, I., Clarke, M., Ruttan, T., Allison, S. L., Schalich, J., Heinz, F. X., Harrison, S. C., Rey, F. A., and Fuller, S. D. (2001). Molecular organization of a recombinant subviral particle from tick-borne encephalitis virus. *Mol. Cell* **7**(3):593–602.

Franklin, K., and Paxinos, G. (1997). "The Mouse Brain in Stereotaxic Coordinates," 1st Ed. Academic Press, Sydney, Australia.

Gao, G. F., Hussain, M. H., Reid, H. W., and Gould, E. A. (1994). Identification of naturally occurring monoclonal antibody escape variants of louping ill virus. *J. Gen. Virol.* **75**(3):609–614.

Gething, M. J., McCammon, K., and Sambrook, J. (1986). Expression of wild-type and mutant forms of influenza hemagglutinin: the role of folding in intracellular transport. *Cell* **46**(6):939–950.

Gorbalenya, A. E., Donchenko, A. P., Koonin, E. V., and Blinov, V. M. (1989). N-terminal domains of putative helicases of flavi- and pestiviruses may be serine proteases. *Nucleic Acids Res.* **17**(10):3889–3897.

Gritsun, T. S., Desai, A., and Gould, E. A. (2001). The degree of attenuation of tick-borne encephalitis virus depends on the cumulative effects of point mutations. *J. Gen. Virol.* **82**(7):1667–1675.

Gualano, R. C., Pryor, M. J., Cauchi, M. R., Wright, P. J., and Davidson, A. D. (1998). Identification of a major determinant of mouse neurovirulence of dengue virus type 2 using stably cloned genomic-length cDNA. *J. Gen. Virol.* **79**(3):437–446.

Guirakhoo, F., Bolin, R. A., and Roehrig, J. T. (1992). The Murray Valley encephalitis virus prM protein confers acid resistance to virus particles and alters the expression of epitopes within the R2 domain of E glycoprotein. *Virology* **191**(2):921–931.

Guirakhoo, F., Heinz, F. X., and Kunz, C. (1989). Epitope model of tick-borne encephalitis virus envelope glycoprotein E: Analysis of structural properties, role of carbohydrate side chain, and conformational changes occurring at acidic pH. *Virology* **169**(1):90–99.

Guirakhoo, F., Heinz, F. X., Mandl, C. W., Holzmann, H., and Kunz, C. (1991). Fusion activity of flaviviruses: Comparison of mature and immature (prM-containing) tick-borne encephalitis virions. *J. Gen. Virol.* **72**(6):1323–1329.

Guirakhoo, F., Hunt, A. R., Lewis, J. G., and Roehrig, J. T. (1993). Selection and partial characterization of dengue 2 virus mutants that induce fusion at elevated pH. *Virology* **194**(1):219–223.

Guirakhoo, F., Weltzin, R., Chambers, T. J., Zhang, Z. X., Soike, K., Ratterree, M., Arroyo, J., Georgakopoulos, K., Catalan, J., and Monath, T. P. (2000). Recombinant chimeric yellow fever-dengue type 2 virus is immunogenic and protective in nonhuman primates. *J. Virol.* **74**(12):5477–5485.

Guirakhoo, F., Zhang, Z. X., Chambers, T. J., Delagrave, S., Arroyo, J., Barrett, A. D., and Monath, T. P. (1999). Immunogenicity, genetic stability, and protective efficacy of a recombinant, chimeric yellow fever-Japanese encephalitis virus (ChimeriVax-JE) as a live, attenuated vaccine candidate against Japanese encephalitis. *Virology* **257**(2):363–372.

Hahn, C. S., Dalrymple, J. M., Strauss, J. H., and Rice, C. M. (1987a). Comparison of the virulent Asibi strain of yellow fever virus with the 17D vaccine strain derived from it. *Proc. Nat. Acad. Sci. USA* **84**(7):2019–2023.

Hahn, C. S., Hahn, Y. S., Rice, C. M., Lee, E., Dalgarno, L., Strauss, E. G., and Strauss, J. H. (1987b). Conserved elements in the 3′ untranslated region of flavivirus RNAs and potential cyclization sequences. *J. Mol. Biol.* **198**(1):33–41.

Hall, R. A., Khromykh, A. A., Mackenzie, J. M., Scherret, J. H., Khromykh, T. I., and Mackenzie, J. S. (1999). Loss of dimerisation of the nonstructural protein NS1 of Kunjin virus delays viral replication and reduces virulence in mice, but still allows secretion of NS1. *Virology* **264**(1):66–75.

Hanley, K. A., Lee, J. J., Blaney, J. E., Murphy, B. R., and Whitehead, S. S. (2002). Paired charge-to-alanine mutagenesis of dengue virus type 4 NS5 generates mutants with temperature-sensitive, host range, and mouse attenuation phenotypes. *J. Virol.* **76**(2):525–531.

Hasegawa, H., Yoshida, M., Shiosaka, T., Fujita, S., and Kobayashi, Y. (1992). Mutations in the envelope protein of Japanese encephalitis virus affect entry into cultured cells and virulence in mice. *Virology* **191**(1):158–165.

Heinz, F. X. (1986). Epitope mapping of flavivirus glycoproteins. *Adv Virus Res.* **31**(1):103–168.

Heinz, F. X., Stiasny, K., Puschner-Auer, G., Holzmann, H., Allison, S. L., Mandl, C. W., and Kunz, C. (1994). Structural changes and functional control of the tick-borne encephalitis virus glycoprotein E by the heterodimeric association with protein prM. *Virology* **198**(1):109–117.

Hiramatsu, K., Tadano, M., Men, R., and Lai, C. J. (1996). Mutational analysis of a neutralization epitope on the dengue type 2 virus (DEN2) envelope protein: Monoclonal antibody resistant DEN2/DEN4 chimeras exhibit reduced mouse neurovirulence. *Virology* **224**(2):437–445.

Holbrook, M. R., Li, L., Suderman, M. T., Wang, H., and Barrett, A. D. (2000). The French neurotropic vaccine strain of yellow fever virus accumulates mutations slowly during passage in cell culture. *Virus Res.* **69**(1):31–39.

Holbrook, M. R., Ni, H., Shope, R. E., and Barrett, A. D. (2001). Amino acid substitution(s) in the stem-anchor region of Langat virus envelope protein attenuates mouse neurovirulence. *Virology* **286**(1):54–61.

Holzmann, H., Heinz, F. X., Mandl, C. W., Guirakhoo, F., and Kunz, C. (1990). A single amino acid substitution in envelope protein E of tick-borne encephalitis virus leads to attenuation in the mouse model. *J. Virol.* **64**(10):5156–5159.

Holzmann, H., Stiasny, K., Ecker, M., Kunz, C., and Heinz, F. X. (1997). Characterization of monoclonal antibody-escape mutants of tick-borne encephalitis virus with reduced neuroinvasiveness in mice. *J. Gen. Virol.* **78**(1):31–37.

Huang, C. H., and Wong, C. (1963). Relation of the peripheral multiplication of Japanese B encephalitis virus to the pathogenesis of the infection in mice. *Acta Virol.* **7**(16):322–330.

Hurrelbrink, R. J., and McMinn, P. C. (2001). Attenuation of Murray Valley encephalitis virus by site-directed mutagenesis of the hinge and putative receptor-binding regions of the envelope protein. *J. Virol.* **75**(16):7692–7702.

Iacono-Connors, L. C., Smith, J. F., Ksiazek, T. G., Kelley, C. L., and Schmaljohn, C. S. (1996). Characterization of Langat virus antigenic determinants defined by monoclonal antibodies to E, NS1 and preM and identification of a protective, non-neutralizing preM-specific monoclonal antibody. *Virus Res.* **43**(2):125–136.

Jennings, A. D., Gibson, C. A., Miller, B. R., Mathews, J. H., Mitchell, C. J., Roehrig, J. T., Wood, D. J., Taffs, F., Sil, B. K., Whitby, S. N., *et al.* (1994). Analysis of a yellow fever virus isolated from a fatal case of vaccine-associated human encephalitis. *J. Infect. Dis.* **169**(3):512–518.

Jennings, A. D., Whitby, J. E., Minor, P. D., and Barrett, A. D. (1993). Comparison of the nucleotide and deduced amino acid sequences of the envelope protein genes of the wild-type French viscerotropic strain of yellow fever virus and the live vaccine strain, French neurotropic vaccine, derived from it. *Virology* **192**(2):692–695.

Jiang, W. R., Lowe, A., Higgs, S., Reid, H., and Gould, E. A. (1993). Single amino acid codon changes detected in louping ill virus antibody-resistant mutants with reduced neurovirulence. *J. Gen. Virol.* **74**(5):931–935.

Johansson, M., Brooks, A. J., Jans, D. A., and Vasudevan, S. G. (2001). A small region of the dengue virus-encoded RNA-dependent RNA polymerase, NS5, confers interaction with both the nuclear transport receptor importin-beta and the viral helicase, NS3. *J. Gen. Virol.* **82**(4):735–745.

Johnson, R. T. (1982). Pathogenesis of CNS infections. *In* "Viral Infections of the Nervous System" (R. T. Johnson, ed.)., Raven Press, New York.

Johnson, R. T., Burke, D. S., Elwell, M., Leake, C. J., Nisalak, A., Hoke, C. H., and Lorsomrudee, W. (1985). Japanese encephalitis: Immunocytochemical studies of viral antigen and inflammatory cells in fatal cases. *Ann. Neurol.* **18**(5):567–573.

Kawano, H., Rostapshov, V., Rosen, L., and Lai, C. J. (1993). Genetic determinants of dengue type 4 virus neurovirulence for mice. *J. Virol.* **67**(11):6567–6575.

Khromykh, A. A., Kenney, M. T., and Westaway, E. G. (1998). trans-complementation of flavivirus RNA polymerase gene NS5 by using Kunjin virus replicon-expressing BHK cells. *J. Virol.* **72**(9):7270–7279.

Khromykh, A. A., Meka, H., Guyatt, K. J., and Westaway, E. G. (2001). Essential role of cyclization sequences in flavivirus RNA replication. *J. Virol.* **75**(14):6719–6728.

Khromykh, A. A., Sedlak, P. L., and Westaway, E. G. (1999). trans-complementation analysis of the flavivirus Kunjin NS5 gene reveals an essential role for translation of its N-terminal half in RNA replication. *J. Virol.* **73**(11):9247–9255.

Koonin, E. V. (1993). Computer-assisted identification of a putative methyltransferase domain in NS5 protein of flaviviruses and lambda 2 protein of reovirus. *J. Gen. Virol.* **74**(4):733–740.

Kozerski, C., Ponimaskin, E., Schroth-Diez, B., Schmidt, M. F., and Herrmann, A. (2000). Modification of the cytoplasmic domain of influenza virus hemagglutinin affects enlargement of the fusion pore. *J. Virol.* **74**(16):7529–7537.

Kuhn, R. J., Zhang, W., Rossman, M. G., Pletnev, S. V., Corver, J., Lenches, E., Jones, C. T., Mukhopadhyay, S., Chipman, P. R., Strauss, E. G., Baker, T. S., and Strauss, J. H. (2002). Structure of dengue virus: implications for flavivirus organization, maturation and fusion. *Cell* **108**(10):717–725.

Kümmerer, B. M., and Rice, C. M. (2002). Mutations in the yellow fever virus nonstructural protein NS2A selectively block production of infectious particles. *J. Virol.* **76**(10):4773–4784.

Labuda, M., Jiang, W. R., Kaluzova, M., Kozuch, O., Nuttall, P. A., Weismann, P., Eleckova, E., Zuffova, E., and Gould, E. A. (1994). Change in phenotype of tick-borne encephalitis virus following passage in *Ixodes ricinus* ticks and associated amino acid substitution in the envelope protein. *Virus Res.* **31**(3):305–315.

Lee, E., and Lobigs, M. (2000). Substitutions at the putative receptor-binding site of an encephalitic flavivirus alter virulence and host cell tropism and reveal a role for glycosaminoglycans in entry. *J. Virol.* **74**(19):8867–8875.

Lee, E., and Lobigs, M. (2002). Mechanism of virulence attenuation of glycosaminoglycan-binding variants of Japanese encephalitis virus and Murray Valley encephalitis virus. *J. Virol.* **76**(10):4901–4911.

Lee, E., Weir, R. C., and Dalgarno, L. (1997). Changes in the dengue virus major envelope protein on passaging and their localization on the three-dimensional structure of the protein. *Virology* **232**(2):281–290.

Li, W., Li, Y., Kedersha, N., Anderson, P., Emara, M., Swiderek, K. M., Moreno, G. T., and Brinton, M. A. (2002). Cell proteins TIA-1 and TIAR interact with the 3′ stem-loop of the West Nile virus complementary minus-strand RNA and facilitate virus replication. *J. Virol.* **76**(23):11989–12000.

Lin, C., Amberg, S. M., Chambers, T. J., and Rice, C. M. (1993). Cleavage at a novel site in the NS4A region by the yellow fever virus NS2B-3 proteinase is a prerequisite for processing at the downstream 4A/4B signalase site. *J. Virol.* **67**(4):2327–2335.

Lindenbach, B. D., and Rice, C. M. (1999). Genetic interaction of flavivirus nonstructural proteins NS1 and NS4A as a determinant of replicase function. *J. Virol.* **73**(6):4611–4621.

Lindenbach, B. D., and Rice, C. M. (2001). Flaviviridae: The viruses and their replication. *In* "Fields Virology" (D. M. Knipe and P. M. Howley, eds.), pp. 991–1041. Lippincott, Williams and Wilkins, Philadelphia.

Lobigs, M., and Garoff, H. (1990). Fusion function of the Semliki Forest virus spike is activated by proteolytic cleavage of the envelope glycoprotein precursor p62. *J. Virol.* **64**(3):1233–1240.

Lobigs, M., Usha, R., Nestorowicz, A., Marshall, I. D., Weir, R. C., and Dalgarno, L. (1990). Host cell selection of Murray Valley encephalitis virus variants altered at an RGD sequence in the envelope protein and in mouse virulence. *Virology* **176**(2):587–595.

Malkova, D., and Frankova, V. (1959). The lymphatic system in the development of experimental tick-borne encephalitis in mice. *Acta Virol.* **3**(2):210–214.

Mandl, C. W., Allison, S. L., Holzmann, H., Meixner, T., and Heinz, F. X. (2000). Attenuation of tick-borne encephalitis virus by structure-based site-specific mutagenesis of a putative flavivirus receptor binding site. *J. Virol.* **74**(20):9601–9609.

Mandl, C. W., Guirakhoo, F., Holzmann, H., Heinz, F. X., and Kunz, C. (1989a). Antigenic structure of the flavivirus envelope protein E at the molecular level, using tick-borne encephalitis virus as a model. *J. Virol.* **63**(2):564–571.

Mandl, C. W., Heinz, F. X., Stockl, E., and Kunz, C. (1989b). Genome sequence of tick-borne encephalitis virus (Western subtype) and comparative analysis of nonstructural proteins with other flaviviruses. *Virology* **173**(1):291–301.

Mandl, C. W., Holzmann, H., Meixner, T., Rauscher, S., Stadler, P. F., Allison, S. L., and Heinz, F. X. (1998). Spontaneous and engineered deletions in the 3′ noncoding region of tick-borne encephalitis virus: Construction of highly attenuated mutants of a flavivirus. *J. Virol.* **72**(3):2132–2140.

Markoff, L., Pang, X., Houng, H-S., Falgout, B., Olsen, R., Jones, E., and Polo, S. (2002). Derivation and characterization of a dengue type 1 host range-restricted mutant virus that is attenuated and highly immunogenic in monkeys. *J. Virol.* **76**(7):3318–3328.

Matusan, A. E., Kelley, P. G., Pryor, M. J., Whisstock, J. C., Davidson, A. D., and Wright, P. J. (2001a). Mutagenesis of the dengue virus type 2 NS3 proteinase and the production of growth-restricted virus. *J. Gen. Virol.* **82**(7):1647–1656.

Matusan, A. E., Pryor, M. J., Davidson, A. D., and Wright, P. J. (2001b). Mutagenesis of the dengue virus type 2 NS3 protein within and outside helicase motifs: Effects on enzyme activity and virus replication. *J. Virol.* **75**(20):9633–9643.

McMinn, P. C., Dalgarno, L., and Weir, R. C. (1996a). A comparison of the spread of Murray Valley encephalitis viruses of high or low neuroinvasiveness in the tissues of Swiss mice after peripheral inoculation. *Virology* **220**(2):414–423.

McMinn, P. C., Lee, E., Hartley, S., Roehrig, J. T., Dalgarno, L., and Weir, R. C. (1995). Murray valley encephalitis virus envelope protein antigenic variants with altered hemagglutination properties and reduced neuroinvasiveness in mice. *Virology* **211**(1):10–20.

McMinn, P. C., and Sammels, L. (1997). The molecular pathogenesis of flavivirus encephalitis. *Microbiol. Aust.* **18**(9):A32.

McMinn, P. C., Weir, R. C., and Dalgarno, L. (1996b). A mouse-attenuated envelope protein variant of Murray Valley encephalitis virus with altered fusion activity. *J. Gen. Virol.* **77**(9):2085–2088.

Men, R., Bray, M., Clark, D., Chanock, R. M., and Lai, C. J. (1996). Dengue type 4 virus mutants containing deletions in the 3′ noncoding region of the RNA genome: Analysis of growth restriction in cell culture and altered viremia pattern and immunogenicity in rhesus monkeys. *J. Virol.* **70**(6):3930–3937.

Monath, T. P. (1986). Pathobiology of the flaviviruses. *In* "The Togaviridae and Flaviviridae" (S. Schlesinger and M. J. Schlesinger, eds.), pp. 375–440. Plenum, New York.

Monath, T. P., Arroyo, J., Levenbook, I., Zhang, Z-X., Catalan, J., Draper, K., and Guirakhoo, F. (2002). Single mutation in the flavivirus envelope protein hinge region increases neurovirulence for mice and monkeys but decreases viscerotropism for monkeys: Relevance to development and safety testing of live, attenuated vaccines. *J. Virol.* **76**(4):1932–1943.

Monath, T. P., Cropp, C. B., and Harrison, A. K. (1983). Mode of entry of a neurotropic arbovirus into the central nervous system: Reinvestigation of an old controversy. *Lab. Invest.* **48**(4):399–410.

Monath, T. P., Soike, K., Levenbook, I., Zhang, Z. X., Arroyo, J., Delagrave, S., Myers, G., Barrett, A. D., Shope, R. E., Ratterree, M., Chambers, T. J., and Guirakhoo, F. (1999). Recombinant, chimaeric live, attenuated vaccine (ChimeriVax) incorporating the envelope genes of Japanese encephalitis (SA14–14–2) virus and the capsid and nonstructural genes of yellow fever (17D) virus is safe, immunogenic and protective in nonhuman primates. *Vaccine* **17**(15–16):1869–1882.

Murray, J. M., Aaskov, J. G., and Wright, P. J. (1993). Processing of the dengue virus type 2 proteins prM and C-prM. *J. Gen. Virol.* **74**(2):175–182.

Murthy, H. M., Clum, S., and Padmanabhan, R. (1999). Dengue virus NS3 serine protease: Crystal structure and insights into interaction of the active site with substrates by molecular modeling and structural analysis of mutational effects. *J. Biol. Chem.* **274**(9):5573–5580.

Muylaert, I. R., Chambers, T. J., Galler, R., and Rice, C. M. (1996). Mutagenesis of the N-linked glycosylation sites of the yellow fever virus NS1 protein: Effects on virus replication and mouse neurovirulence. *Virology* **222**(1):159–168.

Nestorowicz, A., Chambers, T. J., and Rice, C. M. (1994). Mutagenesis of the yellow fever virus NS2A/2B cleavage site: Effects on proteolytic processing, viral replication, and evidence for alternative processing of the NS2A protein. *Virology* **199**(1):114–123.

Ni, H., Chang, G. J., Xie, H., Trent, D. W., and Barrett, A. D. (1995). Molecular basis of attenuation of neurovirulence of wild-type Japanese encephalitis virus strain SA14. *J. Gen. Virol.* **76**(2):409–413.

Ni, H. L., and Barrett, A. D. T. (1998). Attenuation of Japanese encephalitis virus by selection of its mouse brain membrane receptor preparation escape variants. *Virology* **241**(1):30–36.

Nitayaphan, S., Grant, J. A., Chang, G. J., and Trent, D. W. (1990). Nucleotide sequence of the virulent SA-14 strain of Japanese encephalitis virus and its attenuated vaccine derivative, SA-14-14-2. *Virology* **177**(2):541–552.

Nowak, T., Farber, P. M., Wengler, G., and Wengler, G. (1989). Analyses of the terminal sequences of West Nile virus structural proteins and of the in vitro translation of these proteins allow the proposal of a complete scheme of the proteolytic cleavages involved in their synthesis. *Virology* **169**(2):365–376.

Pletnev, A. G. (2001). Infectious cDNA clone of attenuated Langat tick-borne flavivirus (strain E5) and a 3′ deletion mutant constructed from it exhibit decreased neuroinvasiveness in immunodeficient mice. *Virology* **282**(2):288–300.

Pletnev, A. G., Bray, M., Hanley, K. A., Speicher, J., and Elkins, R. (2001). Tick-borne Langat/mosquito-borne dengue flavivirus chimera, a candidate live attenuated vaccine for protection against disease caused by members of the tick-borne encephalitis virus complex: Evaluation in rhesus monkeys and in mosquitoes. *J. Virol.* **75**(17):8259–8267.

Pletnev, A. G., Bray, M., Huggins, J., and Lai, C. J. (1992). Construction and characterization of chimeric tick-borne encephalitis/dengue type 4 viruses. *Proc. Nat. Acad. Sci. USA* **89**(21):10532–10536.

Pletnev, A. G., Bray, M., and Lai, C. J. (1993). Chimeric tick-borne encephalitis and dengue type 4 viruses: Effects of mutations on neurovirulence in mice. *J. Virol.* **67**(8):4956–4963.

Pletnev, A. G., and Men, R. (1998). Attenuation of the Langat tick-borne flavivirus by chimerization with mosquito-borne flavivirus dengue type 4. *Proc. Nat. Acad. Sci. USA* **95**(4):1746–1751.

Proutski, V., Gould, E. A., and Holmes, E. C. (1997). Secondary structure of the 3′ untranslated region of flaviviruses: Similarities and differences. *Nucleic Acids Res.* **25**(6):1194–1202.

Pryor, M. J., Gualano, R. C., Lin, B., Davidson, A. D., and Wright, P. J. (1998). Growth restriction of dengue virus type 2 by site-specific mutagenesis of virus-encoded glycoproteins. *J. Gen. Virol.* **79**(11):2631–2639.

Pryor, M. J., and Wright, P. J. (1993). The effects of site-directed mutagenesis on the dimerization and secretion of the NS1 protein specified by dengue virus. *Virology* **194**(2):769–780.

Pryor, M. J., and Wright, P. J. (1994). Glycosylation mutants of dengue virus NS1 protein. *J. Gen. Virol.* **75**(5):1183–1187.

Pugachev, K. V., Nomokonova, N. Y., Dobrikova, E., and Wolf, Y. I. (1993). Site-directed mutagenesis of the tick-borne encephalitis virus NS3 gene reveals the putative serine protease domain of the NS3 protein. *FEBS Lett.* **328**(1–2):115–118.

Qiao, H., Pelletier, S. L., Hoffman, L., Hacker, J., Armstrong, R. T., and White, J. M. (1998). Specific single or double proline substitutions in the "spring-loaded" coiled-coil region of the influenza hemagglutinin impair or abolish membrane fusion activity. *J. Cell Biol.* **141**(6):1335–1347.

Rauscher, S., Flamm, C., Mandl, C. W., Heinz, F. X., and Stadler, P. F. (1997). Secondary structure of the 3′-noncoding region of flavivirus genomes: Comparative analysis of base pairing probabilities. *RNA* **3**(7):779–791.

Reid, H. W., and Doherty, P. C. (1971). Louping-ill encephalomyelitis in the sheep. I. The relationship of viraemia and the antibody response to susceptibility. *J. Comp. Pathol.* **81**(4):521–529.

Rey, F. A., Heinz, F. X., Mandl, C., Kunz, C., and Harrison, S. C. (1995). The envelope glycoprotein from tick-borne encephalitis virus at 2 Å resolution. *Nature* **375**(6529):291–298.

Ryan, M. D., Monaghan, S., and Flint, M. (1998). Virus-encoded proteinases of the Flaviviridae. *J. Gen. Virol.* **79**(5):947–959.

Ryman, K. D., Ledger, T. N., Campbell, G. A., Watowich, S. J., and Barrett, A. D. T. (1998). Mutation in a 17D-204 vaccine substrain-specific envelope protein epitope alters the pathogenesis of yellow fever virus in mice. *Virology* **244**(1):59–65.

Ryman, K. D., Xie, H., Ledger, T. N., Campbell, G. A., and Barrett, A. D. (1997). Antigenic variants of yellow fever virus with an altered neurovirulence phenotype in mice. *Virology* **230**(2):376–380.

Salminen, A., Wahlberg, J. M., Lobigs, M., Liljestrom, P., and Garoff, H. (1992). Membrane fusion process of Semliki Forest virus. II. Cleavage-dependent reorganization of the spike protein complex controls virus entry. *J. Cell Biol.* **116**(2):349–357.

Sanchez, I. J., and Ruiz, B. H. (1996). A single nucleotide change in the E protein gene of dengue virus 2 Mexican strain affects neurovirulence in mice. *J. Gen. Virol.* **77**(10):2541–2545.

Schlesinger, J. J., Chapman, S., Nestorowicz, A., Rice, C. M., Ginocchio, T. E., and Chambers, T. J. (1996). Replication of yellow fever virus in the mouse central nervous system: Comparison of neuroadapted and non-neuroadapted virus and partial sequence analysis of the neuroadapted strain. *J. Gen. Virol.* **77**(6)**:**1277–1285.

Shi, P. Y., Brinton, M. A., Veal, J. M., Zhong, Y. Y., and Wilson, W. D. (1996). Evidence for the existence of a pseudoknot structure at the 3′ terminus of the flavivirus genomic RNA. *Biochemistry* **35**(13)**:**4222–4230.

Stadler, K., Allison, S. L., Schalich, J., and Heinz, F. X. (1997). Proteolytic activation of tick-borne encephalitis virus by furin. *J. Virol.* **71**(11)**:**8475–8481.

Steinhauer, D. A., Martin, J., Lin, Y. P., Wharton, S. A., Oldstone, M. B., Skehel, J. J., and Wiley, D. C. (1996). Studies using double mutants of the conformational transitions in influenza hemagglutinin required for its membrane fusion activity. *Proc. Nat. Acad. Sci. USA* **93**(23)**:**12873–12878.

Stiasny, K., Allison, S. L., Marchler-Bauer, A., Kunz, C., and Heinz, F. X. (1996). Structural requirements for low-pH-induced rearrangements in the envelope glycoprotein of tick-borne encephalitis virus. *J. Virol.* **70**(11)**:**8142–8147.

Stiasny, K., Allison, S. L., Schalich, J., and Heinz, F. X. (2002). Membrane interactions of the tick-borne encephalitis virus fusion protein E at low pH. *J. Virol.* **76**(8)**:**3784–3790.

Sumiyoshi, H., Tignor, G. H., and Shope, R. E. (1995). Characterization of a highly attenuated Japanese encephalitis virus generated from molecularly cloned cDNA. *J. Infect. Dis.* **171**(5)**:**1144–1151.

Swanson, L. W., Köhler, C., and Björklund, A. (1987). Integrated systems of the CNS. *In* "Handbook of Chemical Neuroanatomy" (A. Björklund and T. Hökfelt, eds.), pp. 125–277. Elsevier, New York.

Ta, M., and Vrati, S. (2000). Mov34 protein from mouse brain interacts with the 3′ noncoding region of Japanese encephalitis virus. *J. Virol.* **74**(11)**:**5108–5115.

Valle, R. P. C., and Falgout, B. (1998). Mutagenesis of the NS3 protease of dengue virus type 2. *J. Virol.* **72**(1)**:**624–632.

van der Most, R. G., Corver, J., and Strauss, J. H. (1999). Mutagenesis of the RGD motif in the yellow fever virus 17D envelope protein. *Virology* **265**(1)**:**83–95.

van der Most, R. G., Murali-Krishna, K., Ahmed, R., and Strauss, J. H. (2000). Chimeric yellow fever/dengue virus as a candidate dengue vaccine: quantitation of the dengue virus-specific CD8 T-cell response. *J. Virol.* **74**(17)**:**8094–8101.

Wallner, G., Mandl, C. W., Ecker, M., Holzmann, H., Stiasny, K., Kunz, C., and Heinz, F. X. (1996). Characterization and complete genome sequences of high- and low-virulence variants of tick-borne encephalitis virus. *J. Gen. Virol.* **77**(5)**:**1035–1042.

Wang, E., Ryman, K. D., Jennings, A. D., Wood, D. J., Taffs, F., Minor, P. D., Sanders, P. G., and Barrett, A. D. (1995). Comparison of the genomes of the wild-type French viscerotropic strain of yellow fever virus with its vaccine derivative French neurotropic vaccine. *J. Gen. Virol.* **76**(11)**:**2749–2755.

Weiner, L. P., Cole, G. A., and Nathanson, N. (1970). Experimental encephalitis following peripheral inoculation of West Nile virus in mice of different ages. *J. Hyg.* **68**(3)**:**435–446.

Wengler, G., Czaya, G., Farber, P. M., and Hegemann, J. H. (1991). In vitro synthesis of West Nile virus proteins indicates that the amino-terminal segment of the NS3 protein contains the active centre of the protease which cleaves the viral polyprotein after multiple basic amino acids. *J. Gen. Virol.* **72**(4)**:**851–858.

Wengler, G., and Wengler, G. (1989). Cell-associated West Nile flavivirus is covered with E+pre-M protein heterodimers which are destroyed and reorganized by proteolytic cleavage during virus release. *J. Virol.* **63**(6)**:**2521–2526.

Westaway, E. G., Mackenzie, J. M., Kenney, M. T., Jones, M. K., and Khromykh, A. A. (1997). Ultrastructure of Kunjin virus-infected cells: Colocalization of NS1 and NS3 with double-stranded RNA, and of NS2B with NS3, in virus-induced membrane structures. *J. Virol.* **71**(9):6650–6661.

Wu, S. J., Grouard-Vogel, G., Sun, W., Mascola, J. R., Brachtel, E., Putvatana, R., Louder, M. K., Filgueira, L., Marovich, M. A., Wong, H. K., Blauvelt, A., Murphy, G. S., Robb, M. L., Innes, B. L., Birx, D. L., Hayes, C. G., and Frankel, S. S. (2000). Human skin Langerhans cells are targets of dengue virus infection. *Nature Med.* **6**(7):816–820.

Xie, H., Ryman, K. D., Campbell, G. A., and Barrett, A. D. (1998). Mutation in NS5 protein attenuates mouse neurovirulence of yellow fever 17D vaccine virus. *J. Gen. Virol.* **79**(8):1895–1899.

You, S., Falgout, B., Markoff, L., and Padmanabhan, R. (2001). In vitro RNA synthesis from exogenous dengue viral RNA templates requires long range interactions between 5′- and 3′-terminal regions that influence RNA structure. *J. Biol. Chem.* **276**(19):15581–15591.

You, S., and Padmanabhan, R. (1999). A novel in vitro replication system for Dengue virus: Initiation of RNA synthesis at the 3′ end of exogenous viral RNA templates requires 5′- and 3′-terminal complementary sequence motifs of the viral RNA. *J. Biol. Chem.* **274**(47):33714–33722.

Yusof, R., Clum, S., Wetzel, M., Murthy, H. M. K., and Padmanabhan, R. (2000). Purified NS2B/NS3 serine protease of dengue virus type 2 exhibits cofactor NS2B dependence for cleavage of substrates with dibasic amino acids in vitro. *J. Biol. Chem.* **275**(14):9963–9969.

Zeng, L., Falgout, B., and Markoff, L. (1998). Identification of specific nucleotide sequences within the conserved 3′-SL in the dengue type 2 virus genome required for replication. *J. Virol.* **72**(9):7510–7522.

ADVANCES IN VIRUS RESEARCH, VOL 60

GENETIC RESISTANCE TO FLAVIVIRUSES

Margo A. Brinton and Andrey A. Perelygin

Biology Department
Georgia State University
Atlanta, Georgia 30303

Resistance to flavivirus-induced disease in mice was first discovered in the 1920s and was subsequently shown to be controlled by the resistant allele of a single dominant autosomal gene. While the majority of current laboratory mouse stains have a homozygous-susceptible phenotype, the resistant allele has been found to segregate in wild mouse populations in many different parts of the world. Resistance is

0065-3527/03 $35.00

flavivirus specific and extends to both mosquito- and tick-borne flaviviruses. Resistant animals are infected productively by flaviviruses but produce lower virus titers, especially in their brains, as compared to susceptible mice. Decreased virus production is observed in resistant animals even during a lethal infection and the times of disease onset and death are also delayed as compared to susceptible mice. An intact immune response is required to clear flaviviruses from resistant mice. The resistant phenotype is expressed constitutively and does not require interferon induction. The *Flv* gene was discovered using a positional cloning approach and identified as *Oas1b*. Susceptible mice produce a truncated Oas1b protein. A C820T transition in the fourth exon of the gene introduced a premature stop codon and was found in all susceptible mouse strains tested. Possible mechanisms by which the product of the resistant allele could confer the resistant phenotype are discussed.

I. Introduction

Numerous studies indicate that the outcome of a virus infection can be influenced by host factors, such as genes, age, sex, nutritional status, and immune competence, as well as by virus virulence factors (Brinton, 1997; Nathanson, 2002). Because many viruses can cause impairment or death of the hosts they infect, they have the potential to exert selective evolutionary pressure on a host population. Virus selective pressure would be expected to preferentially increase the frequency of those host alleles that fortuitously reduce the deleterious effects of a virus infection within a host population. Coevolution of viruses with their hosts toward less deleterious infections ensures the survival of both host and virus (Brinton, 1997; Brinton and Nathanson, 1981).

The majority of studies on host genetic resistance to viruses have been done in mice and chickens because of the availability of inbred populations (Bang, 1978). Identified murine genes controlling resistance to different types of viruses have been reviewed previously (Bang, 1978; Brinton, 1997; Brinton and Nathanson, 1981; Brinton *et al.*, 1984; Casanova *et al.*, 2002; Haller, 1981; Skamene, 1985; Skamene *et al.*, 1980). Virus resistance can be controlled by a single locus, two loci, or more than two loci. In some cases, the dominant allele confers the resistance phenotype, whereas in other cases it confers the susceptible phenotype. Each of the reported virus resistance loci identified in mice appears to be unique and usually controls resistance to a single group of viruses. Only a few of the murine virus resistance genes have mapped to the major histocompatibility locus on chromosome 17. Some of the genes conferring resistance to virus infections regulate innate immunity, and some regulate specific immune responses (Brinton, 1997; Casanova *et al.*, 2002; Urosevic and Shellam, 2002). Others, such as the single locus that controls the outcome of flavivirus infections in mice, function at the intracellular level. The flavivirus resistance gene was named *Flv* in 1989 (Green, 1989).

II. Initial Discoveries of Genetic Resistance to Flaviviruses in Mice

Resistance to flavivirus-induced disease in mice was discovered in the 1920s and rediscovered multiple times in the 1930s because it was not appreciated at the time that all of the viruses used in these early studies belonged to the genus *Flavivirus* in the family *Flavivirdae*. In the early 1920s, Leslie Webster used selective breeding to establish strains of mice from wild progenitors that were either resistant (BR) or susceptible (BS) to a bacterial infection (paratyphoid) (Webster, 1923) and subsequently established mouse strains that were resistant (VR) or susceptible (VS) to an encephalitis virus, louping ill virus (Webster, 1933). Bacterial resistance and viral resistance segregated independently, as indicated by the establishment of BSVS, BSVR, BRVR, and BRVS mouse strains (Webster, 1937). Following the 1933 St. Louis encephalitis virus (SLEV) outbreak in St. Louis, Missouri, susceptibility to SLEV was tested in these mouse strains (Webster, 1936; Webster and Clow, 1936; Webster and Johnson, 1941). The mice used in this study had been inbred for 12 generations by brother–sister matings. A 95% mortality was observed for the VS strains, whereas the VR strains showed 15% mortality. The progeny of matings between resistant and susceptible mice showed 31% mortality. These results indicated that resistance to flavivirus-induced morbidity and mortality was inherited as a dominant trait. The VR and VS mouse strains were subsequently shown to be resistant or susceptible, respectively, to Russian spring–summer encephalitis virus (Casals and Schneider, 1943), indicating that this resistance extended to tick-borne as well as to mosquito-borne flaviviruses.

Differential susceptibility between two partially inbred mouse strains, Swiss and Det, to virulent yellow fever virus (YFV) was first observed by Sawyer and Lloyd (1931) and was subsequently tested by Lynch and Hughes (1936). Swiss mice displayed 100% mortality after intracerebral inoculation, whereas the Det strain showed 77% mortality. F_1 progeny showed 83.3% mortality, and data from the analysis of F_2 animals indicated that susceptibility to this virus was controlled by host genetic factors (Lynch and Hughes, 1936). The relatively high mortality rates observed in the "resistant" Det and F_1 populations were most likely due to the presence of a significant proportion of homozygous-susceptible individuals within these partially inbred populations. In the early 1950s, it was demonstrated that all individuals of a partially inbred strain of mice housed at the Rockefeller Institute's Division of Animal and Plant Pathology at Princeton (PRI) were completely resistant to intracerebral inoculation with the attenuated 17D YFV vaccine (Sabin, 1952a, 1952b). Breeding studies with PRI mice clearly indicated that flavivirus resistance to virus-induced disease was controlled by a dominant allele of a single autosomal gene. It was also shown that young mice (1 to 3 weeks of age) were more susceptible than adults and that intracerebral inoculation of more virulent flaviviruses, such as French neurotropic YFV or Japanese B encephalitis virus (JEV), could partially overcome this resistance as measured by an animal mortality end point. However, higher doses were required to kill resistant mice as compared to susceptible mice.

Among laboratory mice, only the BSVR, BRVR, PRI, and Det strains have the resistant phenotype (Sabin, 1952a, 1952b; Sawyer and Lloyd, 1931; Webster, 1933, 1937). Brinton Darnell *et al.* (1974) demonstrated that adult animals from laboratory mouse strains, such as A/J, AKR/J, BALB/cJ, B10D2/J, C3H/He, C57BL/6J, C57BL/10J, DBA2/J, SJL/J, Swiss/J, and SWR/J, are susceptible to flavivirus-induced disease [88 to 100% mortality after intracerebral inoculation of 17D YFV (Table I)]. Sangster *et al.* (1993) subsequently confirmed that A/J, AKR/Lac, BALB/cJLac, C57BL/10ScSn, DBA/2J, and SJL/J mice are susceptible to 17D YFV, as well as to Murray Valley encephalitis virus (MVEV), and also showed that the CBA, CE, I/M, LP.RIII, NZW, and RIII inbred strains are susceptible to these viruses (Table I). The origins of most of the current laboratory mouse strains are not clearly documented, and it is thought that they may contain a "mosaic of genomes" from *Mus musculus* subspecies *domesticus*, *musculus*, *castaneus*, and *molossinus* (Bonhomme and Guenet, 1989; Bonhomme *et al.*, 1987; Potter, 1978). Also, most of the current inbred mouse strains were derived from a small number of progenitors. The flavivirus susceptibility of the majority of inbred mouse strains suggested that the resistant allele might be rare in wild *Mus* populations. However, data from a number of studies showed that this is not the case.

TABLE I
FLAVIVIRUS SUSCEPTIBILITY OF LABORATORY AND WILD MICE

Type of mouse	Number tested	Challenge virus	Mortality (%)
A/J	10[a]	17D-YF	100
AKR/J	8[a]	17D-YF	88
BALB/cJ	10[a]	17D-YF	100
B10D2/J	26[a]	17D-YF	100
C3H/He	38[a]	17D-YF	100
C3H/RV	34[a]	17D-YF	0
C57BL/6J	9[a]	17D-YF	100
C57BL/10J	5[a]	17D-YF	100
DBA/2J	5[a]	17D-YF	100
SJL/J	10[a]	17D-YF	90
Swiss/J	5[a]	17D-YF	100
SWR/J	10[a]	17D-YF	90
F_1 (*M.m.domesticus* × C3H/He)	91[a]	17D-YF	45
F_2 (F_1 × C3H/He)	41[a]	17D-YF	17
M.m.domesticus			
Devonshire, CA	5[a]	17D-YF	0
LaPuenta, CA	5[a]	17D-YF	0
Maryland	10[a]	17D-YF	20
Soledad, CA	5[a]	17D-YF	0
CASA/Rk	13[b]	MVE-OR2	15
	14[b]	17D-YF	0
CAST/Ei	5[b]	MVE-OR2	0
	3[b]	17D-YF	0
MOLD/Rk	20[b]	MVE-OR2	100
	26[b]	17D-YF	4
C3H/HeJ	9[b]	MVE-OR2	100
	10[c]	MVE-OR155	100
	44[c]	MVE-OR156	95
C3H/RV	14[b]	MEV-OR2	0
	10[c]	MEV-OR155	10
	8[c]	MVE-OR156	0
M.caroli, Thailand	16[c]	MVE-OR156	25
M.cookii, Thailand	12[c]	MVE-OR156	0
M.m.domesticus (CLA), USA	20[c]	MVE-OR156	100
M.m.domesticus (JJD), USA	9[c]	MVE-OR155	44
M.m.molossinus (MOLO), Japan	5[c]	MVE-OR155	100

(continues)

TABLE I (*continued*)

Type of mouse	Number tested	Challenge virus	Mortality (%)
M.m.musculus	9[c]	17D-YF	0
Denmark	20[c]	MVE-OR156	60
Czechoslovakia	10[c]	17D-YF	90
M.spicilegus, Yugoslavia	17[c]	MVE-OR156	0
M.spretus, Spain	20[c]	MVE-OR156	80

[a] From Brinton Darnell *et al.* (1974a).
[b] From Sangster *et al.* (1993).
[c] From Sangster *et al.* (1998).

III. Flavivirus Genetic Resistance in Wild Mice

Both resistant and susceptible alleles of the flavivirus resistance gene were found to be segregating in populations of wild mice in the United States and Australia. When wild *M. musculus domesticus* trapped in Maryland/Virginia or California in the early 1970s were tested for their resistance to an intracerebral inoculation of 17D YFV, the majority of these mice were resistant (Table I). Wild mice trapped in Maryland/Virginia and then barrel bred randomly for three to five generations were mated with susceptible inbred C3H/HeJ mice, and the F_1 and F_2 progeny were tested for virus resistance. Results indicated that these wild populations contained homozygous-resistant, homozygous-susceptible, and heterozygous-resistant individuals and confirmed that resistance was controlled by a single dominant autosomal allele (Brinton Darnell *et al.*, 1974). When wild *M. musculus domesticus* trapped in several different locations in Australia were inoculated intracerebrally with MVEV, about 18 to 25% of the mice survived (Sangster and Shellam, 1986). Thirty-five to 50% of the progeny of matings between wild survivors of MVEV infection and homozygous-susceptible C3H/HeJ mice were resistant, suggesting that the parental wild survivors were heterozygous for the resistant allele. Antiflavivirus antibodies were not found in sera obtained prior to virus inoculation from either the United States or Australian wild mice (Brinton Darnell *et al.*, 1974; Sangster and Shellam, 1986).

In a subsequent study, large numbers of wild *M. musculus domesticus* from Australia and the laboratory-bred progeny of these mice, as well as different species and subspecies of *Mus* from around the world, were tested for resistance to flavivirus-induced disease using different strains of MVEV and 17D YFV as the test viruses (Sangster *et al.*, 1998). The resistant allele of the *Flv* locus was shown to be distributed broadly among *M. musculus domesticus* populations in Australia and was also demonstrated in populations of *M. musculus musculus* from Denmark (Skive), *M. musculus molossinus* (MOLO) from Japan, *M. spretus* from Spain, *M. spicilegus* from Yugoslavia, and *M. caroli* and *M. cookii* from Thailand (Sangster *et al.*, 1998) (Table I). Interestingly, all of the Czechoslovakian *M. musculus musculus* (strain CZI-O) tested were susceptible to flavivirus infection. These studies showed that heterozygous animals did not survive intracerebral challenge with two virulent strains of MVEV (OR155 and OR156), but did survive challenge with the attenuated MVEV OR2. However, the presence of a single copy of the resistant allele in heterozygotes delayed both the time of disease onset and death as compared to homozygous-susceptible animals.

Inbred strains derived from wild individuals of different species within the *Mus* complex have also been tested for flavivirus resistance (Sangster and Shellam, 1986; Sangster *et al.*, 1993). Although the CASA/Rk and CAST/Ei strains derived from *M. musculus castaneus* displayed resistance characteristics that were the same as those of mice carrying the PRI *Flv*-resistant allele, the resistant allele in these mice was designated Flv^r-like. The resistant phenotype in the MOLD/Rk strain derived from *M. musculus molossinus* differed from that of Flv^r. These mice did not develop disease after an intracerebral inoculation of the attenuated 17D YFV vaccine, but did after an intracerebral inoculation with MVEV (Sangster *et al.*, 1993). The resistant allele in these mice was designated minor resistance allele, Flv^{mr}. Another study "rediscovered" inherited resistance to flaviviruses in inbred strains of mice derived from wild ancestors of *M. musculus domesticus* (WMP/Pas), *M. musculus musculus* (MAI/Pas, MBT/Pas, PWK/Pas), and *M. spretus* (SEG/Pas, STF/Pas) by testing them with West Nile virus (WNV) (Mashimo *et al.*, 2002). These authors reported that resistance to WNV was controlled by a new gene, *Wnv*, rather than by *Flv*, which is the case.

The worldwide distribution of mice that carry the resistant allele of the *Flv* gene, as well as the presence of this allele in divergent mouse species, suggest that it was not acquired recently. The high frequency of the resistant allele in wild *Mus* populations suggests that it may have conferred a selective advantage. The flavivirus that exerted the pressure for the selection of this allele is not known but could have been a progenitor of current flaviviruses that existed in the distant past.

IV. Development of Congenic Flavivirus-Resistant and -Susceptible Mouse Strains

To create a congenic pair of resistant and susceptible strains, susceptible inbred C3H/He mice were crossed with the homozygous-resistant PRI mice. All F_1 animals were heterozygotes and survived intracerebral inoculation with 17D YFV (Groschel and Koprowski, 1965). These F_1 animals were backcrossed to C3H/He mice, the progeny were tested for virus resistance, and survivors were backcrossed again to C3H/He mice. This backcross protocol was continued through the eighth generation. Eighth generation carriers of the selected Flv^r allele were mated to each other, and homozygous-resistant progeny were identified by analysis of the virus resistance of their offspring. From these mice the homozygous (Flv^r/Flv^r) inbred C3H/RV strain that is congenic with C3H/He was established (Groschel and Koprowski, 1965). This strain has since been renamed C3H.PRI-Flv^r (Green, 1989).

As a result of backcrossing, rapid allelic homogenization would be expected to occur in regions of the genome that were not linked to the Flv^r allele, as only the Flv region would be maintained in the heterozygous state throughout the breeding scheme by repeated selection of progeny for virus resistance. However, due to linkage with the selected Flv^r allele, flanking chromosomal regions from the parental PRI strain would also be retained. The lengths of such flanking regions can vary greatly because of the random distribution of crossover sites. For N backcross generations (if $N > 5$), the average length of the flanking region has been estimated to be $200/N$ (Silver, 1995). Although the average size of flanking regions decreases slowly in subsequent backcross generations, these regions are predicted to be no more than 25 centiMorgans (cM) in length by the eighth backcross generation. The length of the donor PRI region present in the C3H.PRI-Flv^r mouse strain genome has been estimated to be 31 cM (Urosevic *et al.*, 1999), which is slightly longer than the theoretical prediction. Because of the long flanking regions from the PRI parent, allelic differences in several additional genes have been demonstrated between congenic C3H.PRI-Flv^r and C3H/He mice (see Section VII).

Using the standard backcross protocol, the minor resistance allele, *Flv*^{mr}, from the MOLD/Rk strain was introduced into the genetic background of the inbred-susceptible C3H/HeJARC strain and then fixed in the homozygous state (*Flv*^{mr}/*Flv*^{mr}) by a series of brother–sister matings generating the congenic strain, C3H.MOLD-Flv^{mr} (Urosevic *et al.*, 1999). An additional congenic strain, C3H.M.domesticus-Flv^{r}, was developed in the C3H/HeJARC background to investigate the characteristics of the *Flv*^{r}-like allele derived from wild Australian *M. musculus domesticus* mice (Urosevic *et al.*, 1999). Sizes of the donor chromosomal regions flanking the selected resistant alleles in the C3H.M.domesticus-Flv^{r} and C3H.MOLD-Flv^{mr} strains have been estimated to be 9 and 11 cM, respectively.

V. Virus Specificity of the *Flv*^{r} Phenotype in Mice

The disease resistance conferred by the Flv^{r} allele is specific for flaviruses and has been demonstrated with a number of different flaviviruses. Mosquito-borne flaviviruses tested include WNV (Brinton Darnell *et al.*, 1974; Goodman and Koprowski, 1962a, 1962b; Hanson *et al.*, 1969; Mashimo *et al.*, 2002; Sabin, 1954), dengue (Sabin, 1954), SLEV (Casals and Schneider, 1943), YFV (Brinton Darnell *et al.*, 1974; Goodman and Koprowski, 1962a, 1962b; Groschel and Koprowski, 1965; Lynch and Hughes, 1936; Perelygin *et al.*, 2002; Sabin, 1952a; Sangster and Shellam, 1986; Sawyer and Lloyd, 1931), JEV (Sabin, 1954), Banzi (Jacoby and Bhatt, 1976), Ilheus (Hanson *et al.*, 1969), MVEV (Sangster and Shellam, 1986), and Kunjin, Alfuy, and Kokobera (Shellam *et al.*, 1998). The tick-borne flaviviruses tested include louping ill (Casals and Schneider, 1943; Webster, 1923) and Russian spring–summer encephalitis viruses (Casals and Schneider, 1943). Flavivirus-resistant and -susceptible mouse strains were shown to be equally susceptible to viruses from other families, such as an arenavirus, lymphocytic choriomeningitis (Sabin, 1954); three bunyaviruses, Rift Valley fever, sandfly fever (Sabin, 1954), and Kununurra (Shellam *et al.*, 1998); a picornavirus, poliovirus (Sabin, 1954); a rhabdovirus, rabies (Sabin, 1954); seven togaviruses, Western equine encephalitis, Eastern equine encephalitis, Venezuelan equine encephalitis (Jacoby and Bhatt, 1976; Sabin, 1954), Chikungunya, Sindbis (Hanson *et al.*, 1969), Semliki Forest virus (Sangster and Shellam, 1986), and Ross River virus (Urosevic and Shellam, 2002); and two herpes viruses, herpes simplex (Sabin, 1954) and murine cytomegalovirus (Shellam *et al.*, 1998). Vesicular stomatitis virus (VSV) and Sindbis virus were shown to grow equally well in resistant and susceptible embryofibroblasts (Brinton Darnell and Koprowski, 1974a; Perelygin *et al.*, 2002) (Fig. 6A).

VI. Characteristics of the *Flv*^{r} Phenotype

A. *Flavivirus Infections in Resistant and Susceptible Animals*

Flv^r/*Flv*^{r} mice are usually resistant to disease induced by flaviviruses, but not to infection by these viruses. The virulence of the infecting flavivirus, as well as the dose and route of inoculation, affects the final outcome of an infection. Although virulent flaviviruses can cause death in resistant mice after intracerebral injection, even during lethal infections, some effects attributable to the resistant allele are evident.

The minimal lethal virus dose required to kill resistant mice is higher than that for susceptible mice. For instance, *Flv*^{r}/*Flv*^{r} mice are completely resistant to intracerebral challenge with undiluted 17D YFV vaccine, whereas congenic susceptible mice show 100% mortality with a 10^{-2} dilution of this virus (Table II). Whereas resistant mice display 33% mortality after intracerebral inoculation with 50,000 plaque-forming units (PFU) of WNV, susceptible mice show 82% mortality after inoculation with 5000 PFU (Table II). Differences in ambient temperature do not alter the resistant phenotype. At least 100,000 times more Banzi virus was required to kill resistant mice as compared to susceptible mice after intraperitoneal inoculation (Jacoby and Bhatt, 1976).

Resistant mice produce lower yields of flaviviruses than susceptible mice. Although a decrease in the growth of all of the flaviviruses tested so far has been observed in resistant mice, the extent of this effect varies depending on the virus. Brain tissue has been the primary focus of *in vivo* virus titration studies. After an intracerebral inoculation of 10^3 LD_{50} of MVEV, virus titers were similar in resistant and susceptible brains on day 2, but peak titers in the brains of susceptible mice were >1000 times higher than in resistant mouse brains (Fig. 1A) (Urosevic *et al.*, 1997b). All of the MVEV-infected susceptible mice died by day 6, but all of the resistant mice survived, and virus titers in resistant mouse brains declined after day 6. The amount of total viral RNA was shown to be significantly higher in susceptible brains as compared to resistant brains (Urosevic *et al.*, 1997b). After an intracerebral inoculation of $10^{5.5}$ PFU of WNV E101, both resistant and susceptible mice died, but peak brain titers were 10,000-fold higher in susceptible brains (Fig. 1B) (Hanson *et al.*, 1969). Peak titers in WNV-infected susceptible

TABLE II
EFFECT OF AMBIENT TEMPERATURE ON THE OUTCOME OF FLAVIVIRUS INFECTIONS[a]

Mouse strain	Virus	Amount of virus[b]	Temperature (°C)	No. dead/total
C3H/He	WNV	5 000[c]	23	6/9
		5 000	35[d]	8/8
C3H/RV	WNV	50,000	23	3/9
		50,000	35	3/9
C3H/He	17D-YFV	10^{-2}[e]	23	6/6
		10^{-2}	35	5/5
		10^{-3}	23	3/4
		10^{-3}	35	3/5
C3H/RV	17D-YFV	Undiluted	23	0/8
		Undiluted	35	0/8
NMRI	17D-YFV	10^{-2}	5	9/10
		10^{-2}	23	9/10
		10^{-2}	35	6/10

[a] Data from Brinton Darnell *et al.* (1974a).
[b] Three- to 6-month-old animals were given an intracerebral injection of 0.03 ml of virus.
[c] Total PFU/animal.
[d] Mice kept at 35°C were acclimated to this temperature 24 h before infection.
[e] Dilution of standard 17D-YFV vaccine.

mice were reached by day 3, and all of the susceptible mice died by day 4. In resistant brains, peak virus titers were observed by day 4 and death occurred on day 7 or 8. The final titer point for resistant mice in Fig. 1B was obtained from moribund animals and showed an increase. None of the resistant mice developed disease after an intraperitoneal injection of $10^{8.5}$ PFU of WNV E101, whereas 50% of susceptible mice died after injection of $10^{7.5}$ PFU by this route (Brinton, 1986).

All susceptible mice died by day 6 after an intracerebral infection with 10^2 $TCID_{50}$ of Banzi virus (Bhatt *et al.*, 1981). Although all of the resistant mice died after the same dose of Banzi, they died by day 8 and brain titers on day 8 were more than 10^2 $TCID_{50}$ lower than those in susceptible mice on day 6. The pathogenesis of Banzi infections in the brains of resistant and susceptible mice was similar, but the onset of pathogenic changes began 2 days later in resistant animals. After an intraperitoneal inoculation of resistant mice with 250 $TCID_{50}$ of Banzi virus and susceptible mice with 100 $TCID_{50}$, virus replication in spleens and thymus was delayed for about a day in

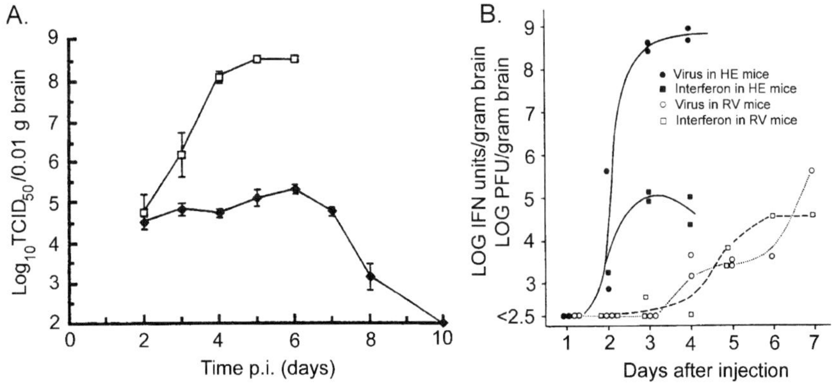

FIG 1. Replication of MVEV and WNV in the brains of resistant C3H.PRI-Flv^r and susceptible C3H/He mice. (A) Mice were infected with 10^3 LD_{50} of the OR2 strain of MVEV by the intracerebral route. Virus titers were determined by $TCID_{50}$ assay and are expressed as $\log_{10}$ $TCID_{50}/0.01$ g of brain tissue. All of the susceptible mice died by 6 days after infection. None of the resistant mice showed signs of disease, and virus was cleared from their brains by 10 days after infection. □, susceptible C3H/He; ◆, resistant C3H.PRI-Flv^r. Data from Urosevic *et al.* (1997b). (B) Resistant (RV;C3H.PRI-Flv^r) and susceptible (He;C3H/He) mice were inoculated intracerebrally with $10^{5.5}$ PFU of WNV, a dose that kills most of the resistant mice. Virus titers were determined by plaque assay on MK_2 cells. After low pH treatment to inactivate WNV, IFN levels were assayed by plaque reduction of vesicular stomatitis virus. Data from Hanson *et al.* (1969).

resistant mice as compared to susceptible mice, but peak titers in these organs were similar in both types of animals (Jacoby and Bhatt, 1976). Virus replication in the brain was first observed on day 3 in susceptible mice and on day 5 in resistant mice. However, peak brain titers differed by $10^{3.5}$ $TCID_{50}$. These data indicate that the spread of flaviviruses is slower in resistant mice.

B. An Intact Immune Response Is Required for Clearance of a Flavivirus Infection in Resistant Mice

The final outcome of a viral infection is the result of a complex interplay between multiple host and viral components (Bang, 1978). Eight-gram homozygous-resistant PRI mice showed 70% mortality after intraperitoneal infection with French neurotropic YFV, whereas 12- to 13-g mice and 20- to 25-g mice displayed 10% and 0% mortality, respectively (Sabin, 1954). C3H.PRI-Flv^r mice did not survive intraperitoneal Banzi virus infections until they were 4 weeks old and were not fully resistant until 8 to 12 weeks of age (Jacoby and Bhatt, 1976). Treatment of resistant C3H.PRI-Flv^r mice with cortisone did not alter their resistance to WNV (Goodman and Koprowski, 1962b). Although sublethal X-irradiation increased mortality rates in resistant mice, it did not increase virus titers in brain tissue (Goodman and Koprowski, 1962b). At higher doses of irradiation both mortality and virus titers increased. Treatment of resistant mice with either 6-thioguanine or endotoxin increased both mortality and virus titers (Goodman and Koprowski, 1962b). Immunosuppression of Banzi virus-infected resistant mice by X-irradiation, cyclophosphamide, or thymectomy decreased their survival time (Bhatt and Jacoby, 1976). For each of these experimental treatments, the mortality rate was higher, the survival time was shorter, and virus titers in brain tissue were higher in susceptible mice as compared to resistant ones.

The kinetics and extent of the antiviral immune response are similar in resistant and susceptible animals. The appearance of a detectable antiviral hemagglutination-inhibiting antibody in the serum of Banzi virus-infected resistant mice was delayed by 3 days as compared to congenic susceptible mice, but similar peak antibody titers were observed in both types of animals by day 8 after infection (Jacoby *et al.*, 1980). After treatment with cycloheximide, the hemagglutination-inhibiting antiviral antibody could not be demonstrated in infected resistant mice and all of the resistant animals died.

Virus-specific T-cell cytotoxic activity was detected by 6 days after Banzi virus infection, peaked at day 8, and was no longer detectable by day 16 (Sheets *et al.*, 1979). No difference in the duration or extent of the cytotoxic T-cell response was observed between congenic-resistant and -susceptible mice. Thymectomized C3H.PRI-Flv^{r} mice produced antiviral antibody but were not protected. However, compared to T-cell-depleted susceptible mice, the spread of virus was slower, peak virus titers in tissues were lower, and time of death was delayed in thymectomized-resistant mice. Adoptively transferred immune spleen cells protected both resistant and susceptible mice (Jacoby *et al.*, 1980). These data indicate that a fully functioning immune response is required for the clearance of flaviviruses from resistant mice and that if viral clearance does not occur, the mice do not survive. However, the humoral and cellular immune responses do not specifically contribute to the Flv-resistant phenotype.

C. *Flavivirus Replication in Resistant and Susceptible Cell Cultures*

Flavivirus resistance is not controlled at the level of virus attachment or entry into cells. Comparable numbers of cells in resistant and susceptible embryofibroblast cultures were infected by WNV as demonstrated

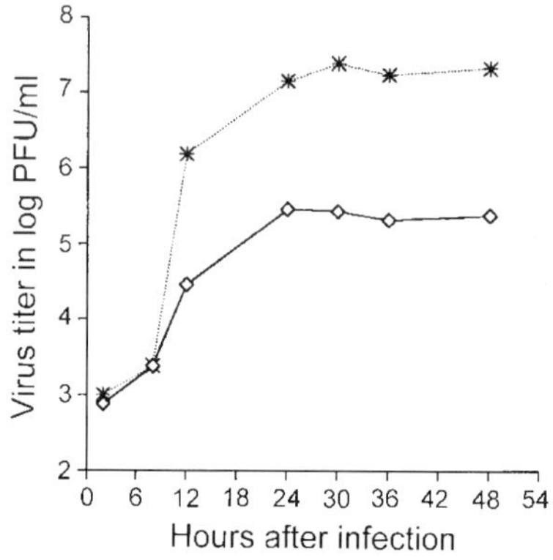

FIG 2. Growth curves of West Nile virus, strain Eg101, in SV40-transformed susceptible C3H/He (✱) and congenic-resistant C3H.PRI-Flv^r (◇) embryofibroblasts. Cells were infected with an MOI of 5, and virus titers were determined by plaque assay on BHK cells.

by immunofluorescence (Brinton Darnell and Koprowski, 1974a). Lower titers of flaviviruses, but not of other types of viruses, are produced by primary cell cultures prepared from the brains, lungs, kidneys, spleens, peritoneal macrophages, and embryofibroblasts of resistant animals (Brinton Darnell and Koprowski, 1974a; Goodman and Koprowski, 1962a, 1962b; Silvia *et al.*, 2001; Vanio, 1963a, 1963b; Webster and Johnson, 1941). A 10- to 100-fold difference was observed in titers produced by resistant and susceptible embryofibroblast cultures after infection with WNV (Fig. 2) (Brinton Darnell and Koprowski, 1974a). This difference was observed with both primary and SV40-transformed embryofibroblast cultures. The maximum titer difference observed in cell cultures was not as great as that observed in brains. Both VSV and Sindbis virus grew equally well in resistant and susceptible cells (Fig. 6A). The titers of MVEV, strain OR2, produced by resistant thioglycollate-elicited peritoneal macrophages were 10-fold lower than those from susceptible macrophages (Silvia *et al.*, 2001). Differential production of Banzi virus by resistant and susceptible brain, spleen, thymus, macrophage, or embryofibroblast cultures was not observed (Bhatt *et al.*, 1981). However, the maximum difference between brain titers in resistant and susceptible mice after a Banzi virus infection was only 10^2 $TCID_{50}$ (Bhatt *et al.*, 1981) as compared to a difference of $>10^5$ PFU/ml in WNV brain titers (Fig. 1B) and a difference of $>10^{3.5}$ $TCID_{50}$ in MVEV brain titers (Fig. 1A).

Analysis of both plus and minus strand viral RNA at early times after infection of resistant and susceptible embryofibroblasts with

WNV (MOI 5) indicated that by 2 h after infection the amounts of plus strand RNA were significantly lower in resistant cells. Interestingly, the amounts of minus strand RNA in the two types of cells were similar (Li and Brinton, unpublished data). While plus strands are released from replication complexes in flavivirus-infected cells, minus strands are found only within replication complexes. Analysis of MVEV RNA in resistant mouse brains also showed that viral plus strand RNA represented a significantly lower proportion of the total viral RNA than it did in susceptible brains (Urosevic *et al.*, 1997b). These data were interpreted as indicating that the *Flv*^{r} allele reduced the efficiency of flavivirus RNA replication (Brinton, 1997; Urosevic *et al.*, 1997b). However, an increased turnover of viral RNA cannot be ruled out.

D. Increased Production of Defective Interfering Particles in Resistant Cells and Mice

When culture fluids from WNV-infected, susceptible C3H/He embryofibroblast cultures were passaged serially at 3-day intervals to fresh confluent-susceptible cultures, virus titers cycled between 10^4 and 10^7 PFU/ml during 11 passages (Brinton Darnell and Koprowski, 1974a). In contrast, titers declined progressively during serial passage of WNV in congenic-resistant C3H.PRI-Flv^{r} cultures, and virus became undetectable between passages 4 and 7 (Brinton Darnell and Koprowski, 1974a). Although the titer of WNV produced during passage decreased as the multiplicity of infection used for the initial infection increased in both resistant and susceptible cultures, this effect was always significantly greater in resistant cell cultures. Virus passaged in the resistant cells, but not virus passaged in the susceptible cells, interfered with the replication of WNV propagated in suckling hamster brain (Brinton, 1983; Brinton Darnell and Koprowski, 1974a). In cross-passage experiments, susceptible cultures could rescue plaquing virus one passage after it had become undetectable in resistant cell culture fluids (Fig. 3) (Brinton, 1983). These data were consistent with the accepted criteria for the demonstration of defective interfering (DI) particle production (Huang and Baltimore, 1970) and suggested that although both types of cell cultures produced DI particles, the resistant cells amplified DI particles more rapidly and efficiently. Homologous interference by DI particles was more efficient in resistant cells than in susceptible cells. Neither differential production of nor interference by DI particles was observed when the rhabdovirus, vesicular stomatitis virus was serially passaged in flavivirus-resistant

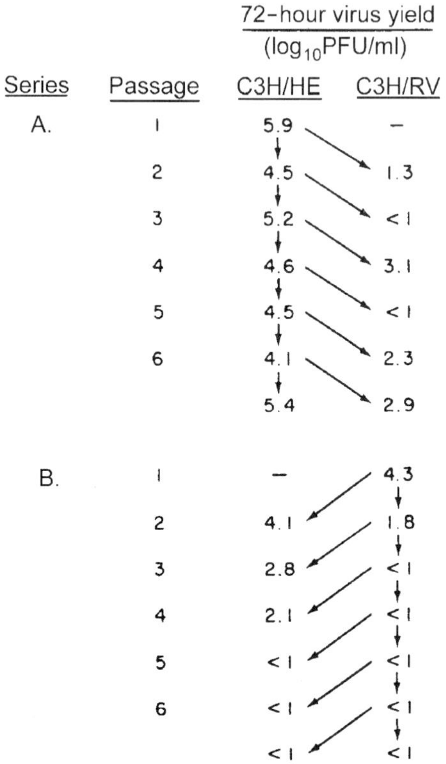

72-hour virus yield
(log10PFU/ml)
Series
Passage
C3H/HE
C3H/RV
A.
1
5.9
–
2
4.5
1.3
3
5.2
< 1
4
4.6
3.1
5
4.5
< 1
6
4.1
2.3
5.4
2.9
B.
1
–
4.3
2
4.1
1.8
3
2.8
< 1
4
2.1
< 1
5
< 1
< 1
6
< 1
< 1
< 1
< 1

FIG 3. Serial undiluted passage of WNV in resistant C3H.PRI-Flv^{r} (C3H/RV) and susceptible C3H/He embryofibroblast cultures. Cells were infected initially at an MOI of 10 with virus that had been plaque purified six times. Culture fluid was harvested 3 days after infection, diluted 1:2 with fresh media, and half of media was transferred to a fresh culture of resistant cells and half to a fresh culture of susceptible cells as indicated by arrows. While susceptible cells supported extended serial passage of WNV, resistant cultures did not. Data from Brinton (1983).

and -susceptible cells (Brinton Darnell and Koprowski, 1974b; Huang, personal communication).

Smith (1981) demonstrated the presence of interfering virus in the brains of adult-resistant C3H.PRI-Flv^{r} mice by 5 days after an intraperitoneal injection of $10^{5.7}$ $TCID_{50}$ of Banzi virus, but not in the brains of similarly infected congenic adult C3H/He mice. Interfering virus was detected in the spleens of both infected resistant and susceptible mice, but was detected later and at lower levels in the susceptible mice as compared to the resistant mice. All of the resistant mice but none of the susceptible mice survived this infection. Even though interfering virus was detected in the brains of weanling or cyclophosphamide-treated adult-resistant mice and the virus titers in the brains of these animals were lower than in comparable susceptible mice, these mice did not survive. As discussed previously, a fully functioning immune response is required to clear virus in resistant mice.

Gradient centrifugation of [^{3}H]uridine-labeled, extracellular WNV particles indicated that a denser fraction enriched for DI particles could be partially resolved (Brinton, 1983; Fig. 4). Only DI particles from enveloped viruses with helical nucleocapsids have been well resolved by gradient centrifugation (Huang and Baltimore, 1970). The majority of the viral RNA in the virion fraction, as well as in the DI-enriched fraction obtained from C3H/He cells by 32 h postinfection, was full-length 40S genome RNA (fraction 25 in Figs. 4D and 4E), but smaller RNAs of various sizes were also observed to be associated with these virus particles. Heterogeneous small RNAs represented a significantly larger proportion of the total RNA in virus particles produced by resistant C3H.PRI-Flv^r cells (Figs. 4I and 4J). Also, the PFU/CPM ratio of virus particles produced by resistant cells was significantly lower than that for particles produced by susceptible cells (Figs. 4C and 4H). Additional studies showed that the proportion of full-length viral RNA in virus particles produced by resistant cells increased proportionally with decreasing MOI of infection and also increased when a six times plaque-purified virus was used to initiate an infection (Figs. 5I and 5N). However, the amount of full-length genomic RNA in virions and infected cells, as well as the titer of virus produced by resistant cells, was always significantly lower as compared to those from susceptible cells. An exception to this was observed only with the 27-h samples obtained after infection with multiply plaque-purified virus at a MOI of 1 (Figs. 5K and 5L).

The observation of multiple sizes of small RNAs in virions is not typical of the DI particle populations described for other types of viruses (Huang and Baltimore, 1970). Usually, only a single species of DI RNA is observed. With further passage, this DI RNA is sometimes replaced by a smaller DI RNA generated by an additional deletion event. The reappearance of multiple-size classes of small RNAs in virions produced by resistant cells between 51 and 72 h after infection with six times plaque-purified WNV (Figs. 5O and 5P) was also unusual, and it was postulated that the original populations of small RNAs had not been eliminated by serial plaque purification in BHK cells (Brinton, 1983). Northern blotting indicated that the small RNAs were virus specific and of positive polarity (Brinton, 1982). An alternative explanation for the production of heterogeneous small RNAs is that they are

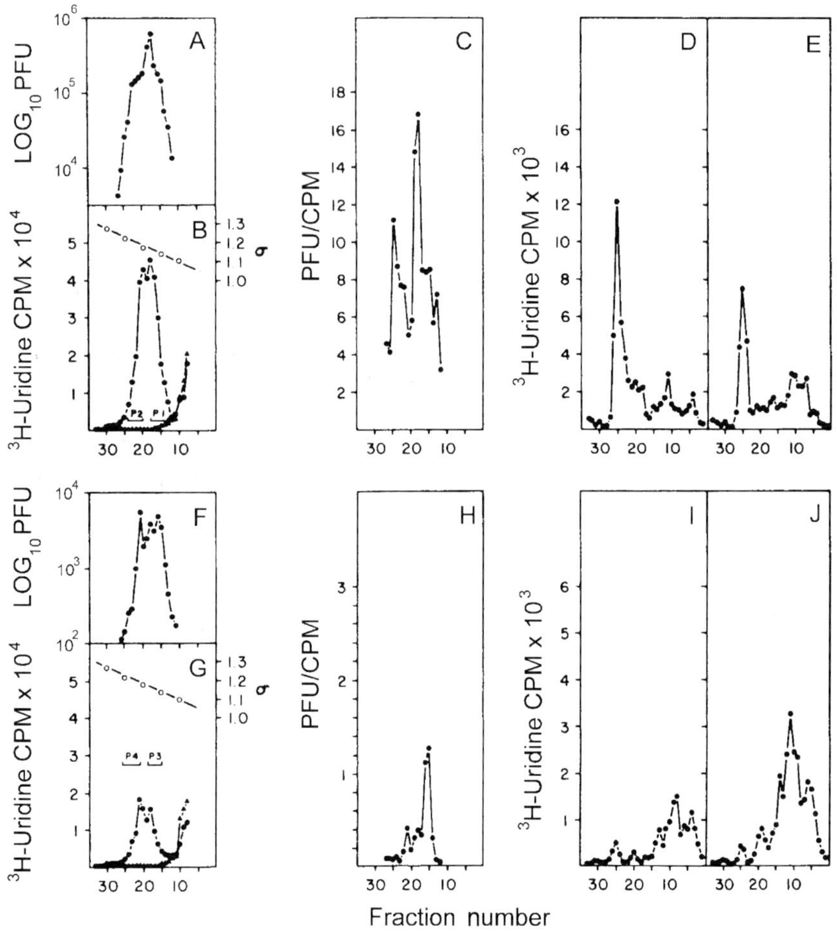

FIG 4. Isopycnic density gradient centrifugation of WNV particles and virion RNA produced by resistant C3H.PRI-Flv^r and susceptible C3H/He cells. Cells were incubated with [^{3}H]uridine from 6 to 32 h after infection with WNV at an MOI of 10. Harvested culture fluids were clarified and then centrifuged on 15 to 45% isopycnic sucrose density gradients. The direction of centrifugation is from right to left. Aliquots of each gradient fraction were analyzed for acid-insoluble radioactivity (B and G) and infectivity (A and F). (A to E) Samples from susceptible cells. (F to J) Samples from resistant cells. The ratio of infectivity to incorporated radioactivity was calculated for each fraction (C and H). The viral RNA in pooled gradient fractions (P1 to P4) was extracted and sedimented on 15 to 30% SDS-sucrose gradients (D, E, I, and J for pools 1 through 4, respectively). ●, virus-infected culture fluids; ▲, uninfected control culture fluid; ○, σ values. Data from Brinton (1983).

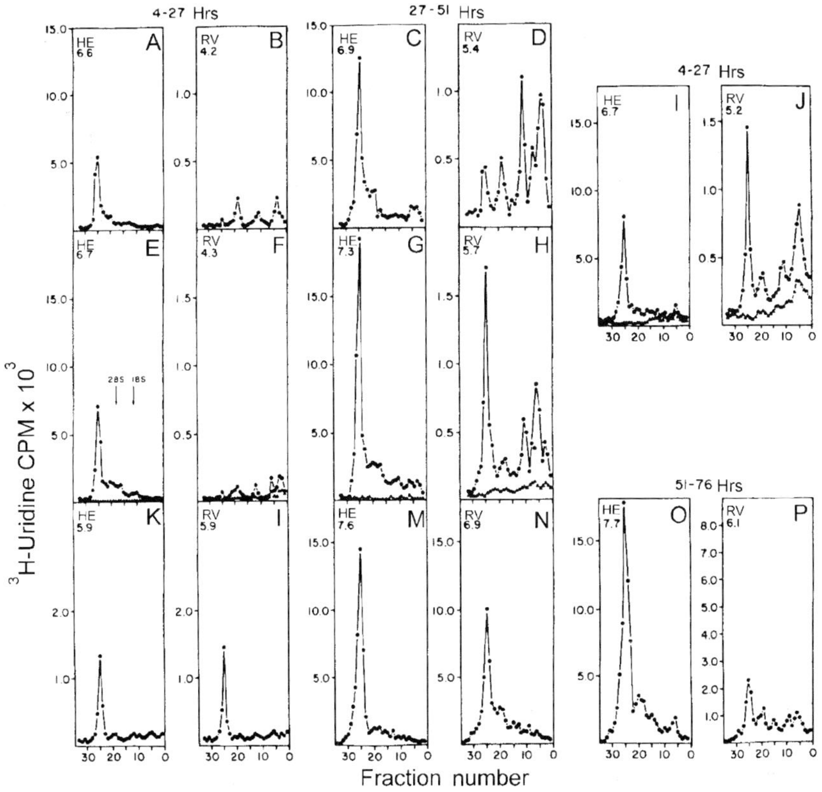

FIG 5. Effect of different multiplicities of infection and serial plaque purification on progeny virus RNA populations. [^{3}H]Uridine-labeled RNA in progeny virus harvested at 27, 51, or 76 h after infection was extracted and centrifuged on SDS-sucrose gradients. The direction of sedimentation is from right to left. RNA in progeny virus produced by resistant (RV) and susceptible (He) cultures infected with virus that had not been plaque purified at a MOI of 10 (A to B), 1 (E to H), or 0.1 (I and J). RNA in progeny virus produced after infection with six times plaque-purified virus at an MOI of 1 (K to P). ●, infected; ▲, uninfected. Data from Brinton (1983).

generated by digestion with a cellular RNase activated by virus infection.

E. *Flavivirus Resistance Does Not Require Induction by Interferon*

In contrast to the Mx gene, which confers resistance to influenza virus-induced disease in mice (Haller *et al.*, 1980), the expression of flavivirus resistance does not require induction by IFN. C3H.PRI-Flv^r

TABLE III
EFFECT OF ANTIBODY TO MOUSE INTERFERON ON THE EXPRESSION OF FLAVIVIRUS RESISTANCE IN MICE[a,b]

Mouse strain	Genotype[c]	Treatment	Mortality	Log_{10} virus titer (PFU per brain)
C3H/RV	R/R	NSG	0/4	2.4
		AIF	0/4	2.8
C3H/He	r/r	NSG	4/4	5.7
		AIF	4/4	6.5

[a] Data from Brinton *et al.* (1982).
[b] Mice were injected with $10^{3.7}$ PFU of 17D-YFV vaccine by the intracerebral route. Sheep normal serum globulin (NSG) or sheep antimouse interferon globulin (AIF) was diluted 1 to 3 with phosphate-buffered saline, and 0.1 ml was injected intravenously just before virus was injected. Results indicate that the anti-IFN antibody did not abrogate flavivirus resistance.
[c] R, resistant; r, susceptible.

mice injected intravenously with sheep anti-α/β interferon (IFN) antibody and then inoculated intracerebrally with 17D-YFV showed no alteration in their flavivirus resistance (Table III) (Brinton *et al.*, 1982). The same treatment of A2G mice that carry the resistant Mx allele rendered them susceptible to influenza virus (Haller *et al.*, 1979). Although treatment with anti-IFN antibody increased the brain titers of YFV in both resistant and susceptible mice as compared to control mice given sheep globulin, the titer of virus remained significantly lower in treated resistant mice as compared to susceptible mice (Table III). When resistant and susceptible mice were injected intravenously with anti-IFN antibody and then inoculated intraperitoneally with 10^7 PFU of WNV (a sublethal dose), none of the resistant mice developed disease symptoms. This was also a sublethal virus dose for untreated susceptible mice, and none of the susceptible mice given normal sheep globulin developed disease. However, three of four susceptible mice given anti-IFN serum developed paralysis and died (mean day of death was 14 days). Treatment of embryofibroblast cultures from C3H.PRI-Flv^r and C3H/He mice with anti-IFN serum prior to infection with WNV resulted in an increase in the yield of virus from both types of cultures (Figs. 6B and 6C). However, the titer of virus produced by the resistant cultures remained lower by 1.0 to 1.5 logs as compared to that from susceptible cells. The lack of a requirement for induction by IFN and the detection of the effect of the *Flv*^{r} allele on viral plus strand levels as early as 2 h after infection (Li

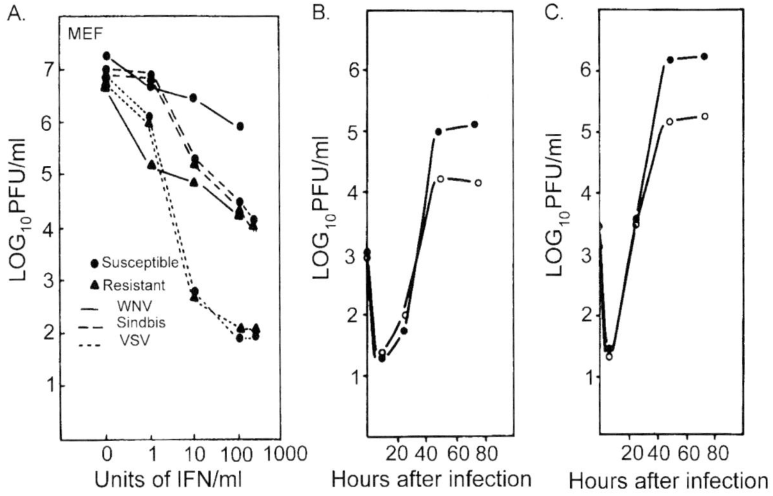

FIG 6. IFN treatment further decreases the yield of virus from resistant cells, but is not needed for expression of the resistant phenotype. (A) Comparison of α/β IFN-mediated suppression of Sindbis virus, vesicular stomatitis virus, and WNV replication in primary cultures of susceptible and resistant mouse embryofibroblasts. Data from Brinton Darnell and Koprowski (1974a). Effect of sheep anti-mouse α/β IFN on the growth of WNV in (B) resistant C3H.PRI-Flv^r and (C) susceptible C3H/He cultures. Anti-IFN globulin (●) or normal serum globulin (○) was added to culture fluids 24 h before infection with WNV at a MOI of 10. Data from Brinton *et al.* (1982).

and Brinton, unpublished data) indicate that the expression of the *Flv*^{r} gene product is constitutive.

Resistant cell cultures and animals do not show either an increased level or a more rapid production of α/β IFNs as compared to susceptible cells or mice. The peak levels of interferon in the brains of susceptible C3H/He mice inoculated intracerebrally with WNV were higher than those produced by resistant brains and were detected 2 days earlier in the susceptible mice (Fig. 1B) (Hanson *et al.*, 1969; Vanio *et al.*, 1961). Analysis of α/β IFNs in concentrated culture media and cell lysates prepared at various times after WNV infection (MOI 20) of resistant and susceptible second passage embryofibroblasts indicated that at 5 h less than 1 unit (U) of IFN was present in all of the samples (Brinton Darnell and Koprowski, 1974a). By 21 h, lysates from resistant cells contained 2 U and media from these cultures contained 1 U of IFN, whereas lysates and media from susceptible cells contained 9 and 10 U, respectively. During serial passage of WNV in mouse embryofibroblast cultures, IFN levels in culture fluids from resistant cells were consistently lower than in those from susceptible cells (Brinton Darnell and Koprowski, 1974a).

Even though the levels of α/β IFNs produced by resistant cells and animals infected with WNV were consistently lower, the inhibitory effect of equivalent amounts of exogenous IFN was slightly greater in resistant cells than in susceptible cells (Hanson *et al.*, 1969; Brinton Darnell and Koprowski, 1974a). When resistant and susceptible cultures were incubated for 24 h with 10, 100, or 1000 U/ml of mouse α/β IFN before infection with WNV, the togavirus, Sindbis, or the rhabdovirus, vesicular stomatitis virus, the yields of the three types of viruses decreased with increasing interferon concentration (Fig. 6A). However, while IFN inhibition of Sindbis and vesicular stomatitis virus was equivalent in resistant and susceptible cells, the yield of WNV from resistant cells, which was already at least 10 times lower than that from susceptible cells, was decreased further by $10^{0.75}$ to 10^{1} PFU/ml at each of the IFN doses tested.

VII. Other Characteristics of Resistant C3H.PRI-Flv^{r} Mice

In addition to a differential susceptibility to flaviviruses, congenic C3H/He and C3H.PRI-Flv^{r} mice also differ in the efficiency of their thermoregulation, in their responses to mosquito bites, and in their sensitivity to LPS.

A. Efficient Thermoregulation

Both C3H.PRI-Flv^{r} mice and the unrelated flavivirus-resistant mouse strain BRVR are more resistant to high and low environmental temperatures than susceptible C3H/He and BSVS mice. This difference was discovered when the heating system in the Wistar Institute animal facility failed in the winter of 1968 and the four strains of mice were subjected to identical extremes of heat and then cold. Ninety-two percent of the C3H.PRI-Flv^{r} survived as compared to only 42% of the C3H/He mice. Eighty-six percent of the BRVR survived, but none of the BSVS mice survived (Koprowski, 1971). These differences were subsequently confirmed experimentally (Lagerspetz *et al.*, 1973). The virus-resistant and -susceptible mice did not differ from each other in their normal body temperatures, in their rate of oxygen consumption at either hot or cold temperatures, or in their response to *l*-noradrenaline. However, heat dissipation after *l*-noradrenaline injection was more efficient in C3H.PRI-Flv^{r} mice than in C3H/He mice, and injections of dopamine, apomorphine, or amantadine hydrochloride caused a larger decrease in colonic temperatures in C3H.PRI-Flv^r than in C3H/He mice. These results suggested that resistant animals have more effective vasomotor control. The observation that virus-resistant mice regulate their body temperatures more effectively in both hot and cold environments suggests that this ability is controlled at the level of the central thermoregulatory region. No effect of ambient temperature on the susceptibility of C3H/He and C3H.PRI-Flv^r mice to either WNV or 17D YFV was observed (Brinton Darnell *et al.*, 1974). The expected phenotype was observed with both strains of mice maintained at either 23 or 35°C (Table II). The normal body temperatures of the two mouse strains were equivalent, and no change in body temperature after flavivirus infection of either strain was observed until C3H/He mice became moribund and then a decrease in body temperature was observed.

B. Mosquito-Stimulated Cytokine Response

Differences in systemic cytokine production were observed between C3H.PRI-Flv^r and C3H/HeJ mice after mosquitoes fed on them. Interferon-γ production was downregulated in C3H/HeJ mice on days 7 and 10 after *Culex pipiens* feeding and on day 7 after *Aedes aegypti* feeding, whereas the Th2 cytokines, IL-4 and IL-10, were upregulated by 4 to 7 days with either type of mosquito (Zeidner *et al.*, 1999). In contrast to the immunosuppressive effect observed in C3H/HeJ mice, systemic interferon-γ and IL-2 were upregulated on days 7 and 10 and IL-4 was downregulated on day 10 after *C. pipiens* had fed on congenic C3H.PRI-Flv^r.

C. Rickettsia tsutsugamushi *Resistance*

C3H.PRI-Flv^r mice were also shown to be resistant to lethal infections with the Gilliam strain of *R. tsutsugamushi*, an obligate intracellular organism, whereas the congenic C3H/He mice were susceptible (Jerrells and Osterman, 1981, 1982). Resistant mice survived intraperitoneal infections with 10^5 50% mouse lethal doses (MID_{50}) of *R. tsutsugamushi*, whereas susceptible mice were killed by 10^3 MID_{50}. Although both resistant and susceptible mouse strains survived low-dose infections, an intravenous injection of an antimacrophage agent, such as silica or carrageenan, changed the infection to a lethal one in susceptible C3H/He mice, but not in C3H.PRI-Flv^r mice. Irradiation of the C3H.PRI-Flv^r also did not increase their susceptibility. However, the combination of irradiation and silica injection did render C3H.PRI-Flv^r

mice susceptible. It had been shown previously that natural resistance in mice to lethal infections of *R. tsutsugamushi* was controlled by a single, autosomal-dominant allele *Ric*^{r} (Groves *et al.*, 1980).

D. Lipopolysaccharide Responsiveness

Although the majority of inbred mouse strains respond to the stimulatory effects of lipopolysaccharide (LPS), C3H/HeJ, C3H/Bts (Glode and Rosenstreich, 1976), C57BL/10ScCR, and C57BL/10ScN (Vogel *et al.*, 1979) mice do not respond. These mice are resistant to LPS-induced septic shock, but susceptible to gram-negative bacterial infections. The point mutation that caused the nonresponsiveness of the C3H/HeJ and C3H/Bts mice occurred between 1960 and 1968, whereas the mutation in C57BL/10ScCR and C57BL/10ScN was estimated to have occurred some time between 1947 and 1961. Mutations responsible for the LPS nonresponsiveness in C3H/HeJ and C57BL/10ScCR mice have been localized to the Toll-like receptor-4 gene (*T1r4*). C3H/HeJ mice synthesize a mutated mRNA, whereas C57BL/10ScCR mice produce no mRNA from this gene (Poltorak *et al.*, 1998; Qureshi *et al.*, 1999). *T1r4* has been mapped to the *Lps* locus on chromosome 4 (Hoshino *et al.*, 1999; Poltorak *et al.*, 1998; Qureshi *et al.*, 1999; Watson and Riblet, 1974; Watson *et al.*, 1977). Reversion of a single nucleotide in exon 3 of the *Lps* locus of C3H/HeJARC mice, which were maintained for many years in Australia, was reported (Silvia and Urosevic, 1999). The C3H/HeJARC LPS responsive strain was used to produce the congenic flavivirus-resistant strains: C3H.M.domesticus-Flv^r and C3H.-MOLD-Flv^{mr}. C3H.PRI-Flv^r-resistant mice were made in the early 1960s using C3H/He mice that were LPS responsive. Analysis of the LPS sensitivity of animals from the three resistant strains, C3H.PRI-Flv^r, C3H.M.domesticus-Flv^r, and C3H.MOLD-Flv^{mr}, indicated that all three were highly responsive to LPS (Goodman and Koprowski, 1962a; Silvia and Urosevic, 1999). Both LPS-sensitive and -resistant C3H mice are equally susceptible to flavivirus-induced disease.

E. Summary

Mapping data clearly demonstrated that the difference in LPS sensitivity between flavivirus-resistant and -susceptible mice was due to mutations in the *Lps* locus located on a different chromosome than the *Flv* locus. However, there have been no studies to date to map the locations of the genes that control thermoregulation or the cytokine response to mosquito feeding. The observation of the same difference in thermoregulation with two sets of unrelated flavivirus-resistant and -susceptible mice suggests that an *Flv*-linked gene controls this trait. The mosquito feeding response has so far only been studied in one strain of flavivirus-resistant mice.

VIII. Mapping the *Flv* Gene to Chromosome 5

The *Flv* locus was predicted to map to mouse chromosome 5 based on coinheritance of its alleles with the *Ric* locus alleles (Jerrells and Osterman, 1981) as described earlier. The *Ric* locus has also been designated as *Eta*-1 and *Spp1*. C3H.PRI-Flv^r mice are homozygous for the *Ric*^{r} allele, whereas C3H/HeJ mice have the *Ric*^{s} allele. The retinal degradation (*rd*) locus was previously mapped close to the *Ric* locus and the *rd* locus was shown to be distal to the *Ric* locus on chromosome 5 (Groves *et al.*, 1980). The resistant C3H.PRI-Flv^r strain carries the wild-type allele at the *rd* locus, whereas susceptible C3H/HeJ mice have the defective allele. The genotype of the *rd* locus in progeny of mating experiments was determined using either histological examination or Southern blot analysis using an *rd*-specific probe. Southern blot analysis of C3H.PRI-Flv^r DNA using an *rd* locus cDNA probe demonstrated MspI fragments of 4.0, 1.9, and 0.7 kb, which were absent in C3H/He genomic DNA (Shellam *et al.*, 1993). Because the flavivirus-resistant and -susceptible mouse strains are congenic, a demonstration of allelic differences between these strains in two closely linked loci, *Ric* and *rd*, suggested that these genes were linked to the *Flv* locus (Shellam *et al.*, 1993). Two sets of three-point linkage analyses in mice were performed to define the map position of the *Flv* locus relative to four markers, *Pgm-1, Eta-1 (Ric), rd,* and *Gus-s*, on chromosome 5 (Sangster *et al.*, 1994). Results of these analyses indicated a gene order of *Pgm-1, Eta-1, rd, Flv, Gus-s* (Fig. 7).

Subsequently, primers for 20 microsatellite markers, which had been mapped to this region of mouse chromosome 5 by the Whitehead Institute, were used for finer mapping of the Flv locus (Urosevic *et al.*, 1993, 1995). Twelve of these microsatellites were genotyped relative to the *Flv* gene in 1325 backcross animals from both (C3H/HeJ × C3H.PRI-Flv^r) F_1 × C3H/HeJ and (BALB/c × C3H.PRI-Flv^r) F_1 × BALB/c matings (Urosevic *et al.*, 1997a). Three of the microsatellite markers, D5Mit408, D5Mit159, and D5Mit242, were tightly linked to the *Flv* locus (Fig. 7) and one of these markers, D5Mit159, showed no recombination with *Flv*, indicating linkage of <0.15 cM. For the other two markers, D5Mit242 and D5Mit408, four and two recombinants with the *Flv* locus (0.30 and 0.15 cM), respectively, were found (Urosevic *et al.*, 1997a).

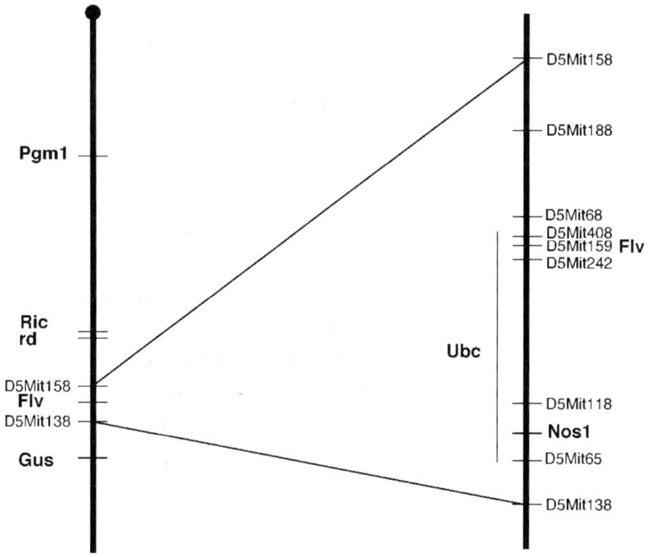

FIG 7. Genetic map of the distal portion of mouse chromosome 5 with an expansion of the *Flv* region. Four loci (left), *Pgm1, Ric, rd,* and *Gus*, were initially used to map the *Flv* locus on chromosome 5. Data from Sangster *et al.* (1994). A set of microsatellite markers was used to fine map the *Flv* locus (right). Data modified from Urosevic *et al.* (1995).

IX. Analysis of Other Known Genes That Map Close to the Flv Locus on Chromosome 5

A. The Nos-1 *Locus*

Nitric oxide (NO) has been reported to have an antiviral effect on many types of viruses (Reiss and Komatsu, 1998), including flaviviruses (Lin *et al.*, 1997). The nitric oxide synthase 1 (*Nos-1*) gene had previously been mapped to the same region of mouse chromosome 5 as the *Flv* locus (Lee *et al.*, 1995; Fig. 7) and was studied as a possible candidate for the *Flv* gene (Silvia *et al.*, 2001). Stimulation of NO in C3H.PRI-Flv^r and C3H/He brain astrocytes or peritoneal macrophage cultures inhibited MVEV replication to the same extent in all cultures (Silvia *et al.*, 2001). The *Nos-1* gene is expressed constitutively only in neurons and endothelial cells, whereas flavivirus resistance has been observed in cells from additional types of tissues. Also, the broad spectrum of NO antiviral activity is not consistent with the virus specificity of *Flv* resistance.

B. *The* Ubc *Locus*

Ubiquitin is a highly conserved, small protein (76 amino acids) that functions as a tag in the selective proteolysis of abnormal or foreign proteins by the 26S proteasome (Ciechanover *et al.*, 1980; Jennissen, 1995). Genes encoding ubiquitin fall into two classes: monomeric loci and polyubiquitin genes. In monomeric loci (*Uba*), a single ubiquitin-coding region is fused to a carboxyl-extension protein. Polyubiquitin genes (such as *Ubb* and *Ubc*) contain multiple tandem ubiquitin-coding regions. Proteins encoded by both monomeric and polyubiquitin genes are processed to ubiquitin monomers, which can then be conjugated with each other or with other proteins. Ubiquitin has been implicated in the degradation of viral proteins in infected cells (Ciechanover *et al.*, 1980; Wilkinson *et al.*, 1980).

Ubc was previously mapped close to the microsatellite marker D5Mit65 (1.52 ± 1.50 cM) on mouse chromosome 5 in an interspecific (*M. musculus* × *M. spretus*) backcross (Klingenspor *et al.*, 1997). The genetic distance between the D5Mit65 marker and *Flv* was estimated to be equal to 3.00 ± 0.15 cM using independent *M. musculus* backcrosses (Urosevic *et al.*, 1997a). Data from these studies suggest that *Ubc* and *Flv* are closely linked (Fig. 7). *Ubc* genomic DNAs amplified by polymerase chain reaction (PCR) from C3H.PRI-Flv^{r} and C3H/He cells were sequenced and compared (Perelygin *et al.*, 2002a). The *Ubc* gene from C3H.PRI-Flv^{r} contained 11 functional monomers, whereas the C3H/He *Ubc* encoded only 9 monomers. Because this difference would not be expected to have a significant effect on ubiquitin function and because ubiquitin is not known to have a flavivirus-specific function, it was not likely that the observed difference in *Ubc* monomer number could confer the flavivirus resistance phenotype (Perelygin *et al.*, 2002a).

X. Positional Cloning of the *Flv* Gene

From microsatellite mapping data (Urosevic *et al.*, 1997a), the *Flv* gene and the D5Mit159 marker were estimated to be less than 0.15 cM apart on mouse chromosome 5 (Fig. 7). The distance between two genes can be expressed either in recombination units, cM (genetic map), or in bp (physical map). Because the current length of the mouse genome is estimated to be about 1360.9 cM (Dietrich *et al.*, 1996) or 2.8 Gb (Gregory *et al.*, 2002), 1 cM of the genetic map is equivalent to about 2.06 Mb in the physical map. Therefore, 0.15 cM was estimated to be equivalent to 309 kb. The probability that a genomic contig of 618 kb, which contained 309 kb downstream and 309 kb upstream of the D5Mit159 microsatellite, would contain the *Flv* locus was estimated to be >95% (Perelygin *et al.*, 2002b).

The *Flv* gene was identified by first finding all of the genes in a region surrounding the D5Mit159 marker on chromosome 5 and then comparing the sequences of the mRNAs transcribed from each of these genes in congenic-resistant and -susceptible animals or cells (Perelygin *et al.*, 2002b). Genomic clones covering a region of >700 kb on murine chromosome 5 were selected from bacterial artificial chromosome (BAC) libraries using (1) a probe designed from the D5Mit159 marker sequence (available from the Whitehead Institute) and (2) additional probes designed from the terminal sequences of the selected BAC clones. Genes within the selected BAC clones were then identified using several methods. Four genes, AF217002, AF217003, AF319547, and AF328927, were found using a cDNA selection protocol (Lovett, 1994), and five genes, AF261233, AF328926, AF418010, AF453830, and *Oas1a*, were found using an exon trapping protocol (Fig. 8). Partial cDNAs obtained were extended to full length, cloned, and sequenced. The map location of these nine genes within this region of chromosome 5 was subsequently confirmed by searching the Celera mouse genome database with the cDNA sequences. Primers were then designed to amplify the cDNAs of 12 additional transcripts (mCT16571, mCT16562, mCT16559, mCT16567, mCT16565, mCT15075, mCT15073, mCT15304, mCT15074, mCT15317, mCT15079, and mCT15087) that were predicted within this region of chromosome 5 by the Celera mouse genome database. One additional gene, *Oas1h*, was mapped into the contig using the cDNA sequence AB067530 submitted to GenBank by Kakuta *et al.* (2002) (Fig. 8).

Based on the prediction that the mouse genome contained ~30,000 genes (now considered a low estimate) and is ~2.8 billion bp in length, it was estimated that a 1 million bp (1 Mb) region of mouse genomic DNA would contain about 11 genes. This is equivalent to an average density of about 22 genes per 1 cM. From these estimates it was predicted that there would be about 7–8 genes in the BAC contig selected. This >700-kb region of chromosome 5 analyzed contained more than three times the number of genes predicted. A total of 22 genes, including 10 loci encoding 2′-5′ oligoadenylate synthetases (*Oas*), were identified in this region of mouse chromosome 5 (Fig. 8; Perelygin *et al.*, 2002b). Two additional *Oas-like* loci were separated from the *Oas2* gene by ~6 Mb of genomic DNA. The sequence redundancy between regions of the *Oas* genes complicated the amplification and

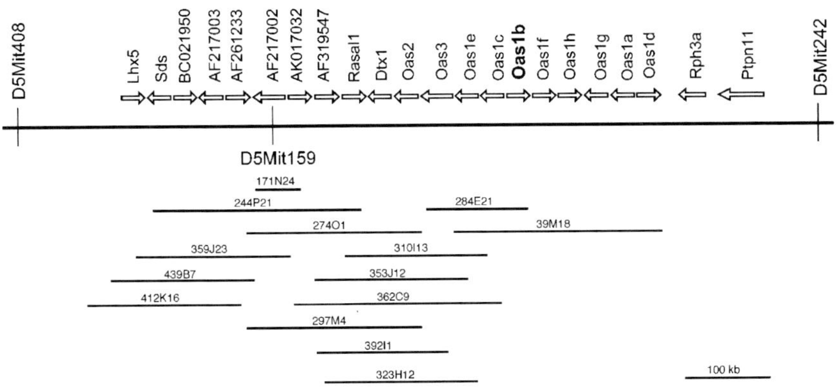

FIG 8. Physical and transcript maps of the *Flv* interval. Genes are represented by their accepted abbreviations or the GenBank accession numbers of their transcripts. The direction of gene transcription is indicated by arrows. The centromere is located on the left side of the region shown. The *Oas1b* (*Flv*) gene is indicated in bold. The flanking microsatellite markers and the D5Mit159 marker are shown. Horizontal bars beneath the genes represent the BAC clones listed by their library names. Data from Perelygin *et al.* (2002b).

identification of the individual cDNAs. Whereas the human *OAS* gene family (four genes including one *OAS-like* gene) had been well characterized, Perelygin *et al.* (2002b) were the first to characterize the mouse *Oas* gene family and to provide a simple nomenclature for the members of this family. Prior to this study, only 5 of the 12 mouse *Oas* genes had been identified. This gene family in mice is significantly larger (12 genes) than the family in humans (4 genes) due to multiple duplications, primarily of the *Oas1* genes (Fig. 11).

Criteria used for the identification of the *Flv* locus were based on previous data obtained on the characteristics of the resistant phenotype. These criteria were (a) that the gene map to a region near the D5Mit159 marker on chromosome 5, (b) that expression of mRNA from the *Flv*^{r} allele be constitutive and not tissue restricted, (c) that the resistant C3H.PRI-Flv^r and susceptible C3H/He alleles differ from each other by a mutation that affects the expression or function of the Flv gene product, and (d) that this mutation shows an absolute correlation with phenotype in all resistant and susceptible mouse strains. A final criterion, that expression of the product of the resistant allele in susceptible animals has a dominant, negative effect on flavivirus replication, is still under investigation.

XI. Identification of the *Flv* Gene as *Oas1b*

Full-length cDNAs for each of the 22 genes identified in the *Flv* region were amplified by RT-PCR from congenic flavivirus-resistant (C3H.PRI-Flv^r) and -susceptible (C3H/He) mouse strains, sequenced, and compared. Sequences of the majority of these genes were either identical or very similar, differing by only a few silent substitutions between the two mouse strains. In contrast, two genes, Na^+/Ca^{2+} exchanger and *Oas1b*, were polymorphic. Na^+/Ca^{2+}-exchanger cDNAs from C3H.PRI-Flv^r and C3H/He mouse strains differed by five nonsynonymous substitutions. However, a random distribution of these substitutions was subsequently observed in two additional susceptible (BALB/c and C57BL/6) and one resistant (BRVR) mouse strains, indicating that none of these substitutions correlated with the susceptible phenotype.

Most of the single base differences found between Oas1b cDNAs of congenic C3H.PRI-Flv^r and C3H/He mouse strains were silent, but a C820T transition resulted in a premature stop codon in the susceptible C3H/He strain. The Oas1b protein produced in the C3H/He cells lacked 30% of its C-terminal sequence as compared to the C3H.PRI-Flv^r protein (Fig. 9). Oas1b proteins from the two mouse strains also differed by two additional nonsynonymous mutations: a threonine-to-alanine substitution at amino acid 65 and an arginine-to-glutamine substitution at amino acid 190 in the C3H/He protein (Perelygin *et al.*, 2002b).

Comparison of *Oas1b* genomic (AF481734) and cDNA (AF328926) sequences revealed the presence of six exons. The individual Oas1b exons from eight additional mouse strains were amplified from genomic DNA and sequenced. The *Oas1b* gene in four resistant mouse strains (BRVR, C3H.PRI-Flv^r, CASA/Rk, and CAST/Ei) encodes an identical full-length protein, but in five susceptible strains (129/SvJ, BALB/c, C3H/He, C57BL/6, and CBA/J) it encodes an identical truncated protein. Although the *Oas1b* cDNA sequence from the MOLD/Rk strain (minor resistant phenotype) also encodes a full-length protein, it differs from the proteins of the other resistant strains by 14 amino acid substitutions (Fig. 9). The MOLD/Rk Oas1b protein sequence contains alanine at position 65, as do the proteins encoded by the susceptible strains. The flavivirus-susceptible phenotype correlated with the C820T transition in the *Oas1b* gene in the 10 mouse strains studied. Based on this consistent correlation, the *Flv* locus was identified as the *Oas1b* gene. According to the Celera mouse database, the *Oas1b* gene is located about 250 kb downstream of the

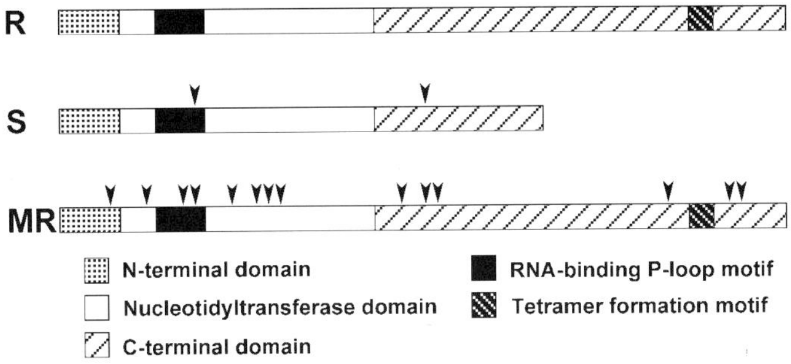

FIG 9. Domain architecture of Oas1b proteins. The N-terminal domain (−30 amino acids) (gray) and the C-terminal domain (hatched) are specific to the Oas protein family according to a ProDom tool analysis. The nucleotidyltransferase domain (Pfam 01909), the P-loop motif, and the CFK tetramerization motif are indicated. R, products of the Flv^r and Flv^r-like alleles; S, product of the Flv^s allele; MR, product of the Flv^{mr} allele. Positions of amino acid substitutions between the Flv^r and the Flv^s or the Flv^{mr} proteins are indicated by arrows. Data from Perelygin *et al.* (2002b).

D5Mit159 microsatellite (Fig. 8). D5Mit159 mapped to an intronic region of the calcium channel gene (GenBank accession number AF217002). Interestingly, one of the first murine Oas sequences reported, designated *Oas1a* (Ichii *et al.*, 1986), turned out to be the *Flv* gene (now called *Oas1b*). The sequence submitted to GenBank in 1986 was obtained from a flavivirus-susceptible mouse.

XII. POSSIBLE MECHANISM OF ACTION OF THE RESISTANT ALLELE OF THE *Oas1b* GENE

There are three types of the 2′-5′ oligoadenylate synthetases in mice. The small 2′-5′ oligoadenylate synthetases (40–47 kDa) are encoded by a cluster of eight genes, *Oas1a* through *Oas1h*, whereas the medium (85–86 kDa) and large (126 kDa) forms of this enzyme are encoded by the *Oas2* and *Oas3* genes, respectively. All 10 of these genes are closely linked to each other and are located in the distal part of mouse chromosome 5 (Fig. 8).

The protein encoded by *Oas1b* is one of the small 2′-5′ oligoadenylate synthetases. All Oas proteins contain three domains (Fig. 9). The central domain has a conserved nucleotidyltransferase fold, whereas the N- and the C-terminal domains are unique and vary significantly among the family members of the different size classes. Functional motifs identified previously in small 2′-5′ oligoadenylate synthetases include an N-terminal LxxxP motif, which is required for 2′-5′

oligoadenylate synthetase activity (Ghosh *et al.*, 1997a), and a P-loop motif (Fig. 9) postulated to be involved in double-stranded RNA (dsRNA) binding (Saraste *et al.*, 1990). A DAD Mg^{2+}-binding motif is also required for normal functioning of murine 2′-5′ oligoadenylate synthetases (Yamamoto *et al.*, 2000). The products of both resistant and susceptible alleles of the *Oas1b* gene have identical LxxxP and DAD motifs. However, the P-loop motif of both alleles of *Oas1b* contains a four amino acid deletion not found in any of the other murine 2′-5′ oligoadenylate synthetases (Perelygin *et al.*, 2002b).

A C-terminal CFK motif (Fig. 9) has been reported to be required for tetramerization of the small form of human 2′-5′ oligoadenylate synthetase (Ghosh *et al.*, 1997b). Homotetramers of small Oas proteins remain inactive until activated by interaction with a dsRNA. After activation, homotetramers catalyze the synthesis of 2′-5′ oligoadenylates (2-5 A) from the ATP (Fig. 10). The 2-5A produced subsequently activates the inactive endoribonuclease, RNase L, by causing it to dimerize. Activated RNase L preferentially cleaves RNAs after single-stranded AU and UU dinucleotides (Wreschner *et al.*, 1981). Susceptible mice encode a C-terminally truncated Oas1b protein that is missing the CFK motif and these proteins should not be able to form homotetramers. Flavivirus susceptibility correlates in all cases tested with the loss of the Oas1b CFK motif. Additional unknown functional motifs may also be located in the C-terminal region. It is not currently known whether the truncated Oas1b proteins are stable in susceptible cells or whether they retain other functions. A recombinant truncated Oas1b protein expressed in *Escherichia coli* did not synthesize 2-5A (Kakuta *et al.*, 2002). Preliminary studies with recombinant full-length Oas1b protein suggest that it also does not have synthetase activity (Perelygin and Brinton, unpublished data).

The protein encoded by *Oas1b* differs from the other mammalian 2′-5′ oligoadenylate synthetases by a four amino acid deletion in the P-loop motif. One hypothesis to explain the flavivirus specificity of the *Flv* resistance is that there is a specific interaction between the Oas1b protein and an as yet unknown double-stranded region within the flavivirus RNA. Oas1b proteins from resistant and susceptible mice differ by a single base substitution in the P-loop domain that could also be involved in altering RNA-binding specificity.

RNase L may only be activated at specific locations within the cell. Previous studies with other viruses have shown that RNase L can preferentially target viral RNA over cell RNA (Li *et al.*, 1998; Nilsen and Baglioni, 1979). Degradation of flavivirus RNA by RNase L may be one mechanism by which the efficiency of flavivirus replication is

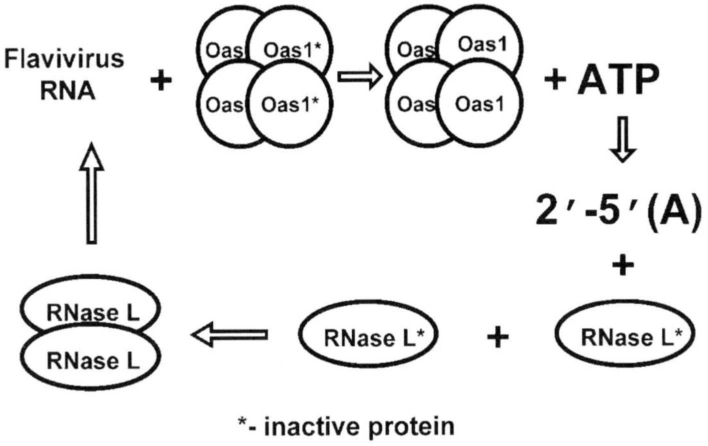

FIG 10. Model of the mechanism of action of Oas1 proteins in flavivirus-infected cells.

reduced. However, this activity may be contributed by other Oas1 gene products, which are active 2-5A synthetases, and does not explain the viral-specific effect of the Flv^r gene product. The Oas1b protein from resistant cells may have additional functions that affect virus replication (Samuel, 2002). To date, there have been few functional studies of murine *Oas* gene products, but data obtained with human *OAS* gene products suggest that they can function in apoptosis, gene regulation, and developmental pathways (Hovnanian *et al.*, 1998). It was reported previously that the half-lives of complexes formed between the WNV 3′ minus SL RNA and cell proteins in resistant cells were longer than those of complexes formed in susceptible cells (Shi *et al.*, 1996). It is currently not known if the products of the different *Oas1b* alleles could cause this effect. However, the product of the *Flv^r* allele may act directly or indirectly to reduce the efficiency of flavivirus plus strand RNA synthesis. The functional significance of *Oas1* gene duplication in mice is currently not known. It is also not known to what extent the other Oas1 proteins contribute to the total antiflaviviral activity mounted by the host cell.

XIII. Conclusions

Several previous observations are consistent with the hypothesis that the *Flv^r* gene product exerts its main effect via RNase L activation. RNase protection experiments indicated that the levels of flavivirus plus strand RNA, but not minus strand RNA, were reduced significantly in WNV-infected resistant cells as compared with infected susceptible cells (Li and Brinton, unpublished data). Flavivirus minus strand RNAs are present in infected cells only within double-stranded replication intermediates and therefore would be resistant to RNase L cleavage, whereas nacent plus strand RNAs are released from the replication complexes and so would be cleavage sensitive. However, the Flv^{r} gene product may act instead at the level of plus strand RNA synthesis. The observation that plus strand RNA levels are significantly lower in resistant cells 2 h after infection (Li and Brinton, unpublished data) suggests that a constitutively expressed gene product is responsible for the decrease in viral RNA. The Oas1b gene product is constitutively expressed (Eskildsen *et al.*, 2002b; Kakuta *et al.*, 2002; Perelygin *et al.*, 2002b).

Although produced by both resistant and susceptible cells, heterogeneous small viral RNAs represent a higher proportion of the total virus RNA in WNV-infected resistant cells than susceptible cells (Brinton, 1983, 1997). These heterogeneous small RNAs are packaged into the virions (Figs. 4 and 5; Brinton, 1983). Virions produced by resistant cells, but not those produced by susceptible cells, have homologous interfering activity and lower RNA to infectivity ratios. The heterogeneous sizes of the small RNAs contained in these virions were not typical of DI RNA populations and suggest that the small RNAs present in these virions may instead be a collection of fragments of the genomic RNA produced by RNase digestion. Flavivirus RNAs contain conserved 3′ and 5′ cyclization sequences that facilitate long-distance 3′–5′ RNA–RNA interactions within the genome RNA (Hahn *et al.*, 1987; Khromykh *et al.*, 2001; Brinton, 2002). Base pairing between the cyclization sequences within 3′ and 5′ RNA fragments could generate paired fragments that are functionally similar to DI RNAs in their ability to interfere with the replication of homologous full-length viral genomes. The lower infectivity of virions produced by resistant cells would be expected to reduce the efficiency of virus spread in the host. The small RNAs observed in virions may be generated by RNase L or another cellular RNase that is activated later in the replication cycle after upregulation of the expression of the *Oas* genes by the viral capsid protein (see later) and/or IFN. Different flavivirus genomes may vary in their susceptibility to RNase L digestion because of differences in the frequency of target sites for RNase L in single-stranded regions. Different strains of hepatitis C virus have been shown to differ in their susceptibility to RNase L digestion (Han and Barton, 2002).

The virus specificity of the resistant phenotype conferred by the *Flv*^{r} allele may be due to preferential activation of the Oas1b protein by flavivirus RNAs. Although the RNAs of different flaviviruses may be able to specifically activate the product of the *Flv*^{r} allele, the efficiency of this activation may vary among flaviviruses. Interestingly, a strain of MVEV with a deletion near the 5′ end of the 3′ NCR replicates in resistant mice more efficiently than other MVEV strains (Urosevic *et al.*, 1999). Sequence differences within the binding domain of the Oas1b protein may also affect specificity. The *Flv*^{mr} allele, which contains additional mutations, conferred resistance to attenuated YFV-17D but not to virulent MVEV.

Although neither IFN treatment nor IFN production is required for the expression of *Flv*^{r} resistance, IFN pretreatment of resistant cells augments the effect of the *Flv*^{r} allele by further decreasing the virus yield (Fig. 6). Studies with dengue virus have shown that pretreatment of susceptible cell cultures with IFN-α or -β efficiently reduced the number of cells infected, as well as the level of viral RNA replication, and that combinations of IFN-α and -γ or IFN-β and -γ were even more effective (Diamond *et al.*, 2000). Further studies with susceptible knockout cells indicated that the antiviral effect of IFN on dengue virus replication was equally efficient in PKR minus/RNase L minus and control cells (Diamond and Harris, 2001). The authors postulated that IFN acts by inhibiting the translation of genomic RNA via an unknown PKR/RNase L-independent mechanism. IFN induces the expression of a large number of genes, but it is not known how many of these gene products are involved in antiviral functions (Samuel, 2001). Studies with knockout flavivirus-susceptible mice that did not express functional PKR, RNase L, or Mx-1 indicated the existence of alternative antiviral pathways of interferon action against both encephalomyocarditis virus (Zhou *et al.*, 1999) and Sindbis virus (Ryman *et al.*, 2002).

In human hepatocytes infected with a hepatitis C replicon, the viral capsid protein has been shown to enter the nucleus and activate the expression of the INF-inducible human *OAS1* gene in a dose-dependent manner (Naganuma *et al.*, 2000). Gene activation by the capsid protein was further enhanced in cells that had been treated with IFN. It is not yet known whether the capsid proteins of viruses in the genus *Flavivirus* can activate *Oas1* genes in a similar manner. Flavivirus infection has also been reported to have additional effects on the host cell. WNV infection of diploid vertebrate cells results in the activation of NF-κB via virus-induced phosphorylation of the inhibitor κB, which in turn increases the transcription of MHC-1 mRNA and the surface expression of MHC-1 (Kesson and King, 2001). WNV infection has also been reported to induce the cellular expression of intracellular adhesion

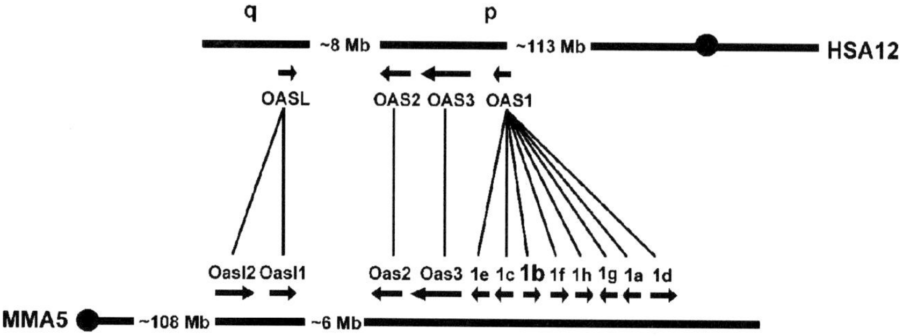

FIG 11. Orthologous relationships between members of human *OAS* and murine *Oas* gene families. The human *OAS1* gene has eight murine orthologs. Of these, the murine *Oas1a* gene is most similar to the human *OAS1*.

molecule-1 (ICAM-1) in an IFN-independent manner (Shen *et al.*, 1995). The interaction between flaviviruses and their host cells is complex, and the Flv^r gene product represents only one component in this interaction. However, variations in a single component, such as in the products of the different alleles of *Oas1b*, can be sufficient to change the outcome of the infection and to tip the balance in favor of either the host or the virus.

Comparison of the mouse and human genomes (Fig. 11) showed that there is not a direct equivalent of the mouse *Oas1b* gene in humans. Although the eight mouse *Oas1* genes are orthologous to the single human *OAS1* gene, the human OAS1 cDNA sequence is more similar to that of murine *Oas1a* than to *Oas1b*. Further study of the mechanism of action of the murine *Flv* gene may nevertheless lead to the development of new antiviral strategies for humans. Variation in human susceptibility to diseases caused by various flaviviruses has been observed in the United States, Cuba, Haiti, Africa, and South America (Centers for Disease Control and Prevention, 2001; Halstead, 1992; Halstead *et al.*, 2001; Luby *et al.*, 1967; Monath and Heinz, 1996; Sabin, 1954). Although reduced immune function is associated with the development of severe disease in humans following a flavivirus infection, some evidence suggests that there may also be a genetic component of human susceptibility. During the spread of WNV in the United States, different bird species have varied widely in their susceptibility to West Nile virus infections (Bernard *et al.*, 2001) and only about 1% of infected humans developed CNS disease (Peterson *et al.*, 2003). It is not currently known whether susceptibility to flavivirus–induced diseases in either humans or birds is genetically controlled and, if so, which gene(s) is responsible.

References

Bang, F. B. (1978). Genetics of resistance of animals to viruses. I. Introduction and studies in mice. *Adv. Virus Res.* **23:**269–348.

Bernard, K. A., Maffei, J. G., Jones, S. A., Kauffman, E. B., Ebel, G. D., *et al.* (2001). West Nile virus infection in birds and mosquitoes, New York state, 2000. *Emerg. Infect. Dis.* **7:**679–685.

Bhatt, P. N., and Jacoby, R. O. (1976). Genetic resistance to lethal flavivirus encephalitis. II. Effect of immunosuppression. *J. Infect. Dis.* **134:**166–173.

Bhatt, P. N., Johnson, E. A., Smith, A. L., and Jacoby, R. O. (1981). Genetic resistance to lethal flavivirus encephalitis. III. Replication of Banzi virus in vitro and in vivo in tissues of congenic susceptible and resistant mice. *Arch. Virol.* **69:**273–286.

Bonhomme, F., and Guenet, J. L. (1989). The wild house mouse and its relatives. *In* "Genetic Variants and Strains of the Laboratory Mouse" (M.F. Lyon and A.G. Searle, eds.), pp. 646–662. Oxford Univ. Press, Oxford, UK.

Bonhomme, F., Guenet, J. L., Dod, B., Moriwaki, K., and Bulfield, G. (1987). The polyphyletic origin of laboratory inbred mice and their rate of evolution. *Biol. J. Linn. Soc.* **30:**51–58.

Brinton, M. A. (1982). Characterization of West Nile virus persistent infections in genetically resistant and susceptible mouse cells. I. Generation of defective nonplaquing virus particles. *Virology* **116:**84–98.

Brinton, M. A. (1983). Analysis of extracellular West Nile virus particles produced by cell cultures from genetically resistant and susceptible mice indicates enhanced amplification of defective interfering particles by resistant cultures. *J. Virol.* **46:**860–870.

Brinton, M. A. (1986). Replication of flaviviruses. *In* "The Togaviridae and Flaviviridae" (S. Schlesinger and M. J. Schlesinger, eds.), pp. 327–374. Plenum Press, New York.

Brinton, M. A. (1997). Host susceptibility to viral disease. *In* "Viral Pathogenesis" (N. Nathanson, ed.), pp. 303–328. Lippincott-Raven, Philadelphia.

Brinton, M. A. (2002). The molecular biology of West Nile virus: A new invader of the Western Hemisphere. *Annu. Rev. Microbiol.* **56:**371–402.

Brinton, M. A. (2001). Host factors involved in West Nile virus replication. *Ann. N.Y. Acad. Sci.* **951:**207–219.

Brinton, M. A., Arnheiter, H., and Haller, O. (1982). Interferon independence of genetically controlled resistance to flaviviruses. *Inf. Imm.* **36:**284–288.

Brinton, M. A., Blank, K. J., and Nathanson, N. (1984). Host genes that influence susceptibility to viral diseases. *In* "Concepts in Viral Pathogenesis" (A.L. Notkins and M.B.A. Oldstone, eds.), pp. 71–78. Springer-Verlag, New York.

Brinton, M. A., and Nathanson, N. (1981). Genetic determinants of virus susceptibility: Epidemiology implications of murine models. *Epidemiol. Rev.* **3:**115–139.

Brinton Darnell, M., and Koprowski, H. (1974a). Genetically determined resistance to infection with group B arboviviruses. II. Increased production of interfering particles in cell cultures from resistant mice. *J. Infect. Dis.* **129:**248–256.

Brinton Darnell, M., and Koprowski, H. (1974b). Increased production of interfering virus particles in cell cultures from mice resistant to group B arbovirus infection. *In* "Mechanisms of Virus Disease" (W. S. Robinson and C. F. Fox, eds.), pp. 147–158. W. A. Benjamin, Menlo Park, Calif..

Brinton Darnell, M., Koprowski, H., and Lagerspetz, K. (1974). Genetically determined resistance to infection with group B arboviruses. I. Distribution of the resistance gene among various mouse populations and characteristics of gene expression in vivo. *J. Infect. Dis.* **129:**240–247.

Casals, J., and Schneider, H. A. (1943). Natural resistance and susceptibility to Russian spring-summer encephalitis in mice. *Proc. Soc. Exp. Biol. Med.* **54:**201–202.

Casanova, J.-L., Schurr, E., Abel, L., and Skamene, E. (2002). Forward genetics of infectious diseases: immunological impact. *Trends Immunol.* **23:**469–472.

Centers for Disease Control and Prevention. (2001). Fever, jaundice, and multiple organ system failure associated with 17D-derived yellow fever vaccination, 1996–2001. MMRW **50:**643–645.

Ciechanover, A., Heller, H., Elias, S., Haas, A. L., and Hershko, A. (1980). ATP-dependent conjugation of reticulocyte proteins with the polypeptide required for protein degradation. *Proc. Natl. Acad. Sci. USA* **77:**1365–1368.

Diamond, M. S., and Harris, E. (2001). Interferon inhibits dengue virus infection by preventing translation of viral RNA through a PKR-independent mechanism. *Virology* **289:**297–311.

Diamond, M. S., Roberts, T. G., Edgil, D., Lu, B., Ernst, J., and Harris, E. (2000). Modulation of dengue virus infection in human cells by alpha, beta, and gamma interferons. *J. Virol.* **74:**4957–4966.

Dietrich, W. F., Miller, J., Steen, R., Merchant, M. A., Damron-Boles, D., Husain, Z., Dredge, R., Daly, M. J., Ingalls, K. A., O'Connor, T. J., *et al.* (1996). A comprehensive genetic map of the mouse genome. *Nature* **380:**149–152.

Eskildsen, S., Hartmann, R., Kjeldgaard, N. O., and Justesen, J. (2002). Gene structure of the murine 2′-5′-oligoadenylate synthetase family. *Cell Mol. Life Sci.* **59:**1212–1222.

Ghosh, A., Desai, S. Y., Sarkar, S. N., Ramaraj, P., Ghosh, S. K., Bandyopadhyay, S., and Sen, G. C. (1997a). Effects of mutating specific residues present near the amino terminus of 2′-5′-oligoadenylate synthetase. *J. Biol. Chem.* **272:**15452–15458.

Ghosh, A., Sarkar, S. N., Guo, W., Bandyopadhyay, S., and Sen, G. C. (1997b). Enzymatic activity of 2′-5′-oligoadenylate synthetase is impaired by specific mutations that affect oligomerization of the protein. *J. Biol. Chem.* **272:**33220–33226.

Glode, L. M., and Rosenstreich, D. L. (1976). Genetic control of B cell activation by bacterial lipopolysaccharide is mediated by multiple distinct genes or alleles. *J. Immunol.* **117:**2061–2066.

Goodman, G. T., and Koprowski, H. (1962a). Macrophages as a cellular expression of inherited natural resistance. *Proc. Natl. Acad. Sci. USA* **48:**160–165.

Goodman, G. T., and Koprowski, H. (1962b). Study of the mechanism of innate resistance to virus infection. *J. Cell. Comp. Physiol.* **59:**333–373.

Green, M. C. (1989). Catalogue of mutant genes and polymorphic loci. *In* "Genetic Variants and Strains of Laboratory Mice" (M. C. Green, ed.), pp. 12–403. Gustav Fischer, Stuttgart.

Gregory, S. G., Sekhon, M., Schein, J., Zhao, S., Osoegawa, K., *et al.* (2002). A physical map of the mouse genome. *Nature* **418:**743–750.

Groschel, D., and Koprowski, H. (1965). Development of a virus-resistant inbred mouse strain for the study of innate resistance to arbo B viruses. *Arch. Gesamte. Virusforsch.* **18:**379–391.

Groves, M. G., Rosenstreich, D. L., Taylor, B. A., and Osterman, J. V. (1980). Host defenses in experimental scrub typhus: Mapping the gene that controls natural resistance in mice. *J. Immunol.* **125:**1398–1399.

Hahn, C. S., Hahn, Y. S., Rice, C. M., Lee, E., Dalgarno, L. *et al.* (1987). Conserved elements in the 3′ untranslated region of flavivirus RNAs and potential cyclization sequences. *J. Mol. Biol.* **198:**33–41.

Haller, O. (ed.) (1981). "Natural Resistance to Tumors and Viruses." Springer-Verlag, Berlin.

Haller, O., Arnheiter, H., Gresser, I., and Lindenmann, J. (1979). Genetically determined, interferon-dependent resistance to influenza virus in mice. *J. Exp. Med.* **149:**601–612.

Haller, O., Arnheiter, H., Lindenmann, J., and Gresser, I. (1980). Host gene influences sensitivity to interferon action selectively for influenza virus. *Nature* **283:**660–662.

Halstead, S. B. (1992). The XXth century dengue pandemic: Need of surveillance and research. *World Health Stat. Q.* **45:**292–298.

Halstead, S. B., Streit, T. G., Lafontant, J. G., Putvatana, R., Russell, K., *et al.* (2001). Haiti: Absence of dengue hemorrhagic fever despite hyperendemic dengue virus transmission. *Am. J. Trop. Med. Hyg.* **65:**180–183.

Han, J. Q., and Barton, D. J. (2002). Activation and evasion of the antiviral 2′-5′ oligoadenylate synthetase/ribonuclease L pathway by hepatitis C virus mRNA. *RNA* **8:**512–525.

Hanson, B., Koprowski, H., Baron, S., and Buckler, C. E. (1969). Interferon-mediated natural resistance of mice to arbo B virus infection. *Microbios* **1B:**51–68.

Hoshino, K., Takeuchi, O., Kawai, T., Sanjo, H., Ogawa, T., Takeda, Y., Takeda, K., and Akira, S. (1999). Cutting edge: Toll-like receptor 4 (TLR4)-deficient mice are hyporesponsive to lipopolysaccharide. *J. Immunol.* **162:**3749–3752.

Hovnanian, A., Rebouillat, D., Mattei, M. G., Levy, E. R., Marie, I., Monaco, A. P., and Hovanessian, A. G. (1998). The human 2′,5′-oligoadenylate synthetase locus is composed of three distinct genes clustered on chromosome 12q24.2 encoding the 100-, 69-, and 40-kDa forms. *Genomics* **52:**267–277.

Huang, A. S., and Baltimore, D. (1970). Defective viral particles and viral disease processes. *Nature* **226:**325–327.

Ichii, Y., Fukunaga, R., Shiojiri, S., and Sokawa, Y. (1986). Mouse 2-5A synthetase cDNA: Nucleotide sequence and comparison to human 2-5A synthetase. *Nucleic Acids Res.* **14:**10117.

Jacoby, R. O., and Bhatt, P. N. (1976). Genetic resistance to lethal flavivirus encephalitis. I. Infection of congenic mice with Banzi virus. *J. Infect. Dis.* **134:**158–165.

Jacoby, R. O., Bhatt, P. N., and Schwartz, A. (1980). Protection of mice from lethal flaviviral encephalitis by adoptive transfer of splenic cells from donors infected with live virus. *J. Infect. Dis.* **141:**617–624.

Jennissen, H. P. (1995). Ubiquitin and the enigma of intracellular protein degradation. *Eur. J. Biochem.* **231:**1–30.

Jerrells, T. R., and Osterman, J. V. (1981). Host defenses in experimental scrub typhus: Inflammatory response of congenic C3H mice differing at the Ric gene. *Infect. Immun.* **31:**1014–1022.

Jerrells, T. R., and Osterman, J. V. (1982). Role of macrophages in innate and acquired host resistance to experimental scrub typhus infection of inbred mice. *Infect. Immun.* **37:**1066–1073.

Kakuta, S., Shibata, S., and Iwakura, Y. (2002). Genomic structure of the mouse 2′,5′-oligoadenylate synthetase gene family. *J. Interferon Cytokine Res.* **22:**981–993.

Kesson, A. M., and King, N. J. (2001). Transcriptional regulation of major histocompatibility complex class I by flavivirus West Nile is dependent on NK-kappa B activation. *J. Infect. Dis.* **15:**947–954.

Khromykh, A. A., Meka, H., Guyatt, K. J., and Westaway, E. G. (2001). Essential role of cyclization sequences in flavivirus RNA replication. *J. Virol.* **75:**6719–6728.

Klingenspor, M., Bodnar, J., Welch, C., Xia, Y. R., Lusis, A. J., and Reue, K. (1997). Localization of ubiquitin gene family members to mouse chromosomes 5, 11, and 18. *Mann. Genome* **8:**789–790.

Koprowski, H. (1971). When a thing's really good it cannot die. *In* "Of Microbes and Life" (J. Monod and E. Borek, eds.), pp. 62–69. Columbia Univ. Press, New York.

Lagerspetz, K. Y. H., Koprowski, H., Darnell, M., and Tarkkonen, H. (1973). Thermoregulation in group B arbovirus-resistant and group B arbovirus-susceptible mice. *Am. J. Physiol.* **225:**532–537.

Lee, C. G. L., Gregg, A. R., and O'Brien, W. E. (1995). Localisation of the neuronal form of nitric oxide synthetase to mouse chromosome 5. *Mamm. Genome* **6:**56–57.

Li, X. L., Blackford, J. A., and Hassel, B. A. (1998). RNase L mediates the antiviral effect of interferon through a selective reduction in viral RNA during encephalomyocarditis virus infection. *J. Virol.* **72:**2752–2759.

Lin, Y. L., Huang, Y. L., Ma, S. H., Yeh, C. T., Chiou, S. Y., Chen, L. K., and Liao, C. L. (1997). Inhibition of Japanese encephalitis virus infection by nitric oxide: Antiviral effect of nitric oxide on RNA virus replication. *J. Virol.* **71:**5227–5235.

Lovett, M. (1994). Direct selection of cDNAs using genomic contigs. *In* "Current Protocols in Human Genetics" (J. Seidman and C. Seidman, eds.), pp. 6.3.1–6.3.15. Wiley, New York.

Luby, J. P., Miller, G., Gardner, P., *et al.* (1967). The epidemiology of St. Louis encephalitis in Houston, Texas. *Am. J. Epidemiol.* **86:**584–597.

Lynch, C. J., and Hughes, T. P. (1936). The inheritance of susceptibility to yellow fever encephalitis in mice. *Genetics* **21:**104–112.

Mashimo, T., Lucas, M., Simon-Chazottes, D., Frenkiel, M. P., Montagutelli, X., Ceccaldi, P. E., Deubel, V., Guenet, J. L., and Despres, P. (2002). A nonsense mutation in the gene encoding 2′-5′-oligoadenylate synthetase/L1 isoform is associated with West Nile virus susceptibility in laboratory mice. *Proc. Natl. Acad. Sci. USA* **99:**11311–11316.

Monath, T. P., and Heinz, F. X. (1996). Flaviviruses. *In* "Virology" (B. N. Fields, D. M. Knipe, R. M. Chanock, J. L. Melnick, T. P. Monath, and B. Roizman, eds.), 3rd Ed. pp. 375–440. Lippincott-Raven, Philadelphia.

Naganuma, A., Nozaki, A., Tanaka, T., Sugiyama, K., Takagi, H., Mori, M., Shimotohno, K., and Kato, N. (2000). Activation of the interferon-inducible 2′-5′-oligoadenylate synthetase gene by hepatitis C virus core protein. *J. Virol.* **74:**8744–8750.

Nathanson, N. (2001). Host susceptibility to viral diseases. *In* "Viral Pathogenesis and Immunity," pp. 158–169. Lippincott Williams and Wilkins, Philadelphia.

Nilsen, T. W., and Baglioni, C. (1979). Mechanism for discrimination between viral and host mRNA in interferon-treated cells. *Proc. Natl. Acad. Sci. USA* **76:**2600–2604.

Perelygin, A. A., Kondraschov, F., Rogozin, I. B., and Brinton, M. A. (2002a). Evolution of the mouse polyubiquitin C gene. *J. Mol. Evol.* **55:**202–210.

Perelygin, A. A., Scherbik, S. S., Zhulin, I. B., Stockman, B. M., Li, Y., and Brinton, M. A. (2002b). Positional cloning of the murine flavivirus resistance gene. *Proc. Natl. Acad. Sci. USA* **99:**9322–9327.

Petersen, L. R., Marfin, A. A., and Gubler, D. J. (2003). West Nile Virus. *JAMA* **290:**524–528.

Poltorak, A., He, X., Smirnova, I., Lui, M.-Y., Van Huffel, C. *et al.* (1998). Defective LPS signaling in C3H/HeJ and C57BL/10ScCr mice: Mutations in the Tlr4 gene. *Science* **282:**2085–2088.

Potter, M. (1978). Comments on the relationship of inbred strains to the genus *Mus*. *In* "Origins of Inbred Mice" (H. C. Morse, ed.), pp. 497–509. Academic Press, New York.

Qureshi, S. T., Lariviere, L., Leveque, G., Clermount, S., Moore, K. J., Gros, P., and Malo, D. (1999). Endotoxin-tolerant mice have mutations in Toll-like receptor 4 (Tlr4). *J. Exp. Med.* **189:**615–625.

Reiss, C. S., and Komatsu, T. (1998). Does nitric oxide play a critical role in viral infections. *J. Virol.* **72:**4547–4551.
Ryman, K. D., White, L. J., Johnson, R. E., and Klimstra, W. B. (2002). Effects of PKR/RNase L-dependent and alternative pathways on alphavirus replication and pathogenesis. *Viral Immunol.* **15:**53–76.
Sabin, A. B. (1952a). Genetic, hormonal and age factors in natural resistance to certain viruses. *Ann. N. Y. Acad. Sci.* **54:**936–944.
Sabin, A. B. (1952b). Nature of inherited resistance to viruses affecting the nervous system. *Proc. Natl. Acad. Sci. USA* **38:**540–546.
Sabin, A. B. (1954). Genetic factors affecting susceptibility and resistance to virus diseases of the nervous system. *Res. Publ. Assoc. Res. Nerv. Ment. Dis.* **33:**57–67.
Samuel, C. E. (2001). Antiviral actions of interferons. *Clin. Microbiol. Rev.* **14:**778–809.
Samuel, C. E. (2002). Host genetic variability and West Nile virus susceptibility. *Proc. Natl. Acad. Sci. USA* **99:**11555–11557.
Sangster, M. Y., Heliams, D. B., Mackenzie, J. S., and Shellam, G. R. (1993). Genetic studies of flavivirus resistance in inbred strains derived from wild mice: Evidence for a new resistance allele at the flavivirus locus (Flv). *J. Virol.* **67:**340–347.
Sangster, M. Y., Mackenzie, J. S., and Shellam, G. R. (1998). Genetically determined resistance to flavivirus infection in wild *Mus musculus domesticus* and other taxonomic groups in the genus. *Mus. Arch. Virol.* **143:**697–715.
Sangster, M. Y., and Shellam, G. R. (1986). Genetically controlled resistance to flaviviruses within the house mouse complex of species. *Curr. Top. Microbiol. Immunol.* **127:**313–318.
Sangster, M. Y., Urosevic, N., Mansfield, J. P., Mackenzie, J. S., and Shellam, G. R. (1994). Mapping the Flv locus controlling resistance to flaviviruses on mouse chromosome 5. *J. Virol.* **68:**448–452.
Saraste, M., Sibbald, P. R., and Wittinghofer, A. (1990). The P-loop: A common motif in ATP- and GTP-binding proteins. *Trends Biochem. Sci.* **15:**430–434.
Sawyer, W. A., and Lloyd, W. (1931). The use of mice in tests of immunity against yellow fever. *J. Exp. Med.* **54:**553–555.
Sheets, P., Schwartz, A., Jacoby, R. O., and Bhatt, P. N. (1979). T cell-mediated cytotoxicity for L292 fibroblasts infected with Banzi virus (Flavivirus). *J. Infect. Dis.* **140:**384–391.
Shellam, G. R., Sangster, M. Y., and Urosevic, N. (1998). Genetic control of host resistance to flavivirus infection in animals. *Rev. Sci. Tech.* **17:**231–248.
Shellam, G. R., Urosevic, N., Sangster, M. Y., Mansfield, J. P., and Mackenzie, J. S. (1993). Characterization of allelic forms at the retinal degeneration (rd) and b-glucuronidase (Gus) loci for the mapping of the flavivirus resistance (Flv) gene on mouse chromosome 5. *Mouse Genome* **91:**572–574.
Shen, J., Devery, J. M., and King, N. J. (1995). Early induction of interferon-independent virus-specific ICAM-1 (CD54) expression by flavivirus in quiescent but not proliferating fibroblasts-implications for virus-host interactions. *Virology* **208:**437–449.
Shi, P.-Y., Li, W., and Brinton, M. A. (1996). Cell proteins bind specifically to the 3′ stem-loop structure of West Nile virus minus-strand RNA. *J. Virol.* **70:**6278–6287.
Silver, L. M. (1995). "Mouse Genetics, Concepts and Applications." Oxford Univ. Press, New York.
Silvia, O. J., Shellam, G. R., and Urosevic, N. (2001). Innate resistance to flavivirus infection in mice controlled by Flv is nitric oxide-independent. *J. Gen. Virol.* **82:**603–607.

Silvia, O. J., and Urosevic, N. (1999). Variations in LPS responsiveness among different mouse substrains of C3H lineage and their congenic derivative sublines. *Immunogenetics* **50:**354–357.

Skamene, E. (ed.) (1985). "Genetic Control of Host Resistance to Infection and Malignancy." A. R. Liss, New York.

Skamene, E., Kongshavn, P. A. L., and Landy, M. (eds.) (1980). "Genetic Control of Natural Resistance to Infection and Malignancy." Academic Press, New York.

Smith, A. L. (1981). Genetic resistance to lethal flavivirus encephalitis: effect of host age and immune status and route of inoculation on production of interfering Banzi virus in vivo. *Am. J. Trop. Med. Hyg.* **30:**1319–1323.

Urosevic, N., Mann, K., Hodgetts, S. I., and Shellam, G. R. (1997a). The use of microsatellites in high-resolution genetic mapping around the mouse flavivirus resistance locus (Flv). *Arbovirus Res. Aust.* **7:**296–299.

Urosevic, N., Mansfield, J. P., Mackenzie, J. S., and Shellam, G. R. (1995). Low resolution mapping around the flavivirus resistance locus (Flv) on mouse chromosome 5. *Mamm. Genome* **6:**454–458.

Urosevic, N., Sangster, M. Y., Mansfield, J. P., Mackenzie, J. S., and Shellam, G. R. (1993). Flavivirus resistance (Flv^r) gene in mice: Mapping studies. *Arbovirus Res. Aust.* **6:**130–134.

Urosevic, N., and Shellam, G. R. (2002). Host genetic resistance to Japanese encephalitis group viruses. *Curr. Top. Microbiol. Immunol.* **267:**153–170.

Urosevic, N., Silva, O. J., Sangster, M. Y., Mansfield, J. P., Hodgetts, S. I., and Shellam, G. R. (1999). Development and characterization of new flavivirus-resistant mouse strains bearing Flv^r-like and Flv^{mr} alleles from wild or wild-derived mice. *J. Gen. Virol.* **80:**897–906.

Urosevic, N., van Maanen, M., Mansfield, J. P., Mackenzie, J. S., and Shellam, G. R. (1997b). Molecular characterization of virus-specific RNA produced in the brains of flavivirus-susceptible and -resistant mice after challenge with Murray Valley encephalitis virus. *J. Gen Virol.* **78:**23–29.

Vanio, T. (1963a). Virus and hereditary resistance in vitro. I. Behavior of West Nile (E-101) virus in the cultures prepared from genetically resistant and susceptible strains of mice. *Ann. Med. Exp. Biol. Fenn.* **41:**(Suppl. 1) 1–24.

Vanio, T. (1963b). Virus and hereditary resistance in vitro. II. Behavior of West Nile (E-101) virus in cultures prepared from challenged resistant, challenged back-cross and nonchallenged susceptible mice. *Ann. Med. Exp. Fenn.* **41**(Suppl. 1), 25–35.

Vanio, R., Gawatkin, R., and Koprowski, H. (1961). Production of interferon by brains of genetically resistant and susceptible mice infected with West Nile virus. *Virology* **14:**385–387.

Vogel, S. N., Marshall, S. T., and Rosenstreich, D. L. (1979). Analysis of the effects of lipopolysaccharide on macrophages: Differential phagocytic responses of C3H/HeN and C3H/HeJ macrophages in vitro. *Infect. Immun.* **25:**328–336.

Watson, J., and Riblet, R. (1974). Genetic control of responses to bacterial lipopolysaccharide in mice. I. Evidence for a single gene that influences mitogenic and immunogenic responses to lipopolysaccharides. *J. Exp. Med.* **140:**1147–1161.

Watson, J., Riblet, R., and Taylor, B. A. (1977). The response of recombinant inbred strains of mice to bacterial lipopolysaccharides. *J. Immunol.* **118:**2088–2093.

Webster, L. T. (1923). Microbic virulence and host susceptibility in mouse typhoid infection. *J. Exp. Med.* **37:**231–244.

Webster, L. T. (1933). Inherited and acquired factors in resistance to infection. I. Development of resistant and susceptible lines of mice through selective breeding. *J. Exp. Med.* **57:**793–817.

Webster, L. T. (1936). Experimental encephalitis (St. Louis type) in mice with high inborn resistance. *J. Exp. Med.* **63:**827–845.

Webster, L. T. (1937). Inheritance of resistance of mice to enteric bacterial and neurotropic virus infections. *J. Exp. Med.* **65:**261–286.

Webster, L. T., and Clow, A. D. (1936). Experimental encephalitis (St. Louis type) in mice with high inborn resistance: A chronic subclinical infection. *J. Exp. Med.* **63:**827–845.

Webster, L. T., and Johnson, M. S. (1941). Comparative virulence of St Louis encephalitis virus cultured with brain tissue from innately susceptible and innately resistant mice. *J. Exp. Med.* **74:**489–494.

Wilkinson, K. D., Urban, M. K., and Haas, A. L. (1980). Ubiquitin is the ATP-dependent proteolysis factor I of rabbit reticulocytes. *J. Biol. Chem.* **255:**7529–7532.

Wreschner, D. H., McCauley, J. W., Skehel, J. J., and Kerr, I. M. (1981). Interferon action-: Sequence specificity of the ppp(A2′p)nA-dependent ribonuclease. *Nature* **289:**414–417.

Yamamoto, Y., Sono, D., and Sokawa, Y. (2000). Effects of specific mutations in active site motifs of 2′,5′-oligoadenylate synthetase on enzymatic activity. *J. Interferon Cytokine Res.* **20:**337–344.

Zeidner, N. S., Higgs, S., Happ, C. M., Beaty, B. J., and Miller, B. R. (1999). Mosquito feeding modulates Th1 and Th2 cytokines in flavivirus-susceptible mice: An effect mimicked by injection of sialokinins, but not demonstrated in flavivirus-resistant mice. *Parasite Immunol.* **21:**35–44.

Zhou, A., Paranjape, J. M., Der, S. D., Williams, B. R. G., and Silverman, R. H. (1999). Interferon action in triply deficient mice reveals the existence of alternative antiviral pathways. *Virology* **258:**435–440.

ADVANCES IN VIRUS RESEARCH, VOL 60

IMMUNOBIOLOGY OF MOSQUITO-BORNE ENCEPHALITIC FLAVIVIRUSES

Arno Müllbacher, Mario Lobigs, and Eva Lee

Division of Immunology and Genetics
John Curtin School of Medical Research
The Australian National University
Canberra City, A.C.T. 2601, Australia

I. Introduction

A. The JEV Serocomplex

The JEV serocomplex comprises Japanese encephalitis virus (JEV), West Nile virus (WNV), Murray Valley encephalitis (MVE) virus, Kunjin (KUN) virus, Alfuy (ALF) virus, and St. Louis encephalitis (SLE) virus on the basis of cross-neutralization using polyclonal antisera (Calisher *et al.*, 1989). Phylogenetic relationship confirms the membership of this serocomplex in general with the exception of SLE, which is grouped with the Ntaya serocomplex on the basis of nucleotide and amino acid sequence relatedness (Kuno *et al.*, 1998).

Members of the JEV serocomplex are all mosquito-borne, and several, including JEV, WNV, MVE, and SLE, are well-known human pathogens. The life cycles of these four flaviviruses critically rely on amplifying hosts, such as pigs and birds, for transmission. The geographical distribution of JEV and WNV has increased over the last

0065-3527/03 $35.00

decade to be currently the most widely spread members of the serocomplex. JEV is medically the most important member, with approximately 40,000 cases reported annually, and is found in most parts of Asia, including east, south, southeast, as well as the Pacific islands and northern Australia in recent years (Mackenzie *et al.*, 2001; Vaughn and Hoke, 1992). WNV is found all over Europe and Africa as well as India; it has recently spread to the Western Hemisphere (involving most of the United States) and is rapidly establishing a foot hold in local migratory bird populations (Marfin *et al.*, 2001; Petersen and Roehrig, 2001). MVE, KUN, and ALF are predominantly found in mainland Australia; epidemic outbreaks of MVE have been reported in Papua New Guinea (Mackenzie *et al.*, 1994; Marshall, 1988). Epidemics of SLE occur with regular frequency in the southern United States, depending on environmental factors.

B. Pathogenesis in Human Infections and in the Mouse Model

The majority of human infections with encephalitic flaviviruses are asymptomatic or give rise to only a mild febrile illness. However, in a small percentage of infected individuals the mild infection turns into a life-threatening encephalitis. Thus, a key question in the pathogenesis of encephalitic flaviviral disease concerns the conditions that allow virus entry from the blood into the central nervous system (CNS). The clinical manifestations of encephalitic flavivirus disease have been reviewed (Monath, 1986; Monath and Heinz, 1996). Host factors appear to play a role in disease susceptibility. The ratio of inapparent-to-apparent infections and case fatality rates are influenced by age, and elderly patients are the most susceptible (Marfin *et al.*, 1993; Nash *et al.*, 2001; Tsai *et al.*, 1998). Underlying medical conditions (e.g., hypertension, diabetes mellitus, and coinfection with a second virus) may predispose infected persons to neurological complications by facilitating passage of the neurotropic viruses across the blood–brain barrier (Marfin *et al.*, 1993; Nash *et al.*, 2001; Wasay *et al.*, 2000). Although considerable variation in virulence properties exist between isolates of SLE (Monath *et al.*, 1980), JEV (Chen *et al.*, 1990), and WNV (Porter *et al.*, 1993) but not MVE (Lobigs *et al.*, 1988), it remains uncertain whether human isolates represent distinct neuroinvasive phenotypes.

Histological examination of fatal human cases of infection with viruses of the JEV serotype show viral antigen in neurons with the greatest involvement in the thalamus and brainstem but relatively little evidence of inflammatory responses (Johnson *et al.*, 1985; Nash *et al.*, 2001; Reyes *et al.*, 1981; Shieh *et al.*, 2000). Macrophages were the predominant inflammatory cells found to invade the brain parenchyma and localized in the vicinity of virus-infected cells; T and B cells were mainly seen in perivascular regions. Viral antigen is progressively cleared from the brain at later stages of disease (Johnson *et al.*, 1985), probably due to the induction of adaptive immune responses but with little evidence of immunopathology in neuronal destruction. A vigorous virus-specific antibody (Ab) response, systemically and within the CNS, appears to be an important factor in recovery from acute encephalitic flavivirus infection (Burke *et al.*, 1985a), whereas elevated levels of tumor necrosis factor in the serum and cerebrospinal fluid (probably reflecting the magnitude of the inflammatory response) is a poor prognostic indicator for recovery (Ravi *et al.*, 1997). The administration of dexamethasone as an immunosuppressant in patients with acute JEV-CNS illness was of no significant benefit, indicating that death resulted from infection of many neurons and disruption of their function (Hoke *et al.*, 1992). Flavivirus persistence in the human CNS has been found in a small proportion of patients, which may account for late sequelae in some patients (Ravi *et al.*, 1993).

In addition to encephalitis, rare human cases of acute hepatitis and pancreatitis have also been reported as a result of WNV infections (Cernescu *et al.*, 2000; Georges *et al.*, 1987; Nash *et al.*, 2001; Perelman and Stern, 1974), but not by infections with other members of the JEV serotype. Clinical infections with the mosquito-borne encephalitic flaviviruses in humans mostly occur in the absence of detectable viremia consistent with the notion that humans are dead-end hosts in the natural transmission cycle. Only WNV has been isolated consistently from sera of human patients (Georges *et al.*, 1987; Lvov *et al.*, 2000; Southam and Moore, 1954). However, given the route of entry and the small virus dose encountered in natural flavivirus infections, as well as the length of the incubation period before symptoms of CNS infection become apparent (1 to 2 weeks), it is most likely that an extraneural phase of virus growth precedes virus entry into the brain. It is not known which tissues are important in supporting peripheral virus growth and whether the magnitude of progeny virus produced in these tissues is a decisive factor that determines the incidence of CNS involvement.

Mice provide an excellent animal model for flaviviral encephalitis in humans. All members of the JEV serocomplex are neurotropic in mice. When directly inoculated into the brain, virus grows to high titers, causing a mainly fatal encephalomyelitis. The disease outcome following extraneural inoculation is influenced by host factors and virulence properties of the viruses. Genetic resistance of wild and inbred strains of mice has been described (Brinton *et al.*, 1982; Darnell *et al.*, 1974; Miura *et al.*, 1990). The flavivirus resistance gene (*Flv*) has recently been identified as an interferon (IFN)-inducible gene encoding 2′-5′-oligoadenylate synthetase (Mashimo *et al.*, 2002; Perelygin *et al.*, 2002). The age of the animals is of particular importance in their susceptibility to encephalitic flavivirus infection; mice, until the age of 3 to 4 weeks, are highly susceptible to a low peripheral virus inoculum (Eldadah *et al.*, 1967; Grossberg and Scherer, 1966; Huang and Wong, 1963; MacDonald, 1952a, 1952b). In these young mice, encephalitic flaviviruses can grow in diverse extraneural tissues and generate viremia (Huang and Wong, 1963; McMinn *et al.*, 1996; Monath *et al.*, 1983). Thus, the kinetics of virus spread into the brain from the site of extraneural inoculation can be followed. Virus is first detected in the olfactory lobe prior to spread to the remainder of the brain, suggesting that in 3-week-old mice neuroinvasion occurs via olfactory neurons (McMinn *et al.*, 1996; Monath *et al.*, 1983). The olfactory neuroepithelium is not protected by the blood-brain barrier and is richly supplied with capillaries having fenestrated endothelia, potentially facilitating infection by virus in the circulation. Brain infection with the encephalitic flaviviruses results in extensive neuronal necrosis in the 3-week-old mice (Matthews *et al.*, 2000), which is apparently exacerbated by neutrophil infiltration, triggering a nitric oxide (NO)-mediated immunopathological reaction (Andrews *et al.*, 1999). Given their susceptibility to peripheral inoculation with neuroinvasive viruses of the JEV serocomplex, 3-week-old mice are routinely used in virulence determination of virus isolates by comparing LD_{50} values after intraperitoneal (i.p.) and intracranial (i.c.) inoculation (Lobigs *et al.*, 1988, 1990; Monath *et al.*, 1980).

In mice older than 3 to 4 weeks of age, extraneural inoculation of viruses of the JEV serocomplex often fails to result in morbidity and mortality over a wide dose range (Eldadah *et al.*, 1967; Grossberg and Scherer, 1966; Huang and Wong, 1963; Licon Luna *et al.*, 2002; MacDonald, 1952a). The erratic course of infection in the older animals often prevents a reliable determination of the LD_{50}. Adult animals are highly susceptible to i.c. injection with small virus doses, leading to high virus titers in the brain and fatal encephalitis. However, virus growth in extraneural tissues is poor and viremia often undetectable (Huang and Wong, 1963; Licon Luna *et al.*, 2002; MacDonald, 1952b). A number of factors could account for age-dependent differences in the susceptibility of mice to flavivirus infection; these include (but not exclusively) (1) higher peripheral virus yield and viremia in younger animals (see earlier), (2) an age-dependent anatomical difference in the permeability of the blood-brain barrier, (3) a stronger tropism of the viruses for developing neurons (Ogata *et al.*, 1991; Oliver and Fazakerley, 1998), and (4) absence, delay, or dysfunction of innate and/or adaptive immune responses (Ridge *et al.*, 1996; Sarzotti *et al.*, 1996; Trgovcich *et al.*, 1999). When peripheral infection of older mice progresses to encephalitis, prominent inflammatory changes are found in the brain with leukocyte infiltrates and perivascular cuffing, and the inflammatory reaction is accompanied by neuronal destruction (Hase *et al.*, 1990).

As in human infections with the encephalitic flaviviruses, the relatively rare stochastic event that permits virus in the circulation to breach the blood-brain barrier appears to be the crucial step leading to a mostly fatal encephalitis in peripherally infected older mice. The mechanism allowing encephalitic flaviviruses to breach the blood-brain barrier remains uncertain, but four candidate routes for CNS invasion have been canvassed (Johnson and Mims, 1968): (1) by the neuronal route after infection of peripheral nerves, (2) by infection of highly susceptible olfactory neurons that are unprotected by the blood-brain barrier (McMinn *et al.*, 1996; Monath *et al.*, 1983), (3) by virus entry into vascular endothelial cells of capillaries in the brain, transcytosis, and release of virus into the brain parenchyma (Liou and Hsu, 1998), or (4) by diffusion of virus between capillary endothelial cells in individuals displaying leakiness of the blood-brain barrier due to factors unrelated or secondary to the viral infection. Numerous factors influencing the integrity of the blood-brain barrier, such as microwave radiation, infection with a second pathogen, hypercarbia, inhalation anesthetics, and the administration of small amounts of detergent or lipopolysaccharide have been shown to increase brain invasion and mortality by the encephalitic flaviviruses (Ben-Nathan *et al.*, 2000; Gupta and Pavri, 1987; Hayashi and Arita, 1977; Kobiler *et al.*, 1989; Lange and Sedmak, 1991; Lustig *et al.*, 1992). The possible contribution of antiviral cytolytic lymphocytes to the fourth mechanism of blood-brain barrier disruption is discussed later.

In summary, many parallels exist between the pathogenesis of neurotropic flaviviruses in humans and adult mice, highlighting the value of this animal model in gaining an understanding of the human disease. These include (1) the poor or undetectable virus growth in extraneural tissues contrasting with efficient virus replication in the CNS, (2) the rare occurrence of virus entry into the brain following peripheral infection, and (3) the lack of a correlation between the magnitude of viremia and brain invasion.

II. Innate Immune Responses

A. *Interferons*

The earliest host responses to viral infection are nonspecific and involve, among other cytokines, the induction of IFN. Type I IFNs, IFN-α and β, are produced by leukocytes and fibroblasts, respectively, in response to infection and activate the transcription of a host of IFN-inducible genes that leads to the induction of antiviral pathways (Goodbourn *et al.*, 2000). IFN-γ is made exclusively by cells of the immune system (NK and T cells) and has important immunoregulatory functions; these include the activation of monocytic cells, the enhancement of expression of a subset of chemokines, and the induction of increased expression of molecules of the MHC class I and class II antigen presentation pathways. Most of the antiviral activity of IFN-γ is mediated by NO radicals synthesized by monocytic phagocytes, following induction of the enzyme NO synthase by the cytokine (Guidotti and Chisari, 2000; Reiss and Komatsu, 1998). The effect of NO on growth of flaviviruses in cultured cells and *in vivo* has been investigated. Inhibition of growth of JEV in IFN-γ-activated mouse macrophage and co-cultured neuroblastoma cells correlated with cellular NO production, and mortality in JEV-infected mice increased when the activity of NO synthase was inhibited (Lin *et al.*, 1997; Saxena *et al.*, 2000, 2001). The antiviral mechanism of NO on JEV replication in cell culture involves a block in viral RNA and protein synthesis (Lin *et al.*, 1997). In contrast, no inhibition of TBE replication by NO was seen in cultures of mouse macrophages and a marginal increase in survival of infected mice was found when a competitive inhibitor of NO production was administered (Kreil and Eibl, 1996). In a study of i.p. infection of 3-week-old mice with MVE, the contribution of NO to viral pathogenesis was also noted (Andrews *et al.*, 1999). Given the close relatedness of MVE and JEV, the difference in disease outcome following NO synthase inhibition is unexpected. It may be reconciled by differences in the animal models used; infection with JEV was in 4- to 6-week-old mice with sublethal virus doses, resulting in less mortality and longer survival times than in the 3-week-old animals lethally infected with MVE. Consistent with a role of NO in protection from encephalitic flavivirus infection, we have found slightly increased mortality in 6-week-old NO synthase knock-out mice infected intravenously (i.v.) with a low dose of MVE relative to wt mice (Lobigs *et al.*, 2003b).

Mice with targeted disruption of the IFN-γ gene have impaired production of NO by monocytes and reduced MHC class II expression (Dalton *et al.*, 1993). These mice were not significantly more susceptible to challenge with dengue virus (Johnson and Roehrig, 1999) or yellow fever virus (Liu and Chambers, 2001) than wild-type mice. We have found that peripheral infection of IFN-γ knockout mice with a small dose of MVE (10^2 PFU, i.v.) gives rise to only a marginal increase in mortality from encephalitis in comparison to that in wild-type mice (Lobigs *et al.*, 2003b). These studies suggest that in adult mice the potentially deleterious inflammatory response resulting from the production of IFN-γ is more than compensated by the antiviral effect of the cytokine, which may involve NO production, class-switching of immunoglobulin isotypes, or a direct antiviral function in the CNS (Binder and Griffin, 2001; Kundig *et al.*, 1993).

The importance of type I IFNs in recovery from infection with the encephalitic flaviviruses has been shown *in vivo* (1) by the therapeutic and prophylactic effects of administration of IFN inducers (Haahr, 1971; Taylor *et al.*, 1980; Vargin *et al.*, 1977) or recombinant IFN (Brooks and Phillpotts, 1999; Pinto *et al.*, 1988), and (2) by uncontrolled growth of MVE and JEV in the brain and extraneural tissues of 6-week-old mice lacking a functional type I IFN receptor (Lee and Lobigs, 2002; Lobigs *et al.*, 2003b).

B. The Natural Killer and Other Cellular Innate Immune Responses

Natural killer (NK) cells are part of the innate immune response and are induced and activated by a variety of infectious organisms, including most viruses. Furthermore, the NK cell response has been implicated as an important component in the early defense in numerous virus infections (Biron *et al.*, 1999). However, few studies have attempted to elucidate the role played by NK cells in flavivirus infections.

Besides their cytolytic function, NK cells are susceptible to and synthesize and regulate a host of cytokines, which influences the adaptive immune response. Class I IFNs induced as a result of virus infections are known to greatly affect NK cell activity by specifically enhancing cytolytic activity of NK cells (Biron *et al.*, 1999). However, due to this extensive network of cytokines, and effector cells, it is difficult to assign host resistance to flavivirus infection to any particular NK cell function. Induction of classical NK cell cytolytic activity as a result of flavivirus infection has been reported in mice (Hill *et al.*, 1993a; Kopecky *et al.*, 1991; MacFarlan *et al.*, 1977; Momburg *et al.*, 2001; Vargin and Semenov, 1986) and humans (Ilyinskikh *et al.*, 1990).

Classical NK cells have been isolated from the brain of mice infected i.c. with WNV (Liu *et al.*, 1989), suggesting that NK cells are recruited and/or activated at the site of flavivirus replication *in vivo*. However, NK cells isolated from WNV-infected brains have been shown to be refractory in killing astrocytes (Liu and Müllbacher, 1988) even prior to MHC class I upregulation due to flavivirus infection (Liu *et al.*, 1989b). To assess if NK cell cytotoxicity plays a role in resistance or recovery from flavivirus infections, we used adult beige mice, which are deficient in cytolytic function of NK cells (Roder *et al.*, 1979). No increase in the susceptibility to MVE infection (Licon Luna R., Lobigs, M., Müllbacher, A., and Lee, E., unpublished observations) was observed. Furthermore, mutant mice defective in either the exocytosis pathway (perforin-deficient mice [perf$^{-/-}$]; Kägi *et al.*, 1994), or the Fas pathway (mutant mice with a defect in the Fas ligand, gld) died with similar frequency and kinetics after high- or low-dose infection with MVE (Licon Luna *et al.*, 2002) or WNV (Wang, Y., Lobigs, M., Lee, E., and Müllbacher, A., unpublished observations) as compared to wild-type mice. These observations do not preclude that in the presence of a functional NK cell response, NK cells are involved in the antiviral response and compensated for by redundant mechanisms in their absence. We have hypothesized (Hill *et al.*, 1993a; Lobigs *et al.*, 2003a; Momburg *et al.*, 2001) that the reduced susceptibility of flavivirus-infected target cells to NK cell lysis, due to flavivirus-mediated MHC class I upregulation (see later), is a strategy employed by flaviviruses to sideline NK-mediated host defense early during infection.

Macrophages and monocytes are essential components in the host defense against most infections, including viral infections (Blanden, 1982). One study (Ben-Nathan *et al.*, 1996), using macrophage depletion in a mouse model investigating WNV-induced encephalitis, found increased neuroinvasion by virus even when nonneurovirulent strains were used. The authors speculated that an elevated viremia due to the absence of macrophages leads to an early invasion of the CNS. However, macrophages have a pleiotropic role in most immune responses. In the case of flavivirus infection the immunobiology and pathogenesis is further complicated by the phenomenon of immune-enhancement. This refers to the observation first made by Hawkes (1964), whereby suboptimal neutralizing Ab concentrations enhanced flavivirus infection, in particular, of Fc receptor-bearing macrophages (Peiris *et al.*, 1981). Although initially demonstrated with MVE and WNV, a possible role of Ab-mediated enhancement of flavivirus infection in pathogenesis has only been invoked in dengue virus etiology (Burke and Monath, 2001).

Nothing is known in regard to the role of other cellular innate immune responses in flavivirus infection, such as $\gamma\delta$ T cells recently identified in response to vaccinia virus (VV) infections (Selin *et al.*, 2001).

III. Adaptive Immune Response

A. *The Humoral Immune Response*

The humoral response in mice inoculated with encephalitic flaviviruses is normally characterized by appearance of neutralizing Ab between 4 to 6 days post-inoculation (Camenga *et al.*, 1974; Monath and Borden, 1971; Webb *et al.*, 1968) with the IgM response predominant in the early phase of infection (Monath and Borden, 1971; Webb *et al.*, 1968). Serum Ab in patients with acute encephalitis comprises mainly virus-specific IgM, with peak levels detectable 7 days after admission in JEV-infected human patients (Burke *et al.*, 1985b). Serum IgM Ab in rhesus macaque monkeys intranasally infected with JEV is detectable from around day 10 post-inoculation, whereas IgG Ab is not detected in the 28-day period after infection (Raengsakulrach *et al.*, 1999).

The importance of humoral response in recovery from encephalitis is demonstrated in studies by Webb *et al.* (1968), Camenga *et al.* (1974), and Mathur *et al.* (1983), all showing that administration of Ab early during infection (within 3 to 6 days) can protect against an otherwise lethal infection; the first two studies using immune-deficient mice (γ-irradiated and cyclophosphamide-treated, respectively) for the challenge, the latter relying on lethal i.c. infection. Similar protection can also be achieved by adoptive transfer of immune splenocytes (Camenga *et al.*, 1974; Jacoby *et al.*, 1980; Mathur *et al.*, 1983), and this protection is abrogated by pretreatment of splenocytes with anti-thymocyte Ab (Jacoby *et al.*, 1980; Mathur *et al.* 1983), demonstrating the importance of T cells in recovery from flavivirus-induced encephalitis (see later). Notably, Mathur *et al.* (1983) reported that deletion of B cells from the splenocytes using anti-mouse IgG plus complement did not impact significantly on the outcome of adoptive transfer, but the efficacy of anti-B cell treatment had not been confirmed in the study. A study (Diamond *et al.*, 2003) of WNV infection in mice genetically deficient in B cells (μMT) or in both B and T cells (*RAG1*) showed significantly greater susceptibility compared to that of congenic wild-type mice; passive transfer of immune splenocytes and B cells to the *RAG1* mice, as well as immune sera to the μMT mice, reduced mortality. The importance of B cells and humoral response in protection from lethal virus challenge is also demonstrated in envelope (E) protein-based DNA vaccination against JEV (Pan *et al.*, 2001), as protection could be adoptively transferred with immune sera but not splenocytes or T cell-enriched splenocytes. Furthermore, gene knockout mice defective in B cells or $CD4^+$ T cells were not protected by DNA vaccination, in contrast to mice defective in $CD8^+$ T cells surviving lethal JEV challenge after DNA immunization (Pan *et al.*, 2001).

The role of T cells in eliciting humoral responses against flaviviruses was investigated in the following studies: T cell depletion by thymectomy and subsequent irradiation did not significantly alter the level of hemagglutination-inhibiting (HI) Ab in mice infected with Banzi virus, in contrast to cyclophosphamide treatment, which suppressed the HI response (Bhatt and Jacoby, 1976); nude mice with no functional T cells showed severely depressed production of JEV-specific Ab (Lad *et al.*, 1993); depressed Ab levels were also reported in infection of athymic mice by the French neurotropic strain of YFV (Bradish *et al.*, 1980), while Hotta *et al.* (1981) reported that nude Balb/c mice show depressed IgM production without any subsequent rise in IgG levels after infection with a mouse-adapted strain of dengue type 1 virus. The depressed Ab levels reported in most of these studies of flavivirus-infected mice lacking functional T cells suggests that B cell activation by flaviviruses is at least partially dependent on T cell help, in contrast to viruses like vesicular somatitis virus and polio virus, which can activate B cells in the absence of T cell help (Bachmann *et al.*, 1993). Hence, in addition to abolishing class- and isotype-switching, absence of helper T cells may impact on the generation of a strong neutralizing IgM response against flaviviruses.

Administration of monoclonal antibodies (mAb) to mice during infection with encephalitic flavivirus SLE (Mathews and Roehrig, 1984), MVE (Hawkes *et al.*, 1988), and JEV (Cecilia *et al.*, 1988) offers significant protection from otherwise lethal i.p. virus challenge. In the SLE study, the efficacy of passive protection by mAb correlated well with *in vitro* neutralization activity of the Ab and animals could be protected by passive transfer of neutralizing mAb up to day 3 post-challenge, prior to neural invasion by the virus. Unlike in the SLE study, nonneutralizing mAbs against MVE can also give significant protection against a lethal i.p. challenge (Hawkes *et al.*, 1988). In the JEV study (Cecilia *et al.*, 1988), a panel of E protein-specific mAbs were tested for neutralization, HI, and passive protection against lethal JEV challenge. The mAbs, which gave significant protection, mostly showed high neutralization activity, except for one that did not neutralize *in vitro*. The contribution of NS1, a cell surface localized nonstructural protein, to protection is demonstrated in active immunization using purified NS1 of YFV, as well as passive transfer of mAbs to NS1 (Schlesinger *et al.*, 1985), both resulting in effective protection against YF17D-induced encephalitis. Comparison of immunization with recombinant VV that expressed JEV or MVE prM and E proteins, NS1 protein, or NS3 protein demonstrated low levels of protection with NS1-based immunization in contrast to high levels of protective immunity elicited by E protein-based immunization (Hall *et al.*, 1996; Konishi *et al.*, 1991). In addition, protective immunity is also elicited by DNA immunization with JEV NS1 gene: immunized mice showed no neutralizing activity but a strong complement-dependent cytolytic activity against JEV-infected cells (Lin *et al.*, 1998).

What is the mechanism of neutralization of flavivirus infectivity by Ab? *In vitro* neutralization of WNV infectivity by Ab has been shown by Gollins and Porterfield (1986) to depend on inhibition of the intraendosomal acid-catalysed fusion step during virus entry. Studies of the entry process of JEV using electron and confocal microscopy techniques showed that neutralizing mAb strongly inhibits JEV-induced fusion and internalization into cells, but not binding of virus to cells (Butrapet *et al.*, 1998). Interestingly, Se-Thoe *et al.* (2000) showed by electron microscopy that direct fusion between virus and plasma membrane normally occurs during DEN-2 entry into LLC-MK2 cells but is inhibited in the presence of a neutralizing mAb or DEN-immune human sera; virus uptake is observed to occur by endocytosis instead. In contrast, Crill and Roehrig (2001) showed, using a panel of mAb against dengue E protein, that neutralization of virus infectivity correlated with inhibition by the mAb of virus adsorption to host cells.

B. T Cell Responses

T cells are of crucial importance for the recovery from most virus infections. Individuals deficient in T cells but not B cells are unable to control infections with viruses that are avirulent in normal individuals (Fulginiti *et al.*, 1968). Only a limited number of studies have addressed the role played by T cells in flavivirus infections. Using T cell-deficient mice (nu/nu), experiments showed that T cells are necessary for recovery (Lad *et al.*, 1993) and protection after i.c. challenge with JEV (Miura *et al.*, 1990). However, because general T cell depletion or lack of T cells will affect the humoral, as well as the cellular, immune defense against flavivirus infections, information gained from such studies are hard to interpret.

The two dominant lineages of peripheral T cells, $CD4^+$ and $CD8^+$ T cells, recognize epitopes consisting of a complex of a short peptide (in case of flavivirus-immune T cells, the peptide is derived from a flavivirus-encoded protein) and a self-MHC class II or class I molecule, respectively. Professional antigen-presenting cells (APC) are required to present peptide/MHC complexes to T cells for activation. All nucleated cells express class I MHC, whereas class II MHC expression is restricted to only some cell types of the reticuloendothelial system. Thus, antigen-primed $CD4^+$ T cells, dendritic cells (DC), macrophages, or B cells, all of which express constitutively high concentrations of class II MHC antigens, efficiently induce cell proliferation (Kulkarni *et al.*, 1991b). For the induction of naïve T cell responses, DC cells may be the only cell type with sufficient qualitative and quantitative stimulatory capacity to trigger activation of both subsets of T cells (Lanzavecchia and Sallusto, 2001).

1. The CD4 T Cell Response

Upon recognition of appropriate MHC class II/peptide complexes, $CD4^+$ T cells become activated, proliferate, and mature into one of two T helper subsets, TH1 or TH2, depending on the stimulatory milieu. In response to most viral infections, the cytolytic function resides predominantly in the $CD8^+$ T cell population. Rarely do $CD4^+$ T cells differentiate into cytolytic effector cells. However, the T cell response against flaviviruses is an exception. In dengue virus infections in humans, the bulk of the cytolytic response resides in the $CD4^+$ T cell population (Kurane *et al.*, 1989). Interestingly, cytolytic $CD4^+$ T cell clones have also been established from peripheral blood monocytes of individuals immunized with JEV vaccine (Aihara *et al.*, 1998; Kurane and Takasaki, 2000). The reason for this unusual phenotype of $CD4^+$ T cells is not clear. It is possible that these clones are a result of *in vitro* restimulation and are thus of limited biological relevance, especially as their *in vivo* function is restricted to infected cells expressing class II MHC antigens. In this context, however, it is noteworthy that WNV-infection of astrocytes *in vitro* induces the expression of MHC class II (Liu *et al.*, 1989b). In case of JEV-immune $CD4^+$ T cell clones, these were derived from individuals immunized with inactivated virus preparations (Aihara *et al.*, 1998; Kurane and Takasaki, 2000). Inactivated virus preparations, with the exception of those produced by gamma irradiation (Müllbacher *et al.*, 1988), are known to induce primarily $CD4^+$ T cell and not classical $CD8^+$ Tc cell responses (Braciale and Yap, 1978). However, using live VV encoding the structural prM and E and nonstructural NS1 genes of JEV as a vaccine did induce $CD8^+$ Tc cells as expected (Konishi *et al.*, 1998). Thus, at present, we do not have a clear understanding of why $CD4^+$ T cells acquire cytolytic potential in response to flavivirus infections. The cytolytic pathway by which these flavivirus-immune $CD4^+$ T cells kill target cells has not been established. It may involve the release of cytokines (IFN-γ or TNF-α) or direct killing via the granule exocytosis or Fas-mediated pathways (see later). The possibility that this constitutes a diversion mechanism of the host's immune responses, such as immune class deviation, evolved by flaviviruses cannot be discounted. Elucidation of the mechanism underlying the phenomenon of cytolytic $CD4^+$ T cell responses as a result of flavivirus infection in humans may be of great importance in attempting to develop prophylactic and therapeutic interventions.

It is generally assumed that $CD4^+$ T cells are important in the host's defenses against flaviviral infections by facilitating a protective Ab response and a cytotoxic T (Tc) cell response. To our knowledge, experiments using $CD4^+$ T cell-depleted or T cell-deficient mice have not been performed to definitively establish the role $CD4^+$ T cells play in recovery from flavivirus infection. T cell help is required to switch B cell immunoglobulin synthesis to the IgG isotype with virus neutralization property (Mathews *et al.*, 1991, 1992; Uren *et al.*, 1987). However, it does not appear to play a significant role in the CNS inflammatory response. Using WNV infection in the mouse model to study leukocyte infiltration of the CNS, we noted that $CD4^+$ T cells were not present in cell isolates from infected brains, whereas NK and $CD8^+$ T cell infiltrates were found (Liu *et al.*, 1989a). These findings have recently been confirmed by immunohistochemical analysis of WNV-infected brain sections (Wang *et al.*, 2003). Furthermore, the CD4/CD8 T cell ratio in spleen changes significantly in favor of CD8 cells as a result of WNV infection (Wang *et al.*, 2003). The reason for this is unknown. It is also not known if flavivirus-immune $CD4^+$ T cells are required for the induction and/or activation of $CD8^+$ T cells.

A number of studies have identified the source of peptide determinants, recognized in association of MHC class II by flavivirus-immune $CD4^+$ T cells. In the majority of mouse haplotypes, immunogenic peptides are derived from the E and NS1 proteins (Mathews *et al.*, 1991; Mathews *et al.*, 1992) and these are also the source of peptides recognized by human $CD4^+$ T cells in response to JEV vaccination (Aihara *et al.*, 1998, Konishi *et al.*, 1995). This of course is not surprising because the E structural protein, together with the cell surface-expressed NS1 protein, is responsible for the bulk of the Ab response to these viruses and thus flavivirus-specific B cells can function as APCs.

The most detailed and extensive analysis of the CD4$^+$ T cell response against JEV serocomplex flaviviruses has been undertaken in the mouse in response to infection with KUN (Kulkarni *et al.*, 1991a, 1991b, 1991c, 1992). This was made possible by the availability of a panel of VV recombinants expressing overlapping sequences of the KUN genome covering the entire viral genome (Hill *et al.*, 1993a; Parrish *et al.*, 1991). KUN-immune CD4$^+$ T cells from H-2^b, H-2^d, and H-2^k mice responded to peptides predominantly encoded within the structural polyprotein region plus the nonstructural proteins, NS1 and NS2. In addition, peptide determinants encoded within the NS4B-NS5 region were also immunogenic in all three haplotypes tested. The limitation of mapping studies using recombinant expression vectors was highlighted when it was found that a VV construct encoding all nonstructural genes of KUN (including the NS4B-NS5 region) did not induce proliferation of T cells from KUN-primed mice, in contrast to a VV recombinant encoding the NS4A to NS5 region only. This may imply that the primary sequence of a protein, when expressed under different temporal and spatial constraints, is not equally immunogenic (Hill *et al.*, 1993a; Kulkarni *et al.*, 1992).

Clones of CD4$^+$ T cells induced in response to either MVE, KUN, or WNV showed extensive cross-reactivities with respect to activation by heterologous flaviviruses (Uren *et al.*, 1987). In a more detailed analysis, synthetic peptides with sequences derived from the E glycoprotein of MVE were also recognized by T cell clones derived from KUN and WNV immunized mice (Mathews *et al.*, 1991). Such cross-reactivity has also been observed with CD4$^+$ T cell clones established from peripheral mononuclear cells of JEV-vaccinated human donors (Aihara *et al.*, 1998). Thus it appears that, like class I MHC-restricted T cell responses (see later), class II MHC-restricted responses are widely cross-reactive within the JEV serocomplex.

2. *The CD8$^+$ T Cell Response*

CD8$^+$ T cells exert their effector function by at least two different mechanisms. One is the synthesis and secretion of cytokines such as IFN-γ (Boehm *et al.*, 1997) and TNF-α (Vassalli, 1992), and the other is cellular cytotoxicity mediated either by the perforin/granzyme exocytosis or the Fas pathway (Kägi *et al.*, 1994; Lowin *et al.*, 1994; Rouvier *et al.*, 1993). These two effector functions do not necessarily reside within the same cell. Recent estimates of Tc cell responses to viruses, using either intracellular IFN-γ or MHC-class I tetramer staining, far outnumber estimates of CD8$^+$ T cells precursors using limiting dilution analysis with a functional target cell lysis readout (Altman *et al.*, 1996; Butz and Bevan, 1998; Regner *et al.*, 2001b). As to cytolytic function, the predominant view at present suggests that target cell lysis by the exocytosis pathway is the primary mechanism by which Tc cells limit viral infections and the putative major role of the Fas pathway is associated with immune regulation (Kägi *et al.*, 1995; Nagata, 1997; Rouvier *et al.*, 1993). This view may be valid for certain virus infections but cannot be generalized, and may not be applicable to the immune response against flavivirus infection (see later).

In mice of different haplotypes, Tc cells are readily induced in the spleen when inoculated with live WNV, KUN, MVE, or JE, by either the i.p. or i.v. route (Gajdosova *et al.*, 1981; Kesson *et al.*, 1987; Lobigs *et al.*, 1994, 1997; Parrish *et al.*, 1991; Regner *et al.*, 2001a). Yet in most mouse strains only one restriction element, either K or D, is associated with any given Tc cell response (Hill *et al.*, 1993a; Hill *et al.*, 1993b). The mechanism responsible for this immune-response gene phenomenon is not understood and may differ from that of self-tolerance creating "holes" in the T-cell receptor (TCR) repertoire, which is responsible for immune-response gene defects in other virus-immune Tc cell responses (Doherty *et al.*, 1978; Hill *et al.*, 1993b; Müllbacher *et al.*, 1983; Zinkernagel *et al.*, 1978). The peak of the Tc cell response against the encephalitic flaviviruses occurs 5–6 days post-infection (Kesson *et al.*, 1987), similar to that observed with other viruses (Zinkernagel and Doherty, 1979). The primary *in vivo* Tc cell response is generally weak, and in most experimental studies *in vitro* restimulation is undertaken to obtain increased cytolytic activity (see later). For target cells to become sensitive to lysis by flavivirus-immune Tc cells, relatively prolonged infection times with flaviviruses are required; in the order of 12 to 24 hours. The reason for this is not understood but suggests that the class I MHC antigen–presenting pathway is inefficient in processing and presentation of flavivirus gene products early after infection. This requirement for long infection time of target cells for lysis to occur is unlikely to reside in the Tc effector population (i.e., low affinity TCRs requiring high antigen density on target cells to trigger the lytic signals) because modification of targets with appropriate peptides resulted in rendering targets susceptible to lysis at peptide concentrations similar to those reported for other viral systems (Regner *et al.*, 2001c). Also, infection of targets with VV recombinants encoding the relevant flavivirus gene products sensitized targets after 1 to 2 hours of infection, strongly inferring that the Tc cells immune to flaviviruses have TCRs with affinities comparable to those generated against other viruses, such as influenza or poxviruses.

The requirement for prolonged infection times for Tc-mediated lysis does not seem to be cell type-dependent. Both tumor cell lines, such as L929, and other methylcholanthrene-induced fibroblasts, as well as primary mouse embryo fibroblasts and freshly obtained mouse macrophages, require long infection times to become susceptible to lysis by flavivirus-immune Tc cells (Kesson *et al.*, 1988).

The memory Tc cell response to JEV serocomplex flaviviruses in mice exhibits some unusual features. The ability of lymphocytes of mice primed previously with a virus to be stimulated *in vitro* to generate potent secondary Tc cells is an indication of Tc cell memory, and lasts for the lifetime of the animal (Müllbacher and Flynn, 1996). Yet, in the case of encephalitic flaviviruses, *in vitro* stimulation of *in vivo*-primed splenocytes to generate secondary virus-immune Tc cells occurs efficiently only during a brief interval after priming (~14 days) (Kesson *et al.*, 1988; Lobigs *et al.*, 1996). This suggests that successful *in vitro* boosting only occurs when Tc cells are still in an activated state and that they later become refractory, or even suppressive, when encountering flavivirus-infected cells. We have speculated that this unresponsiveness of memory flavivirus-immune Tc cells to viral restimulation *in vitro* is due to the increased MHC-class I expression on flavivirus-infected stimulator cells (Lobigs *et al.*, 1996). The virus-mediated upregulation of MHC class I results in a significant increased presentation of both viral and self-peptides, where the latter may be above the threshold required for activation of self-reactive Tc cells (Blanden *et al.*, 1987). Therefore, it is conceivable that flavivirus-mediated upregulation of MHC class I leads to transient T cell autoimmunity, followed by downregulation of both autoimmunity and virus-specific Tc cell memory responses. Consistent with this interpretation is our observation that restimulation of flavivirus-primed splenocytes after lengthy periods following priming generates Tc cells that exhibit high lysis of mock-infected target cells (Kesson *et al.*, 1988; Lobigs *et al.*, 1996).

Long-term primed splenocytes are readily boosted *in vitro* to generate potent Tc cells when restimulation occurs with peptide-pulsed stimulator cells rather than flavivirus-infected cells (Regner *et al.*, 2001a). This may be important in terms of recombinant vaccination strategies aimed at inducing anti-flaviviral Tc cell immunity. An example is shown in Fig. 1, where a DNA vaccine encoding the MVE nonstructural polyprotein region from NS2B to the C terminus of NS5 was used to vaccinate CBA/H (H-2^k) mice. When immune splenocytes were stimulated *in vitro* with MVE or the two immunodominant K^k-restricted peptide determinants, MVE_{1785} and MVE_{1971} (Regner

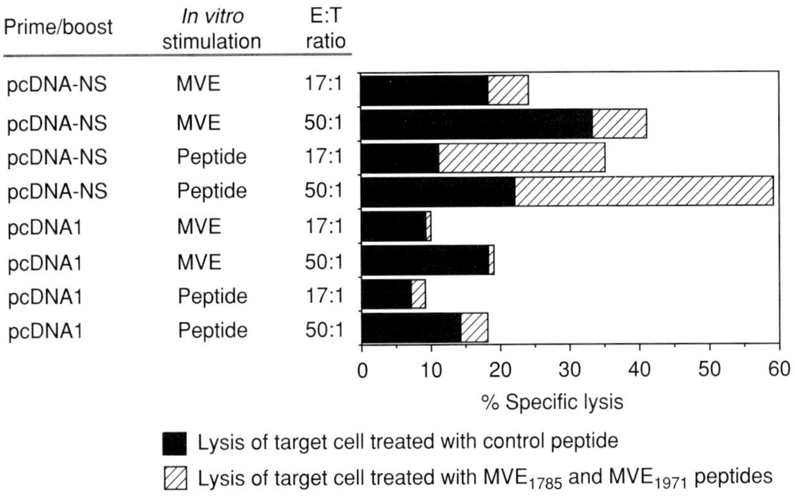

FIG 1. DNA vaccine-induced Tc cell response against MVE: restimulation, *in vitro*, with virus induces high anti-self cytotoxicity. CBA/H mice were primed, and boosted 3 weeks later, with pcDNA1.NS plasmid DNA (Lobigs *et al.*, 1994) or pcDNA1 control DNA using a gene gun as described (Colombage *et al.*, 1998). Two weeks after boosting, splenocytes were stimulated *in vitro* for 5 days with MVE or peptides MVE_{1785} plus MVE_{1971} as described (Regner *et al.*, 2001c); the effector cells were used on mouse L929 cells (H-2^k) treated with the 2 MVE-derived peptides or the K^k-restricted influenza virus control peptide NP_{50-57} at the indicated effector:target (E:T) cell ratios in a ^{51}Cr-release assay as described (Regner *et al.*, 2001c).

et al., 2001c), we found that stimulation with MVE but not with the peptides resulted in high lysis of control peptide-treated target cells. Thus, the ratio of "nonspecific" to viral peptide-specific target cell killing was >2-fold greater for the virus-stimulated in comparison to the peptide-stimulated effectors (80% and 37%, respectively, at E:T ratio of 50:1). Accordingly, persistent memory flavivirus-immune Tc cells are clearly generated as a consequence of a primary flavivirus infection or immunization with a recombinant subunit vaccine. However, the protective value of vaccination-induced Tc cell immunity against the encephalitic flaviviruses is relatively poor, given the high self-reactive component in the memory Tc cell response against the viruses. We also found that restimulation of MVE-primed splenocytes with peptide or virus differentially induced cytotoxicity and IFN-γ production. MVE-infected stimulators only induced the former, while stimulators modified with a immunodominant H-2K^k-restricted peptide induced both cytotoxicity and intracellular IFN-γ (Regner *et al.*, 2001b). To what extent IFN-γ synthesis is involved in regulation of memory Tc cell activation, protection, or immunopathogenesis is not known. Our own observation, using IFN-γ knockout mice, suggests that the lack of IFN-γ does not impair Tc cell memory formation or activation in response to pox and influenza viruses (Müllbacher *et al.*, unpublished data). Furthermore, IFN-γ knockout mice do not seem to be significantly more susceptible to MVE than are wild-type mice (Lobigs *et al.*, 2003b). However, in the context of dengue virus infections it has been suggested that IFN-γ may be involved in flavivirus-induced immunopathology (Kurane *et al.*, 1989). Thus, additional work on the role of IFN-γ in the immunobiology of flavivirus infections is warranted.

Similar to the $CD4^+$ T cell response, the viral genes coding for the antigenic peptide determinants from KUN recognized by Tc cells have been identified using VV–KUN recombinants (Hill *et al.*, 1993a; Parrish *et al.*, 1991). With the exception of the H-2^b haplotype, the dominant determinant in H-2^k, H-2^d, H-2^q, and H-2^s haplotypes could be mapped to the NS3/NS4A region. The same was also true for Tc cells from WNV-primed animals (Hill *et al.*, 1993a). In addition, the source of the K^k–restricted peptide determinant in the Tc cell response against MVE was mapped to the NS3 protein (Lobigs *et al.*, 1994) and that of the MVE-specific D^k- and D^d-restricted responses to the NS4B and NS2B-3 proteins, respectively (Lobigs *et al.*, 1997). Using K^k-specific peptide motif prediction, two peptides, MVE_{1785} (REHSGNEI) and MVE_{1971} (DEGEGR VI), encoded within the NS3 region, were identified to be immunogenic and were recognized by MVE and WNV-immune Tc cells (Lobigs *et al.*, 1994). A thorough analysis of the H-2^k MVE-immune Tc cell response against these peptide determinants revealed some interesting properties. First, the two peptides had different binding affinities for the H-$2K^k$ molecules as assessed by MHC-class I stabilization assays and half-life of peptide/MHC-class I complexes, with MVE_{1785} having lower affinity than MVE_{1971}. Yet, despite the lower binding affinity of MVE_{1785}, more MVE-immune Tc cell clones responded to this peptide than the higher affinity MVE_{1971} peptide as revealed by limiting dilution analysis. Furthermore, the lower affinity peptide was more efficient in triggering intracellular IFN-γ synthesis (Regner *et al.*, 2001c). These findings strongly suggest that the TCR repertoire in naïve animals is a limiting factor in the immune response rather than peptide affinity. Thus, vaccines designed on the basis of peptide affinity will not necessarily lead to the most effective immune response.

Using the same two peptides mentioned previously, and corresponding peptides based on sequence alignments for closely related (WNV,

KUN, JEV) and more distant (YFV, DEN) flaviviruses, in studies on cytotoxicity and intracellular IFN-γ synthesis, an extraordinary extent of cross-reactivity was revealed (Regner *et al.*, 2001a). In the extreme case, MVE_{1785}–immune Tc cell recognized target cells treated with a peptide (DEN_{1971}), which shared only a single amino acid (one of two anchor residues) with the MVE_{1785} peptide. No cross-reactivity was found with targets modified with more similar K^k-binding self-peptides. Even more surprising was that K^k-restricted flavivirus-immune Tc cells did not cross-react on target cells infected with other viruses including alpha- and bunyawera-viruses (Müllbacher *et al.* 1986; Regner *et al.*, 2001a). Target cells infected with VV recombinants encoding the NS3 to NS5 gene region of hepatitis C virus, a member of the Flaviviridae family, were similarly not recognized by either MVE- or WNV-immune Tc cells (our unpublished observation). The reason for this selective cross-reactivity of the Tc cell response within the flaviviruses is not clear; however, all cross-reactive peptides are derived from the NS3 protein, which is a protein with highly conserved function and thus structure, although with quite variable amino acid sequence.

Studies by Takada *et al.* (2000) investigating the Tc cell response to JEV also found that, depending on the haplotype of mice used, peptide determinants derived from different viral gene products are recognized by flavivirus-immune Tc cells. The consequences that follow from these studies are that effective vaccines suitable to protect outbred population should ideally encompass the whole virus genome.

IV. Virus-Host Interplay in Immunity and Pathogenesis

A. *Upregulation of MHC Class I Antigen Presentation: A Mechanism for Immune Escape?*

Many viruses interfere with MHC class I antigen processing and presentation (Ploegh, 1998; Yewdell and Bennink, 1999). The effect is almost always a block-in or downregulation of the MHC class I pathway. This is thought to constitute a strategy of immune evasion preventing the recognition of virus-infected cells by Tc cells.

The MHC class I pathway functions in the surveillance of the intracellular pool of peptides to allow detection of "foreign" peptide determinants by $CD8^+$ T cells (Fig. 2). The peptides are 8 to 10 amino acids in length and are derived from degradation of proteins of host or pathogen origin. They are predominantly of cytosolic origin and are generated by cellular proteases of which the proteosome, a multi-subunit and

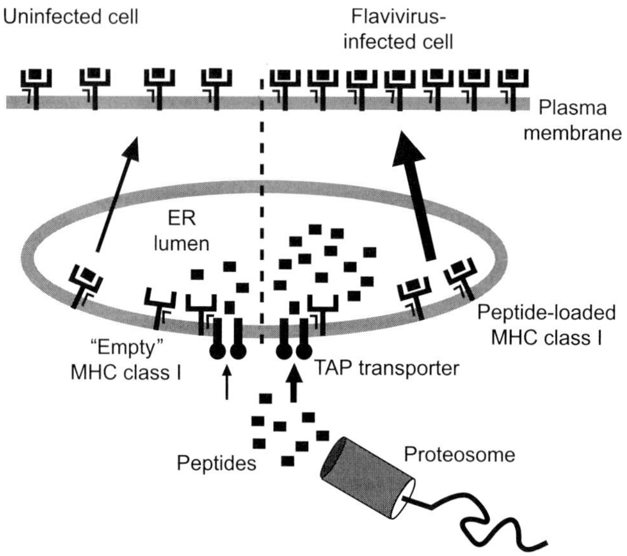

FIG 2. Schematic diagram of the MHC class I antigen processing and presentation pathway showing upregulation of TAP-dependent peptide transport into the ER and MHC class I cell surface expression as a result of flavivirus infection.

multi-catalytic protease abundant in the nucleus and cytoplasm, is most important (Tanaka and Kasahara, 1998; York and Rock, 1996). In order to assemble with the MHC class I restriction elements, most cytoplasmic peptides are transported into the lumen of the endoplasmic reticulum (ER) by the *t*ransporter *a*ssociated with antigen *p*rocessing (TAP), an ER-resident heterodimeric ATP-binding cassette transmembrane transporter (Elliott, 1997; Momburg and Hämmerling, 1998). Binding of peptides to MHC class I molecules is facilitated by a number of chaperones (tapasin, calreticulin, ERp57), which associate with TAP in the peptide loading complex (Cresswell *et al.*, 1999). Peptide-loaded MHC class I molecules leave the ER and travel via the exocytic pathway to the plasma membrane. MHC class I–restricted antigen presentation is regulated by IFN-γ, which increases expression of the class I molecules *per se*, and that of accessory molecules of the class I pathway (TAP and proteosome subunits) by transcriptional regulation (Momburg and Hämmerling, 1998; Tanaka and Kasahara, 1998). The induction of TAP expression by IFN-γ is more rapid than that of MHC class I molecules (Epperson *et al.*, 1992; Min *et al.*, 1996). This is consistent with the view that the constitutive level of TAP expression is insufficient to support inducible increases of MHC class I.

In contrast to the inhibitory effect of many viruses on MHC class I antigen presentation, upregulation of MHC class I expression has been observed in cells infected with flaviviruses. This phenomenon has been documented in different cell types (fibroblasts, trophoblasts, myoblasts, astrocytes, macrophages, B cells, and endothelial cells) of different species origin (human, mouse, and hamster) (Bao *et al.*, 1992; King and Kesson, 1988; King et al., 1989; Liu *et al.*, 1989b; Lobigs *et al.*, 1996; Müllbacher and Lobigs, 1995; Shen *et al.*, 1997). It is induced by flaviviruses from different serocomplexes, including yellow fever virus, dengue virus, and members of the JEV serotype (Lobigs *et al.*, 1996). Several lines of evidence suggest that flavivirus-induced MHC class I upregulation is not the consequence of a virus-induced IFN response: (1) the effect was only partially abrogated by treatment of infected cells with Ab against IFNs (King and Kesson, 1988); (2) WNV infection induced MHC class I expression in trophoblasts, a cell refractory to IFN-mediated induction of MHC class I (King *et al.*, 1989); (3) the cell surface expression of recombinant mouse MHC class I molecules, devoid of promoter sequences for IFN-induced transcription factors, was upregulated following flavivirus infection but not by treatment of cells with poly(IC), a strong IFN inducer (Müllbacher and Lobigs, 1995). Furthermore, the increased biosynthesis of MHC class I molecules was discounted as a mechanism of flavivirus-mediated MHC class I upregulation (Momburg *et al.*, 2001; Müllbacher and Lobigs, 1995).

The mechanism of flavivirus-mediated upregulation of MHC class I cell surface expression involves the increase of peptide supply into the lumen of the ER for assembly with the class I restriction elements (see Fig. 2) (Momburg *et al.*, 2001; Müllbacher and Lobigs, 1995). This effect of flavivirus infection is not seen in cells devoid of functional TAP but is apparent in cells with very low peptide transporter activity due to a mutation in the TAP2 subunit (RMA-S; Yang *et al.*, 1992) or developmental downregulation of the transporter activity (Syrian hamster BHK and NIL-2 cells; Lobigs *et al.*, 1995, 1999), suggesting TAP-dependence of the process (Momburg *et al.*, 2001; Müllbacher and Lobigs, 1995). Recently we have shown in *in vitro* peptide transport experiments that the transport activity of TAP is augmented by up to 50% at the early phase of infection with different flaviviruses, although TAP expression was not increased, and the substrate specificity of the transporter remained unaltered (Momburg *et al.*, 2001). This is the first demonstration of transient upregulation of TAP-dependent peptide import into the lumen of the ER as a consequence of a viral infection. The flaviviral gene products responsible for this effect as well as the detailed biochemical mechanism, remain to be resolved.

The biological function of upregulation of MHC class I restricted antigen presentation in flavivirus-infected cells is not immediately apparent in terms of an advantageous strategy for the virus. The increase in MHC class I cell surface expression results in augmented lysis of flavivirus-infected cells by allo-reactive and virus-specific (influenza virus and VV when used in double infections with a flavivirus) effector Tc cells (Liu *et al.*, 1989a; Müllbacher and Lobigs, 1995). However, consistent with the finding that increased cell surface class I expression reduces susceptibility to NK cell-mediated lysis (Lanier, 1998), it was found that flaviviruses are poor inducers of NK cells and that flavivirus infection reduces susceptibility of cells to NK cell lysis (Momburg *et al.*, 2001). Thus, an escape from early NK cell lysis may be crucial in the flavivirus lifecycle for the production of a viremia of sufficient magnitude and duration to allow transmission of the virus from an infected host to an arthropod vector during a blood meal.

B. Contribution of Cellular Immune Responses to Encephalitic Flavivirus Pathogenesis

Infection of the CNS by flaviviruses is accompanied by inflammatory responses which can increase the disease severity. This is reflected in prolonged survival and/or reduced mortality of animals with flaviviral CNS infection upon immunosuppression with cyclophosphamide or T cell depletion (Camenga and Nathanson, 1975; Hirsch and Murphy, 1968; Semenov *et al.,* 1975a, 1975b). Multiple effector functions of inflammatory brain infiltrates may exacerbate flaviviral encephalitis. The cytolytic effector functions of NK and Tc cells (mediated by perforin-dependent granule exocytosis or Fas ligand engagement [Henkart, 1994; Kägi *et al.*, 1994]) can accelerate the time to death from encephalitis in 6-week-old mice infected peripherally with a high dose of MVE. This conclusion was drawn from experiments using mice deficient in perforin, granzyme, or Fas ligand expression, and was more accentuated in mice lacking both Fas ligand and perforin (Licon Luna *et al.*, 2002). T lymphocyte-mediated cytotoxicity effected through the granule exocytosis and Fas pathways is thought to be involved in the killing of infected astrocytes and neurons, respectively (Medana *et al.*, 2000, 2001). Alternatively, the inhibition of production of reactive oxygen and nitrogen intermediates by inflammatory cells (predominantly neutrophils and macrophages) in the brain of MVE- or TBE-infected mice was found to reduce disease severity (Andrews *et al.*, 1999;

Kreil and Eibl, 1996). However, the exacerbating contribution of the immunopathological response to the disease outcome is not always apparent. Survival following intracerebral WNV infection was unaffected by immunosuppression (Camenga and Nathanson, 1975), and cyclophosphamide treatment increased the incidence of fatal encephalitis in mice peripherally infected with WNV (Camenga *et al.*, 1974). This is corroborated in our studies using β2-microglobulin (β-2 m) knockout mice deficient in $CD8^+$ T cells (Wang *et al.*, 2003). Mortality of β-2 m knockout mice given low doses of WNV (10^3 PFU) i.v. was 80% compared to 30% of wt mice, given the same dose. Also, in 4-week-old B and T cell-deficient SCID mice, no difference in time to death was seen following brain infection with WNV (Halevy *et al.*, 1994). The latter results suggest that WNV-induced cytopathology, rather than associated immunopathology in the CNS, was the key factor in mortality of infected mice.

In addition to the potentially detrimental role of the inflammatory process in the CNS during encephalitic flavivirus infection, it appears that the cellular immune response may contribute at a second stage to the disease process (e.g., in the events leading to neuroinvasion). We have shown that the cytolytic effector functions of NK/Tc cells play a role in entry of MVE into the brain of 6-week-old mice infected i.v. with a low virus dose. In this model for flavivirus encephalitis, mortality of wild-type animals is approximately 50% despite low or undetectable virus growth in extraneural tissues. In contrast, mice deficient in either the granule exocytosis- or Fas-mediated pathways of cytotoxicity showed reduced mortality following peripheral MVE infection, and mice deficient in both effector pathways were even less susceptible (Licon Luna *et al.*, 2002). Based on these data, we propose that cytotoxic lymphocytes can play a role in brain invasion of neurotropic viruses by killing infected endothelial cells lining brain capillaries, thus allowing virus access into the brain parenchyma. This pathway for neuroinvasion is disrupted in mice deficient in the cytolytic effector function of NK/Tc cells, as well as mice lacking $CD8^+$ T cells and expression of their restriction elements (β2-microglobulin k/o mice; Wang *et al.*, 2003; our unpublished data).

References

Aihara, H., Takasaki, T., Matsutani, T., Suzuki, R., and Kurane, I. (1998). Establishment and characterization of Japanese encephalitis virus-specific, human CD4(+) T-cell clones: flavivirus cross-reactivity, protein recognition, and cytotoxic activity. *J. Virol.* **72**(10):8032–8036.

Altman, J. D., Moss, P. A. H., Goulder, P. J. R., Barouch, D. H., McHeyzer-Williams, M. G., Bell, J. I., McMichael, A. J., and Davis, M. M. (1996). Phenotypic analysis of antigen-specific T lymphocytes [published erratum appears in *Science* (1998) Jun 19; 280(5371):1821]. *Science.* **274**(5284)**:**94–96.

Andrews, D. M., Matthews, V. B., Sammels, L. M., Carrello, A. C., and McMinn, P. C. (1999). The severity of Murray Valley encephalitis in mice is linked to neutrophil infiltration and inducible nitric oxide synthase activity in the central nervous system. *J. Virol.* **73**(10)**:**8781–8790.

Bachmann, M. F., Kundig, T. M., Kalberer, C. P., Hengartner, H., and Zinkernagel, R. M. (1993). Formalin inactivation of vesicular stomatitis virus impairs T-cell but not T-help-independent B-cell responses. *J. Virol.* **67**(7)**:**3917–3922.

Bao, S., King, N. J., and Dos Remedios, C. G. (1992). Flavivirus induces MHC antigen on human myoblasts: a model of autoimmune myositis? *Muscle Nerve* **15:**1271–1277.

Ben-Nathan, D., Huitinga, I., Lustig, S., van Rooijen, N., and Kobiler, D. (1996). West Nile virus neuroinvasion and encephalitis induced by macrophage depletion in mice. *Arch. Virol.* **141:**459–469.

Ben-Nathan, D., Kobiler, D., Rzotkiewicz, S., Lustig, S., and Katz, Y. (2000). CNS penetration by noninvasive viruses following inhalational anesthetics. *Ann. N. Y. Acad. Sci.* **917:**944–950.

Bhatt, P. N., and Jacoby, R. O. (1976). Genetic resistance to lethal flavivrus encephalitis. II. Effect of immunosuppression. *J. Infect. Dis.* **134**(2)**:**166–173.

Binder, G. K., and Griffin, D. E. (2001). Interferon-gamma-mediated site-specific clearance of alphavirus from CNS neurons. *Science* **293:**303–306.

Biron, C. A., Nguyen, K. B., Pien, G. C., Cousens, L. P., and Salazar-Mather, T. P. (1999). Natural killer cells in antiviral defense: function and regulation by innate cytokines. *Annu. Rev. Immunol.* **17:**189–220.

Blanden, R. V. (1982). Role of macrophages in elimination of viral infection. *In* "Self-Defense Mechanisms: Role of Macrophages" (D. Mizune, Z. A. Cohn, K. Takeya, and N. Ishida, eds.), pp. 269–278. University of Tokyo Press, Tokyo.

Blanden, R. V., Hodgkin, P. D., Hill, A., Sinickas, V. G., and Müllbacher, A. (1987). Quantitative considerations of T cell activation and self-tolerance. *Immunol. Rev.* **98:**75–93.

Boehm, U., Klamp, T., Groot, M., and Howard, J. C. (1997). Cellular responses to interferon-γ. *Ann. Rev. Immunol.* **15:**749–795.

Braciale, T. J., and Yap, K. L. (1978). Role of viral infectivity in the induction of influenza virus-specific cytotoxic T cells. *J. Exp. Med.* **147:**1236–1252.

Bradish, C. J., Fitzgeorge, R., and Titmuss, D. (1980). The responses of normal and athymic mice to infections by togaviruses: strain differentiation in active and adoptive immunization. *J. Gen. Virol.* **46**(2)**:**255–265.

Brinton, M. A., Arnheiter, H., and Haller, O. (1982). Interferon independence of genetically controlled resistance to flaviviruses. *Infect. Immun.* **36:**284–288.

Brooks, T. J., and Phillpotts, R. J. (1999). Interferon-alpha protects mice against lethal infection with St. Louis encephalitis virus delivered by the aerosol and subcutaneous routes. *Antiviral. Res.* **41**(1)**:**57–64.

Burke, D. S., Lorsomrudee, W., Leake, C. J., Hoke, C. H., Nisalak, A., Chongswasdi, V., and Laorakpongse, T. (1985a). Fatal outcome in Japanese encephalitis. *Am. J. Trop. Med. Hyg.* **34:**1203–1210.

Burke, D. S., and Monath, T. P. (2001). Flaviviruses. *In* "Fields Virology" (D. M. Knipe and P. M. Howley, eds.), Vol. 1, pp. 1043–1126. Lippincott Williams & Wilkins, Philadelphia.

Burke, D. S., Nisalak, A., Ussery, M. A., Laorakpongse, T., and Chantavibul, S. (1985b). Kinetics of IgM and IgG responses to Japanese encephalitis virus in human serum and cerebrospinal fluid. *J. Infect. Dis.* **151**(6):1093–1099.

Butrapet, S., Kimura-Kuroda, J., Zhou, D. S., and Yasui, K. (1998). Neutralizing mechanism of a monoclonal antibody against Japanese encephalitis virus glycoprotein E. *Am. J. Trop. Med. Hyg.* **58**(4):389–398.

Butz, E. A., and Bevan, M. J. (1998). Massive expansion of antigen-specific $CD8^+$ T cells during an acute virus infection. *Immunity* **8**(2):167–175.

Calisher, C. H., Karabatsos, N., Dalrymple, J. M., Shope, R. E., Porterfield, J. S., Westaway, E. G., and Brandt, W. E. (1989). Antigenic relationships between flaviviruses as determined by cross-neutralization tests with polyclonal antisera. *J. Gen. Virol.* **70**(Pt 1):37–43.

Camenga, D. L., and Nathanson, N. (1975). An immunopathologic component in experimental togavirus encephalitis. *J. Neuropathol. Exp. Neurol.* **34**(6):492–500.

Camenga, D. L., Nathanson, N., and Cole, G. A. (1974). Cyclophosphamide-potentiated West Nile viral encephalitis: relative influence of cellular and humoral factors. *J. Infect. Dis.* **130**(6):634–641.

Cecilia, D., Gadkari, D. A., Kedarnath, N., and Ghosh, S. N. (1988). Epitope mapping of Japanese encephalitis virus envelope protein using monoclonal antibodies against an Indian strain. *J. Gen. Virol.* **69**(Pt 11):2741–2747.

Cernescu, C., Nedelcu, N. I., Tardei, G., Ruta, S., and Tsai, T. F. (2000). Continued transmission of West Nile virus to humans in southeastern Romania, 1997–1998. *J. Infect. Dis.* **181**(2):710–712.

Chen, W. R., Tesh, R. B., and Rico-Hesse, R. (1990). Genetic variation of Japanese encephalitis virus in nature. *J. Gen. Virol.* **71**(Pt 12):2915–2922.

Colombage, G., Hall, R., Pavy, M., and Lobigs, M. (1998). DNA-based and alphavirus-vectored immunisation with prM and E proteins elicits long-lived and protective immunity against the flavivirus, Murray Valley encephalitis virus. *Virology* **250**(1):151–163.

Cresswell, P., Bangia, N., Dick, T., and Diedrich, G. (1999). The nature of the MHC class I peptide loading complex. *Immunol. Rev.* **172**:21–28.

Crill, W. D., and Roehrig, J. T. (2001). Monoclonal antibodies that bind to domain III of dengue virus E glycoprotein are the most efficient blockers of virus adsorption to Vero cells. *J. Virol.* **75**(16):7769–7773.

Dalton, D. K., Pitts-Meek, S., Keshav, S., Figari, I. S., Bradley, A., and Stewart, T. A. (1993). Multiple defects of immune cell function in mice with disrupted interferon-gamma genes. *Science* **259**(5102):1739–1742.

Darnell, M. B., Koprowski, H., and Lagerspetz, K. (1974). Genetically determined resistance to infection with group B arboviruses. I. Distribution of the resistance gene among various mouse populations and characteristics of gene expression in vivo. *J. Infect. Dis.* **129**(3):240–247.

Diamond, M. S., Shrestha, B., Marri, A., Mahan, D., and Engle, M. (2003). B cells and antibody play critical roles in the immediate defense of disseminated infection by West Nile encephalitis virus. *J. Virol.* **77**(4):2578–2586.

Doherty, P. C., Biddison, W. E., Bennink, J. R., and Knowles, B. B. (1978). Cytotoxic T-cell responses in mice infected with influenza and vaccinia viruses vary in magnitude with H-2 genotype. *J. Exp. Med.* **148**:534–543.

Eldadah, A. H., Nathanson, N., and Sarsitis, R. (1967). Pathogenesis of West Nile Virus encephalitis in mice and rats. I. Influence of age and species on mortality and infection. *Am. J. Epidemiol.* **86**(3):765–775.

Elliott, T. (1997). Transporter associated with antigen processing. *Adv. Immunol.* **65:**47–109.

Epperson, D. E., Arnold, D., Spies, T., Cresswell, P., Pober, J. S., and Johnson, D. R. (1992). Cytokines increase transporter in antigen processing-1 expression more rapidly than HLA class I expression in endothelial cells. *J. Immunol.* **149:**3297–3301.

Fulginiti, V. A., Kempe, C. H., Hathaway, W. E., Pearlman, D. S., Sieber, O. F., Eller, J. J., Joyner, J. J., and Robinson, A. (1968). Progressive vaccinia in immunologically deficient individuals. *In* "Immunologic Deficiency Diseases in Man" (D. Bergsma, ed.), Vol. IV, pp. 129–144. The National Foundation-March of Dimes, White Plains N. Y.

Gajdosova, E., Oravec, C., and Mayer, V. (1981). Cell-mediated immunity in flavivirus infections. I. Induction of cytotoxic T lymphocytes in mice by an attenuated virus from the tick-born encephalitis complex and its group-reactive character. *Acta Virol.* **25:**10–18.

Georges, A. J., Lesborges, J. L., Georges-Courbot, M. C., Meunies, D. M. Y., and Gonzales, J. P. (1987). Fatal hepatitis from West Nile virus. *Ann. Inst. Pasteur/Virol.* **138:**237–244.

Gollins, S. W., and Porterfield, J. S. (1986). A new mechanism for the neutralization of enveloped viruses by antiviral antibody. *Nature* **321**(6067)**:**244–246.

Goodbourn, S., Didcock, L., and Randall, R. E. (2000). Interferons: cell signalling, immune modulation, antiviral response and virus countermeasures. *J. Gen. Virol.* **81** (Pt 10)**:** 2341–2364.

Grossberg, S. E., and Scherer, W. F. (1966). The effect of host age, virus dose and route of inoculation on inapparent infection in mice with Japanese encephalitis virus. *Proc. Soc. Exp. Biol. Med.* **123:**118–124.

Guidotti, L. G., and Chisari, F. V. (2000). Cytokine-mediated control of viral infections. *Virology* **273**(2)**:**221–227.

Gupta, A. K., and Pavri, K. M. (1987). Alteration in immune response of mice with dual infection of *Toxocara canis* and Japanese encephalitis virus. *Trans. R. Soc. Trop. Med. Hyg.* **81:**835–840.

Haahr, S. (1971). The influence of Poly I:C on the course of infection in mice inoculated with West Nile virus. *Arch. Gesamte. Virusforsch.* **35**(1)**:**1–9.

Halevy, M., Akov, Y., Ben-Nathan, D., Kobiler, D., Lachmi, B., and Lustig, S. (1994). Loss of active neuroinvasiveness in attenuated strains of West Nile virus: pathogenicity in immunocompetent and SCID mice. *Arch. Virol.* **137**(3–4)**:**355–370.

Hall, R. A., Brand, T. N., Lobigs, M., Sangster, M. Y., Howard, M. J., and Mackenzie, J. S. (1996). Protective immune responses to the E and NS1 proteins of Murray Valley encephalitis virus in hybrids of flavivirus-resistant mice. *J. Gen. Virol.* **77**(Pt 6)**:**1287–1294.

Hase, T., Dubois, D. R., and Summers, P. L. (1990). Comparative study of mouse brains infected with Japanese encephalitis virus by intracerebral or intraperitoneal inoculation. *Int. J. Exp. Pathol.* **71:**857–869.

Hawkes, R. A. (1964). Enhancement of the infectivity of arboviruses by specific antisera produced in domestic fowls. *Aust. J. Exp. Biol. Med. Sci.* **42:**465–482.

Hawkes, R. A., Roehrig, J. T., Hunt, A. R., and Moore, G. A. (1988). Antigenic structure of the Murray Valley encephalitis virus E glycoprotein. *J. Gen. Virol.* **69**(Pt 5)**:**1105–1109.

Hayashi, K., and Arita, T. (1977). Experimental double infection of Japanese encephalitis virus and herpes simplex virus in mouse brain. *Jpn. J. Exp. Med.* **47:**9–13.

Henkart, P. A. (1994). Lymphocyte-mediated cytotoxicity: two pathways and multiple effector molecules. *Immunity* **1:**343–346.

Hill, A. B., Lobigs, M., Blanden, R. V., Kulkarni, A. B., and Müllbacher, A. (1993a). The cellular immune response to flaviviruses. *In* "Viruses and the Cellular Immune Response" (D. B. Thomas, ed.), pp. 363–388. Marcel Dekker, New York.

Hill, A. B., Müllbacher, A., and Blanden, R. V. (1993b). Ir genes, peripheral cross-tolerance and immunodominance in MHC class I restricted T cell responses: an old quagmire revisited. *Immunol. Rev.* **133:**75–91.

Hirsch, M. S., and Murphy, F. A. (1968). Effects of anti-lymphoid sera on viral infections. *Lancet* **2**(7558)**:**37–40.

Hoke, C. H., Jr., Vaughn, D. W., Nisalak, A., Intralawan, P., Poolsuppasit, S., Jongsawas, V., Titsyakorn, U., and Johnson, R. T. (1992). Effect of high-dose dexamethasone on the outcome of acute encephalitis due to Japanese encephalitis virus. *J. Infect. Dis.* **165**(4)**:**631–637.

Hotta, H., Murakami, I., Miyasaki, K., Takeda, Y., Shirane, H., and Hotta, S. (1981). Inoculation of dengue virus into nude mice. *J. Gen. Virol.* **52**(Pt 1)**:**71–76.

Huang, C. H., and Wong, C. (1963). Relation of the peripheral multiplication of Japanese B encephalitis virus to the pathogenesis of the infection in mice. *Acta. Virol.* **7:**322–330.

Ilyinskikh, N. N., Zagromov, E. J., and Lepekhin, A. V. (1990). The analysis of some indices of immune response, DNA repair, and micronuclei content in cells from tick-borne encephalitis patients. *Acta Virol.* **34**(6)**:**554–562.

Jacoby, R. O., Bhatt, P. N., and Schwartz, A. (1980). Protection of mice from lethal flaviviral encephalitis by adoptive transfer of splenic cells from donors infected with live virus. *J. Infect. Dis.* **141**(5)**:**617–624.

Johnson, A. J., and Roehrig, J. T. (1999). New mouse model for dengue virus vaccine testing. *J. Virol.* **73**(1)**:**783–786.

Johnson, R. T., Burke, D. S., Elwell, M., Leake, C. J., Nisalak, A., Hoke, C. H., and Lorsomrudee, W. (1985). Japanese encephalitis: immunocytochemical studies of viral antigen and inflammatory cells in fatal cases. *Ann. Neurol.* **18**(5)**:**567–573.

Johnson, R. T., and Mims, G. A. (1968). Pathogenesis for viral infections of the nervous system. *N. Engl. J. Med.* **278:**84–92.

Kägi, D., Ledermann, B., Bürki, K., Zinkernagel, R. M., and Hengartner, H. (1995). Lymphocyte-mediated cytotoxicity in vitro and in vivo: Mechanisms and significance. *Immunol. Rev.* **146:**95–115.

Kägi, D., Vignaux, F., Ledermann, B., Burki, K., Depraetere, V., Nagata, S., Hengartner, H., and Golstein, P. (1994). Fas and perforin pathways as major mechanisms of T cell-mediated cytotoxicity. *Science* **265:**528–530.

Kesson, A. M., Blanden, R. V., and Müllbacher, A. (1987). The primary in vivo murine cytotoxic T-cell response to the flavivirus, West Nile. *J. Gen. Virol.* **68:**2001–2006.

Kesson, A. M., Blanden, R. V., and Müllbacher, A. (1988). The secondary in vitro murine cytotoxic T-cell response to the flavivirus, West Nile. *Immunol. Cell Biol.* **66:**23–32.

King, N. J. C., and Kesson, A. M. (1988). Interferon-independent increases in class I major histocompatibility complex antigen expression follow flavivirus infection. *J. Gen. Virol.* **69:**2535–2543.

King, N. J. C., Maxwell, L. E., and Kesson, A. M. (1989). Induction of class I major histocompatibility complex antigen expression by West Nile virus on γ interferon-refractory early murine trophoblast cells. *Proc. Natl. Acad. Sci. USA* **86:**911–915.

Kobiler, D., Lustig, S., Gozes, Y., Ben-Nathan, D., and Akov, Y. (1989). Sodium dodecyl-sulphate induces a breach in the blood-brain barrier and enables a West Nile virus variant to penetrate into mouse brain. *Brain Res.* **496:**314–316.

Konishi, E., Kurane, I., Mason, P. W., Shope, R. E., Kanesa-Thasan, N., Smucny, J. J., Hoke, C. H., Jr., and Ennis, F. A. (1998). Induction of Japanese encephalitis virus-specific cytotoxic T lymphocytes in humans by poxvirus-based JE vaccine candidates. *Vaccine* **16**(8)**:**842–849.

Konishi, E., Kurane, I., Mason, P. W., Innis, B. L., and Ennis, F. A. (1995). Japanese encephalitis virus-specific proliferative responses of human peripheral blood T lymphocytes. *Am. J. Trop. Med. Hyg.* **53:**278–283.

Konishi, E., Pincus, S., Fonseca, B. A., Shope, R. E., Paoletti, E., and Mason, P. W. (1991). Comparison of protective immunity elicited by recombinant vaccinia viruses that synthesize E or NS1 of Japanese encephalitis virus. *Virology* **185**(1)**:**401–410.

Kopecky, J., Tomkova, E., and Vlcek, M. (1991). Immune response of the long-tailed field mouse (*Apodemus sylvaticus*) to tick-borne encephalitis virus infection. *Folia Parasitol.* **38**(3)**:**275–282.

Kreil, T. R., and Eibl, M. M. (1996). Nitric oxide and viral infection: NO antiviral activity against a flavivirus in vitro, and evidence for contribution to pathogenesis in experimental infection in vivo. *Virology.* **219**(1)**:**304–306.

Kulkarni, A. B., Müllbacher, A., and Blanden, R. V. (1991a). Effect of high ligand concentration on West Nile virus-specific T cell proliferation. *Immunol. Cell Biol.* **69:**27–38.

Kulkarni, A. B., Müllbacher, A., and Blanden, R. V. (1991b). Functional analysis of macrophages, B cells and splenic dendritic cells as antigen-presenting cells in West Nile virus-specific murine T lymphocyte proliferation. *Immunol. Cell Biol.* **69:**71–80.

Kulkarni, A. B., Müllbacher, A., and Blanden, R. V. (1991c). In-vitro T cell proliferative response to the flavivirus West Nile. *Viral Immunol.* **4:**73–82.

Kulkarni, A. B., Müllbacher, A., Parrish, C. R., Westaway, E. G., Coia, G., and Blanden, R. V. (1992). Analysis of the murine MHC class II-restricted T cell response to the flavivirus Kunjin using vaccinia expression. *J. Virol.* **66:**3583–3592.

Kundig, T. M., Hengartner, H., and Zinkernagel, R. M. (1993). T cell-dependent IFN-gamma exerts an antiviral effect in the central nervous system but not in peripheral solid organs. *J. Immunol.* **150**(6)**:**2316–2321.

Kuno, G., Chang, G. J., Tsuchiya, K. R., Karabatsos, N., and Cropp, C. B. (1998). Phylogeny of the genus Flavivirus. *J. Virol.* **72**(1)**:**73–83.

Kurane, I., Meager, A., and Ennis, F. A. (1989). Dengue virus-specific human T cell clones. Serotype cross-reactive proliferation, interferon gamma production, and cytotoxic activity. *J. Exp. Med.* **170**(3)**:**763–775.

Kurane, I., and Takasaki, T. (2000). Immunogenicity and protective efficacy of the current inactivated Japanese encephalitis vaccine against different Japanese encephalitis virus strains. *Vaccine* **18** (Suppl 2)**:**33–35.

Lad, V. J., Gupta, A. K., Goverdhan, M. K., Ayachit, V. L., Rodrigues, J. J., and Hungund, L. V. (1993). Susceptibility of BL6 nude (congenitally athymic) mice to Japanese encephalitis virus by the peripheral route. *Acta Virol.* **37**(4)**:**232–240.

Lange, D. G., and Sedmak, J. (1991). Japanese encephalitis virus (JEV): potentiation of lethality in mice by microwave radiation. *Bioelectromagnetics* **12:**335–348.

Lanier, L. L. (1998). NK cell receptors. *Annu. Rev. Immunol.* **16:**359–393.

Lanzavecchia, A., and Sallusto, F. (2001). Regulation of T cell immunity by dendritic cells. *Cell* **106**(3)**:**263–266.

Lee, E., and Lobigs, M. (2002). Mechanism of virulence attenuation of glycosaminoglycan-binding variants of Japanese encephalitis virus and Murray Valley encephalitis virus. *J. Virol.* **76**(10)**:**4901–4911.

Licon Luna, R. M., Lee, E., Müllbacher, A., Blanden, R. V., Langman, R., and Lobigs, M. (2002). Lack of both Fas ligand and perforin protects from flavivirus-mediated encephalitis in mice. *J. Virol.* **76:**3202–3211.

Lin, Y. L., Chen, L. K., Liao, C. L., Yeh, C. T., Ma, S. H., Chen, J. L., Huang, Y. L., Chen, S. S., and Chiang, H. Y. (1998). DNA immunization with Japanese encephalitis virus nonstructural protein NS1 elicits protective immunity in mice. *J. Virol.* **72**(1)**:**191–200.

Lin, Y. L., Huang, Y. L., Ma, S. H., Yeh, C. T., Chiou, S. Y., Chen, L. K., and Liao, C. L. (1997). Inhibition of Japanese encephalitis virus infection by nitric oxide: antiviral effect of nitric oxide on RNA virus replication. *J. Virol.* **71**(7)**:**5227–5235.

Liou, M. L., and Hsu, C. Y. (1998). Japanese encephalitis virus is transported across the cerebral blood vessels by endocytosis in mouse brain. *Cell Tissue Res.* **293:**389–394.

Liu, T., and Chambers, T. J. (2001). Yellow fever virus encephalitis: properties of the brain-associated T-cell response during virus clearance in normal and gamma interferon-deficient mice and requirement for $CD4^+$ lymphocytes. *J. Virol.* **75**(5)**:**2107–2118.

Liu, Y., Blanden, R. V., and Müllbacher, A. (1989a). Identification of cytolytic lymphocytes in West Nile virus infected murine central nervous system. *J. Gen. Virol.* **70:**565–574.

Liu, Y., King, N., Kesson, A., Blanden, R. V., and Müllbacher, A. (1989b). Flavivirus infection up-regulates the expression of class I and class II major histocompatibility antigens on and enhances T cell recognition of astrocytes in vitro. *J. Neuroimmunol.* **21:**157–168.

Liu, Y., and Müllbacher, A. (1988). Astrocytes are not susceptible to lysis by natural killer cells. *J. Neuroimmunol.* **19:**101–110.

Lobigs, M., Arthur, C., Müllbacher, A., and Blanden, R. V. (1994). The flavivirus nonstructural protein, NS3, is a dominant source of cytotoxic T cell peptide determinants. *Virology* **202:**195–201.

Lobigs, M., Blanden, R. V., and Müllbacher, A. (1996). Flavivirus-induced up-regulation of MHC class I antigens: implications for the induction of $CD8^+$ T-cell-mediated autoimmunity. *Immunol. Rev.* **152:**5–19.

Lobigs, M., Marshall, I. D., Weir, R. C., and Dalgarno, L. (1988). Murray Valley encephalitis virus field strains from Australia and Papua New Guinea: studies on the sequence of the major envelope protein gene and virulence for mice. *Virology* **165:**245–255.

Lobigs, M., Müllbacher, A., Blanden, R. V., Hämmerling, G. J., and Momburg, F. (1999). Antigen presentation in Syrian hamster cells: substrate selectivity of TAP controlled by polymorphic residues in TAP1 and differential requirements for loading of H2 class I molecules. *Immunogenetics* **49:**931–941.

Lobigs, M., Müllbacher, A., and Pavy, M. (1997). The $CD8^+$ T cell response to flavivirus infection. *In* "Arbovirus research in Australia" (B. Kay, M. D. Brown, and J. G. Askov, eds.), Vol. 7, pp. 160–165. QIM, Brisbane.

Lobigs, M., Müllbacher, A., and Regner, M. (2003a). MHC class I up-regulation by flaviviruses: Immune interaction with unknown advantage to host or pathogen. *Immunol. Cell Biol.* **81:**217–223.

Lobigs, M., Müllbacher, A., Wang, Y., Pavy, M., and Lee, E. (2003b). Role of type I and type II interferon responses in recovery from infection with an encephalitic flavivirus. *J. Gen. Virol.* **84:**567–572.

Lobigs, M., Rothenfluh, H. S., Blanden, R. V., and Müllbacher, A. (1995). Functionally polymorphic MHC class I peptide transporters in the MHC class I monomorphic Syrian hamster. *Immunogenetics* **42:**398–407.

Lobigs, M., Usha, R., Nestorowicz, A., Marshall, I. D., Weir, R. C., and Dalgarno, L. (1990). Host cell selection of Murray Valley encephalitis virus variants altered at an RDG sequence in the envelope protein and in mouse virulence. *Virology* **176:**587–595.

Lowin, B., Hahne, M., Mattmann, C., and Tschopp, J. (1994). Cytolytic T-cell cytotoxicity is mediated through perforin and Fas lytic pathways. *Nature* **370:**650–652.

Lustig, S., Danenberg, H. D., Kafri, Y., Kobiler, D., and Ben-Nathan, D. (1992). Viral neuroinvasion and encephalitis induced by lipopolysaccharide and its mediators. *J. Exp. Med.* **176:**707–712.

Lvov, D. K., Butenko, A. M., Gromashevsky, V. L., Larichev, V. P., Gaidamovich, S. Y., Vyshemirsky, O. I., Zhukov, A. N., Lazorenko, V. V., Salko, V. N., Kovtunov, A. I., Galimzyanov, K. M., Platonov, A. E., Morozova, T. N., Khutoretskaya, N. V., Shishkina, E. O., and Skvortsova, T. M. (2000). Isolation of two strains of West Nile virus during an outbreak in southern Russia, 1999. *Emerg. Infect. Dis.* **6**(4)**:**373–376.

MacDonald, F. (1952a). Murray Valley encephalitis infection in the laboratory mouse. I. Influence of age on the susceptibility of infection. *Aust. J. Exp. Biol. Med. Sci.* **30:**319–324.

MacDonald, F. (1952b). Murray Valley encephalitis infection in the laboratory mouse. II. Multiplication of virus inoculated intramuscularly. *Aust. J. Exp. Biol. Med. Sci.* **30:**325–332.

MacFarlan, R. I., Burns, W. H., and White, D. O. (1977). Two cytotoxic cells in peritoneal cavity of virus-infected mice: antibody-dependent macrophages and nonspecific killer cells. *J. Immunol.* **119:**1569–1574.

Mackenzie, J. S., Chua, K. B., Daniels, P. W., Eaton, B. T., Field, H. E., Hall, R. A., Halpin, K., Johansen, C. A., Kirkland, P. D., Lam, S. K., McMinn, P., Nisbet, D. J., Paru, R., Pyke, A. T., Ritchie, S. A., Siba, P., Smith, D. W., Smith, G. A., van den Hurk, A. F., Wang, L. F., and Williams, D. T. (2001). Emerging viral diseases of Southeast Asia and the Western Pacific. *Emerg. Infect. Dis.* **7**(3)**:**497–504.

Mackenzie, J. S., Lindsay, M. D., Coelen, R. J., Broom, A. K., Hall, R. A., and Smith, D. W. (1994). Arboviruses causing human disease in the Australasian zoogeographic region. *Arch Virol.* **136**(3–4)**:**447–467.

Marfin, A. A., Bleed, D. M., Lofgren, J. P., Olin, A. C., Savage, H. M., Smith, G. C., Moore, P. S., Karabatsos, N., and Tsai, T. F. (1993). Epidemiologic aspects of a St. Louis encephalitis epidemic in Jefferson County, Arkansas, 1991. *Am. J. Trop. Med. Hyg.* **49**(1)**:**30–37.

Marfin, A. A., Petersen, L. R., Eidson, M., Miller, J., Hadler, J., Farello, C., Werner, B., Campbell, G. L., Layton, M., Smith, P., Bresnitz, E., Cartter, M., Scaletta, J., Obiri, G., Bunning, M., Craven, R. C., Roehrig, J. T., Julian, K. G., Hinten, S. R., and Gubler, D. J. (2001). Widespread West Nile virus activity, eastern United States, 2000. *Emerg. Infect. Dis.* **7**(4)**:**730–735.

Marshall, I. D. (1988). Murray Valley and Kunjin encephalitis. *In* "The Arboviruses: Epidemiology and Ecology" (T. P. Monath, ed.), Vol. 3, pp. 151–189. CRC Press, Boca Raton, Fla.

Mashimo, T., Lucas, M., Simon-Chazottes, D., Frenkiel, M. P., Montagutelli, X., Ceccaldi, P. E., Deubel, V., Guenet, J. L., and Despres, P. (2002). A nonsense mutation in the gene encoding 2′-5′-oligoadenylate synthetase/L1 isoform is associated with West Nile virus susceptibility in laboratory mice. *Proc. Natl. Acad. Sci. USA* **99**(17)**:**11311–11316.

Mathews, J. H., Allan, J. E., Roehrig, J. T., Brubaker, J. R., Uren, M. F., and Hunt, A. R. (1991). T-helper cell and associated antibody response to synthetic peptides of the E glycoprotein of Murray Valley encephalitis virus. *J. Virol.* **65**(10)**:**5141–5148.

Mathews, J. H., and Roehrig, J. T. (1984). Elucidation of the topography and determination of the protective epitopes on the E glycoprotein of Saint Louis encephalitis virus by passive transfer with monoclonal antibodies. *J. Immunol.* **132**(3)**:**1533–1537.

Mathews, J. H., Roehrig, J. T., Brubaker, J. R., Hunt, A. R., and Allan, J. E. (1992). A synthetic peptide to the E glycoprotein of Murray Valley encephalitis virus defines multiple virus-reactive T- and B-cell epitopes. *J. Virol.* **66**(11)**:**6555–6562.

Mathur, A., Arora, K. L., and Chaturvedi, U. C. (1983). Host defence mechanisms against Japanese encephalitis virus infection in mice. *J. Gen. Virol.* **64**(Pt 4)**:**805–811.

Matthews, V., Robertson, T., Kendrick, T., Abdo, M., Papadimitriou, J., and McMinn, P. (2000). Morphological features of Murray Valley encephalitis virus infection in the central nervous system of Swiss mice. *Int. J. Exp. Pathol.* **81**(1)**:**31–40.

McMinn, P. C., Dalgarno, L., and Weir, R. C. (1996). A comparison of the spread of Murray Valley encephalitis viruses of high or low neuroinvasiveness in the tissues of Swiss mice after peripheral inoculation. *Virology* **220**(2)**:**414–423.

Medana, I., Li, Z., Flugel, A., Tschopp, J., Wekerle, H., and Neumann, H. (2001). Fas ligand (CD95L) protects neurons against perforin-mediated T lymphocyte cytotoxicity. *J. Immunol.* **167**(2)**:**674–681.

Medana, I. M., Gallimore, A., Oxenius, A., Martinic, M. M., Wekerle, H., and Neumann, H. (2000). MHC class I-restricted killing of neurons by virus-specific $CD8^+$ T lymphocytes is effected through the Fas/FasL, but not the perforin pathway. *Eur. J. Immunol.* **30**(12)**:**3623–3633.

Min, W., Pober, J. S., and Johnson, D. R. (1996). Kinetically coordinated induction of TAP1 and HLA class I by IFN gamma. *J. Immunol.* **156:**3174–3183.

Miura, K., Onodera, T., Nishida, A., Goto, N., and Fujisaki, Y. (1990). A single gene controls resistance to Japanese encephalitis virus in mice. *Arch. Virol.* **112**(3–4)**:**261–270.

Momburg, F., and Hämmerling, G. J. (1998). Generation and TAP-mediated transport of peptides for major histocompatibility complex class I molecules. *Adv. Immunol.* **68:**191–256.

Momburg, F., Müllbacher, A., and Lobigs, M. (2001). Modulation of transporter associated with antigen processing (TAP)-mediated peptide import into the endoplasmic reticulum by flavivirus infection. *J. Virol.* **75:**5663–5671.

Monath, T. P. (1986). Pathobiology of the flaviviruses. *In* "The Togaviridae and Flaviviridae" (S. Schlesinger and M. Schlesinger, eds.), pp. 375–440. Plenum Press, New York.

Monath, T. P., and Borden, E. C. (1971). Effects of thorotrast on humoral antibody, viral multiplication, and interferon during infection with St. Louis encephalitis virus in mice. *J. Infect. Dis.* **123**(3)**:**297–300.

Monath, T. P., Cropp, C. B., Bowen, G. S., Kemp, G. E., Mitchell, C. J., and Gardner, J. J. (1980). Variation in virulence for mice and rhesus monkeys among St. Louis encephalitis virus strains of different origin. *Am. J. Trop. Med. Hyg.* **29**(5)**:**948–962.

Monath, T. P., Cropp, C. B., and Harrison, A. K. (1983). Mode of entry of a neurotropic arbovirus into the central nervous system: Reinvestigation of an old controversy. *Lab. Invest.* **48:**399–410.

Monath, T. P., and Heinz, F. X. (1996). Flaviviruses. *In* "Fields Virology" (B. N. Fields, D. M. Knipe, P. M. Howley, R. M. Chanock, J. L. Melnick, T. P. Monath, B. Roizman, and S. E. Strauss, eds.), 3rd ed., pp. 961–1034. Lippincott-Raven, Philadelphia.

Müllbacher, A., Ada, G. L., and Tha Hla, R. (1988). Gamma-irradiated influenza A virus can prime for a cross-reactive and cross-protective immune response against influenza A virus. *Immunol. Cell Biol.* **66:**153–157.

Müllbacher, A., Blanden, R. V., and Brenan, M. (1983). Neonatal tolerance to MHC antigens alters Ir gene control of the cytotoxic T cell response to vaccinia virus. *J. Exp. Med.* **157:**1324–1338.

Müllbacher, A., and Flynn, K. (1996). Aspects of cytotoxic T cell memory. *Immunol. Rev.* **150:**113–127.

Müllbacher, A., and Lobigs, M. (1995). Up-regulation of MHC class I by flavivirus-induced peptide translocation into the endoplasmic reticulum. *Immunity* **3:**207–214.

Müllbacher, A., Marshall, I. D., and Ferris, P. (1986). Classification of Barmah Forest virus as an alphavirus using cytotoxic T cell assays. *J. Gen. Virol.* **67:**295–299.

Nagata, S. (1997). Apoptosis by death factor. *Cell* **88**(N3)**:**355–365.

Nash, D., Mostashari, F., Fine, A., Miller, J., O'Leary, D., Murray, K., Huang, A., Rosenberg, A., Greenberg, A., Sherman, M., Wong, S., and Layton, M. (2001). The outbreak of West Nile virus infection in the New York City area in 1999. *N. Engl. J. Med.* **344**(24)**:**1807–1814.

Ogata, A., Nagashima, K., Hall, W. W., Ichikawa, M., Kimura-Kuroda, J., and Yasui, K. (1991). Japanese encephalitis virus neurotropism is dependent on the degree of neuronal maturity. *J. Virol.* **65:**880–886.

Oliver, K. R., and Fazakerley, J. K. (1998). Transneuronal spread of Semliki Forest virus in the developing mouse olfactory system is determined by neuronal maturity. *Neuroscience* **82:**867–877.

Pan, C. H., Chen, H. W., Huang, H. W., and Tao, M. H. (2001). Protective mechanisms induced by a Japanese encephalitis virus DNA vaccine: requirement for antibody but not CD8(+) cytotoxic T-cell responses. *J. Virol.* **75**(23)**:**11457–11463.

Parrish, C. R., Coia, G., Hill, A. B., Müllbacher, A., Westaway, E. G., and Blanden, R. V. (1991). Preliminary analysis of murine cytotoxic T cell responses to the proteins of the flavivirus Kunjin using vaccinia virus expression. *J. Gen. Virol.* **72:**1645–1653.

Peiris, J. S. M., Gordon, S., Unkeless, J. C., and Porterfield, J. S. (1981). Monoclonal anti-Fc receptor IgG blocks antibody enhancement of viral replication in macrophages. *Nature* **289:**189–191.

Perelman, A., and Stern, J. (1974). Acute pancreatitis in West Nile Fever. *Am. J. Trop. Med. Hyg.* **23**(6)**:**1150–1152.

Perelygin, A. A., Scherbik, S. V., Zhulin, I. B., Stockman, B. M., Li, Y., and Brinton, M. A. (2002). Positional cloning of the murine flavivirus resistance gene. *Proc. Natl. Acad. Sci. USA* **99**(14)**:**9322–9327.

Petersen, L. R., and Roehrig, J. T. (2001). West Nile virus: a reemerging global pathogen. *Emerg. Infect. Dis.* **7**(4)**:**611–614.

Pinto, A. J., Morahan, P. S., and Brinton, M. A. (1988). Comparative study of various immunomodulators for macrophage and natural killer cell activation and antiviral efficacy against exotic RNA viruses. *Int. J. Immunopharmacol.* **10**(3)**:**197–209.

Ploegh, H. L. (1998). Viral strategies of immune evasion. *Science* **280:**248–253.

Porter, K. R., Summers, P. L., Dubois, D., Puri, B., Nelson, W., Henchal, E., Oprandy, J. J., and Hayes, C. G. (1993). Detection of West Nile virus by the polymerase chain reaction and analysis of nucleotide sequence variation. *Am. J. Trop. Med. Hyg.* **48**(3)**:**440–446.

Raengsakulrach, B., Nisalak, A., Gettayacamin, M., Thirawuth, V., Young, G. D., Myint, K. S., Ferguson, L. M., Hoke, C. H., Jr., Innis, B. L., and Vaughn, D. W. (1999). An intranasal challenge model for testing Japanese encephalitis vaccines in rhesus monkeys. *Am. J. Trop. Med. Hyg.* **60**(3)**:**329–337.

Ravi, V., Desai, A. S., Shenoy, P. K., Satishchandra, P., Chandramuki, A., and Gourie-Devi, M. (1993). Persistence of Japanese encephalitis virus in the human nervous system. *J. Med. Virol.* **40**(4)**:**326–329.

Ravi, V., Parida, S., Desai, A., Chandramuki, A., Gourie-Devi, M., and Grau, G. E. (1997). Correlation of tumor necrosis factor levels in the serum and cerebrospinal fluid with clinical outcome in Japanese encephalitis patients. *J. Med. Virol.* **51:**132–136.

Regner, M., Lobigs, M., Blanden, R. V., Milburn, P., and Müllbacher, A. (2001a). Antiviral cytotoxic T cells cross-reactively recognize disparate peptide determinants from related viruses but ignore more similar self- and foreign determinants. *J. Immunol.* **166**(6)**:**3820–3828.

Regner, M., Lobigs, M., Blanden, R. V., and Müllbacher, A. (2001b). Effector cytolytic function but not IFN-gamma production in cytotoxic T cells triggered by virus-infected target cells in vitro. *Scand. J. Immunol.* **54**(4)**:**366–374 [Erratum *Scand. J. Immunol.* **54:**640–642.].

Regner, M., Müllbacher, A., Blanden, R. V., and Lobigs, M. (2001c). Immunogenicity of two peptide determinants in the cytolytic T-cell response to flavivirus infection: inverse correlation between peptide affinity for MHC class I and T-cell precursor frequency. *Viral Immunol.* **14**(2)**:**135–149.

Reiss, C. S., and Komatsu, T. (1998). Does nitric oxide play a critical role in viral infections? *J. Virol.* **72**(6)**:**4547–4551.

Reyes, M. G., Gardner, J. J., Poland, J. D., and Monath, T. P. (1981). St. Louis encephalitis: Quantitative histologic and immunofluorescent studies. *Arch. Neurol.* **38:**329–334.

Ridge, J. P., Fuchs, E. J., and Matzinger, P. (1996). Neonatal tolerance revisited: turning on newborn T cells with dendritic cells. *Science* **271:**1723–1726.

Roder, J. C., Lohmann-Matthes, M. L., Domzig, W., and Wigzell, H. (1979). The beige mutation in the mouse. II. Selectivity of the natural killer (NK) cell defect. *J. Immunol.* **123**(5)**:**2174–2181.

Rouvier, R., Luciani, M.-F., and Golstein, P. (1993). Fas involvement in $Ca2^{+}$-independent T-cell-mediated cytotoxicity. *J. Exp. Med.* **177:**195–200.

Sarzotti, M., Robbins, D. S., and Hoffman, P. M. (1996). Induction of protective CTL responses in newborn mice by a murine retrovirus. *Science* **271:**1726–1728.

Saxena, S. K., Mathur, A., and Srivastava, R. C. (2001). Induction of nitric oxide synthase during Japanese encephalitis virus infection: evidence of protective role. *Arch. Biochem. Biophys.* **391**(1)**:**1–7.

Saxena, S. K., Singh, A., and Mathur, A. (2000). Antiviral effect of nitric oxide during Japanese encephalitis virus infection. *Int. J. Exp. Pathol.* **81**(2)**:**165–172.

Schlesinger, J. J., Brandriss, M. W., and Walsh, E. E. (1985). Protection against 17D yellow fever encephalitis in mice by passive transfer of monoclonal antibodies to the nonstructural glycoprotein gp48 and by active immunization with gp48. *J. Immunol.* **135**(4)**:**2805–2809.

Se-Thoe, S. Y., Ling, A. E., and Ng, M. M. (2003). Alteration of virus entry mode: a neutralisation mechanism for Dengue-2 viruses. *J. Med. Virol.* **62:**364–376.

Selin, L. K., Santolucito, P. A., Pinto, A. K., Szomolanyi-Tsuda, E., and Welsh, R. M. (2001). Innate immunity to viruses: Control of vaccinia virus infection by gammadelta T cells. *J. Immunol.* **166**(11)**:**6784–6794.

Semenov, B. F., Khozinsky, V. V., and Vargin, V. V. (1975a). The damaging action of cellular immunity in flavivirus infections of mice. *Med. Biol.* **53**(5)**:**331–336.

Semenov, B. F., Vargin, V. V., Zschiesche, W., and Veckenstedt, A. (1975b). Influence of certain immunodepressants on experimental flavivirus and enterovirus infections in mice. *Intervirology* **5**(3–4)**:**220–224.

Shen, J., T-To, S. S., Schrieber, L., and King, N. J. (1997). Early E-selectin, VCAM-1, ICAM-1, and late major histocompatibility complex antigen induction on human endothelial cells by flavivirus and comodulation of adhesion molecule expression by immune cytokines. *J. Virol.* **71:**9323–9332.

Shieh, W. J., Guarner, J., Layton, M., Fine, A., Miller, J., Nash, D., Campbell, G. L., Roehrig, J. T., Gubler, D. J., and Zaki, S. R. (2000). The role of pathology in an investigation of an outbreak of West Nile encephalitis in New York, 1999. *Emerg. Infect. Dis.* **6**(4)**:**370–372.

Southam, C. M., and Moore, A. E. (1954). Induced virus infections in man by the Egypt isolates of West Nile virus. *Am. J. Trop. Med. Hyg.* **3:**19–50.

Takada, K., Masaki, H., Konishi, E., Takahashi, M., and Kurane, I. (2000). Definition of an epitope on Japanese encephalitis virus (JEV) envelope protein recognized by JEV-specific murine $CD8^+$ cytotoxic T lymphocytes. *Arch. Virol.* **145**(3)**:**523–534.

Tanaka, K., and Kasahara, M. (1998). The MHC class I ligand-generating system: roles of immunoproteasomes and the interferon-gamma-inducible proteasome activator PA28. *Immunol. Rev.* **163:**161–176.

Taylor, J. L., Schoenherr, C., and Grossberg, S. E. (1980). Protection against Japanese encephalitis virus in mice and hamsters by treatment with carboxymethylacridanone, a potent interferon inducer. *J. Infect. Dis.* **142**(3)**:**394–399.

Trgovcich, J., Aronson, J. F., Eldridge, J. C., and Johnston, R. E. (1999). TNFalpha, interferon, and stress response induction as a function of age-related susceptibility to fatal Sindbis virus infection of mice. *Virology* **263:**339–348.

Tsai, T. F., Popovici, F., Cernescu, C., Campbell, G. L., and Nedelcu, N. I. (1998). West Nile encephalitis epidemic in southeastern Romania. *Lancet* **352**(9130)**:**767–771.

Uren, M. F., Doherty, P. C., and Allan, J. E. (1987). Flavivirus-specific murine $L3T4^+$ T cell clones: induction, characterization and cross-reactivity. *J. Gen. Virol.* **68**(Pt 10)**:** 2655–2663.

Vargin, V. V., and Semenov, B. F. (1986). Changes of natural killer cell activity in different mouse lines by acute and asymptomatic flavivirus infections. *Acta. Virol.* **30**(4)**:**303–308.

Vargin, V. V., Zschiesche, W., and Semenov, B. F. (1977). Effects of tilorone hydrochloride on experimental flavivirus infections in mice. *Acta. Virol.* **21**(2)**:**114–118.

Vassalli, P. (1992). The pathophysiology of tumor necrosis factor. *Annu. Rev. Immunol.* **10:**411–452.

Vaughn, D. W., and Hoke, C. H., Jr. (1992). The epidemiology of Japanese encephalitis: prospects for prevention. *Epidemiol. Rev.* **14:**197–221.

Wang, Y., Lobigs, M., Lee, E., and Müllbacher, A. (2003). $CD8^+$ T cells mediate recovery and immunopathology in West Nile virus encephalitis. *J. Virol.* Accepted.

Wasay, M., Diaz-Arrastia, R., Suss, R. A., Kojan, S., Haq, A., Burns, D., and Van Ness, P. (2000). St. Louis encephalitis: a review of 11 cases in a 1995 Dallas, Tex, epidemic. *Arch. Neurol.* **57**(1)**:**114–118.

Webb, H. E., Wight, D. G., Wiernik, G., Platt, G. S., and Smith, C. E. (1968). Langat virus encephalitis in mice. II. The effect of irradiation. *J. Hyg. (Lond).* **66**(3)**:**355–364.

Yang, Y., Früh, K., Chambers, J., Waters, J. B., Wu, L., Spies, T., and Peterson, P. A. (1992). Major histocompatibility complex (MHC)-encoded HAM2 is necessary for antigenic peptide loading onto class I MHC molecules. *J. Biol. Chem.* **267:**11669–11672.

Yewdell, J. W., and Bennink, J. R. (1999). Mechanisms of viral interference with MHC class I antigen processing and presentation. *Annu. Rev. Cell Dev. Biol.* **15:**579–606.

York, I. A., and Rock, K. L. (1996). Antigen processing and presentation by the class I major histocompatibility complex. *Annu. Rev. Immunol.* **14:**369–396.

Zinkernagel, R. M., Althage, A., Cooper, S., Kreeb, G., Klein, P. A., Sefton, B., Flaherty, L., Stimpfling, J., Shreffler, D., and Klein, J. (1978). *Ir*-genes in *H*-2 regulate generation of anti-viral cytotoxic T cells: Mapping to *K* or *D* and dominance of unresponsiveness. *J. Exp. Med.* **148:**592–605.

Zinkernagel, R. M., and Doherty, P. C. (1979). MHC-restricted cytotoxic T cells: studies on the biological role of polymorphic major transplantation antigens determining T-cell restriction-specificity, function, and responsiveness. *Adv. Immunol.* **117:**51–177.

ADVANCES IN VIRUS RESEARCH, VOL 60

IMMUNE MODULATION BY FLAVIVIRUSES

Nicholas J. C. King,* Bimmi Shrestha,† and Alison M. Kesson‡

*Department of Pathology
Institute of Biomedical Research
School of Medical Sciences
University of Sydney 2006
New South Wales, Australia
†Department of Medicine
Washington University School of Medicine
St. Louis, Missouri 63110
‡Department of Pediatrics and Child Health
Department of Virology & Microbiology
The Children's Hospital at Westmead
University of Sydney 2006
New South Wales, Australia

Flaviviruses cause pleomorphic disease with significant morbidity and mortality worldwide. Interestingly, in contrast to most viruses, which subvert or avoid host immune systems, members of the neurotropic Japanese encephalitis serocomplex cause functional changes

0065-3527/03 $35.00

associated with increased efficacy of the immune response. These viruses induce increased cell surface expression of immune recognition molecules, including class I and II major histocompatibility complex (MHC) and various adhesion molecules. Increases are functional: infected cells are significantly more susceptible to both virus- and MHC-specific cytotoxic T cell lysis. Induced changes are modulated positively or negatively by Th1 and Th2 cytokines, as well as by cell cycle position and adherence status at infection. Infection also increases costimulatory molecule expression on Langerhans cells in the skin. Local interleukin-1β production causes accelerated migration of phenotypically altered Langerhans cells to local draining lymph nodes, where initiation of antiviral immune responses occur. The exact mechanism(s) of upregulation is unclear, but changes are associated with NF-κB activation and increased *MHC* and *ICAM-1* gene transcription, independently of interferon (IFN) or other proinflammatory cytokines. Increased MHC and adhesion molecule expression may contribute to the pathogenesis of flavivirus encephalitis. Results from a murine model of flavivirus encephalitis developed in this laboratory suggest that fatal disease is immunopathological in nature, with IFN-γ playing a crucial role. We hypothesize that these viruses may decoy the adaptive immune system into generating low-affinity T cells, which clear virus poorly, as part of their survival strategy. This may enable viral growth and immune escape in cycling cells, which do not significantly upregulate cell surface molecules.

I. Introduction

The incidence of infectious virus disease remains a serious health and economic issue worldwide. Despite some important successes, preventive vaccination for many viruses is still not effective (McDade and Hughes, 2000). Thus critical understanding of both host–virus interactions and ways in which viruses modulate and manipulate host immune defense systems remains pivotal to the design of rational intervention in clinical disease.

Flaviviridae are a family of positive, single-stranded RNA viruses, comprising some 70 members of which, up until recently, the best known were perhaps dengue and yellow fever. Most flaviviruses are arthropod borne, with vertebrates constituting the intermediate host essential for virus survival. These viruses cause a wide range of disease with significant morbidity and mortality in both humans and animals.

West Nile virus (WNV) and the Australasian encephalitic group, which includes Murray Valley encephalitis (MVE), Kunjin, and Japanese encephalitis (JE), form the Japanese encephalitis antigenic complex (Calisher *et al.*, 1989). These, together with other regional neurotropic flaviviruses, such as St. Louis and tick-borne encephalitis viruses, cause neuropathological illness ranging from malaise through paralysis, coma, and death. The global economic cost of these infections is therefore extremely high. The prospect of outbreaks of infection with these viruses in nonendemic areas has increased markedly as global warming extends the environment permissive for the mosquito vector and world travel increases the size of the host population pool (Hannah *et al.*, 1995; Hubalek and Halouzka, 1999; Ritchie *et al.*, 1997). Thus, for example, while WNV encephalitis is endemic in many developing countries, periodic epidemic outbreaks occur elsewhere—most notably in New York, where WNV encephalitis was reported for the first time in August 1999. This outbreak also highlighted the high mortality of epidemic WNV encephalitis at some 10% of reported clinical cases (Anderson *et al.*, 1999; Briese *et al.*, 1999; Lanciotti *et al.*, 1999).

II. Strategies for Virus Survival in the Immune-Competent Host

The survival of viruses, as obligate intracellular pathogens that infect complex multicellular hosts, is driven principally by three primary but interrelated imperatives. The first is the standard biological requirement to replicate, the second is to be transmitted to the next host, and the third is to continue to use the host pool *ad infinitum*. Each of these poses a separate set of impediments to virus survival. If any of these imperatives is not met consistently, the virus will become extinct.

In the first and second requirements, the immediate outcome of infection is the result of a race between virus replication/transmission and the generation of an eradicating response by the host, which comprises both innate and adaptive immune components. The virus life cycle must therefore be sufficiently rapid to enable replication and transmission of large numbers of progeny virions before the immune response becomes effective.

In the third requirement, available host pools are limited both by finite ecological niche size and the time it takes for reproduction of the host to occur. This is because in simple terms, the outcome of any pathogenic infection for the host is lifetime immunity or death. Either outcome effectively reduces the potential host pool for the virus.

Lifetime immunity of the host pool is the basis and goal for the successful eradication of a virus species by vaccination.

The reproduction time of humans is nominally 12–14 years, but this is the minimum time; values vary enormously, depending on cultural, economic, and other factors. Because infection tends to generate lifetime host immunity, a virus must time its replication and transmission to spread through the host pool more slowly than the reproduction time of the host in order to survive. It follows that the longer the reproduction time of the host, the bigger the host pool that is required for virus survival. Increasing world travel has allowed once discrete communities to mix, effectively increasing the size of the human host pool. This expansion in the host pool is illustrated by the emergence of WNV encephalitis on the eastern seaboard of the United States for the first time in August 1999, with consequences that have yet to be fully evaluated (Kramer and Bernard, 2001).

Although interrelated and often interactive, viral strategies for successfully accomplishing these three survival imperatives are generally different from one another and few subserve more than one imperative. It can be argued that the number of virus strategies that address each requirement reflects the complexity of the problem for virus survival. Thus the imperative for successful virus replication in the face of surveillance by the innate and adaptive immune systems is subserved by a range of virus-mediated mechanisms as multifarious as those mounted by the immune system to prevent it.

A. Host Responses

The host has, at its disposal, a complex network in which numerous cell types and soluble factors of the immune system participate. There are two major parts to this network, which both complement and interact with one another. The first is innate immunity, responsible for rapid antimicrobial responses. This is a nonspecific response, established early on in the evolution of antimicrobial defenses. It is the most immediately mobilized response and protects the host within hours of infection, principally through activities of neutrophils and macrophages and their products. Inactivation of these early responses renders the host demonstrably more susceptible to various microbial pathogens and may result in overwhelming infection and death. Increasing evidence points to the importance of the innate immune response in viral infections. A functional innate response may eradicate some infections, but if not, usually prevents them from overwhelming the host. This allows sufficient time for generation of the slower antigen-specific, adaptive immune response, which makes up the second part of this network. The adaptive immune response comprises elements of both humoral and cellular immune responses, prosecuted by B and T cells, respectively. Effector B cells (plasma cells) produce specific antibodies as part of the humoral response. The cellular immune response produces cytotoxic T cells (CTL), which kill virus-infected cells, and T helper (Th) cells, which produce a spectrum of cytokines. The combination of cytokines produced by Th cells influences the differentiation of both B and T cells down specific response pathways, depending on the infecting organism. Th1 responses are associated with the production of proinflammatory cytokines, such as tumor necrosis factor (TNF) and interferon (IFN)-γ (type II interferon) and generally occur in virus infections. Th2 responses, however, tend to occur in parasitic infections and are associated with the production of cytokines such as interleukin (IL)-4 and IL-10, which are anti-inflammatory. These cells play a central role in the generation and maintenance of specific immunological memory, critical to the accelerated abrogative responses associated with a secondary challenge by the same virus (Abbas *et al.*, 2000).

One of the first responses to an infection involves the recruitment and activation of neutrophils, macrophages, and natural killer (NK) cells. Cytokines such as TNF, IL-1, IL-12, IL-18, type I, and type II IFN, produced in this response, further activate both phagocytic cells and various leukocyte subsets that go on to play key roles in the subsequent development of antigen-specific antiviral immunity (Karupiah *et al.*, 2000).

An early cellular response to viral infection is the production of type I IFN, IFN-α/β, by infected cells. These are directly and primarily antiviral in that they have a direct effect on viral replication within the cell, but they are also indirectly antiviral because they increase cell surface class I major histocompatibility complex (MHC) antigen expression on infected cells. A complex of viral peptide and MHC is specifically recognized by T cells via the T cell receptor (TcR). Thus increased cell surface MHC expression makes recognition by T cells more likely, as cell surface MHC concentration is a crucial factor determining the efficiency of induction and execution of the cellular immune response (Goldstein and Mescher, 1987; King *et al.*, 1985, 1986; O'Neill and Blanden, 1979; Shimonkevitz *et al.*, 1985). Type I IFN also help to direct the adaptive immune response (Le Bon *et al.*, 2001; Mattei *et al.*, 2001; Tough *et al.*, 1996). Type II IFN — IFN-γ — secreted by activated T cells and NK cells, is not only antiviral, but also potently upregulates both class I and II MHC expression (MHC I and II), which further facilitates T-cell recognition of the infected cells (Stark *et al.*, 1998). Moreover, IFN-γ enhances the activation status of macrophages and neutrophils, inducing increased and prolonged expression of nitric oxide synthase 2, which catalyzes the production of the radical gas nitric oxide (NO) from L-arginine. NO is strongly antimicrobial and, via formation of various reactive nitrogen intermediates, including peroxynitrite, formed by the interaction of NO with superoxide anion radicals, is thought to target several cellular metabolic enzymes and redox-sensitive pathways, as well as DNA (Fierro *et al.*, 1999; Nathan, 1992). Superoxide is formed from the reduction of molecular oxygen by the enzyme NADPH oxidase in IFN-γ-activated macrophages and neutrophils (Cassatella *et al.*, 1990). They quickly form reactive oxygen species, which are themselves toxic in much the same way as reactive nitrogen species.

The actions of both these reactive species are not specific for pathogens. While both are constitutively produced at low levels and required for normal cellular physiology, sustained high levels result in fatal cell damage (Karupiah *et al.*, 2000).

Thus there is clearly cross-talk and feedback regulation between innate and adaptive immune response compartments, which collaborate in a finely tuned and efficient manner in the destruction and final eradication of the invading pathogen. Disruption or dysregulation of elements within these compartments may therefore lead either to reduction in the host response with resultant overwhelming virus replication or to an overexuberant and ongoing immune response, resulting in immunopathology and autoimmunity. In both cases, excessive tissue destruction results in increased host morbidity and mortality.

B. Interaction with the Vertebrate Host

Virus-evolved strategies to ensure survival and replication in the host in the face of these complex concerted defensive networks usually fall into one of three scenarios. These may involve direct interaction with the immune system of the host, subversion of elements of the immune response, or evasion of the immune response. Such viral strategies are well described. For example, in the first scenario, HIV interacts directly with the immune system of the host by using CD4 and chemokine receptors to infect T cells and macrophages, which express these molecules (Cohen *et al.*, 2000; Murphy, 2001). Once infected, T cells die. HIV thus specifically incapacitates the immune response against it.

Host ligand or receptor components may also be mimicked by soluble viral gene products produced by infected cells. This subverts specific elements of the immune response to reduce host antiviral responses. Examples include soluble mediator mimicry, where homologues of immunosuppressive-soluble mediators, such as cytokines, chemokines, and growth factors, are encoded by virus or receptor mimicry, where receptor homologues for antiviral soluble mediators are encoded. Large DNA viruses such as poxviruses and herpesviruses have genomes that encode several of these molecules, presumably acquired from the host in the process of coevolution (Lalani *et al.*, 2000; McFadden and Murphy, 2000; Murphy, 2001; Nash *et al.*, 1999). Viruses may also evade the immune response using immunological "self"-mimicry, where the viral peptide and MHC are seen by the immune system as "self" (Mullbacher and Blanden, 1979). They may further nonspecifically or specifically decrease the cell surface levels of MHC molecule expression on the infected cells. Thus infected cells essentially become immunologically silent, where the reduced MHC expression reduces the probability of recognition by, and activation of, T cells. This is a typical outcome of infection by vaccinia, some herpesviruses (Yewdell and Bennink, 1999), and several subgenera of adenoviruses (Pääbo *et al.*, 1986a,b; Wold and Gooding, 1991). Any of these scenarios may result in a reduction in the efficacy of the antiviral response, which allows the virus to gain a replicative foothold in the host more readily.

Many of these strategies have arisen through continuing changes in the virus genome, afforded by the luxury of coevolution in a monogamous virus–host relationship. However, in the case of arthropod-borne flaviviruses, the requirement for replication in both warm-blooded vertebrate and cold-blooded invertebrate hosts may impose constraints on genomic variation. This may explain their high degree of genetic stability compared to many RNA viruses. Moreover, compared to the large DNA viruses, the small, single-stranded genome has little space for genes other than those that are already present and essential to survival in either or both hosts. Thus flaviviruses are likely to have strategies that may enable evasion or subversion of the immune response in other ways.

III. Flavivirus Upregulates Cell Surface Recognition Molecules

Most known or suspected viral strategies for evading or modulating the immune response, as well as the counteracting host responses for subverting them, fall within the framework of our understanding of virus–host interactions and are thus logical, if not predictable.

However, some virus-mediated effects on the host have no obvious advantage to the virus and some would appear to be frankly disadvantageous in the context of this framework. Several flaviviruses, principally those of the JE serocomplex, seem to directly produce effects that would on first appraisal appear to be suicidal for the virus.

In the last decade we have used WNV (*Sarafend*) as a prototype for the JE serocomplex. We have shown significant upregulation variously, of the key immune recognition molecules, MHC I and II (Argall *et al.*, 1991; Bao *et al.*, 1992; Douglas *et al.*, 1994; Johnston *et al.*, 1996; King and Kesson, 1988; King *et al.*, 1989; Liu *et al.*, 1989; Shen *et al.*, 1995a, 1997), intercellular adhesion molecule-1 (ICAM-1; CD54), vascular cellular adhesion molecule-1 (VCAM-1; CD106), E-selectin (CD62E) (Shen *et al.*, 1995b, 1997), P-selectin (CD62P) (N. J. C. King, unpublished), and B7-1 (CD80) (Johnston *et al.*, 1996), on the surface of infected cells *in vitro* in a variety of primary cells isolated from mouse, rat, and human. In all cases we have shown this for WNV, but many of these molecules are also demonstrably upregulated by Kunjin, JE, and MVE viruses.

As can be seen from the aforementioned list, most of these molecules all fall into the immunoglobulin superfamily. However, not all members of this family are upregulated by flavivirus infection. For example, the TcR is not. Members of other families such as CD44, from the homing cellular adhesion molecule family, leukocyte function antigen-1 (LFA-1, CD11a/CD18), Mac-1 (CD11b/CD18), CD11c/CD18, and VLA-4 (CD49d/CD29), members of the integrin family, are similarly not upregulated by WNV infection (N. J. C. King, unpublished). This suggests a specific targeting of immune recognition molecules by these viruses (Fig. 1).

A. MHC I and II

In the case of MHC I, upregulation is first detectable as early as 6 h postinfection in some cells, but continues to increase until the cell succumbs to the infection (King and Kesson, 1988; Kesson and King, 2001). Because WNV does not significantly depress cellular metabolism initially, infected cells generally survive for at least 48 h, frequently up to 5 days and, in some cases, almost indefinitely, depending on the cell type. From a practical point of view, this gives a substantial temporal window in which to study such virus–host cell interactions. The increased expression of MHC I is functional (Douglas *et al.*, 1994; King *et al.*, 1993). Fibroblasts infected by WNV become

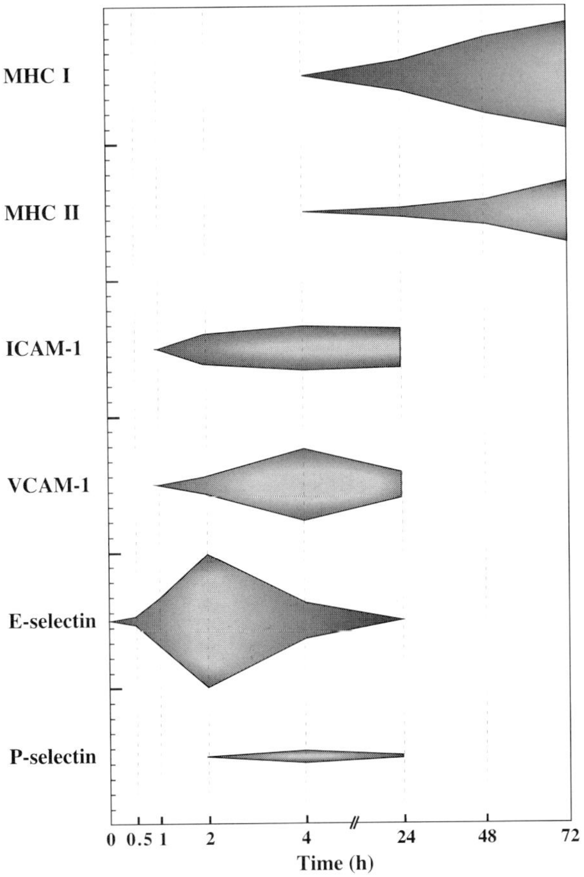

FIG 1. The kinetics of expression of various cell surface molecules after infection with WNV. The graph shows the modulating amplitude of expression (abcissa) over the time period shown (ordinate), represented as a bilateral deviation from the baseline expression. Values for expression have been normalized in all antigens to represent -fold increases over baseline expression shown at the midline of each graph.

increasingly susceptible to both WNV-specific CTL and MHC I-specific CTL between 16 and 96 h (Kesson and King, 2001). Furthermore, WNV-infected target cells coinfected with influenza can display influenza virus peptide to be killed more efficiently by influenza virus-specific CTL (Liu *et al.*, 1989). This indicates that the virus peptide is not limiting and that it is the WNV-induced increase in cell surface molecule expression that makes killing of WNV-infected targets more efficient with time.

The upregulation of MHC II by WNV occurs in a more restricted cell range than the more ubiquitous upregulation of MHC I. Such cells include murine macrophages (Shen *et al.*, 1995a), astrocytes (Liu *et al.*, 1989), Langerhans cells (Johnston *et al.*, 1996), and rat Schwann cells (Argall *et al.*, 1991), as well as human myoblasts (Bao *et al.*, 1992) and endothelial cells (Shen *et al.*, 1997). MHC II is expressed constitutively at high concentrations on macrophages and Langerhans cells, and these cells are capable of presenting antigens efficiently to initiate a cellular immune response. The WNV-induced MHC II is also functional; it can trigger cytokine release from MHC II-specific effector T cells (Liu *et al.*, 1989), whereas WNV-infected Langerhans cells can present WNV antigens to initiate an antiviral CTL response (N. J. C. King, unpublished).

B. Adhesion Molecules

The WNV-induced upregulation of adhesion molecules, ICAM-1, VCAM-1, and E-selectin, accompanying MHC upregulation is of considerable interest. The rolling receptor, E-selectin, on the vascular endothelial surface, slows the flow of individual leukocytes carrying sLe^X. Subsequent firm attachment of leukocytes occurs via ICAM-1 and VCAM-1, expressed in high concentrations on the activated endothelial cell surface, through integrin receptors, LFA-1 or Mac-1 and VLA-4, respectively, expressed on leukocytes. Together, these molecules are part of the complex array of cell surface molecules that modulate the temporal homing, endothelial attachment, and subsequent diapedesis of leukocyte subsets from the blood into local sites of infection (Springer, 1994; Wang and Springer, 1998). However, these molecules also increase the avidity of interaction between T cells and infected cells where the recognition by the TcR and its cognate ligand, MHC+virus peptide, is involved. Thus ICAM-1 may lower the threshold for T-cell activation (Kuhlman *et al.*, 1991).

The kinetics of upregulation of ICAM-1, VCAM-1, and E-selectin by WNV are quite different. On human umbilical vein endothelial cells (HUVEC), E-selectin increases within 30 min, peaking at 2 h, and is decreased substantially by 4 h. Increases in ICAM-1 and VCAM-1 are both detectable by 2 h, peak at 4 h, and while ICAM-1 is still highly expressed at 24 h, VCAM-1 expression is reduced almost to baseline levels by this time (Shen *et al.*, 1997). Curiously, a quantal effect of

WNV-induced adhesion molecule upregulation was observed with ICAM-1. In fibroblasts, no increases were seen until the multiplicity of infection reached 5 plaque-forming units (pfu) per cell, with 25–50 pfu/cell producing peak responses at 2, 4, and 24 h. This apparent on–off response is unlike the smoother sigmoidal dose–response effect seen with increasing ICAM-1 expression with titrated concentrations of IFN-α/β or γ. It is possible that MHC I responds in the same quantal way, but this response is smoothed by IFN-β feedback to these cells in infected cultures, whereas the contribution by IFN-β to ICAM-1 upregulation in infected cells is relatively small (Shen *et al.*, 1995b).

C. Adherence Status

Monocytes normally circulate in the bloodstream in a relatively nonactive state and become highly activated when required. Activating stimuli usually result in endothelial attachment and diapedesis of monocytes into the tissues at sites of inflammation to become fully activated, functional macrophages. In culture, attachment to plastic and/or exposure to cytokines such as IFN-γ produces changes in morphology and behavior consistent with activation in these cells (Eierman *et al.*, 1989; Friedman and Beller, 1987; Haskill *et al.*, 1988; King *et al.*, 1991). This dependence on adherence for a full response is also found in WNV infection. Nonadherent monocytes or fibroblasts can be infected readily under nonadherent conditions, but do not upregulate MHC I or ICAM-1 significantly until attachment occurs. If attachment is only allowed to occur 24 or 48 h after WNV infection, MHC I and ICAM-1 are induced to greater levels than those seen in cells infected under adherent conditions from the start. The production of NO by monocytes in response to WNV infection is also similarly influenced by adherence status. This has obvious implications for the consequences of infection of monocytes during viremia in flavivirus disease. Thus, it could be argued, infected monocytes might readily facilitate the transmission of virus to other cells upon attachment to endothelial cells and subsequent migration into surrounding tissues. Although NO is obviously antiviral for some viruses (Karupiah *et al.*, 1993, 2000), including JE virus (Lin *et al.*, 1997), it does not seem to be antiviral for tick-borne encephalitis virus (Kreil and Eibl, 1996) or WNV (N. J. C. King, unpublished). This raises the question of whether the induction of NO by WNV is a host response that can be negotiated successfully by some flaviviruses or is perhaps induced specifically by WNV to some replicative advantage.

D. Cytokine Modulation of West Nile Virus (WNV)-Induced Upregulation

In general, WNV seems to produce responses similar to those of the antiviral Th1 cytokines. However both Th1 and Th2 cytokines can modulate the upregulation of adhesion molecules by WNV. In particular, the proinflammatory Th1 cytokine, TNF, synergizes with WNV to upregulate adhesion molecule expression more than either agent alone. Interestingly, P-selectin is briefly upregulated by WNV infection on some endothelial cells (N. J. C. King, unpublished), although not on HUVEC. Notwithstanding, the combination of WNV and TNF over 24 h in HUVEC produces an increase that is greater than with TNF alone (Shen *et al.*, 1997).

In contrast, the Th2 cytokine, IL-4, downregulates the early increase in E-selectin produced by WNV, although in combination with WNV it delays the decrease seen by 24 h in VCAM-1 expression with WNV infection alone. Similar to the synergy seen with TNF, IL-4 upregulates the expression of P-selectin in concert with WNV by 24 h to a greater extent than IL-4 alone.

The comodulation of adhesion molecules by WNV and cytokines together has a number of important implications because of the likely recognition of, and/or interaction with, WNV-infected cells by leukocytes producing these cytokines. We have speculated that cerebrovascular endothelial cells might become infected during WNV viremia *in vivo*, resulting in more avid attachment of leukocytes to these cells. This may enhance the transmission of WNV into the brain to cause the hallmark encephalitis of WNV infection, either through breakdown of the blood–brain barrier, via the accompanying inflammatory cytokine release, as in JE (Mathur *et al.*, 1992), or through infection of leukocytes in contact with infected endothelium, which then diapedese into the brain (Shen *et al.*, 1997). If circulating monocytes infected during viremia, as suggested earlier, come into contact with infected endothelium, this may allow such a Trojan horse effect to occur even more readily. Moreover, the modulation of adhesion by NO (Kubes *et al.*, 1991, 1993; De Caterina *et al.*, 1995) may positively influence this still further. Despite the fact that endothelial cells and monocyte/macrophages may be infected readily *in vitro* (Shen *et al.*, 1995a, 1997), we have seen neither significant endothelial infection nor any infected leukocytes in the brain at any time during the pathogenesis of WNV encephalitis (Shrestha *et al.*, 2002). However, it should be kept in mind that WNV, although neurotropic, causes a primary systemic disease and that such cell–cell interactions may be important in WNV transmission within the host and/or in the host response to WNV infection outside the central nervous system.

E. Cell Cycle

Initial work in which upregulation of MHC I was shown suggested that this response may depend on the stage of cell cycle in which infection took place (King and Kesson, 1988). Subsequent experiments clearly demonstrated that the WNV-induced upregulation of MHC I and ICAM-1 is cell cycle dependent (Douglas *et al.*, 1994; Shen *et al.*, 1995b). Thus fibroblasts in G_0 respond to WNV infection with significantly greater upregulation of MHC I than cycling cells. Moreover, ICAM-1 is only upregulated in G_0; cycling fibroblasts do not upregulate ICAM-1 in response to WNV infection. We have shown that on a cell-for-cell basis, infected cells in G_0 produce the same amount of IFN-β as infected cycling cells. At the same time, cycling cells increase MHC-I expression in response to IFN-β to a greater extent than G_0 cells. Cycling fibroblasts are also more susceptible to WNV infection; that is, a significantly greater proportion of cycling cells are infectable by WNV than can be infected in G_0 under the same conditions. Experiments inactivating IFN-β in WNV-infected cultures showed that the contribution by IFN-β to MHC I upregulation was greatest in cycling cells; thus IFN-independent upregulation was greater in cells in G_0. Perhaps not surprisingly, because metabolism is most active during the cell cycle, cycling cells produce significantly more virus on a cell-for-cell basis than cells in G_0 (Kesson *et al.*, 2002). Taken together, this indicates that differential upregulation of MHC I and ICAM-1 seen in cells infected at different stages of the cell cycle is not a cellular response to differences in type I IFN concentrations produced by infected cells.

IV. Mechanisms of Upregulation

Although the production of type I IFN undoubtedly contributes to the upregulation of MHC I, WNV itself directly increases MHC I expression independently of IFN in most, if not all, infected cells. Thus the complete neutralization of supernatant IFN-α/β by the polyclonal antibody in WNV-infected fibroblast cultures over 24 h is still followed by significant upregulation of MHC I. Supernatants from WNV-infected cultures thus treated and that have been UV irradiated to inactivate WNV do not upregulate cell surface molecules on fresh, uninfected cells, confirming the effective neutralization of IFN-α/β in WNV-infected cultures. Fibroblasts from IFN type I receptor gene knockout mice (King and Kesson, unpublished results), which do not respond to IFN, and murine trophoblasts (King *et al.*, 1989), which neither make nor respond to IFN, nevertheless significantly upregulate MHC I after WNV infection. Experiments using high concentrations of alternative cytokines, regarded as possible candidates for induction by WNV in infected cells, including TNF, IL-1α/β, or IL-4, to upregulate the same cell surface molecules upregulated during WNV infection, show no upregulation, significantly less upregulation, or upregulation with completely different kinetics than those occurring with WNV infection. The possibility cannot be excluded that a soluble mediator(s) induced by WNV infection, including one or more chemokines, indirectly upregulates these various cell surface molecules. However, such soluble factor candidates are likely to be limited to short-lived and long-acting and/or highly sensitive to inactivation by UV irradiation (King and Kesson, 1988; Shen *et al.*, 1995b, 1997).

A. *Transcription*

The mechanism of the interaction between flavivirus and the vertebrate host cell to cause these direct and dramatic increases in the expression of cell surface molecules is presently unknown. Evidence suggests that flavivirus-induced upregulation of MHC I antigen expression involves an increase in peptide supply to the endoplasmic reticulum (Momburg *et al.*, 2001; Müllbacher and Lobigs, 1995). However, this does not explain how flaviviruses are able to upregulate non-MHC I molecules and argues for a more common mechanism. Data from our laboratory has demonstrated that the increase in expression of MHC I and ICAM-1 proteins occurs as a direct result of an increase in transcription of the *MHC I* and *ICAM-1* genes (Kesson and King, 2001; Kesson *et al.*, 2002). Experiments designed to block transcription with actinomycin D are associated with a failure to increase the expression of MHC I when the cells are infected with WNV (King and Kesson, 1988).

The increase in mRNA is not purely a result of autocrine stimulation of WNV-infected cells by WNV-induced IFN-β production by these cells, as treatment of cells with polyclonal anti-IFN-α/β antibody (anti-IFN-α/β) during infection of the cells with WNV is also associated with a significant increase in MHC I mRNA. Whether, in addition to this, the mRNA is stabilized by viral infection to prolong its half-life, thereby further augmenting the induced cell surface expression, is not known. The upregulation of cell surface molecules and their mRNA implies that translation may also be increased, although this has not been shown.

B. WNV Induces Nuclear Factor-κB Activation

The genes for cell surface recognition molecules, which increase in expression as a result of WNV infection, are all targets for the transcription factor, nuclear factor-κB (NF-κB). NF-κB is an inducible eukaryotic transcription factor belonging to the highly conserved *rel*-related family present in an inactive cytoplasmic form in almost all cells. Members of this family control the transcriptional activity of various promoters of a wide range of proteins involved in inflammatory and immune responses, including the promoters of MHC I, MHC II, ICAM-1, and VCAM-1. The NF-κB/Rel family of proteins binds an array of homologous decanucleotide sequences (5′-GGGRHTYYCC-3′) of differing affinities with resultant transcription of a number of genes involved in the inflammatory and/or immune responses. In vertebrates, five distinct DNA-binding subunits are currently known that might heterodimerize extensively, thereby forming complexes with distinct transcriptional activity, DNA sequence specificity, and cell type- and cell stage-specific distribution. NF-κB/Rel family members of the first group include NF-κB1 (p50) and NF-κB2 (p52), which are synthesized as precursor proteins of 105 (p105) and 100 (p100) kDa, respectively. The second group includes Rel A (p65), Rel B, and c-Rel, which are synthesized as mature proteins. The activity of NF-κB/Rel proteins is regulated by their subcellular localization. In unstimulated cells, NF-κB complexes are retained in the cytoplasm by the interaction with a second family of proteins with regulatory/inhibitory functions, inhibitor-κB (I-κB). Uncoupling of an I-κB protein from NF-κB generally requires phosphorylation, ubiquitination, and rapid degradation in proteosomes (Baeuerle and Henkel, 1994). Upon cell stimulation by a large variety of pathological conditions, NF-κB is released from its inhibitory subunit (Baeuerle and Baltimore, 1988). The activated nuclear factor can then translocate to the nucleus where it binds its cognate DNA sequences, initiating transcription of numerous target genes.

The importance of NF-κB in signaling transcription of *MHC I* and *ICAM-1* genes strongly suggests that this transcription factor may be the common link that controls the flavivirus upregulation of these cell surface recognition molecules. Data show that NF-κB is activated in WNV-infected cells but not in IFN-β-treated cells, as demonstrated by electromobility shift assays and transactivation assays, using an NF-κB reporter plasmid. Supershift analysis of WNV-induced NF-κB activation revealed a double NF-κB complex, a p65/p50 heterodimer, and a faster migrating p50 homodimer. Activation of NF-κB and subsequent upregulation of cell surface expression of MHC I are partially abrogated by the protein kinase C inhibitor H-7, presumably by blocking phosphorylation of I-κB and subsequent activation of NF-κB. In addition, salicylates, which are also implicated in blocking I-κB phosphorylation, were also effective in blocking WNV-induced activation of NF-κB and subsequent MHC I upregulation (Kesson and King, 2001). Moreover, WNV infection of RAW-Luc cells incubated in the presence of a competing oligonucleotide containing the NF-κB site resulted in a significant reduction in luciferase production and a decrease in MHC I and ICAM-1 upregulation compared with cells with the irrelevant oligonucleotide (A. M. Kesson, unpublished). These data suggest that activation of the transcription factor NF-κB is a critical common step in the pathway of flavivirus-induced upregulation of cell surface recognition molecules. How flavivirus activates NF-κB and what the implications are for interaction with the host cell have yet to be determined (Fig. 2).

V. Models of WNV Disease Pathogenesis

Upregulated cell surface immune recognition molecule expression in response to WNV *in vitro* suggested that infection *in vivo* would cause an enhanced immune response. It seemed reasonable to suppose that this would start at the level of induction and be reflected in disease pathogenesis. Most flaviviruses are transmitted via the skin, with the first line of defense presumably being Langerhans cells (LC) in the epidermis. After an intervening viremia, these viruses spread systemically to the central nervous system to cause encephalitis. Data from a model of MVE virus in young outbred mice suggest that disease pathogenesis is associated with a vigorous immune response (Andrews *et al.*, 1999). Therefore, we have developed two *in vivo* models of WNV infection specifically to investigate the parameters involved in the usual initiation of immune responses by LC against WNV and to determine the immune parameters associated with the development of immunopathology. The first is a model of WNV ear skin infection to approximate the normal context of WNV infection, and the second is a model of WNV encephalitis pathogenesis. Both models are in adult inbred mice. The use of inbred mice allows for immunological manipulation and comparison between defined strains, whereas in adult mice, the blood–brain barrier and adaptive immune systems are completely mature.

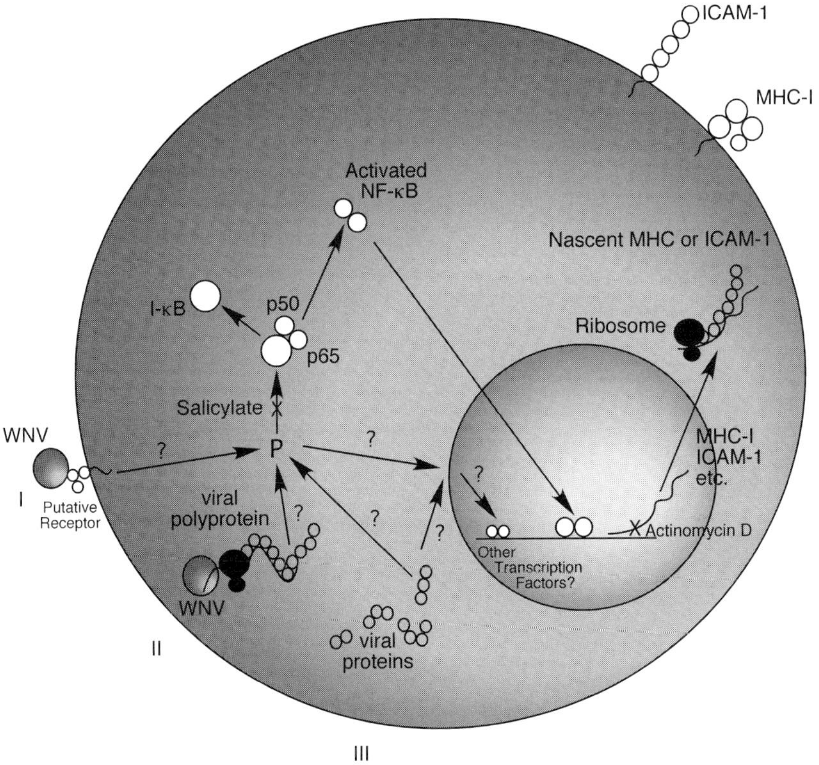

FIG 2. Diagram depicting the possible pathways of action by WNV on the cell. Virus entry via the presumptive receptor is followed by uncoating and translation of polyprotein and generation of viral proteins. The possible inductive effects of these processes on a phosphorylation event, which results in activation of NF-κB and subsequent transcription and translation of various genes, including MHC I and ICAM-I, are indicated by arrows. Processes involved in cell surface expression of translated protein are not shown, but may also be affected by WNV (Momburg *et al.*, 2001; Müllbacher and Lobigs, 1995). Unknown possible pathways are indicated by question marks.

A. A Skin Model of WNV Infection

1. Langerhans Cell Changes

Langerhans cells are dendritic cells situated at regular intervals between keratinocytes in the epidermal skin layer. Although comprising less than 3% of epidermal cells in the skin, they form a network throughout the epidermis in apparent contact with one another via their dendrites. LC are thought to be critical in the initiation of immune responses to foreign antigens in the skin. With the introduction of foreign antigen into the skin, the expression of various cell surface molecules on LC increases, including MHC I and MHC II, ICAM-1, B7-1, and B7-2, and they reduce their expression of E-cadherin (Jakob and Udey, 1998). They migrate from the epidermis to the draining lymph nodes, changing from an antigen-processing to an antigen-presenting phenotype. Upregulation of the costimulatory molecules, B7-1 and B7-2, facilitates the initiation of cellular immune responses in the draining lymph nodes against the foreign antigen brought in and presented by the migrating LC (Kimber *et al.*, 1998, 2000). Much of our knowledge of LC is based on work with contact sensitizers, but clearly the skin immune system is important in epidermally acquired virus infections. Moreover, it is likely that live, replicating viruses, with or without strategies that may prevent, delay, or pervert the initiation of an immune response to such infections, produce effects that are different from those observed in generating contact sensitivity. We hypothesized that WNV would cause an enhanced expression of molecules associated with the antigen-presenting phenotype and enhanced migration to the draining lymph node. We found a significantly increased expression of MHC I and II, ICAM-1, and B7-1 on LC by 16 h after intradermal WNV infection, which continued to increase until 48 h. This was compared to intradermal infection with the alphavirus, Semliki forest virus (SFV), which caused increased expression of these molecules on LC between 8 and 32 h later than that seen in WNV infection. Infiltrating neutrophils and macrophages were visible in the dermis by 24 h in both infections (i.e., 8 h later than the first observed increases in LC expression after WNV infection). Moreover, *in vitro* infection of epidermal cell suspensions with WNV showed that increased antigen expression still occurred on WNV-infected LC in the absence of infiltrating leukocytes, whereas none occurred with SFV infection (Johnston *et al.*, 1996).

2. Migration of Langerhans Cells

Subsequent studies in this model showed more accelerated LC migration to draining lymph nodes in intradermal WNV infection *in vivo* than in comparable SFV infection. Thus peak emigration of LC from the skin and peak LC immigration into the draining lymph node occur 24 h earlier in WNV than in SFV infection. Moreover, the total cell numbers accumulating in the lymph node are greater in WNV than in SFV infection at early time points, suggesting that lymph node shutdown and the processes leading to initiation of an anti-WNV immune response also occur with accelerated kinetics. Most interestingly, however, if either virus is UV inactivated before injection, none of these changes can be detected (Johnston *et al.*, 2000). Thus the presence of a foreign protein is not sufficient in itself to begin the processes associated with initiation of an adaptive antiviral immune response, indicating that this process is highly regulated and suggesting a level of discrimination (Matzinger, 1994) not shown by contact sensitivity studies.

Studies with contact sensitivity models have shown the requirement for TNF and IL-1β in LC migration (Cumberbatch *et al.*, 1997). In contrast, we have shown that migration is dependent on IL-1β and not TNF. We have found that the usual migration of LC in response to WNV is unaltered in TNF gene knockout mice, but is significantly inhibited by neutralizing IL-1β using systemic anti-IL-1β antibody. Inhibition of IL-1β blocks not only the migration of LC to the draining lymph nodes, but also abrogates the changes associated with lymph node shutdown in these mice (Byrne *et al.*, 2001).

Clearly, it is of interest to ascertain whether accelerated LC migration in the context of WNV-induced increased cell surface molecule expression leads to more efficient antigen presentation and initiation of anti-WNV immune responses *in vivo*. Current work in this laboratory is investigating the role of LC in the generation of anti-WNV CTL in this model.

B. A Model of WNV Encephalitis: The Generation of Immunopathology

Most models of flavivirus encephalitis to date have involved the empirical use of neonatal or young (3–4 week old) mice. In such young mice encephalitis develops more readily and is usually more severe than in older mice, often with 100% mortality. Some important observations have been made in these models and, more recently, it has become clear that innate immune responses, usually considered to be most important in bacterial infection, may play an important role in the immunopathological generation of viral encephalitis. In a young mouse model of MVE in Swiss outbred mice (Andrews *et al.*, 1999), the use of aminoguanidine to abrogate the production of NO increased survival significantly. A similar effect was observed in tick-borne encephalitis virus infection (Kreil and Eibl, 1996). In the case of MVE, NO was thought to be produced by neutrophils, which infiltrate the brain in this model, and the depletion of neutrophils reduced mortality even more. This suggests that these cells might use additional modalities, such as superoxide, to potentiate tissue damage in this model.

These data highlighted the immunopathological nature of disease in MVE. The role of TNF was thought to be pivotal in this scenario (1) to activate neurovascular endothelium and increase adhesion molecule expression and (2) to attract neutrophils via induction of the neutrophil-attracting chemokine N51/KC (CXCL1). Upregulation of nitric oxide synthase 2 mRNA was also seen. As mentioned previously, this enzyme is inducible in neutrophils by IFN-γ (Fierro *et al.*, 1999) and may reach sustained levels of expression upon induction in other leukocyte types. IFN-γ, which also induces a range of chemokines that attract various leukocytes, thus has a large number of downstream effects. The combination of recruitment and activation of cells of the innate immune system by cytokines of the adaptive immune system to cause immunopathology further emphasizes the close collaboration of these two systems.

As indicated earlier, cytokines and soluble products of the immune response, such as superoxide and NO, are not in themselves targeted to virus-infected cells. These molecules and their subsequent metabolites thus also interact with critical host-related components within their radius of effect. Under circumstances of dysregulated expression, an inevitable side effect of their role in the clearance and control of virus is collateral damage to the host. This can be a major cause of pathology in viral disease (Karupiah *et al.*, 2000).

Because of the importance of IFN-γ in viral clearance, we developed a model to investigate the requirement of this cytokine in WNV encephalitis using C57BL/6 IFN-γ gene knockout (B6.IFN-$\gamma^{-/-}$) and the corresponding C57BL/6 wild-type (B6.WT) mice as controls. We used an intraperitoneal dose of WNV that produced mortality in the mid-range in wild-type mice so that increased survival or increased mortality, depending on the experimental protocols employed, would be obvious. We found that mortality of the B6.IFN-$\gamma^{-/-}$ group was reduced to less than one third of the mortality of the B6.WT groups. The kinetics of virus infection was similar in both groups: WNV was first detectable in the brainstem on day 6 after intraperitoneal inoculation, spreading caudal to rostral, with similar peak numbers of infected neurons occurring on days 7 and 8 postinfection in both groups. However, twice as many B6.IFN-$\gamma^{-/-}$ as B6.WT mice had detectable WNV in neurons between days 6 and 11.

The route of WNV entry into the brain is unclear. In contrast to the MVE (McMinn *et al.*, 1996) and St. Louis encephalitis (Monath *et al.*, 1983) models, we observed no infection of olfactory neurons in adult or young mice. As mentioned previously, there was little or no detectable neurovascular endothelial infection *in vivo*, although this may have occurred outside the temporal window of our assays. Current data cannot discount the possibility of endothelial transcytosis of virus into the brain, such as occurs in suckling mice in JE (Liou and Hsu, 1998). However, WNV detection in the anterior horn motor neurons of the spinal cord before virus was detectable in the brainstem, shown by our studies, suggests retrograde spread along motor nerves from the periphery as a possible means of WNV entry into the central nervous system.

Leukocyte infiltration into the brain parenchyma, comprising neutrophils and mononuclear cells, was fourfold higher in B6.WT than B6.IFN-$\gamma^{-/-}$ mice at day 7 postinfection. This differential occurred even though there was significantly greater vascular endothelial ICAM-1 and VCAM-1 expression in the brains of B6.IFN-$\gamma^{-/-}$ mice than in B6.WT mice (Fig. 3).

In both groups, microglial activation with migratory and phagocytic phenotypes was clearly observed from day 7 onward in brain sections, with clustering into nodules around infected neurons. Numbers of activated microglia in B6.IFN-$\gamma^{-/-}$ mice were threefold higher than in B6.WT mice at the peak of activation. However, neuronal death was detectable in less than 0.1% of infected neurons, as determined by terminal deoxynucleotidyl–transferase-mediated dUTP-biotin nick end labeling (TUNEL) in both groups. Finally, rechallenge with WNV at 50-fold higher doses in survivors from both strains showed no clinical or histological evidence of disease (Shrestha *et al.*, 2002).

In our model, therefore, WNV (*Sarafend*) encephalitis in the B6.WT exhibits the histopathological and clinical features of human disease (Sampson *et al.*, 2000). This is clearly relevant to the study of WNV in the murine model. There are also several important findings from the B6.IFN-$\gamma^{-/-}$ mouse. These are (1) an increase in survival, without any decrease in viral load; (2) a reduction in leukocyte infiltration into the brain, although critical leukocyte adhesion molecules were markedly upregulated; (3) significantly increased microglial activation with nodule formation; and (4) effective immunity with virus clearance, all in the complete absence of IFN-γ. These results indicate that IFN-γ, or its induced downstream soluble factors, is crucial for full leukocyte recruitment into the brain in WNV encephalitis and that massively upregulated adhesion molecule expression on neurovascular endothelium, independent of IFN-γ, is not sufficient to recruit the usual complement of leukocytes in the absence of IFN-γ. Microglial activation, migration, and nodule formation in response to neuronal infection within the brain are also clearly independent of IFN-γ. The greater activation of microglia in the face of lower infiltration of leukocytes makes it unlikely that the activating stimulus comes from infiltrating leukocytes. Moreover, intranasal inoculation with WNV in this model produces microglial activation 48 h before leukocyte infiltration, suggesting that WNV infection of neurons alone may be responsible for this stimulus (Shrestha *et al.*, 2002).

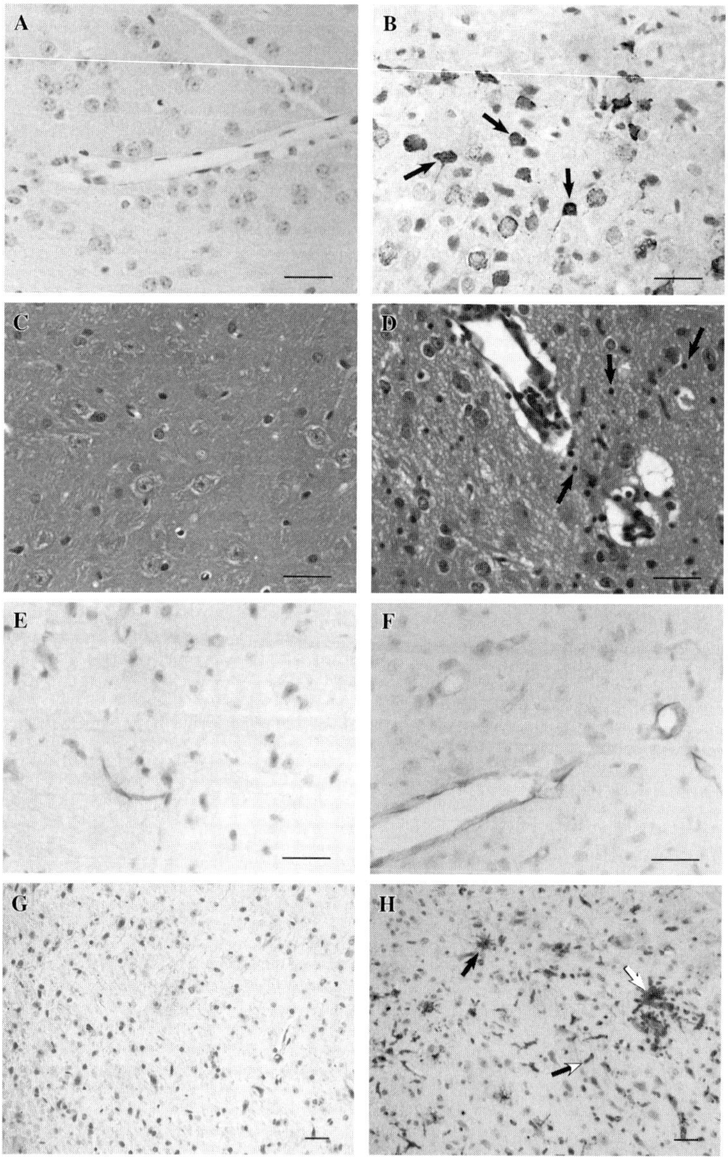

FIG 3. Sagittal sections of perfused, fixed brain from mock-infected (A) and day 9 WNV-infected (B) B6.IFN-$\gamma^{-/-}$ mice labeled for WNV antigen using immunoperoxidase-labeled rat anti-WNV and 3,3-diaminobenzidine tetrahydrochloride (DAB) substrate with hematoxylin as a counterstain. WNV antigen (brown) was first detected in brainstem neurons on day 6 postinfection in both strains and was found only in the cytoplasm and processes of neurons (solid arrows). Hematoxylin and eosin stained sagittal sections

While these findings emphasize the overlap in function between the various soluble mediators produced in response to virus infection in the absence of IFN-γ, in the case of WNV encephalitis, leukocyte recruitment is clearly not as effective. This is similar to the findings in a murine model of yellow fever encephalitis in the same mouse strains, although the discrepancies reported were not as great (Liu and Chambers, 2001). Full recruitment of leukocytes in these models may therefore require IFN-γ-dependent chemotactic signals (Liu *et al.*, 2000, 2001).

These data also demonstrate the acquisition of an efficient adaptive immune response with clearance of WNV, as well as the generation of immunological memory in the absence of IFN-γ. Whether there is a change from a Th1 to a Th2 spectrum of immune response in the absence of IFN-γ is not clear but is the subject of further investigation in this laboratory.

Finally, and most importantly, these data suggest that IFN-γ may cause fatal immunopathology directly or indirectly in WNV encephalitis.

VI. The Paradox of Cell Surface Molecule Upregulation, Immunopathology, and Virus Survival

The survival strategies of viruses, as discussed earlier in this chapter, are aimed at abrogating, subverting, or avoiding any or all of the immune responses pitched against them. The presumed point of this eternal dance is to buy time to increase the chances of transmission from one host to the next. In general, there is nothing to be gained from killing the host, since it will then not reproduce. Over time this will limit numbers of available hosts and may eventually result in virus

of perfused, fixed brains from mock-infected (C) and day 8 WNV-infected (D) B6.WT mice. Infected brain shows leukocytes infiltrating into brain parenchyma (solid arrows). Sagittal sections of perfused, fixed brains from mock-infected (E) and day 5 WNV-infected (F) B6.IFN-$\gamma^{-/-}$ mice labeled for ICAM-1 using immunoperoxidase-labeled rat anti-ICAM-1 and DAB substrate with hematoxylin as a counterstain. Increased ICAM-1 expression (brown), as measured by intensity of staining and number of vessels stained, was first detectable on day 3 in both strains. Sagittal sections of perfused, fixed brains from mock-infected (G) and day 5 WNV-infected (H) B6.WT mice labeled for activated microglia using immunoperoxidase-labeled B4 isolectin and DAB as substrate with hematoxylin as a counterstain. Activation of microglia was first detectable on day 7 and can be seen readily with dendritic (solid arrow) and migratory rod (open arrow) morphologies, which cluster to form nodules around infected neurons (white arrow). Scale bar: 30 μm throughout. (See Color Insert.)

extinction. All our work to date points toward WNV infection enhancing both the stimulus to, and the response by, the immune system. This is the opposite of what we have come to expect of virus survival strategies. Thus, in the light of current thinking, the MHC- and adhesion molecule-upregulating effect of WNV appears paradoxical, as this would make infected cells more likely to become targets of virus-specific, T-cell-mediated lysis. Indeed, as mentioned earlier, infected cells become increasingly susceptible to CTL lysis *in vitro* over time (Kesson and King, 2001) and this too is cell cycle dependent. Thus cells infected in G_0 increase MHC I by 5- to 6-fold more than cells infected in other phases of the cell cycle, which results in a 10-fold difference in the susceptibility to WNV-specific CTL lysis (Douglas *et al.*, 1994). The obvious functionality of upregulated MHC I and the fact that the virus peptide is not limiting implies that the mechanisms for antigen loading of MHC I are comodulated with MHC expression during WNV infection. This is supported by experiments showing that peptide transport by the transporter associated with antigen processing (TAP) is increased in WNV-infected cells and depends on virus macromolecular synthesis. This study showed that synthesis of the TAP protein was not increased (Momburg *et al.*, 2001). However, we have shown that mRNA for TAP-1 and LMP-2 is upregulated during WNV infection (Arnold *et al.*, 2001). This is of some interest, as these genes share an NF-κB site within their bidirectional promoter (Wright *et al.*, 1995). The issues of increased functional MHC expression with concomitant effective virus peptide loading are critically important for T-cell recognition and may give a clue as to what may be happening to advantage the virus in this apparent game of "immunological chicken."

A number of possibilities exist to explain this apparent paradox. One possibility is that antiviral CTL lysis releases infectious virions from infected cells, thus enabling the efficient spread of virus progeny. While this may occur, the time taken for the generation of virus-specific CTL, which peak at day 5 postinfection (Kesson *et al.*, 1987), would significantly delay the release of virions by this strategy. In the meantime, apart from innate antiviral mechanisms, which may be operative, other contemporaneously generated modalities of the adaptive immune response, such as the antiviral antibody response, would have developed a maximal neutralizing effect, mopping up any cell-free infectious virions, thus abrogating all advantage of this for the virus. In addition, significant upregulation of MHC I expression occurs in infected cells before completion of the virus replication cycle and it seems likely that many cells would be lysed before virions were completely mature. Because infected cells are not metabolically severely inhibited by flaviviruses, it is likely that there is more survival advantage for the virus in having several rounds of virus replication accompanied by continual release from one infected cell than the CTL-mediated release of a comparatively small number of flaviviruses after a single replication cycle.

A more likely possibility is that the upregulation of MHC I is associated with inhibition of NK cell lysis (Lopez-Botet *et al.*, 2000; Müllbacher and King, 1989). This has been shown *in vitro* (Momburg *et al.*, 2001). Thus *in vivo* it is possible that subversion of part of the innate immune response occurs in which NK cells may be inhibited from killing flavivirus-infected cells during the early phase of infection.

VII. Flavivirus Modulation of Adaptive Immune Responses: A Hypothesis

Virus-specific CTL populations contain many clones with a range of affinities, including affinities for allo-MHC and heterologous virus antigens (Nahill and Welsh, 1993; Zheng and Liu, 1997), but not completely unrelated antigens (Zarozinski and Welsh, 1997). Langerhans cells, which present antigen to T cells, increase both MHC I and II, as well as adhesion molecules and costimulatory molecules, in an accelerated manner and more than comparable viruses in response to WNV infection *in vitro* and *in vivo* (Johnston *et al.*, 1996). Increased cell surface MHC I or II concentrations on infected antigen-presenting cells would enhance the avidity of the T-cell–antigen-presenting interaction. Nonspecific accessory molecules such as ICAM-1, interacting with LFA-1 on the T cell, can also stabilize T-cell–target interactions, lowering the threshold for T-cell activation (Kuhlman *et al.*, 1991). By increasing the avidity of interaction between these cells, increased expression on WNV-infected, antigen-presenting cells would also trigger functional recognition by low-affinity T cells. This would enable the functional recruitment of both $CD4^+$ or $CD8^+$ T cells, which were previously below the recognition threshold (Goldstein and Mescher, 1987; Kuhlman *et al.*, 1991; Shimonkevitz *et al.*, 1985), and would result in a larger range of T-cell clones being activated. By definition, this range of clones would have varying affinity for MHC I+virus peptide, but would also include a proportion of low-affinity allo-reactive and self-reactive clones. Self-reactive CTL clones could kill both infected and uninfected high MHC I-expressing target cells. Furthermore, because MHC I-specific CTLs can recognize MHC I without the requirement for peptide specificity (Müllbacher *et al.*, 1991), this means that these low-affinity, cross-reactive CTLs would have a high probability of recognizing and killing uninfected, high MHC I-expressing target cells.

Experiments *in vitro* suggest that cross-reactive killing by WNV-specific CTL populations occurs on uninfected, high MHC I-expressing L929 cell targets, compared with lower MHC I-expressing MEF (Kesson, 1989). These targets are not strictly comparable, and we have conducted experiments that show that "WNV-specific" CTL populations lyse IFN-γ-treated, uninfected MEF expressing high MHC I concentrations more efficiently than uninfected, untreated MEF (A. M. Kesson and N. J. C. King, unpublished results). Moreover, we have shown that specific CD8–MHC I interactions are important to stabilize the recognition of infected cells by low-affinity flavivirus-specific CTLs (Kesson, 1989). High-affinity CTLs make less use than low-affinity CTLs of accessory molecules such as CD8 to stabilize the interaction between MHC I and TcR. Stabilization by CD8 of the interaction between MHC I and TcR is required when CTL affinity is low (Goldstein and Mescher, 1987; Shimonkevitz *et al.*, 1985).

Upregulation of cell surface molecules by WNV therefore, although counterintuitive, appears to enhance the host cellular immune response against WNV. However, significant upregulation of these molecules would especially occur in cells infected in G_0, whereas infected cells in other phases of the cycle would be less easily recognized. We speculate that since most cells *in vivo* are in G_0 and WNV replicates better in cycling cells than in cells in G_0 (Kesson *et al.*, 2002), a minority of cells in other phases of the cycle may escape detection long enough to allow sufficient replication of the virus for transmission to take place to an invertebrate host. The high degree of incapacitating illness may serve the function of facilitating increased access to the host for arthropod vectors during this period. Interestingly, experiments using high- and low-affinity virus-specific T cells have shown far better virus clearance by high-affinity T cells *in vivo*, whereas even in high numbers, low-affinity antiviral T cells clear virus comparatively poorly (Derby *et al.*, 2001). With time, however, the eventual proliferation of high-affinity T-cell clones would clear virus from both high and low MHC-expressing infected cells, resolving the infection and leaving the host immune.

In vivo, high MHC I-expressing uninfected targets could be expected to occur in the vicinity of virus-infected cells due to the release of IFN-γ from T-cell–target interactions. These cells would be most susceptible to lysis or cytokine damage by cross-reactive low-affinity self-reactive $CD8^+$ and $CD4^+$ clones, respectively. If virus-specific T-cell populations included low-affinity self-reactive clones that were able to lyse or damage these uninfected cells, this would result in destruction of uninfected tissue. In the case of an encephalitic virus infection, this would potentially increase the damage to the brain, exacerbating the encephalitic syndrome. It has been shown that a wide range of TcR specificities in encephalitic viral disease increases pathology. Conversely, restriction of the range of TcR specificities reduces pathology significantly (Doherty *et al.*, 1994). In our model, encephalitis in the B6.WT mouse is associated with significant infiltration of leukocytes into the brain, whereas the absence of IFN-γ, normally produced by T cells, is associated with reduced leukocyte infiltration, reduced morbidity, and significantly increased survival. This suggests that an overvigorous immune response may be responsible for the pathogenesis of disease, particularly since neutralizing immunity is established in both strains of mice. Cross-reactive damage, whether CTL or cytokine mediated, is more likely to occur in supporting cells in the brain rather than neurons. Supporting cells, such as astrocytes and microglia, readily upregulate MHC and adhesion molecules to high levels in response to IFN-γ and other cytokines. In genetically susceptible individuals, if the normal damping down and eventual abrogation of a particular antiviral response were poor or absent, the continued propagation of self-reactive clones even after the virus was eradicated could lead to autoimmune disease. Why this does not happen more frequently is presumably testimony to the tight controls that must exist to prevent it.

VIII. Maternal Tolerance: A Model of Embryonic Implantation

Along these lines, it has become possible to use WNV as a tool to explore the question of tolerance in a highly specialized and restricted environment, namely that of the implanting embryo. The mechanism by which the semiallogeneic fetus survives without rejection by the maternal immune system has fascinated and confounded immunologists for decades. Because of the theoretical contribution of "foreign" paternal genes, much work has concentrated on the role of paternally derived placental trophoblast cells located at the materno–fetal interface, in direct contact with maternal blood and tissue. In particular, the importance of the lack of expression of cell surface MHC on these cells has been the subject of much debate. We have investigated the use of WNV for several years as a prelude to investigating this problem. The trophectoderm and trophoblast cell layers in direct contact with the maternal circulation usually express little or no MHC I, but can be induced to do so by WNV (King *et al.*, 1989). Moreover, WNV-infected outgrown trophoblast cells with induced MHC I expression can be recognized and killed by WNV-specific and paternal MHC I-specific CTLs (King *et al.*, 1993). We have therefore developed a murine model in which WNV-infected blastocysts are implanted into pseudopregnant mothers to evaluate the effect of localized, upregulated MHC I and adhesion molecule expression in the context of virus infection. This will enable us not only to explore the parameters for successful implantation of the embryo, but also to determine how the maternal immune system deals with virus infection during gestation in the face of paternal-specific systemic and localized immunosuppression.

During pregnancy, paternal MHC I, which is foreign to the maternal immune system, is systemically ignored by downregulating T cells recognizing it (Tafuri *et al.*, 1995). Consistent with our hypothesis, we speculate that since paternal MHC is upregulated by WNV (King *et al.*, 1989), this would trigger recognition *in vivo* by low-affinity anti-WNV T cells, which are not downregulated. These could cross-reactively recognize paternal MHC, to essentially break tolerance to paternal MHC. The upregulation of nonallelic (i.e., nonspecific) molecules such as ICAM-1 and VCAM-1, which also occurs on these cells (N. J. C. King, unpublished), would further increase the avidity of cellular interaction, making this more likely. Because there is no requirement for continuing tolerance to paternal MHC after pregnancy (paternal MHC is now absent), such cross-reactive antipaternal MHC responses may not be as well controlled (i.e., suppressed) as in the maintenance of tolerance to maternal self. Subsequent exposure to paternal MHC through a second pregnancy may reactivate the antiviral memory cells cross-reactive with paternal MHC to destroy the embryo and might constitute a basis for recurrent spontaneous abortions. It may also explain how infection with the closely related JE virus early in gestation causes significant abortion and/or fetal resorption in humans, mice, and pigs (Chaturvedi *et al.*, 1980; Mathur *et al.*, 1983; Monath, 1986; Sugamata and Miura, 1982).

IX. Conclusions

The mechanism(s) by which the interaction between flavivirus and the vertebrate host cell causes direct and dramatic increases in various critical immune recognition molecules expression is presently unknown. These phenomena, however, clearly highlight the specific influence of this virus on the target molecules for CTL recognition. While much work has been done that shows antiviral T cell responses may be "self" cross-reactive, we believe this is the first instance where a virus itself may actually alter the host immune response to augment a low-affinity "self" cross-reactive outcome.

We would argue that this may decoy the immune system into preferentially recognizing infected cells in G_0 with high MHC, allowing infected cycling cells with relatively low MHC, which produce virus more efficiently and which do not upregulate these cell surface molecules to the same extent, to escape detection until significant numbers of high-affinity antiviral CTLs are produced. This has the advantage of gaining significant time for continued viremia, thus increasing the chances of virus transmission. This occurs without the death of the host, which becomes immune and protected against further infection. Clearly, however, a significant amount of further experimental data will be required to confirm this hypothesis.

Acknowledgments

The authors acknowledge the support for this work, variously, of the National Health and Medical Research Council of Australia, the Clive and Vera Ramaciotti Foundation, The Medical Foundation of the University of Sydney, the Watson Munroe Fund, and the Children's Hospital at Westmead Fund.

References

Abbas, A. K., Lichtman, A. H., and Pober, J. S. (2000). "Cellular and Molecular Immunology." Saunders, Philadelphia.

Anderson, J. F., Andreadis, T. G., Vossbrinck, C. R., Tirrell, S., Wakem, E. M., French, R. A., Garmendia, A. E., and Van Kruiningen, H. J. (1999). Isolation of West Nile virus from mosquitoes, crows, and a Cooper's hawk in Connecticut. *Science* **286:**2331–2333.

Andrews, D. M., Matthews, V. B., Sammels, L. M., Carrello, A. C., and McMinn, P. C. (1999). The severity of Murray Valley encephalitis in mice is linked to neutrophil infiltration and inducible nitric oxide synthase activity in the central nervous system. *J. Virol.* **73:**8781–8790.

Argall, K. G., Armati, P. J., King, N. J. C., and Douglas, M. W. (1991). The effects of West Nile virus on major histocompatibility complex class I and II molecule expression by Lewis rat Schwann cells in vitro. *J. Neuroimmunol.* **35:**273–284.

Arnold, S., Hall, R. A., Khromykh, A., Kesson, A. M., King, N. J. C., and Sedger, L. (2001). Mechanisms of flavivirus-mediated regulation of the class I antigen presentation pathway. 1st Australian Virology Group Meeting, Fraser Island, Queensland, Australia.

Baeuerle, P. A., and Baltimore, D. (1988). I kappa B: A specific inhibitor of the NF-kappa B transcription factor. *Science* **242:**540–546.

Baeuerle, P. A., and Henkel, T. (1994). Function and activation of NF-kappa B in the immune system. *Annu. Rev. Immunol.* **12:**141–179.

Bao S., King, N. J. C., and Dos Remedios, C. G. (1992) Flavivirus induces MHC antigen on human myoblasts: A model of autoimmune myositis? *Muscle Nerve* **15:** 1271–1277.

Briese, T., Jia, X. Y., Huang, C., Grady, L. J., and Lipkin, W. I. (1999). Identification of a Kunjin/West Nile-like flavivirus in brains of patients with New York encephalitis. *Lancet* **354:**1261–1262.

Byrne, S. N., Halliday, G. M., Johnston, L. J., and King, N. J. C. (2001). Interleukin-1β but not tumor necrosis factor is involved in West Nile virus-induced Langerhans cell migration from the skin in C57BL/6 mice. *J. Invest. Dermatol.* **117:**702–709.

Calisher, C. H., Karabatsos, N., Dalrymple, J. M., Shope, R. E., Porterfield, J. S., Westaway, E. G., and Brandt, W. E. (1989). Antigenic relationships between flaviviruses as determined by cross-neutralization tests with polyclonal antisera. *J. Gen. Virol.* **70:**37–43.

Cassatella, M., Bazzoni, F., Flynn, R., Dusi, S., Trinchieri, G., and Rossi, F. (1990). Molecular basis of interferon-gamma and lipopolysaccharide enhancement of phagocyte respiratory burst capability: Studies on the gene expression of several NADPH oxidase components. *J. Biol. Chem.* **265:**20241–20246.

Chaturvedi, U. C., Mathur, A., Chandra, A., Das, S. K., Tandon, H. O., and Singh, U. K. (1980). Transplancental infection with Japanese encephalitis virus. *J. Infect. Dis.* **141:**712–715.

Cohen, O., Cicala, C., Vaccarezza, M., and Fauci, A. S. (2000). The immunology of human immunodeficiency virus infection. *In* "Mandel Douglas and Bennett's Principles and Practice of Infectious Diseases" (G. L. Mandell, J. E. Bennett, and R. Dolin, eds.), Vol. 1, pp. 1374–1397. Churchill Livingstone, New York.

Cumberbatch, M., Dearman, R. J., and Kimber, I. (1997). Langerhans cells require signals from both tumour necrosis factor-alpha and interleukin-1-beta for migration. *Immunology* **92:**388–395.

De Caterina, R., Libby, P., Peng, H. B., Thannickal, V. J., Rajavashisth, T. B., Gimbrone, M. A. J., Shin, W. S., and Liao, J. K. (1995). Nitric oxide decreases cytokine-induced endothelial activation: Nitric oxide selectively reduces endothelial expression of adhesion molecules and proinflammatory cytokines. *J. Clin. Invest.* **96:**60–68.

Derby, M. A., Alexander-Miller, M. A., Tse, R., and Berzofsky, J. A. (2001). High-avidity CTL exploit two complementary mechanisms to provide better protection against viral infection than low-avidity CTL. *J. Immunol.* **166:**1690–1697.

Doherty, P. C., Hou, S., Evans, C. F., Whitton, J. L., Oldstone, M. B., and Blackman, M. A. (1994). Limiting the available T cell receptor repertoire modifies acute lymphocytic choriomeningitis virus-induced immunopathology. *J. Neuroimmunol.* **51:**147–152.

Douglas, M. W., Kesson, A. M., and King, N. J. C. (1994). CTL recognition of West Nile virus-infected fibroblasts is cell cycle dependent and is associated with virus-induced increases in class I MHC antigen expression. *Immunology* **82:**561–570.

Eierman, D. F., Johnson, C. E., and Haskill, J. S. (1989). Human monocyte inflammatory mediator gene expression is selectively regulated by adherence substrates. *J. Immunol.* **142:**1970–1976.

Fierro, I. M., Nascimento-DaSilva, V., Arruda, M. A., Freitas, M. S., Plotkowski, M. C., Cunha, F. Q., and Barja-Fidalgo, C. (1999). Induction of NOS in rat blood PMN in vivo and in vitro: Modulation by tyrosine kinase and involvement in bactericidal activity. *J. Leukocyte Biol.* **65:**508–514.

Friedman, A., and Beller, D. I. (1987). The effect of adherence on the *in vitro* induction of cytocidal activity by macrophages. *Immunology* **61:**469–474.

Goldstein, S. A. N., and Mescher, M. F. (1987). Cytotoxic T cell activation by class I protein on cell-size artificial membranes: Antigen density and Lyt-2/3 function. *J. Immunol.* **138:**2034–2043.

Hannah, J., Ritchie, S., Loewenthal, M., Tiley, S., Phillips, D., Broom, A., and Smith, D. (1995). Probable Japanese encephalitis acquired in the Torres Strait. *Commun. Dis. Intell.* **19:**206–208.

Haskill, S., Johnson, C., Eierman, D., Becker, S., and Warren, K. (1988). Adherence induces selective mRNA expression of monocyte mediators and proto-oncogenes. *J. Immunol.* **140:**1690–1694.

Hubalek, Z., and Halouzka, J. (1999). West Nile fever: A re-emerging mosquito-borne viral disease in Europe. *Emerg. Infect. Dis.* **5:**643–650.

Jakob, T., and Udey, M. C. (1998). Regulation of E-cadherin-mediated adhesion in Langerhans cell-like dendritic cells by inflammatory mediators that mobilize Langerhans cells in vivo. *J. Immunol.* **160:**4067–4073.

Johnston, L. J., Halliday, G. M., and King, N. J. C. (2000). Langerhans cells migrate to local lymph nodes following cutaneous infection with an arbovirus. *J. Invest. Dermatol.* **114:**560–568.

Johnston, L. J., Halliday, G. M., and King, N. J. C. (1996). Phenotypic changes in Langerhans cells after infection with arboviruses: A role in the immune response to epidermally acquired viral infection? *J. Virol.* **70** 4761–4766.

Karupiah, G., Hunt, N. H., King, N. J. C., and Chaudhri, G. (2000). NADPH oxidase, Nrampl and nitric oxide synthase 2 in host antimicrobial responses. *Rev. Immunogenet.* **2:**387–415.

Karupiah, G., Xie, Q. W., Buller, R. M., Nathan, C., Duarte, C., and MacMicking, J. D. (1993). Inhibition of viral replication by interferon-gamma-induced nitric oxide synthase. *Science* **261:**1445–1448.

Kesson, A. M. (1989). "T Lymphocyte Response to Flavivirus Infection." Department of Microbiology, John Curtin School of Medical Research. A.C.T., Australian National University.

Kesson, A. M., Blanden, R. V., and Müllbacher, A. (1987). The primary in vivo murine cytotoxic T cell response to the flavivirus West Nile. *J. Gen. Virol.* **68:**2001–2006.

Kesson, A. M., Cheng, Y., King, N. J. C. (2002) Regulation of immune recognition molecules by flavivirus, West Nile. *Viral Immunology* **15:**273–283.

Kesson, A. M., and King, N. J. C. (2001). Transcriptional regulation of major histocompatibility complex class I by flavivirus West Nile is dependent on NF-κB activation. *J. Infect. Dis.* **184:**947–954.

Kimber, I., Cumberbatch, M., Dearman, R. J., Bhushan, M., and Griffiths, C. E. M. (2000). Cytokines and chemokines in the initiation and regulation of epidermal Langerhans cell mobilisation. *Br. J. Dermatol.* **142:**401–412.

Kimber, I., Dearman, R. J., Cumberbatch, M., and Huby, R. J. D. (1998). Langerhans cells and chemical allergy. *Curr. Opin. Immunol.* **10:**614–619.

King, N. J. C., Maxwell, L. E., and Kesson, A. M. (1989). Induction of class I major histocompatibility complex antigen expression by West Nile virus on gamma interferon-refractory early murine trophoblast cells. *Proc. Natl. Acad. Sci. USA* **86:**911–915.

King, N. J. C., Müllbacher, A., and Blanden, R. V. (1986). Relationship between surface H-2 concentration, size of different target cells, and lysis by cytotoxic T cells. *Cell. Immunol.* **98:**525–532.

King, N. J. C., Müllbacher, A., Tian, L., Rodger, J. C., Lidbury, B., and Hla, R. T. (1993). West Nile virus infection induces susceptibility of in vitro outgrown murine blastocysts to specific lysis by paternally directed allo-immune and virus-immune cytotoxic T cells. *J. Reprod. Immunol.* **23:**131–144.

King, N. J. C., Sinickas, V. G., and Blanden, R. V. (1985). H-2K and H-2D antigens are independently regulated in mouse embryo fibroblasts. *Exp. Clin. Immunogenet.* **2:**206–214.

King, N. J. C., Ward, M. H., and Holmes, K. T. (1991). Magnetic resonance studies of murine macrophages: Proliferation is not a prerequisite for acquisition of an "activated" high resolution spectrum. *FEBS Lett.* **287:**97–101.

King, N. J. C., and Kesson, A. M. (1988). Interferon-independent increases in class I major histocompatibility complex antigen expression follow flavivirus infection. *J. Gen. Virol.* **69:**2535–2543.

Kramer, L. D., and Bernard, K. A. (2001). West Nile virus in the Western Hemisphere. *Curr. Opin. Infect. Dis.* **14:**519–525.

Kreil, T. R., and Eibl, M. M. (1996). Nitric oxide and viral infection: NO antiviral activity against a flavivirus in vitro, and evidence for contribution to pathogenesis in experimental infection in vivo. *Virology* **219:**304–306.

Kubes, P., Suzuki, M., and Granger, D. N. (1991). Nitric oxide: An endogenous modulator of leukocyte adhesion. *Proc. Natl. Acad. Sci. USA* **88:**4651–4655.

Kubes, P. R., Kanwar, S., Niu, X. F., and Gaboury, J. P. (1993). Nitric oxide synthesis inhibition induces leukocyte adhesion via superoxide and mast cells. *FASEB J.* **7:**1293–1299.

Kuhlman, P., Moy, V. T., Lollo, B. A., and Brian, A. A. (1991). The accessory function of murine intercellular adhesion molecule-1 in T lymphocyte activation: Contributions of adhesion and co-activation. *J. Immunol.* **146:**1773–1782.

Lalani, A. S., Barrett, J. W., and McFadden, G. (2000). Modulating chemokines: More lessons from viruses. *Immunol. Today* **21:**100–106.

Lanciotti, R. S., Roehrig, J. T., Deubel, V., Smith, J., Parker, M., Steele, K., Crise, B., Volpe, K. E., Crabtree, M. B., Scherret, J. H., Hall, R. A., MacKenzie, J. S., Cropp, C. B., Panigrahy, B., Ostlund, E., Schmitt, B., Malkinson, M., Banet, C., Weissman, J., Komar, N., Savage, H. M., Stone, W., McNamara, T., and Gubler, D. J. (1999). Origin of the West Nile virus responsible for an outbreak of encephalitis in the northeastern United States. *Science* **286:**2333–2337.

Le Bon, A., Schiavoni, G., D'Agostino, G., Gresser, I., Belardelli, F., and Tough, D. F. (2001). Type I interferons potently enhance humoral immunity and can promote isotype switching by stimulating dendritic cells in vivo. *Immunity* **14:**461–470.

Lin, Y. L., Huang, Y. L., Ma, S. H., Yeh, C. T., Chiou, S. Y., Chen, L. K., and Liao, C. L. (1997). Inhibition of Japanese encephalitis virus infection by nitric oxide: Antiviral effect of nitric oxide on RNA virus replication. *J. Virol.* **71:**5227–5235.

Liou, M. L., and Hsu, C. Y. (1998). Japanese encephalitis virus is transported across the cerebral blood vessels by endocytosis in mouse brain. *Cell Tissue Res.* **293:**389–394.

Liu, M. T., Armstrong, D., Hamilton, T. A., and Lane, T. E. (2001). Expression of Mig (monokine induced by interferon-gamma) is important in T lymphocyte recruitment and host defense following viral infection of the central nervous system. *J. Immunol.* **166:**1790–1795.

Liu, M. T., Chen, B. P., Oertel, P., Buchmeier, M. J., Armstrong, D., Hamilton, T. A., and Lane, T. E. (2000). The T cell chemoattractant IFN-inducible protein 10 is essential in host defense against viral-induced neurologic disease. *J. Immunol.* **165:**2327–2330.

Liu, T., and Chambers, T. J. (2001). Yellow fever virus encephalitis: Properties of the brain-associated T-cell response during virus clearance in normal and gamma interferon-deficient mice and requirement for CD4(+) lymphocytes. *J. Virol.* **75:**2107–2118.

Liu, Y., King, N. J. C., Kesson, A., Blanden, R. V., and Müllbacher, A. (1989). Flavivirus infection up-regulates the expression of class I and class II major histocompatibility antigens on and enhances T cell recognition of astrocytes in vitro. *J. Neuroimmunol.* **21**:157–168.

Lopez-Botet, M., Bellon, T., Llano, M., Navarro, F., Garcia, P., and de Miguel, M. (2000). Paired inhibitory and triggering NK cell receptors for HLA class I molecules. *Hum. Immunol.* **61**:7–17.

Mathur, A., Arora, K. L., and Chaturvedi, U. C. (1983). Host defence mechanisms against Japanese encephalitis virus infection in mice. *J. Gen. Virol.* **64**:805–811.

Mathur, A., Khanna, N., and Chaturvedi, U. C. (1992). Breakdown of blood-brain barrier by virus-induced cytokine during Japanese encephalitis virus infection. *Int. J. Exp. Pathol.* **73**:603–611.

Mattei, F., Schiavoni, G., Belardelli, F., and Tough, D. F. (2001). IL-15 is expressed by dendritic cells in response to type I IFN, double-stranded RNA, or lipopolysaccharide and promotes dendritic cell activation. *J. Immunol.* **167**:1179–1187.

Matzinger, P. (1994). Tolerance, danger, and the extended family. *Annu. Rev. Immunol.* **12**:991–1045.

McDade, J. E., and Hughes, J. M. (2000). New and emerging infectious diseases. *In* "Principles and Practice of Infectious Diseases" (G. L. Mandell, J. E. Bennett, and R. Dolin, eds.), Vol. 1, pp. 178–183. Churchill Livingstone, Philadelphia.

McFadden, G., and Murphy, P. M. (2000). Host-related immunomodulators encoded by poxviruses and herpesviruses. *Curr. Opin. Microbiol.* **3**:371–378.

McMinn, P. C., Dalgarno, L., and Weir, R. C. (1996). A comparison of the spread of Murray Valley encephalitis viruses of high or low neuroinvasiveness in the tissues of Swiss mice after peripheral inoculation. *Virology* **220**:414–423.

Momburg, F., Müllbacher, A., and Lobigs, M. (2001). Modulation of transporter associated with antigen processing (TAP)-mediated peptide import into the endoplasmic reticulum by flavivirus infection. *J. Virol.* **75**:5663–5671.

Monath, T. P. (1986). Pathobiology of the Flaviviruses. *In* "The Togaviridae and Flaviviridae" (S. Schlesinger and M. J. Schlesinger, eds.), pp. 375–440. Plenum Press, New York.

Monath, T. P., Cropp, C. B., and Harrison, A. K. (1983). Mode of entry of a neurotropic arbovirus into the central nervous system: Reinvestigation of an old controversy. *Lab. Invest.* **48**:399–410.

Müllbacher, A., and Blanden, R. V. (1979). H-2-linked control of cytotoxic T-cell responsiveness to alphavirus infection: Presence of H-2Dk during differentiation and stimulation converts stem cells of low responder genotype to T cells of responder phenotype. *J. Exp. Med.* **149**:786–790.

Müllbacher, A., Hill, A. B., Blanden, R. V., Cowden, W. B., King, N. J., and Hla, R. T. (1991). Alloreactive cytotoxic T cells recognize MHC class I antigen without peptide specificity. *J. Immunol.* **147**:1765–1772.

Müllbacher, A., and King, N. J. C. (1989). Target cell lysis by natural killer cells is influenced by beta 2-microglobulin expression. *Scand. J. Immunol.* **30**:21–29.

Müllbacher, A., and Lobigs, M. (1995). Up-regulation of MHC class I by flavivirus-induced peptide translocation into the endoplasmic reticulum. *Immunity* **3**:207–214.

Murphy, P. M. (2001). Viral exploitation and subversion of the immune system through chemokine mimicry. *Nature Immunol.* **2**:116–122.

Nahill, S. R., and Welsh, R. M. (1993). High frequency of cross-reactive cytotoxic T lymphocytes elicited during the virus-induced polyclonal cytotoxic T lymphocyte response. *J. Exp. Med.* **177**:317–327.

Nash, P., Barrett, J., Cao, J. X., Hota-Mitchell, S., Lalani, A. S., Everett, H., Xu, X. M., Robichaud, J., Hnatiuk, S., Ainslie, C., Seet, B. T., and McFadden, G. (1999). Immunomodulation by viruses: The myxoma virus story. *Immunol. Rev.* **168:**103–120.

Nathan, C. (1992). Nitric oxide as a secretory product of mammalian cells. *FASEB J.* **6:**3051–3064.

O'Neill, H. C., and Blanden, R. V. (1979). Quantitative differences in the expression of parentally-derived H-2 antigens in F1 hybrid mice affect T-cell responses. *J. Exp. Med.* **149:**724–731.

Pääbo, S., Nilsson, T., and Peterson, P. A. (1986a). Adenoviruses of subgenera B, C, D, and E modulate cell-surface expression of major histocompatibility complex class I antigens. *Proc. Natl. Acad. Sci. USA* **83:**9665–9669.

Pääbo, S., Weber, F., Nilsson, T., Schaffner, W., and Peterson, P. A. (1986b). Structural and functional dissection of an MHC class I antigen-binding adenovirus glycoprotein. *EMBO J.* **5:**1921–1927.

Ritchie, S. A., Phillips, D., Broom, A., Mackenzie, J., Poidinger, M., and van den Hurk, A. (1997). Isolation of Japanese encephalitis virus from *Culex annulirostris* in Australia. *Am. J. Trop. Med. Hyg.* **56:**80–84.

Sampson, B. A., Ambrosi, C., Charlot, A., Reiber, K., Veress, J. F., and Armbrustmacher, V. (2000). The pathology of human West Nile Virus infection. *Hum. Pathol.* **31:**527–531.

Shen, J., Devery, J. M., and King, N. J. C. (1995a). Adherence status regulates the primary cellular activation responses to the flavivirus West Nile. *Immunology* **84:**254–264.

Shen, J., Devery, J. M., and King, N. J. C. (1995b). Early induction of interferon-independent virus-specific ICAM-1 (CD54) expression by flavivirus in quiescent but not proliferating fibroblasts: Implications for virus-host interactions. *Virology* **208:**437–449.

Shen, J., T-To, S. S., Schrieber, L., and King, N. J. C. (1997). Early E-selectin, VCAM-1, ICAM-1, and late major histocompatibility complex antigen induction on human endothelial cells by flavivirus and comodulation of adhesion molecule expression by immune cytokines. *J. Virol.* **71:**9323–9332.

Shimonkevitz, R., Luescher, B., Cerottini, J. C., and MacDonald, H. R. (1985). Clonal analysis of cytolytic T lymphocyte-mediated lysis of target cells with inducible antigen expression: correlation between antigen density and requirement for Lyt-2/3 function. *J. Immunol.* **135:**892–899.

Shrestha, B., Chaudhri, G., Karupiah, G., Kril, J., and King, N. J. C. (2002). Interferon-γ exacerbates disease in a murine model of West Nile virus encephalitis. Submitted for publication.

Springer, T. A. (1994). Traffic signals for lymphocyte recirculation and leukocyte emigration: The multisteps paradigm. *Cell* **76:**301–314.

Stark, G. R., Kerr, I. M., Williams, B. R. G., Silverman, R. H., and Schreiber, R. D. (1998). How cells respond to interferons. *Annu. Rev. Biochem.* **67:**227–264.

Sugamata, M., and Miura, T. (1982). Japanese encephalitis virus infection in fetal mice at different stages of pregnancy. I. Stillbirth. *Acta Virol.* **26:**279–282.

Tafuri, A., Alferink, J., Moller, P., Hammerling, G. J., and Arnold, B. (1995). T cell awareness of paternal alloantigens during pregnancy. *Science* **270:**630–633.

Tough, D. F., Borrow, P., and Sprent, J. (1996). Induction of bystander T cell proliferation by viruses and type I interferon in vivo. *Science* **272:**1947–1950.

Wang, J., and Springer, T. A. (1998). Structural specializations of immunoglobulin superfamily members for adhesion to integrins and viruses. *Immunol. Rev.* **163:**197–215.

Wold, W. S. M., and Gooding, L. (1991). Region E3 of adenovirus: A cassette of genes involved in host surveillance and virus cell interactions. *Virology* **184:**1–8.

Wright, K. L., White, L. C., Kelly, A., Beck, S., Trowsdale, J., and Ting, J. P.-Y. (1995). Coordinate regulation of the human TAP-1 and LMP-2 genes from a shared bidirectional promoter. *J. Exp. Med.* **181:**1495–1471.

Yewdell, J., and Bennink, J. (1999). Mechanism of viral interference with MHC class I antigen processing and presentation. *Annu. Rev. Cell Dev. Biol.* **15:**579–606.

Zarozinski, C. C., and Welsh, R. M. (1997). Minimal bystander activation of CD8 T cells during the virus-induced polyclonal T cell response. *J. Exp. Med.* **185:**1629–1639.

Zheng, P., and Liu, Y. (1997). Costimulation by B7 modulates specificity of cytotoxic T lymphocytes: A missing link that explains some bystander T cell activation. *J. Exp. Med.* **186:**1787–1791.

MECHANISMS OF DENGUE VIRUS-INDUCED CELL DEATH

Marie-Pierre Courageot, Adeline Catteau, and Philippe Desprès

Unité Postulante des Interactions Moléculaires Flavivirus-Hôtes
Virology Department
Pasteur Institute
75724 Paris Cedex 15, France

The outcome of virus infection depends on viral and host factors. The interactions between flaviviruses and their target cells must be investigated if we are to understood the pathogenicity of these RNA viruses. Host cells are thought to respond to viral infection by initiation of apoptotic cell death. Apoptosis is an active process of cellular self-destruction with distinctive morphological and biochemical features. There is mounting evidence that dengue (DEN) virus can trigger the host cell to undergo apoptosis in a cell-dependent manner. Virally induced apoptosis contributes directly to the cytopathogenic effects of DEN virus in cultured cells. The induction of apoptosis involves the activation of intracellular signaling systems. Although the underlying molecular processes that trigger apoptosis are not well characterized, our knowledge regarding the cellular mechanisms and viral determinants of the outcome of DEN virus infection of target cells is improving. The cellular factors that regulate cell death, such as Bcl-2 family members, can modulate the outcome of DEN virus infection in

0065-3527/03 $35.00

cultured cells. Apoptosis inhibitors delay DEN virus-induced apoptosis, thereby providing a suitable environment for the virus. During DEN virus infection, cell death is also modulated by the virulence of the infecting strains. The purpose of this review is to present recent information on the cellular mechanisms and viral proteins associated with apoptosis in response to DEN virus. This knowledge may provide new insights into the viral pathogenicity.

I. Introduction

Apoptosis is an active process of cell death that is characterized by a number of distinct morphological features and biochemical processes (Fiers *et al.*, 1999; Kimura *et al.*, 2000; Majno and Joris, 1995). The apoptotic process can be initiated by extracellular agents such as hormones, cytokines, killer cells, and a variety of chemical, physical, and microbial agents. Cellular perturbations caused by viral replication can inadvertently trigger signaling pathways or specific sensors that initiate apoptosis, which protects the host from the virus. Viruses can also induce the transcription of a number of genes, including those involved in antiviral defense. These target genes then induce an "antiviral state" in host cells or make them more susceptible to apoptosis, which dramatically limits the extent of viral infection (Goodbourn *et al.*, 2000). Some viruses, in particular those with a large genome such as herpesviruses and poxviruses, have elaborated a variety of strategies to block their host's apoptotic response (Shen and Shenk, 1995; Tschopp *et al.*, 1998).

Interactions between the virus and its target cells are of particular importance for the understanding of the pathogenicity of flaviviruses. For instance, dengue (DEN), Japanese encephalitis, Langat, and West Nile viruses are capable of inducing apoptosis *in vitro* and *in vivo* (Avirutnan *et al.*, 1998; Couvelard *et al.*, 1999; Desprès *et al.*, 1996, 1998; Duarte dos Santos *et al.*, 2000; Hilgard *et al.*, 2000; Jan *et al.*, 2000; Liao *et al.*, 1997, 1998, 2001; Lin *et al.*, 2000b; Marianneau *et al.*, 1997, 1999; Parquet *et al.*, 2001; Pridhod'ko *et al.*, 2001; Su *et al.*, 2001, 2002). Virally induced apoptosis can contribute directly to the pathophysiological manifestations consecutive to flavivirus infection.

The clinical features of DEN hemorrhagic fever (DHF) include liver involvement (Rothman and Ennis, 1999). The target cells for DEN virus in the human liver are hepatocytes and Kupffer cells. Apoptotic cell death following infection should cause liver failure (An *et al.*, 1999;

Couvelard *et al.*, 1999; Feldmann, 1997; Marianneau *et al.*, 1997, 1998a–d; Rosen *et al.*, 1989). Apoptosis is the mechanism by which DEN virus infection causes neuronal cell death in the mouse central nervous system (CNS) (Desprès *et al.*, 1998). Detailed studies of the interactions between the virus and its target cells have led to the identification of viral factors and key effectors molecules that may influence the pathogenicity of DEN virus. The purpose of this review is to summarize what is presently known about the molecular mechanisms associated with apoptosis in response to DEN virus infection.

II. Apoptosis and Necrosis

Cell death can take the form of necrosis or apoptosis. Necrosis is defined as cell death caused by ischemia, pathological environments, and various cytotoxic agents that increase the permeability of the plasma membrane. Ischemic cell death can be completed within 24 h. During necrosis, cells become round and collapse, the cytoplasm swells, the mitochondria dilate, other organelles dissolve, the DNA breaks down in a nonspecific fashion, and the plasma membrane ruptures, spilling the cell contents into the surrounding medium (Fig. 1). As a result, necrosis may cause an inflammatory response.

Apoptosis plays an essential role in development, morphogenesis, tissue remodeling, and immune regulation, but it is also involved in many diseases. Apoptosis is induced following the activation of intracellular signaling pathways. A cell undergoing apoptosis generally shrinks, resulting in cytoplasm condensation, fragmentation of the cell nucleus and chromatin condensation, protein cross-linking, outer membrane leaflet inversion and exposure of phosphatidylserines, and membrane blebbing (Fig. 1). The organelles retain their integrity, and the plasma membrane does not rupture. The plasma membrane disassembles into membrane-enclosed vesicles (apoptotic bodies), which prevent the release of cellular compounds into the extracellular medium. These apoptotic bodies may contain cytoplasm, pyknotic nuclear fragments, and densely packed cell organelles. Apoptosis can be completed within 30 to 60 min. The biochemical hallmarks of apoptosis include the activation of endonucleases, the degradation of DNA into oligonucleosomal fragments, and the activation of a family of cysteine proteases called caspases (cysteinyl aspartate-specific proteases). Biochemical methods have shown that the DNA is specifically cleaved between nucleosomes, resulting in segments of approximately 185 bp that are generally recognized as the hallmark of apoptosis (Fig. 1).

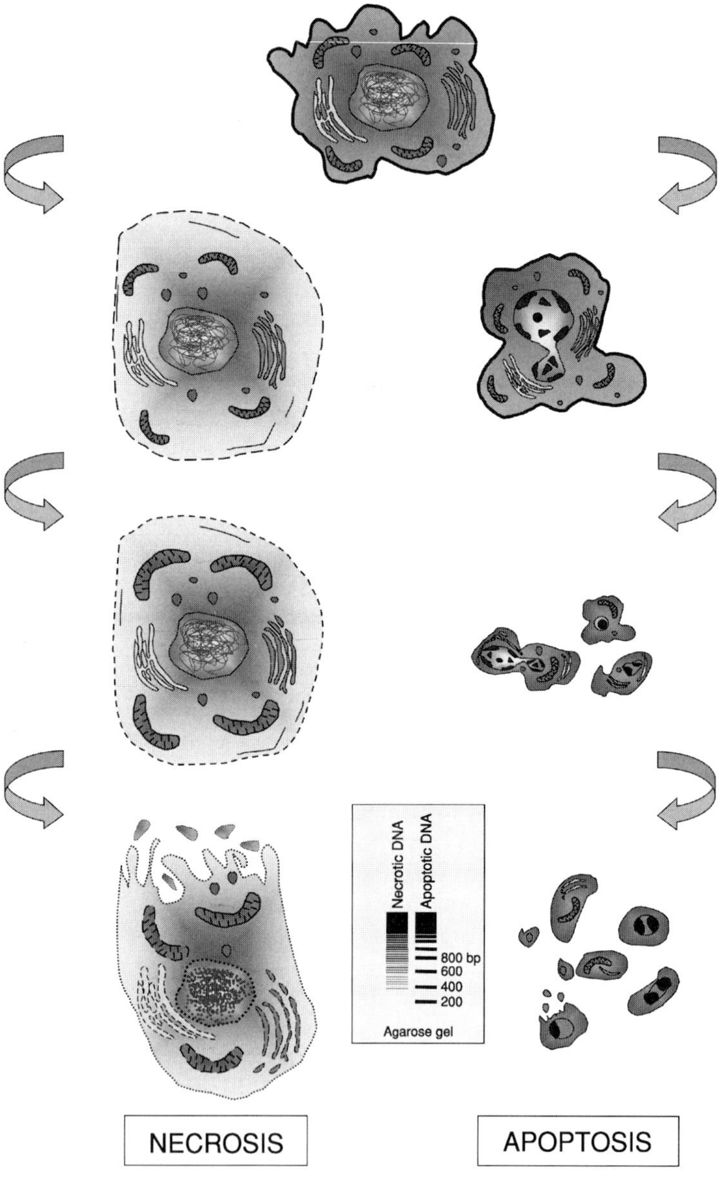

FIG 1. Schematic diagram of the morphological changes that occur during necrosis and apoptosis. A normal cell is shown at the top. (Inset) Gel showing DNA degradation products in the late stages of cell death. (See Color Insert.)

Apoptosis culminates in the engulfment of apoptotic bodies by adjacent macrophages or parenchyma cells. When misregulated, apoptosis can contribute to various diseases, including cancer and autoimmune and neurodegenerative diseases.

III. Molecular Machinery of Apoptosis

The apoptotic process can be divided into three phases: initiation (the cell receives a stimulus that may activate the apoptotic cascade), execution (the point after which death signals become irreversible), and degradation (the cell presents the biochemical and morphological hallmarks of the final stages of apoptosis) (Hengartner, 2000; Slee *et al.*, 1999).

A. Initiation Phase

The apoptotic signal is generated directly or indirectly from receptor-mediated stimuli and apoptosis-promoting inducers that provoke generalized cellular damages, such as cytotoxic drugs, radiation, heat shock, survival factor deprivation, and other cellular stresses (Fig. 2).

1. Interferon-Mediated Signaling Pathway

Type I interferons (IFNs) are produced in direct response to microbial infection and consist of the products of the IFN-α multigene family, which are predominantly synthesized by leukocytes. IFN-α induces a proapoptotic state that ensures that the target cell is triggered to undergo apoptosis (Goodbourn *et al.*, 2000) (Fig. 2a).

The interferon-induced protein kinase, PKR (protein kinase RNA dependent), a serine/threonine kinase with multiple roles in the control of transcription and translation, is associated with the endoplasmic reticulum (ER) membrane within the cell. PKR is activated by binding to double-stranded (ds) RNA, which is frequently present in the cytosol of virally infected cells, via a mechanism involving autophosphorylation (Everett and McFadden, 1999). Calcium depletion from the ER may also activate PKR (Srivastava *et al.*, 1995). When PKR is activated, it phosphorylates eIF-2α, which in turn inhibits protein synthesis (Fig. 2a). PKR also activates the interferon regulatory factor 1 and nuclear transcription factor NF-κB, which are implicated in the transcriptional regulation of dsRNA-activated genes. Phosphorylated PKR helps eliminate the pathogenic agent by mediating apoptosis (Jagus *et al.*, 1999; Lee *et al.*, 1997).

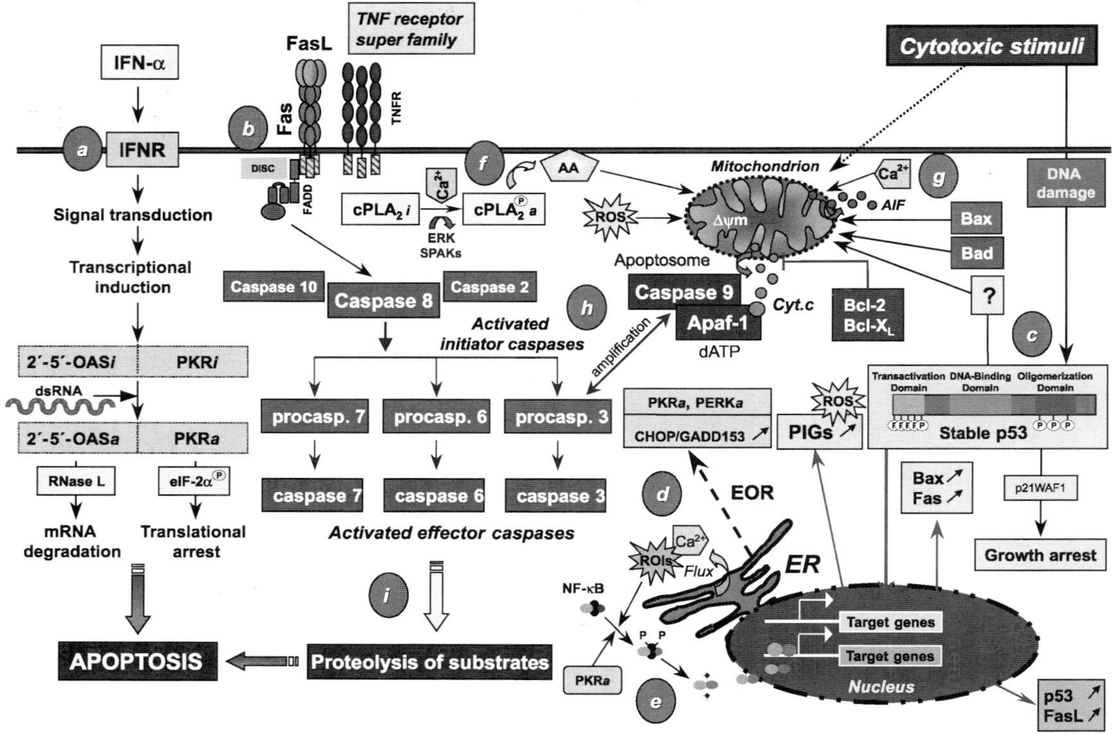

Cytotoxic stimuli
TNF receptor super family
FasL
Fas
TNFR
IFN-α
IFNR
DISC
FADD
Mitochondrion
Ca2+
AA
cPLA2 i
cPLA2 a
ERK
SPAKs
ROS
Δψm
AIF
Bax
Bad
DNA damage
Signal transduction
Transcriptional induction
Apoptosome
Caspase 10
Caspase 8
Caspase 2
Caspase 9
Apaf-1
dATP
Cyt.c
Bcl-2
Bcl-XL
?
Activated initiator caspases
amplification
2´-5´-OASi
PKRi
dsRNA
2´-5´-OASa
PKRa
procasp. 7
procasp. 6
procasp. 3
caspase 7
caspase 6
caspase 3
Activated effector caspases
RNase L
eIF-2α
mRNA degradation
Translational arrest
PKRa, PERKa
CHOP/GADD153
PIGs
Transactivation Domain
DNA-Binding Domain
Oligomerization Domain
Stable p53
Bax
Fas
p21WAF1
Growth arrest
EOR
ROIs
Flux
ER
NF-κB
Target genes
Target genes
Nucleus
p53
FasL
APOPTOSIS
Proteolysis of substrates
PKRa
a
b
c
d
e
f
g
h
i

IFN-induced 2′, 5′-oligoadenylate synthetases also play a major role in apoptosis by activating RNase L (Castelli *et al.*, 1998; Rusch *et al.*, 2000) (Fig. 2a). Functional RNase L may cause dysfunctional changes in the mitochondria, resulting in an apoptotic cascade.

2. Death Receptor-Dependent Signaling Pathway

Death receptors belong to the tumor necrosis factor (TNF) receptor (TNFR) gene superfamily (Fig. 2b). The most well-characterized death receptors are Fas and TNFR-1. Fas L (CD95L) binds to Fas (CD95), and TNF and lymphotoxin α bind to TNFR-1. The ligands mediate their apoptotic effects in an autocrine or paracrine manner via their respective cell surface membrane death receptors (Ashkenazi and Dixit, 1998; Krammer, 2000). The ligation of the homotrimeric molecule, FasL, to its receptor leads to the clustering of Fas and to the formation of a death-inducing signaling complex (DISC). This complex consists of the cytoplasmic region of Fas, the intracellular adaptor molecule, FADD (Fas-associating protein with death domain), and procaspase-8 (Fig. 2b). Upon the recruitment of FADD, caspase-8 is activated through self-cleavage, leading to the initiation of the caspase cascade, which is critical for transduction of the apoptotic signal (Ashkenazi and Dixit, 1998; Krammer, 2000).

3. p53-Dependent Signaling Pathway

Transcriptional factor p53, normally a very short-lived protein, can induce growth arrest and apoptosis in response to intracellular disruptions resulting from DNA damage, metabolite deprivation, heat shock, hypoxia, and activated cellular or viral oncoproteins (May and May, 1999; Sheikh and Fornace, 2000) (Fig. 2c).

p53 is phosphorylated by several protein kinases, including DNA-PK, a kinase that is induced by DNA damage. It also interacts with p300, a histone acetyltransferase that seems to acetylate p53 and to enhance transactivation activity. The p53 protein contains an

---

FIG 2. Intracellular pathways involved in apoptosis. These pathways are mediated through the interferon system (a), the death receptor-dependent system (b), p53 (c), endoplasmic reticulum stress (d), nuclear transcription factor NF-κB (e), cytosolic phospholipase A_2 (f), death-inducing factors converging onto mitochondria (g), and caspase (h). *i*, inactive; *a*, active; AA, arachidonic acid; AIF, apoptosis-inducing factor; *Cyt.c*; cytochrome *c*; IFN, interferon; EOR, endoplasmic reticulum overload; 2′-5′OAS, 2′,5′-oligoadenylate synthetase; $cPLA_2$, cytosolic phospholipase A_2; PIGs, p53-induced genes; PKR, protein kinase RNA dependent; RNase L, ribonuclease L; ROIs, radical oxygen intermediates; ROS, reactive oxygen species; dsRNA, double-stranded RNA. (See Color Insert.)

amino-terminal transactivation domain, a central DNA-binding domain, and a carboxyl-terminal oligomerization domain (Fig. 2c). p53 mediates its death-inducing activity by modulating the expression of its downstream target genes. The central domain facilitates sequence-specific DNA binding to p53 response elements within the regulatory regions of a number of p53-induced genes (PIGs). Tumor suppressor p53 causes apoptosis when it is activated and stabilized (Gottlieb and Oren, 1998). The polyproline region located between the transactivation and the DNA-binding domains of p53 may be necessary to induce apoptosis (Sakamuro *et al.*, 1997).

Multiple mechanisms are believed to be involved in p53-mediated apoptosis (McKenzie *et al.*, 1999) (Fig. 2c). PIGs may function in concert to mediate some of the p53-induced apoptotic effects. These PIGs include genes with proven roles in apoptotic pathways, such as members of the Bcl-2 family, Fas/FasL, KILLER/DR5, GADD45 (the growth arrest and DNA damage-inducible gene number 45), and genes that are involved in the control of redox states (RCGs). The expression of GADD45 is induced by a variety of genotoxic and nongenotoxic stresses. GADD45 then triggers apoptosis by activating the p38 mitogen-activated protein (p38MAP) kinase.

The activation of RCGs results in the production of reactive oxygen species (ROS), which, in turn, causes a collapse of mitochondrial inner transmembrane potential ($\Delta\Psi m$), leading to the rapid release of apoptotic cascade activators (Li *et al.*, 1999). However, p53 also appears to induce apoptosis via mechanisms that do not require its transcriptional activity. For example, p53 can contribute to apoptosis by direct signaling at the mitochondria, thereby amplifying the transcription-dependent apoptosis of p53 (Marchenko *et al.*, 2000).

4. ER Stress-Dependent Signaling Pathway

The ER, a unique oxidizing compartment for the folding of membrane and secretory proteins, is sensitive to environmental changes. The retention of overexpressed membrane proteins by the ER induces a signal pathway termed the ER overload response (EOR) (Kaufman, 1999; Pahl and Baeuerle, 1997; Pahl *et al.*, 1996) (Fig. 2d). It is thought that the EOR increases the expression of novel genes, the products of which might restore proper ER function. Transcriptional induction in response to the EOR may require NF-κB activation (Pahl and Baeuerle, 1995). Calcium and reactive oxygen intermediates (ROIs) might act as second messengers in response to the EOR.

Presumably under conditions of severe ER stress to which the cell cannot adapt, signals are transduced from the ER to the cytoplasm and the nucleus, leading to apoptosis induction. The molecular signaling mechanisms that link ER stress to apoptosis remain largely unknown. Caspase-12 activity is thought to be essential for ER stress-induced apoptosis (Yoneda *et al.*, 2001). Studies suggest that two death-inducing signals are generated in response to perturbations that induce ER stress. The first is the transcriptional induction and p38MAP kinase-mediated activation of transcription factor CHOP/GADD 153 (GEBP homologous protein growth arrest and DNA damage-inducible gene number 153), which regulates cell growth negatively and may cause death in response to ER stress (Zinszner *et al.*, 1998). The second death-inducing signal is activation of the ER stress signaling kinases [PKR and PKR-ER- related kinase (PERK)], which may mediate eIF-2α phosphorylation (Kaufman, 1999). PERK is an eIF-2α kinase located in the ER. It is selectively activated to prevent further mRNA translation. It is largely unknown how ER stress-mediated eIF-2α phosphorylation and the subsequent inhibition of translation initiation activate the apoptotic pathway.

5. *NF-κB-Mediated Apoptotic Pathway*

The transcription factor Rel/NF-κB is a member of the ubiquitous family of Rel-related transcription factors that serve as critical regulators of the inducible expression of many genes (Baeuerle, 2000) (Fig. 2e). Rel/NF-κB either exists in the form of a homodimer or forms a heterodimer with members of a family of structurally related proteins (Mercurio and Manning, 1999). Five NF-κB family proteins have been identified in mammalian cells: RelA (p65), c-Rel, RelB, NF-κB1 (p50/p105), and NF-κB2 (p52/p100). The classic NF-κB dimer contains RelA and NF-κB1. NF-κB exists in the cytoplasm in an inactive form associated with inhibitory proteins termed IκBs, the most important of which are probably IκBα, IκBβ, and IκBϵ. Rel/NF-κB is activated through the signal-induced proteolytic degradation of IκB by components of the ubiquitin-proteasome system. The freed Rel/NF-κB complex can then enter the nucleus, where it binds to a set of related DNA target sites, collectively called the κB sites. These κB sites regulate gene expression directly. A large variety of stimuli can activate Rel/NF-κB. Certain cell types respond to oxidative stress by upregulating Rel/NF-κB activity (Li and Karin, 1999; Mercurio and Manning, 1999; Okamoto *et al.*, 2000).

The transcriptional regulation of target genes such as FasL, some members of the Bcl-2 family, and tumor suppressor p53 is believed to be the mechanism by which Rel/NF-κB promotes apoptosis (Barkett and Gimore, 1999) (Fig. 2e). Functional Rel/NF-κB could also play a pivotal role in apoptosis by making the cell more sensitive to apoptosis-inducing stimuli.

6. *Cytosolic Phospholipase A_2-Dependent Signaling Pathway*

Phospholipases generate important secondary messengers in several cellular processes, including apoptotic cell death. Cytosolic phospholipase A_2 ($cPLA_2$) has been implicated in the receptor-mediated release of arachidonic acid (AA) from membrane phospholipids, playing a key role in many signal transduction reactions (Fig. 2d). AA can activate membrane-associated NADPH oxidase to generate ROS (Cui and Douglas, 1997). AA may also cause dysfunctional changes in the mitochondrial membrane potential, resulting in activation of the apoptotic cascade.

$cPLA_2$ is an important mediator of ischemic, oxidative injuries, and apoptosis in the brain (Hornfelt *et al.*, 1999; Sapirstein and Bonventre, 2000). It is activated by increasing the concentration of intracellular calcium and by phosphorylation. The activation of extracellular signal-regulated kinase (ERK) or stress-activated protein kinases (SPAKs) primes $cPLA_2$ activation (Buschbeck *et al.*, 1999; Lin *et al.*, 1993; Scorrano *et al.*, 2001).

7. *Oxidative Stress-Mediated Signaling Pathway*

Oxidants are capable of causing oxidative damage to macromolecules. This leads to lipid peroxidation, the oxidation of amino acid side chains, the formation of protein–protein cross-links, the oxidation of polypeptide backbones resulting in protein fragmentation, DNA damage, and DNA strand breaks. It is thought that oxidants promote the cross-linking or aggregation of activating signaling components. Changes in the intracellular redox status elicited by the generation of oxidants can trigger selective gene expression, enzyme activation, and DNA synthesis, and can alter the progression of the cell cycle (Baeuerle, 2000; Giron-Calle and Forman, 2000). Markers of oxidative damage include the transcription factors p53 and NF-κB (Baeuerle, 2000; Fiers *et al.*, 1999; Li *et al.*, 1999; Okamoto *et al.*, 2000; Sen, 2000) (Fig. 2e).

A state of moderately increased levels of intracellular oxidants is referred to as oxidative stress. One mechanism by which oxidative stress may cause apoptosis is by increasing the cytosolic concentration of Ca^{2+}. Cells respond to increases in the internal Ca^{2+} concentration by activating a variety of Ca^{2+}-binding proteins, including phospholipases C, A_2, and D, and by activating Ca^{2+}-dependent forms of protein kinase C (Giron-Calle and Forman, 2000). These proteins are modulators of a multitude of cellular reactions, cytoskeletal dynamics, and signal transduction pathways (Hofmann, 1999).

B. Execution Phase

1. Mitochondria

During apoptosis, mitochondria play an essential role in transducing stimuli (Desagher and Martinou, 2000; Green and Kroemer, 1998) (Fig. 2g). Numerous proapoptotic stimuli provoke changes in the permeability of the mitochondrial outer membrane. As a result, the $\Delta\Psi m$ collapses. An irreversible change in mitochondrial permeability may be part of the triggering mechanism for apoptosis.

Mitochondria in cells are committed to apoptosis by the release of apoptogenic molecules, such as apoptosis-inducing factor (AIF) and cytochrome *c*, which are normally confined to the mitochondrial intermembrane space. Cytosolic cytochrome *c* regulates the activities of Apaf-1, a molecule that can promote the clustering of caspase-9 by interacting with CARD-CARD (caspase activation and recruitment domain) in the presence of ATP/dATP (Kumar, 1999) (Fig. 2g). In apoptosomes, Apaf-1 activates caspase-9, driving the next phase of the caspase cascade.

2. Caspases

Cysteine proteases of the caspase family are the central executioners of apoptosis (Green and Kroemer, 1998; Los *et al.*, 2001; Slee *et al.*, 1999) (Fig. 2h). Most caspases are constitutively expressed as inactive proforms (zymogens) (Kumar, 1999). Upon receiving an apoptotic signal, these proforms undergo proteolytic processing to generate the active enzyme, a heterotetramer composed of two heterodimers derived from two precursor molecules. Caspases are generally activated by other caspases. Caspases (the mammalian caspase family contains 14 members) can be described either as upstream/signaling/initiator caspases or as downstream/effector caspases. Upstream caspases include caspases-2, -8, and -10. Downstream caspases (caspases-3, -6, and -7) are thought to lead to apoptosis by cleaving several vital proteins (Fig. 2i). IFN induces caspases-1, -3, and -8, and caspase-12 is essential for ER stress-induced apoptosis.

C. Degradation Phase

The caspase activation events driven by caspase-9 appear to be the simultaneous activation of caspases-3 and -7 (Fig. 2h). Caspase-3 can drive the activation of caspases-2 and -6, followed by the activation of caspases-8 and -10. This caspase cascade may serve as an amplification step that activates the full complement of caspases that are required to dismantle the cell in the appropriate manner.

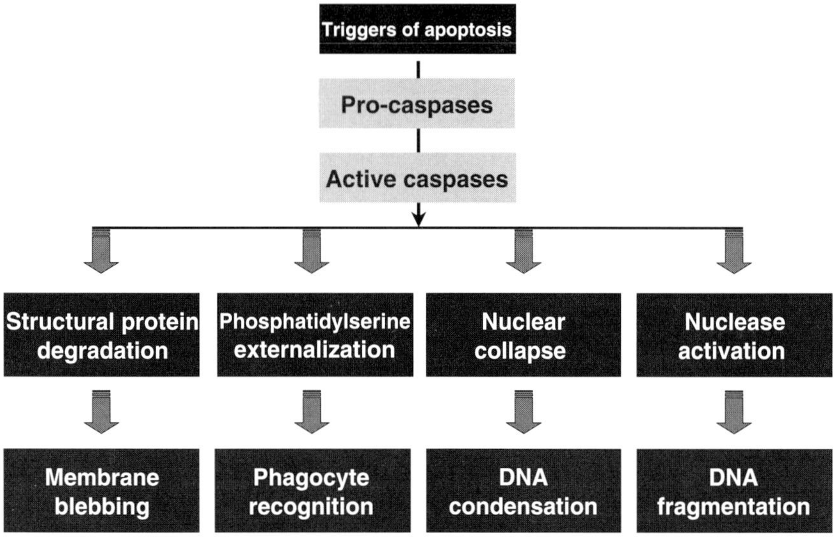

FIG 3. Schematic representation showing how caspases execute the events that lead to the apoptotic death of a cell. (See Color Insert.)

Active caspases have the ability to cleave their key intracellular substrates following an aspartate residue, resulting in the morphological and biochemical changes associated with apoptosis (Degen *et al.*, 2000; Nicholson, 1999). A fraction of the cellular proteome is cleaved by caspases (Fig. 3). Caspases contribute to apoptosis by inactivating or deregulating proteins involved in DNA repair [poly (ADP ribose) polymerase and DNA-PK], mRNA splicing (U1-70K), and DNA replication (140-kDa subunit of replication factor C) (Nicholson, 1999). Caspase-3 activates the DNase that degrades chromosomal DNA [caspase-activated DNase (CAD)] by cleaving and releasing its chaperone-like inhibitor (ICAD), leading to entry of the DNase into the nucleus and cleavage of DNA between nucleosomes. The Acinus protein is also activated by caspase-3 and induces chromatin condensation. Caspases play a key proteolytic role in reorganizing cell structure indirectly by cleaving several proteins involved in cytoskeleton regulation, including gelsolin, spectrin, focal adhesion kinase, and p21-activated kinase 2.

D. Bcl-2 Family Members

The Bcl-2 family of cytoplasmic proteins plays a central role in the regulation of apoptosis (Adams and Cory, 2001). Its interacting

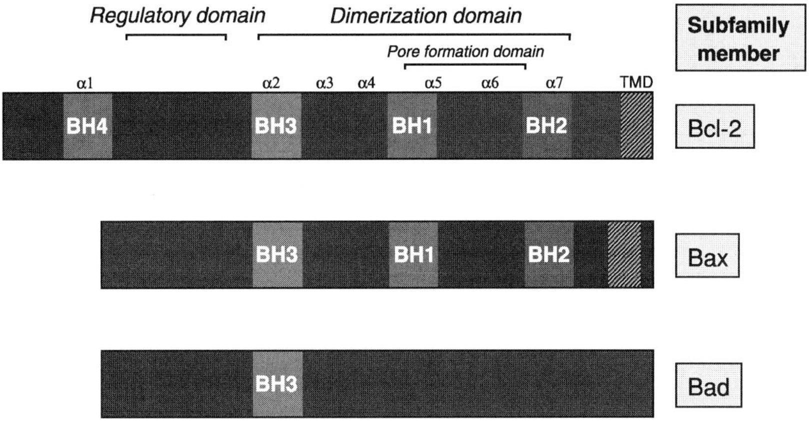

FIG 4. Three members of the Bcl-2 family. Sizes of the domains are approximate and the figure is not drawn to scale. α, α helix; BH, Bcl-2 homology; TMD, transmembrane domain. (See Color Insert.)

proapoptotic and antiapoptotic members integrate with diverse upstream survival and distress signals to determine whether cell death occurs.

Fifteen members of the Bcl-2 family have been identified in mammals so far. All contain at least one of the four conserved regions called Bcl-2 homology domains (BH1–BH4) (Fig. 4). These motifs are formed by α helices and enable the different members of the family to form either homodimers or heterodimers and to regulate each other. *bcl-2*-related proteins have either antiapoptotic or proapoptotic functions. *bcl-2* and several close relatives (such as *bcl-*X_L) inhibit apoptosis, whereas structurally similar relatives (such as *bax*) lead to cell death.

bcl-2 and *bax* share three of the four conserved BH sequence motifs and assume a similar conformation. *bcl-*X_S, which lacks the BH1 and BH2 domains of *bcl-*X_L, functions as a proapoptotic regulator in some systems. Thus, both long and short forms of *bcl-X* proteins differently modulate the apoptotic pathways in response to death stimuli. BH3-only proteins, such as *bad*, seem to be sentinels for cellular damage and critical triggers of apoptosis (Figs. 2g and 4).

p53 and NF-κB may regulate the expression of Bcl-2 genes. The *bax* gene is a target of p53 and is activated transcriptionally via the p53 response elements located within the *bax* gene promoter. *bcl-*X_L gene expression is also upregulated by genotoxic stress in a p53-dependent manner.

It is still not known how Bcl-2 family members function biochemically. In normal cells, *bcl-2* is exclusively membrane associated, and *bcl-*X_L and *bcl-*X_S are present in both soluble and membrane-bound forms. The principal mechanism by which Bcl-2 family members regulate apoptosis is probably by controlling the release of cytochrome *c* and the $\Delta\Psi$m loss due to mitochondrial alterations (Pinton *et al.*, 2001). The antiapoptotic mediators, *bcl-2* and *bcl-*X_L, are produced in many cell types subjected to various cytotoxic treatments. This prevents the release of cytochrome *c* from mitochondria, the activation of caspases, and cell death (Fig. 2g). *bcl-*X_L also appears to be able to exert its protective function by sequestering AIF. Conversely, *bcl-*X_S antagonizes the protective effect of *bcl-*X_L. The death promoter *bax* exists predominantly in the cytosol. The oligomerization of *bax* is required to induce its mitochondrial translocation (Fig. 2g). The oligomeric forms of *bax* might damage mitochondria by pore formation, which triggers the release of apoptotic factors.

IV. Virus-Induced Apoptosis

Viruses must be able to replicate despite defense mechanisms if they are to establish infection in cells. Growing evidence shows that apoptosis can play a key role in the innate response to viral infection (Barber, 2001; Hardwick, 1998; Shen and Shenk, 1995). Cellular apoptosis following viral infection may be a defense mechanism that rapidly eliminates cells in which virus is replicating, thus preventing the release of viral progeny. However, apoptosis can be harmful to the host when nonrenewable cells are infected.

Many viruses have been found to induce the apoptosis of infected cells (Courageot and Desprès, 2000a; O'Brien, 1998; Roulston *et al.*, 1999). The presence of progeny virions in apoptotic bodies may promote their own dissemination and also protect them from contact with antivirus antibodies. Viruses have evolved mechanisms that delay cell death or block apoptosis, thereby facilitating persistent infections (O'Brien, 1998). A block in premature death may enable the virus progeny to spread to neighboring cells while evading host inflammatory responses and protecting the progeny virus from host enzymes (Everett and McFadden, 1999; Griffin and Hardwick, 1997; Teodoro and Branton, 1997; Tschopp *et al.*, 1998). Virally induced apoptosis may act as an initiating stimulus in autoimmunity. Immunization with apoptotic cell fragments concentrating viral and self-antigens might result in the autoimmune targeting of specific self-molecules (Rosen *et al.*, 1995).

V. Apoptotic Cell Death in Response to Dengue Virus Infection

Apoptosis is involved in cell death in response to DEN virus infection in human liver cells, neuronal cells, and endothelial cells *in vitro* and *in vivo* (Table I). Thus, apoptotic cell death may be considered an important feature of DEN pathogenesis.

TABLE I
Cell Types Known to Undergo DEN Virus-Induced Apoptosis

Cell type	Virus	Reference
Liver		
Human hepatocytes	DEN-2	Couvelard *et al.* (1999)
Human Kupffer cells	DEN-1	Marianneau *et al.* (1999)
Hepatoma cell lines		
HepG2	DEN-1	Marianneau *et al.* (1997); Duarte dos Santos *et al.* (2000)
HA22T	DEN-2	Lin *et al.* (2000b)
HuH7	DEN-2	Hilgard *et al.* (2000); Lin *et al.* (2000b)
PLC/PRF5	DEN-2	Lin *et al.* (2000b)
Hep3B	DEN-2	Lin *et al.* (2000b)
Brain		
Mouse neurons	DEN-1	Desprès *et al.*, (1998)
Neuroblastoma cell lines		
Mouse cells	DEN-1	Desprès *et al.* (1996);
Neuro 2a		Duarte dos Santos *et al.* (2000)
N18	DEN-1, 2, 3	Su *et al.* (2001)
Human cells		
SK-N-SH	DEN-2	Jan *et al.* (2000)
NT-2	DEN-2	Jan *et al.* (2000)
Vascular endothelium		
Human endothelial cell line		
ECV 304	DEN-2	Avirutnan *et al.* (1998)
Epithelium		
Hamster kidney cell line		
BHK-21	DEN-1, 2, 3	Jan *et al.* (2000); Su *et al.* (2001)

A. Dengue Virus-Induced Apoptosis in Vivo

The clinical features of severe DEN disease include hemorrhagic diathesis, liver involvement, and encephalopathy. Elevated serum transaminase concentrations in DEN patients indicate the possible involvement of DEN virus infection on liver function. The liver synthesizes most of the hemostatic factors. In DHF, liver involvement is a characteristic sign that the disease may be fatal. Hepatic injury is characterized by centrolobular and midzonal necrosis with hyperplasia of Kupffer cells, the macrophage-like cells residing in the liver. As DEN virus antigens have been detected in hepatocytes and the virus has been recovered from liver specimens at autopsy, the infection of liver cells is one of the pathognomonic features of DEN. The most characteristic sign of DEN-mediated liver injury is the presence of acidophilic or Councilman bodies, which are thought to be apoptotic bodies. *In situ* TUNEL (terminal deoxynucleotidyl transferase-mediated dUTP nick end labeling) was used to detect apoptotic DNA degradation in liver tissue sections from a patient who was diagnosed with DEN at autopsy (Couvelard *et al.*, 1999). Human liver cells undergoing apoptosis were either infected with DEN virus or located close to these cells.

It was questioned whether virally induced apoptosis contributes directly to the pathogenicity of DEN virus infection *in vivo*. The mouse CNS was used as a model system to study the involvement of apoptosis in the pathophysiological response to DEN virus infection. Several lines of evidence suggested that the principal target cells for DEN virus in the mouse CNS are neurons and that their destruction is responsible for mouse death. Thus, DEN virus-induced neuronal death in the CNS might be detrimental to the host.

Newborn mice are poorly sensitive to wild-type strains of DEN virus following intracerebral inoculation (i.c.). Neuroadaptation is critical to the emergence of DEN virus variants with increased neurovirulence in newborn mice. Serial passaging of DEN-1 virus strain FGA/89 (isolated in French Guiana from patient with DF in 1989) in mouse brain and a mosquito cell line was used to select a highly neurovirulent variant (FGA/NA d1d), which replicates more efficiently in the CNS (Desprès *et al.*, 1993, 1998).

Two-day-old Swiss mice inoculated (i.c.) with a lethal dose of neurovirulent variant FGA/NA d1d developed clinically apparent encephalitis 10 days after infection and did not survive for more than 18 days. By day 9 of infection, *in situ* hybridization and an indirect fluorescence-antibody assay on infected brain sections showed that cortical neurons and pyramidal neurons in the hippocampus are the major target cells of the DEN virus in the CNS (Desprès *et al.*, 1998). Brain tissue sections were assayed for apoptosis by the TUNEL method, and TUNEL-stained cells were essentially detected in the cortex and hippocampus. A correlation existed between the distribution of apoptotic cells and the distribution of cells in which the DEN virus was replicating, suggesting that neurons undergoing apoptosis were either infected with the DEN-1 virus or located close to infected cells. Lack of a significant early inflammatory response was characteristic of the apoptotic process in infected tissues. These data suggest that DEN virus infection in the mouse CNS is associated with neuronal damage characterized by apoptotic DNA degradation, which occurs before the mice show any clinical signs of encephalitis (Desprès *et al.*, 1998). This may account for the tissue damage resulting from DEN virus infection in the CNS. However, the neurovirulence of the mouse-passaged FGA/NA d1d variant was essentially restricted to newborn mice. It is not known whether the mature nervous system inhibits apoptosis to protect adult mice against fatal DEN virus infection (Johnston *et al.*, 2001).

B. *Dengue Virus-Induced Apoptosis* in Vitro

1. *Neuronal Cells*

a. Mouse Neuroblastoma Cells Virus-induced neuronal death in the mouse CNS has been implicated as a mechanism for cytopathology in response to DEN-1 virus infection (Desprès *et al.*, 1998). DEN-1 virus induces apoptosis in primary cell cultures derived from mouse brain (Courageot and Desprès, 2000). The replication of DEN virus in the murine neuroblastoma cell lines Neuro 2a and N18 was analyzed to determine the susceptibility of mouse neurons to DEN virus infection (Desprès *et al.*, 1996; Liao *et al.*, 2001; Su *et al.*, 2001).

Human isolates of DEN-1 can induce extensive apoptosis in Neuro 2a cells due to the accumulation of viral proteins within the cell (Desprès *et al.*, 1996) (Fig. 5). The proportion of infected cells in an apoptotic state increases in a virus dose-dependent manner, suggesting that apoptosis is a direct result of virus infection. The intracellular synthesis of viral proteins seems to be essential to activate apoptosis in Neuro 2a cells. Cell growth arrest in the G1 phase has been shown to be associated with the occurrence of apoptosis in N18 cells infected with DEN-2 virus (Su *et al.*, 2001). The causal relationship between G1 growth arrest and apoptosis during DEN virus infection remains elusive.

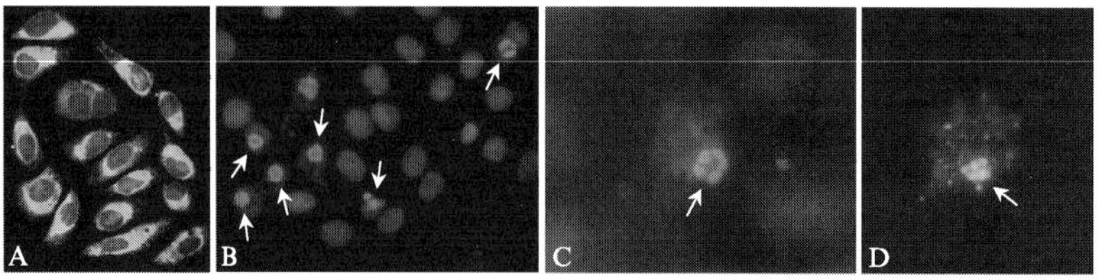

FIG 5. Apoptotic cell death in DEN virus-infected Neuro 2a cells as observed by fluorescence microscopy. Paraformaldehyde-fixed cells were permeabilized with Triton X-100 and assayed for the presence of DEN antigens (A, C) at 20 h postinfection and for apoptosis (B to D) after 30 h. Intracellular DEN antigens were detected by immunofluorescence assay (A and C). The condensation of chromatin into several dense masses was visualized with the DNA intercalator propidium iodide (B and C). Apoptotic DNA degradation was demonstrated in infected cells by the TUNEL method (D). Magnification: ×200. (See Color Insert.)

It has been proposed that apoptosis is triggered by interaction of the DEN envelope glycoproteins with the ER membranes rather than by the release of the virus (Desprès *et al.*, 1996). Overloading of viral proteins in the ER membranes may disrupt the ER membrane and thus cause the release of calcium and the generation of ROI (Fig. 2d). Interestingly, DEN-1 virus-infected Neuro 2a cells show enhanced apoptosis in response to the NF-κB inducer thapsigargin, which inhibits the ER-localized Ca^{2+}-dependent ATPase that promotes calcium uptake by the ER (Desprès *et al.*, 1996) (Fig. 2e). However, antioxidants with different inhibition mechanisms do not block DEN virus-induced apoptosis (Desprès *et al.*, 1996; Lin *et al.*, 2000b). Nonsteroidal anti-inflammatory drugs, salicylates, inhibit DEN-2 virus replication and virus-induced apoptosis in cultured cells (Liao *et al.*, 2001). This appears to be due to the activation of p38 MAP kinase rather than to the suppression of NF-κB activation. To date, the DEN virus has not been found to activate NF-κB to trigger apoptosis in mouse neuroblastoma cell lines (Desprès *et al.*, 1996; Liao *et al.*, 2001). The induction of transcription factor *CHOP/GADD 153* can promote apoptosis in response to ER stress (Fig. 2d). Future experiments should determine whether the DEN virus triggers cells to undergo apoptosis through *CHOP/GADD 153* (Su *et al.*, 2002).

The enforced expression of the antiapoptotic mediator *bcl-2* does not prevent DEN virus-induced apoptosis in N18 and Neuro 2a cells, but does prevent cell death in fibroblast-like cells (Kalinina and Desprès, unpublished results; Marianneau *et al.*, 1998a; Su *et al.*, 2001) (Fig. 2g). DEN-virus infected N18 cells express elevated amounts of *bax*, which makes the antiapoptotic function of *bcl-2* less potent in protecting against apoptotic cell death. In contrast, the enforced expression of the antiapoptotic mediator *bcl-*X_L, which is expressed predominantly in the nervous system, delays apoptosis in DEN virus-infected N18 cells without restricting virus replication. In addition, the proapoptotic regulator *bcl-*X_S potentiates DEN-2 virus-induced apoptosis (Su *et al.*, 2001). Thus, DEN virus-induced apoptosis in mouse neuroblastoma cells is influenced by the long and short forms of *bcl-X* proteins.

b. Human Neuroblastoma Cells Human isolates of DEN-2 virus can cause apoptosis in human neuroblastoma cell lines SK-N-SH and NT-2 (Jan *et al.*, 2000; Su *et al.*, 2001). Caspase-3 activation, mitochondrial release of cytochrome *c*, and generation of intracellular superoxide anion are essential to trigger infected SK-N-SH cells to undergo apoptosis (Jan *et al.*, 2000). The infection of SK-N-SH cells with DEN-2 virus activates $cPLA_2$ to generate AA (Jan *et al.*, 2000). The apoptotic effect can be derived from the action of AA in infected cells (Fig. 2f). The mechanism by which DEN-2 virus infection mediates the activation of $cPLA_2$ is not understood. The transcriptional activity of NF-κB is also involved in DEN-2 virus-induced apoptosis of human neuroblastoma cells. Cellular concentrations of p53, one of the target genes for NF-κB, are increased during the early stages of DEN virus infection (Jan *et al.*, 2000). Genes mediated by p53 generally appear to be necessary for the induction of apoptosis (Fig. 2c). However, transcriptional activation by p53 may be dispensable for the induction of apoptosis. It remains to be determined whether the death-promoting activity of p53 is a critical event in DEN virus-induced apoptosis in SK-N-SH cells.

2. Liver Cells

The presence of viral antigens within human hepatocytes from patients with dengue fever suggests that these cells are susceptible to infection by the DEN virus. Human hepatoma cell lines, HepG2, HuH7, PLC/PRF5, and Hep3B, can support the productive replication of the DEN virus (Duarte dos Santos *et al.*, 2000; Hilgard *et al.*, 2000; Lin *et al.*, 2000b; Marianneau *et al.*, 1996, 1997). The infection of liver cells with the DEN virus induces oxidative stress (Lin *et al.*, 2000a) and apoptosis (Duarte dos Santos *et al.*, 2000; Hilgard *et al.*, 2000; Lin *et al.*, 2000b; Marianneau *et al.*, 1996, 1997). The transcriptional function of NF-κB may play a very important role in DEN-1 virus-induced apoptosis in HepG2 cells because double-stranded oligodeoxynucleotide decoys containing the NF-κB-binding sequence have very efficient protective effects (Marianneau *et al.*, 1997). NF-κB is activated when DEN proteins begin to accumulate in HepG2 cells. The analysis of DEN-1 protein processing in infected HepG2 cells showed that most viral envelope glycoproteins are degraded or aggregate, suggesting that viral morphogenesis is an inefficient process in this cell type (Duarte dos Santos *et al.*, 2000; Marianneau *et al.*, 1997, 1999). The overloading of DEN viral aggregates in ER membranes might account for the induction of apoptosis in human hepatoma cells. It should be of interest to determine whether an unproductive pathway of DEN proteins in the ER causes oxidative damage that activates NF-κB. An endocytotic trafficking mutant (Trfl) isolated from HuH-7 cells was found to be resistant to DEN-1 virus–induced cell death (Hilgard *et al.*, 2000; Xiaoying *et al.*, 2001). Protection was linked to a low level of casein kinase (CK) 2 expression in Trfl cells. The C-terminal regulatory domain of p53 is phosphorylated by CK2 (Fiscella *et al.*, 1994). It is not yet known whether the CK2-mediated phosphorylation of p53 is required for DEN virus-induced apoptosis.

The DEN virus can enter human Kupffer cells *ex vivo*, but the resulting infection is inefficient and no viral progeny has been observed (Marianneau *et al.*, 1999). However, the DEN virus triggers Kupffer cells to undergo apoptosis. As healthy Kupffer cells ingest infected apoptotic bodies, these macrophage-like cells may exert antiviral activities by eliminating apoptotic liver cells (Marianneau *et al.*, 1999). The phagocytic activity of Kupffer cells in the human liver may protect against DEN virus infection by eliminating any infected hepatocytes that undergo apoptosis. However, the mechanisms of host defense involving phagocytosis associated with the production of cytokines may cause local tissue injury or a transient imbalance in homeostasis, leading to more deleterious events (Marianneau *et al.*, 1999). The inefficient clearance or impaired phagocytosis of apoptotic cells might also play a role in the development of autoimmunity.

3. *Endothelial Cells*

Short-lived plasma leakage is a hallmark of severe DEN disease. It is thought that a breakdown of the endothelial barrier might contribute to local vascular leakage. Endothelial cells are permissive to DEN virus infection *in vitro* (Avirutnan *et al.*, 1998). Interestingly, the DEN-2 virus can trigger human umbilical cord vein endothelial (ECV304) cells to undergo apoptosis after 2 or 3 days of infection. Whether vascular leakage could derive from apoptosis in the capillary endothelial cells containing replicating DEN virus particles is an interesting issue that remains to be investigated.

During the early stages of infection, NF-κB is activated in infected endothelial cells. Further experiments should determine whether the DEN virus triggers endothelial cells to undergo apoptosis through the transcriptional activity of NF-κB.

C. Viral Determinants That Influence Dengue Virus-Induced Apoptosis

Little is known about how replication of the DEN virus triggers apoptosis in infected cells. The low-passaged DEN-1 virus isolate FGA/89 and its mouse neurovirulent variant FGA/NA d1d differ in their capacity to induce apoptosis in mouse neuroblastoma cells (Desprès *et al.*, 1998). The genetic determinants that affect neurovirulence of the DEN-1 virus have been determined. FGA/NA d1d has three amino acid substitutions in the envelope E protein (E-196, E-365, and E-405) (Fig. 6). The conservative $M_{196}V$ and $V_{365}I$ substitutions map to the interfaces between the structural domains of the E proteins. The nonconservative $T_{405}I$ substitution maps to a predicted

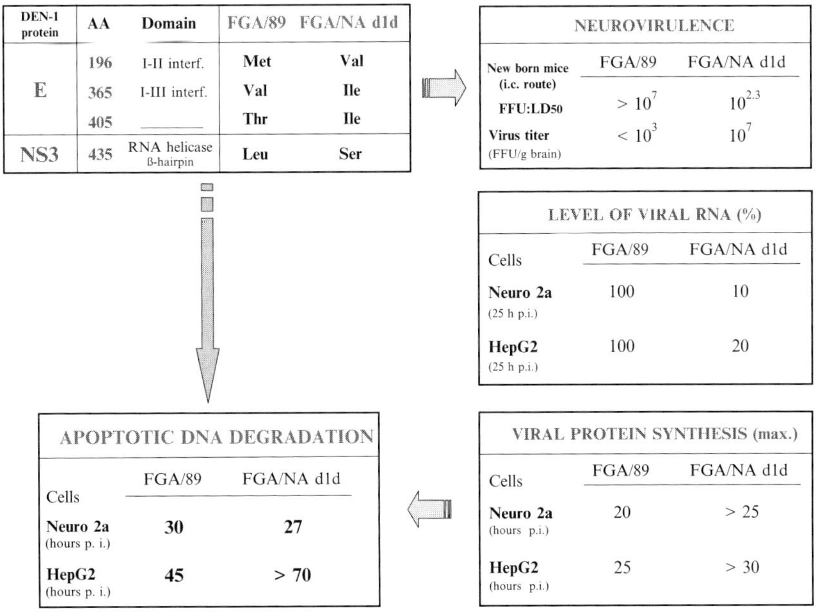

FIG 6. Comparative analysis between DEN-1 virus strain FGA/89 and its neurovirulent variant FGA/NA d1d (Duarte dos Santos *et al.*, 2000). interf., interface; FFU, focus-forming unit; LD_{50}, 50% lethal dose; p.i., postinfection; i.c., intracerebral. (See Color Insert.)

amphipathic α helix of the stem region, next to the membrane anchor region of the E protein. All three amino acid substitutions are located at regions that are likely to be important for conformation changes during the virus life cycle (Duarte dos Santos *et al.*, 2000). FGA/NA d1d has a greater capacity than FGA/89 to produce virus progeny. The identified amino acid substitutions in the E protein may account for the differences in DEN-1 virus production by facilitating virus assembly. The sequence of the NS protein-coding genes only revealed one amino acid substitution in the viral RNA helicase NS3 (NS3-435) (Duarte dos Santos *et al.*, 2000) (Fig. 6). The nonconservative $L_{435}S$ substitution maps to a long β hairpin that emerges from the RNA-binding domain and joins the C-terminal, α-helical domain of the NS3 protein. The connection between both domains may be important for the NTPase/helicase activity of NS3. Viral RNA synthesis is delayed in Neuro 2a and HepG2 cells in which FGA/NA d1d is replicating. The amino acid substitution in the NS3 helicase domain could alter the efficiency of replication complexes to catalyze the synthesis of viral RNA.

The apoptosis-inducing activity of FGA/NA d1d in target cells was compared with that of the parental strain. Following FGA/NA d1d infection, the apoptotic process progresses more rapidly in Neuro 2a cells but is delayed in HepG2 cells (Fig. 6). These data suggest that the determinants identified in the E and NS3 proteins affect the killing kinetics of the DEN-1 virus. Changes in viral growth may account for the differences in the induction of apoptosis in a cell-specific manner (Fig. 6). The induction of apoptosis can also be attributed to distinct apoptotic pathways that are activated in response to virus replication. In conclusion, differences in the DEN virus life cycle may affect virus-mediated cytotoxicity in target cells *in vivo* such as mouse neurons and human hepatocytes (Courageot and Desprès, 2003; Duarte dos Santos *et al.*, 2000).

D. Viral Factors That Contribute to Dengue Virus-Induced Apoptosis

The effect of the DEN virus on apoptosis needs to be studied to unravel the apoptotic pathway. DEN virus-infected neuroblastoma cells underwent apoptosis when viral glycoproteins began to accumulate (Desprès *et al.*, 1996; Duarte dos Santos *et al.*, 2000). As mentioned earlier, changes in the efficiency of viral morphogenesis affect the induction of apoptosis (Duarte dos Santos *et al.*, 2000). Studies in DEN virus-infected Neuro 2a cells indicated that ER α-glucosidase inhibition has a major effect on the productive folding and assembly pathways of DEN glycoproteins prM and E (Courageot *et al.*, 2000). Interestingly, the inhibition of glucose trimming enhances apoptosis in DEN virus-infected CNS (Courageot and Desprès, unpublished results). Although the precise biochemical mechanisms remain to be determined, EOR is probably one of the major signaling pathways leading to the induction of apoptosis in response to DEN virus infection (Desprès *et al.*, 1996; Duarte dos Santos *et al.*, 2000; Marianneau *et al.*, 1997, 1998a–c).

It remains to be determined whether DEN virus-mediated apoptosis can be ascribed to specific viral products. Several lines of evidence suggest that envelope glycoproteins have a major role in the apoptosis triggered by flavivirus replication (Desprès *et al.*, 1996; Duarte dos Santos *et al.*, 2000; Marianneau *et al.*, 1997; Prikhod'ko *et al.*, 2001). The cytotoxic activity of DEN-1 envelope glycoproteins was studied in the stable N2aprM+E cell line, a mouse neuroblastoma cell clone that drives the expression of FGA/89 prM plus E under the control of an inducible, exogenous promoter (Courageot *et al.*, 2000). The expression of recombinant viral products is itself sufficient to induce apoptosis (Courageot and Desprès, unpublished results). Data imply that prM and E are required for an apoptotic response in DEN virus-infected cells.

The induction of apoptosis was mapped to the DEN envelope glycoproteins. We addressed the question of whether their toxic effects are linked to the presence of death-mediating sequences. Our investigation linked translocation of the M ectodomain in the ER lumen to the induction of apoptosis (Catteau *et al.*, 2003a, 2003b). The intraluminal ectodomain of the DEN M protein is highly proapoptotic when transported in the secretory pathway. Further studies are required to determine how the M ectodomain induces an apoptotic cascade. The 40 amino acid sequence of the M ectodomain is remarkably conserved in the four serotypes of DEN virus (80 to 85% identity). We are currently trying to determine whether the induction of apoptosis by the intraluminal ectodomain of the M protein serves an infectious process or is an inescapable consequence of its expression.

VI. Concluding Remarks

Advances in cellular biology have increased our understanding of the mechanisms of DEN virus-induced cell death that determine the outcome of virus infection. Cytotoxicity seems to result from the inappropriate occurrence of apoptosis, which may contribute to the clinical manifestations associated with DEN virus infection. Different serotypes of DEN virus can provoke apoptosis in cultured cell systems. DEN virus replication appears to be a prerequisite for the triggering of the apoptotic process in infected cells. However, the very early stages of DEN virus replication might also enhance apoptosis (Jan *et al.*, 2000).

Growing evidence shows that DEN virus infection activates biochemically distinct apoptotic pathways. In human hepatocytes, neurons, and endothelial cells, DEN virus infection activates NF-κB, and products of the NF-κB gene seem to be essential for the induction of apoptosis. Target genes of NF-κB, such as FasL and p53, may be key mediators of apoptosis in response to DEN virus (Fig. 2e). The transcription-dependent growth arrest of p53 could contribute to DEN virus-induced apoptosis by regulating the expression of its downstream genes, such as RCGs, Fas, and Bax (Fig. 2c). During DEN virus replication, protein phosphorylation steps seem to be essential for the induction of apoptosis. As for other apoptotic pathways, several caspases could be activated in response to DEN virus infection.

Little is currently known about how replication of the DEN virus triggers apoptosis in host cells. The study of viral factors triggering apoptosis revealed that the envelope E protein and the viral RNA helicase domain of the NS3 protein possess determinants that may modulate cell death by altering virus assembly and replicative function (Duarte dos Santos *et al.*, 2000). This observation agrees with our preliminary hypothesis that the early steps of virus assembly may be at the origin of an ER stress, leading to apoptosis. Improvements in our understanding of DEN virus-induced apoptosis have led us to recognize the intraluminal ectodomain of the M ectodomain as a candidate viral inducer of apoptosis. Although the exact role of virion-associated M protein in cell death-inducing activity of DEN virus has not been determined, it is thought that the apoptotic effect is mediated by the interaction of the M ectodomain with signaling molecules during its transport in the secretory pathway (Catteau *et al.*, 2003a, 2003b).

Interference with apoptosis can prolong the life of DEN virus-infected cells, which may contribute to viral persistence. Persistent infection has been established in target cells expressing inhibitors of apoptosis. The ectopic expression of antiapoptotic proteins *bcl-2* and *bcl-*X_L can restrain the apoptotic process in a cell-dependent manner, subsequently facilitating the establishment of persistent DEN virus infection in surviving cells (Su *et al.*, 2001). The establishment of a persistent infection was associated with the abnormal expression of the NS1 glycoprotein (Liao *et al.*, 1997, 1998). The capacity of cellular antiapoptotic genes to block virus-induced cell death supports the attractive hypothesis that the DEN virus has the ability to establish persistent infection *in vivo*. Whether the establishment of chronic infection controlled by the host immune response is the source of progressive late DEN disease is an interesting issue that remains to be investigated.

Acknowledgments

The authors thank Vincent Deubel, Philippe Marianneau, Pierre-Emmanuel Ceccaldi, Claudia Nunes Duarte dos Santos, Felix Rey, Marie Flamand, Marie-Thérèse Drouet, and Marie-Pascale Frenkiel. Research programs were supported in part by grants from the CNRS Interdisciplinaire de Recherche Environnement Vie et Société (program 95N82/0134), DSP/STTC (program 99.34.031), and CAPES/COFECUB (program 254/98). M-P. C. and A. C. are funded by scholarship funds from the French Ministère de l'Education Nationale, de la Recherche et de la Technologie.

References

Adams, J. M., and Cory, S. (2001). Life-or-death decisions by the Bcl-2 protein family. *Trends Biochem. Sci.* **26:**61–66.

An, J., Kimura-Kuroda, J., Hibarayashi, Y., and Yasui, K. (1999). Development of a novel mouse model for dengue virus infection. *Virology* **263:**70–77.

Ashkenazi, A., and Dixit, V. M. (1998). Death receptors: Signaling and modulation. *Science* **281:**1305–1308.

Avirutnan, P., Malasit, P., Seliger, B., Bhakdi, S., and Husmann, M. (1998). Dengue virus infection of human endothelial cells leads to chemokine production, complement activation, and apoptosis. *J. Immunol.* **161:**6338–6346.

Baeuerle, P. (2000). Reactive oxygen species as costimulatory signals of cytokine-induced NF-κB activation pathways. *In* "Antioxidant and Redox Regulation of Genes" (C. K. Sien, H. Sies, and P. A. Baeuerle, eds.), pp. 181–219. Academic Press, San Diego.

Barber, G. N. (2001). Host defense, viruses and apoptosis. *Cell Death Differ.* **8:**113–126.

Barkett, M., and Gimore, T. D. (1999). Control of apoptosis by Rel/NF-κB transcription factors. *Oncogene* **18:**6910–6924.

Buschbeck, M., Ghomashchi, F., Gelb, M. H., Watson, S. P., and Borsch-Haubold, A. G. (1999). Stress stimuli increase calcium-induced arachidonic acid release through phosphorylation of cytosolic phospholipase A2. *Biochem. J.* **344:**359–366.

Castelli, J., Wood, K. A., and Youle, R. J. (1998). The 2-5A system in viral infection and apoptosis. *Biomed. Pharmacoth.* **52:**386–390.

Catteau, A., Kalinina, O., Wagner, M.-C., Deubel, V., Courageot, M.-P., and Despres, P. (2003a). Dengue M ectodomain contains a proapoptotic sequence referred to as *ApoptoM*. *J. Gen. Virol.* **84:**2781–2793.

Catteau, A., Roué, G., Yuste, V. J., Susin, S. A. and Despreš, P. (2003b). Expression of dengue *ApoptoM* sequence results in disruption of mitochodrial potential and caspase activation. Biochemie. (in press).

Courageot, M.-P., and Desprès, P. (2000). Viroses humaines et apoptose. *Ann. Inst. Pasteur/Actualités* **11:**79–93.

Courageot, M.-P., and Desprès, P. (2003). La pathogénicité du virus de la dengue et la mort cellulaire par apoptose. *Virologie* **5:**397–407.

Courageot, M.-P., Frenkiel, M.-P., Duarte dos Santos, C., Deubel, V., and Desprès, P. (2000). α-glucosidase inhibitors reduce dengue virus production by affecting the initial steps of virion morphogenesis in the endoplasmic reticulum. *J. Virol.* **74:**564–572.

Couvelard, A., Marianneau, P., Bedel, C., Drouet, M.-T., Vachon, F., Hénin, D., and Deubel, V. (1999). Report of a fatal case of dengue infection with hepatitis: Demonstration of dengue antigens in hepatocytes and liver apoptosis. *Hum. Pathol.* **30:**1106–1110.

Cui, X.-L., and Douglas, J. G. (1997). Arachidonic acid activates c-jun N-terminal through NADPH oxidase in rabbit proximal tubular epithelial cells. *Proc. Natl. Acad. Sci. USA* **94:**3771–3776.

Degen, W. G. J., Pruijn, G. J. M., Raats, J. M. H., and van Venrooij, W. J. (2000). Caspase-dependent cleavage of nucleic acids. *Cell Death Differ.* **7:**616–627.

Desagher, S., and Martinou, J.-C. (2000). Mitochondria as the central control point of apoptosis. *Trends Cell Biol.* **10:**369–377.

Desprès, P., Flamand, M., Ceccaldi, P.-E., and Deubel, V. (1996). Human isolates of dengue virus type-1 induce apoptosis in mouse neuroblastoma cells. *J. Virol.* **70:**4090–4096.

Desprès, P., Frenkiel, M.-P., Ceccaldi, P.-E., Duarte dos Santos, C., and Deubel, V. (1998). Apoptosis in the mouse central nervous system in response to infection with mouse-neurovirulent dengue viruses. *J. Virol.* **72:**823–829.

Desprès, P., Frenkiel, M.-P., and Deubel, V. (1993). Differences between cell membrane fusion activities of two dengue type-1 isolates reflect modifications of viral structure. *Virology* **196:**209–219.

Duarte dos Santos, C. N., Frenkiel, M.-P., Courageot, M.-P., Rocha, C. S., Vazeille-Falcoz, M.-C., Wien, M., Rey, F., Deubel, V., and Desprès, P. (2000). Determinants in the envelope E protein and viral RNA helicase NS3 that influence the induction of apoptosis in response to infection with dengue type-1 virus. *Virology* **74:**564–572.

Everett, H., and McFadden, G. (1999). Apoptosis: An innate immune response to virus infection. *Trends Microbiol.* **7:**160–165.

Feldmann, G. (1997). Liver apoptosis. *J. Hepatol.* **26:**1–11.

Fiers, W., Beyaert, R., Declercq, W., and Vandeanbeele, P. (1999). More than one way to die: Apoptosis and reactive oxygen damage. *Oncogene* **18:**7719–7730.

Fiscella, M., Zambrano, N., Ullrich, S. J., Unger, T., Lin, D., Cho, B., Mercer, W. E., Anderson, C. W., and Appella, E. (1994). The carboxy-terminal serine 392 phosphorylation site of human p53 is not required for wild-type activities. *Oncogene* **9:**3249–3257.

Giron-Calle, J., and Forman, H. J. (2000). Cell Ca^{2+} in signal transduction: Modulation in oxidative stress. *In* "Antioxidant and Redox Regulation of Genes" (C. K. Sien, H. Sies, and P. A. Baeuerle, eds.), pp. 105–127. Academic Press, San Diego.

Goodbourn, S., Didcock, L., and Randall, R. E. (2000). Interferons: Cell signalling, immune modulation, antiviral responses and virus countermeasures. *J. Gen. Virol.* **81:**2341–2364.

Gottlieb, T. M., and Oren, M. (1998). p53 and apoptosis. *Semin. Cancer Biol.* **8:**359–368.

Green, D., and Kroemer, G. (1998). The central executioners of apoptosis: Caspases or mitochondria. *Trends Cell Biol.* **8:**267–271.

Griffin, D. E., and Hardwick, J. M. (1997). Regulators of apoptosis on the road to persistent alphavirus infection. *Annu. Rev. Microbiol.* **51:**565–592.

Hardwick, J. M. (1998). Viral interference with apoptosis. *Semin. Cell Dev. Biol.* **9:**339–349.

Hengartner, M. O. (2000). The biochemistry of apoptosis. *Nature* **407:**770–776.
Hilgard, P., Czaja, M. J., and Stockert, R. J. (2000). Resistance of the casein kinase deficient cell line Trf1 to dengue virus-induced cell death. *Hepatology*, Meeting October 2000, 215A.
Hofmann, K. (1999). The modular nature of apoptosis signaling proteins. *Cell. Mol. Life Sci.* **55:**1123–1138.
Hornfelt, M., Edstrom, A., and Ekstrom, P. A. R. (1999). Upregulation of cytosolic phospholipase A2 correlates with apoptosis in mouse superior cervical and dorsal ganglia neurons. *Neurosci. Lett.* **265:**87–90.
Jagus, R., Joshi, B., and Barber, G. N. (1999). PKR, apoptosis and cancer. *Int. J. Biochem. Cell Biol.* **31:**123–138.
Jan, J.-T., Chen, B.-H., Ma, S.-H., Liu, C.-I., Tsai, H.-P., Wu, H.-C., Jiang, S.-Y., Yang, K.-D., and Shiao, M.-F. (2000). Potential dengue virus-triggered apoptotic pathway in human neuroblastoma cells: Arachidonic acid, superoxide anion, and NF-κB are sequentially involved. *J. Virol.* **74:**8680–8691.
Johnston, C., Jiang, W., Chu, T., and Levine, B. (2001). Identification of genes involved in the host response to neurovirulent alphavirus infection. *J. Virol.* **75:**10431–10445.
Kaufman, R. J. (1999). Stress signaling from the lumen of the endoplasmic reticulum: Coordination of gene transcriptional and translational controls. *Genes Dev.* **13:**1211–1233.
Kimura, K., Sasano, H., Shimosegawa, T., Mochizuki, S., Nagura, H., and Toyota, T. (2000). Ultrastructure of cells undergoing apoptosis. *Vitam. Horm.* **58:**257–266.
Krammer, P. H. (2000). CD95's deadly mission in the immune system. *Nature* **407:**789.
Kumar, S. (1999). Mechanisms mediating caspase activation in cell death. *Cell Death Differ.* **6:**1060–1066.
Lee, S. B., Rodriguez, D., Rodriguez, J. R., and Esteban, M. (1997). The apoptosis pathway triggered by the interferon-induced protein kinase PKR requires the third basic domain, initiates upstream Bcl-2, and involves ICE-like proteases. *Virology* **231:**81–88.
Li, N., and Karin, M. (1999). Is NF-κB the sensor of oxidative stress? *FASEB J.* **13:**1137–1143.
Li, P.-F., Dietz, R., and von Harsdof, R. (1999). p53 regulates mitochondrial membrane potential through reactive oxygen species and induces cytochrome *c*-independent apoptosis blocked by Bcl-2. *EMBO J.* **18:**6027–6036.
Liao, C.-J., Lin, Y.-L., Shen, S.-C., Shen, J.-Y., Su, H.-L., Huang, Y.-L., Ma, S.-H., Sun, Y.-C., Chen, K.-P., and Chen, L.-K. (1998). Antiapoptotic but not antiviral function of human *bcl-2* assists establishment of Japanese encephalitis virus persistence in cultured cells. *J. Virol.* **72:**9844–9854.
Liao, C.-L., Lin, Y.-J., Wang, J. J., Huang, Y.-L., Yeh, C.-T., Ma, S.-H., and Chen, L.-K. (1997). Effect of enforced expression of human *bcl-2* on Japanese encephalitis virus-induced apoptosis. *J. Virol.* **71:**5963–5971.
Liao, C.-L., Lin, Y.-L., Wu, B.- C., Tsao, C.-H., Wang, M.-C., Liu, C.-I., Huang, Y.-L., Chen, J.-H., Wang, J.-P., and Chen, L. K. (2001). Salicylates inhibit flavivirus replication independently of blocking nuclear factor kappa B activation. *J. Virol.* **75:**7828–7839.
Lin, L. L., Wartmann, M., Lin, A. Y., Knoff, J. L., Seth, A., and Davis, R. J. (1993). $cPAL_2$ is phosphorylated and activated by MAP kinase. *Cell* **72:**269–278.
Lin, Y.-L., Liu, C.-C., Chuang, J. I., Lei, H.-T., Yeh, T.-M., Lin, Y.-S., Huan, Y.-H., and Liu, H.-S. (2000a). Involvement of oxidative stress, NF-IL-6, and RANTES expression in dengue-2-virus-infected human liver cells. *Virology* **276:**114–126.

Lin, Y.-L., Liu, C.-C., Lei, H.-Y., Yeh, T. M., Lin, Y. S., Chen, R. M.-Y., and Liu, H.-S. (2000b). Infection of five human liver cell lines by dengue-2 virus. *J. Med. Virol.* **60:**425–431.

Los, M., Stroh, C., Jänicke, R. U., Engels, H. H., and Schulze-Osthoff, K. (2001). Caspases: More than just killers? *Trends Immunol.* **22:**31–34.

Majno, G., and Joris, I. (1995). Apoptosis, oncosis, and necrosis. *Am. J. Pathol.* **146:**3–15.

Marchenko, N. D., Zaika, A., and Moll, U. M. (2000). Death signal-induced localization of p53 protein to mitochondria. *J. Biol. Chem.* **275:**16202–16212.

Marianneau, P., Cardona, A., Edelman, L., Deubel, V., and Desprès, P. (1997). Dengue virus replication in human hepatoma cells activates NF-kappa B which in turn induces apoptotic cell death. *J. Virol.* **71:**3244–3249.

Marianneau, P., Flamand, M., Courageot, M.-P., Deubel, V., and Desprès, P. (1998a). La mort cellulaire par apoptose en réponse à l'infection par le virus de la dengue: Quelles conséquences dans la pathogénie virale? *Ann. Biol. Clin.* **56:**395–405.

Marianneau, P., Flamand, M., Deubel, V., and Desprès, P. (1998b). Apoptotic cell death in response to dengue virus infection: The pathogenesis of dengue hemorrhagic fever revisited. *Clin. Diagn. Virol.* **10:**113–119.

Marianneau, P., Flamand, M., Deubel, V., and Desprès, P. (1998c). Induction of programmed cell death (apoptosis) by dengue virus *in vitro* and *in vivo*. *Acta Cient. Venez.* **49:**13–17.

Marianneau, P., Mégret, F., Olivier, R., Morens, D. M., and Deubel, V. (1996). Dengue-1 virus binding to human hepatoma HepG2 and simian Vero cell surfaces differs. *J. Gen. Virol.* **77:**2547–2554.

Marianneau, P., Steffan, A. M., Royer, C., Drouet, M.-T., Jaeck, D., Kirn, A., and Deubel, V. (1999). Infection of primary cultures of human Kupffer cells by dengue virus: No viral progeny synthesis but cytokine production is evident. *J. Virol.* **73:**5201–5206.

Marianneau, P., Steffan, A. M., Royer, C., Drouet, M.-T., Kirn, A., and Deubel, V. (1998d). Differing infection of dengue and yellow fever viruses in a human hepatoma cells. *J. Infect. Dis.* **178:**1270–1278.

May, P., and May, E. (1999). Twenty years of p53 research: Structural and functional aspects of the p53 protein. *Oncogene* **18:**7621–7636.

McKenzie, P. P., Guichard, S. M., Middlemas, D. S., Ashmun, R. A., Danks, M. K., and Harris, L. C. (1999). Wild-type p53 can induce p21 and apoptosis in neuroblastoma cells but the DNA damage-induced G_1 checkpoint function is attenuated. *Clin. Cancer Res.* **5:**4199–4207.

Mercurio, F., and Manning, A. M. (1999). Multiple signals converging on NF-κB. *Curr. Opin. Cell Biol.* **11:**226–232.

Nicholson, D. W. (1999). Caspase structure, proteolytic substrates, and function during apoptotic cell death. *Cell Death Differ.* **6:**1028–1042.

O'Brien, V. (1998). Viruses and apoptosis. *J. Gen. Virol.* **79:**1833–1845.

Okamoto, T., Tetsuka, T., Yoshida, S., Kawabe, T. (2000) Redox regulation of NF-κB. *In* "Antioxident and REdox Regulation of Genes" (C. K. Sien, H. Sies, and P. A. Baeuerle, eds.), pp.203–219. Academic Press, San Diego.

Pahl, H. L., and Baeuerle, P. A. (1995). A novel signal transduction pathway from the endoplasmic reticulum to the nucleus is mediated by transcription factor NF-κB. *EMBO J.* **14:**2580–2588.

Pahl, H. L., and Baeuerle, P. A. (1997). The ER-overload response: Activation of NF-κB. *Trends Biochem. Sci.* **22:**63–67.

Pahl, H. L., Sester, M., Burgert, H. G., and Baeuerle, P. A. (1996). Activation of the transcription factor NF-κB by the adenovirus E3/19K protein requires its ER retention. *J. Cell Biol.* **132:**511–522.
Parquet, M. C., Kumatori, A., Hasebe, F., Morita, K., and Igarashi, A. (2001). West Nile virus-induced bax-dependent apoptosis. *FEBS Lett.* **500:**17–24.
Pinton, P., Ferrari, D., Rapizzi, E., Di Virgilio, F., Pozzan, T., and Rizzuto, R. (2001). The Ca^{2+} concentration of the endoplasmic reticulum is a key determinant of ceramide-induced apoptosis: Significance for the molecular mechanism of Bcl-2 action. *EMBO J.* **20:**2690–2701.
Pridhod'ko, G., Pridhod'ko, E., Cohen, J. I., and Pletnev, A. G. (2001). Infection with Langat flavivirus or expression of the envelope protein induces apoptotic cell death. *Virology* **286:**328–335.
Rosen, A., Casciola-Rosen, L., and Ahearn, J. (1995). Novel packages of viral and self-antigens are generated during apoptosis. *J. Exp. Med.* **181:**1557–1561.
Rosen, L., Khin, M. M., and U, T. (1989). Recovery of virus from the liver of children with fatal dengue: Reflections on the pathogenesis on the disease and its possible analogy with that yellow fever. *Res. Virol.* **140:**351–360.
Rothman, A. L., and Ennis, F. A. (1999). Immunopathogenesis of dengue hemorrhagic fever. *Virology* **257:**1–6.
Roulston, A., Marcellus, R. C., and Branton, P. E. (1999). Viruses and apoptosis. *Annu. Rev. Microbiol.* **53:**577–628.
Rusch, L., Zhou, A., and Silverman, R. H. (2000). Caspase-dependent apoptosis by 2′-5′-oligoadenylate activation of RNase L is enhanced by IFN-β. *J. Interferon Cytokine Res.* **20:**1091–1100.
Sakamuro, D., Sabbatini, P., White, E., and Prendergast, G. C. (1997). The polyproline region of p53 is required to activate apoptosis but not growth arrest. *Oncogene* **15:**887–898.
Sapirstein, A., and Bonventre, J. V. (2000). Phospholipase A2 in ischemic and toxic brain injury. *Neurochem. Res.* **25:**745–753.
Scorrano, L., Penzo, D., Petronilli, V., Pagano, F., and Bernardi, P. (2001). Arachidonic acid causes cell death through the mitochondrial permeability transition: Implications for tumor necrosis factor-α apoptotic signaling. *J. Biol. Chem.* **276:**12035–12040.
Sen, C. K. (2000). Oxidants and antioxidants in apoptosis. *In* "Antioxident and Redox Regulation of Genes" (C. K. Sien, H. Sies, and P. A. Baeuerle, eds.), pp. 221–243. Academic Press, San Deigo.
Sheikh, M. S., and Fornace, A. J., Jr. (2000). Role of p53 family members in apoptosis. *J. Cell. Physiol.* **182:**171–181.
Shen, Y., and Shenk, T. E. (1995). Viruses and apoptosis. *Curr. Opin. Genet. Dev.* **5:**105–111.
Slee, E. A., Adrain, C., and Martin, S. J. (1999). Serial killers: Ordering caspase activation events in apoptosis. *Cell Death Differ.* **6:**1067–1074.
Srivastava, S. P., Davies, M. V., and Kaufman, R. J. (1995). Calcium depletion from the endoplasmic reticulum activates the double-stranded RNA-dependent protein kinase (PKR) to inhibit protein synthesis. *J. Biol. Chem.* **273:**2416–2423.
Su, H.-L., Liao, C.-L., Lin, Y.-L. (2002). Japaneese encephalitis virus infection initiates endoplasmic reticulum stress and an unfolded protein response. *J. Virol.* **76:**4162–4171.
Su, H.-L., Lin, Y.-L., Tsao, C.-H., Chen, L.-K., Liu, Y.-T., and Liao, C.-L. (2001). The effect of human *bcl-2* and *bcl-X* genes on dengue virus-induced apoptosis in cultured cells. *Virology* **282:**141–153.

Teodoro, J. G., and Branton, P. E. (1997). Regulation of apoptosis by viral gene products. *J. Virol.* **71:**1739–1746.
Tschopp, J., Thome, M., Hofmann, K., and Mein, E. (1998). The fight of viruses against apoptosis. *Curr. Opin. Genet. Dev.* **8:**82–87.
Xiaoying, S., Potvin, B., Huang, T., Hilgard, P., Spray, D. C., Suadicani, S. P., Wolkoff, A. W., Stanley, P., and Stockert, R. J. (2001). A novel casein kinase 2 alpha-subunit regulates membrane protein traffic in the human hepatoma cell line HuH-7. *J. Biol. Chem.* **276:**2075–2082.
Yoneda, T., Imaizumi, K., Oono, K., Gomi, F., Katayama, T., and Tohyama, M. (2001). Activation of caspase-12, an endoplasmic reticulum (ER) resident caspase, through tumor necrosis factor receptor-associated factor 2-dependent mechanism in response to the ER stress. *J. Biol. Chem.* **276:**13935–13940.
Zinszner, H., Kuroda, M., Wang, X. Z., Batchvarova, N., Lightfoot, R. T., Remotti, H., Stevens, J. L., and Ron, D. (1998). CHOP is implicated in programmed cell death in response to impaired function of the endoplasmic reticulum. *Genes Dev.* **12:**982–995.

ADVANCES IN VIRUS RESEARCH, VOL 60

DYNAMICS OF FLAVIVIRUS INFECTION IN MOSQUITOES

Laura D. Kramer and Gregory D. Ebel

Arbovirus Laboratories
Wadsworth Center
New York State Department of Health
Slingerlands, New York 12159

Mosquito-borne flaviviruses are emerging as the cause of some of the most serious and widespread arthropod-borne viral diseases in the world. Flavivirus outbreaks are influenced by intrinsic (e.g., viral strain, vector competence, host susceptibility) and extrinsic (e.g., temperature, rainfall, human land use) factors that affect mosquito biology in complex ways. The concept of vectorial capacity organizes and integrates these factors, enabling a clearer understanding of their complex interrelationships, how they affect transmission of

0065-3527/03 $35.00

vector-borne disease, and how they impact human health. This review focuses on the components of vectorial capacity, providing an update on our current understanding of how selected aspects of mosquito biology, such as longevity, feeding behavior, oviposition habits, and nutrition of adult and immature stages, impact flavivirus transmission cycles and human disease. The influence of extrinsic factors, such as temperature, rainfall, seasonal and multiyear weather patterns, and human behavior that affects mosquito biology, and therefore flavivirus transmission, is explored. Mechanisms of flaviviral perpetuation over adverse seasons and years are addressed. This review also discusses vector competence, recent advances in mosquito genetics, and vector control as they relate to flavivirus transmission and human health.

I. Mosquito Vectors Associated with Flaviviruses

The mosquito-borne flaviviruses comprise two distinct epidemiological groups: (i) neurotropic viruses, often associated with encephalitic disease in humans or livestock, *Culex* spp. mosquito vectors, and bird reservoirs, and (ii) nonneurotropic viruses, associated with hemorrhagic disease in humans, *Aedes* spp. mosquito vectors, and primate hosts (Gaunt *et al.*, 2001). Excellent and extensive reviews have been published identifying mosquito vectors in the transmission cycles of both these epidemiological groups, including viruses in group (i): the Japanese encephalitis antigenic complex: St. Louis encephalitis virus (SLE) (Day, 2001; Mitchell *et al.*, 1980; Tsai and Mitchell, 1989), Japanese encephalitis virus (JEV) (Burke and Leake, 1988), Murray Valley (MVE) and Kunjin (KUN) encephalitis viruses (Marshall, 1988), and West Nile virus (WNV) (Bernard *et al.*, 2001; Hayes, 1989); and viruses in group (ii): dengue (DEN) (Gubler, 1988) and yellow fever (YF) (Monath, 1988).

Flaviviruses within the Japanese encephalitis antigenic complex have been isolated from a variety of vectors, but are associated most commonly with *Culex* (*Culex*) species mosquitoes. *Culex annulirostris* Skuse is the major vector to humans of MVE and KUN viruses in Australia. *Culex australicus* Dobrotworsky and Drummond, a non-anthropophilic species, is the predominant mosquito found from spring to early summer, but is replaced by *C. annulirostris* from midsummer to autumn when most human infections occur (Dhileepan, 1996; McDonald, 1980). The former is believed to play a major role in early season amplification of MVE and KUN, particularly among nonhuman vertebrate hosts (Russell, 1995), and the latter appears to be responsible for most human infections.

The major Asian vector of JE is *Culex tritaeniorhynchus* Giles (Burke and Leake, 1988). This mosquito is ornithophilic, feeds repeatedly at night, and disperses widely after feeding (Scherer *et al.*, 1959). *C. annulirostris* is the morphological and biological counterpart of *C. tritaeniorhynchus* and appears to be the predominant JE vector in Australia (Ritchie *et al.*, 1997).

Mosquitoes of the genus *Culex* have also been implicated as important vectors of WNV in North America (Centers for Disease Control and Prevention, 2000d), Europe (Hubalek and Halouzka, 1999; Hubalek *et al.*, 1999; Savage *et al.*, 1999), and Africa (Centers for Disease Control and Prevention, 2000a; Miller *et al.*, 2000; Traore-Lamizana *et al.*, 1994). In the northeastern United States, WNV was isolated mainly from *Culex pipiens* complex mosquitoes during 1999 (Nasci *et al.*, 2001b), 2000 (Bernard *et al.*, 2001; Kulasekera *et al.*, 2001), and 2001–2003 (L. D. Kramer, unpublished data; CDC, unpublished data). WNV was also detected in overwintering *Culex* species in New York City (Centers for Disease Control and Prevention, 2000c; Nasci *et al.*, 2001a). *Culex tarsalis* Coquillett is the predominant vector in Western US and Culex quinequefasciatus Say in the Southern Region. The precise role of most other potential vector species in either enzootic transmission of WNV or transmitting the infection to human beings remains unclear. *Culex salinarius* Coquillett was found frequently infected with WNV in New York (peak minimal infection rate of 9 per 1000 in Staten Island during the 2000 transmission season) (Kulasekera *et al.*, 2001). This species, which feeds indiscriminately and aggressively, may act as a bridge vector, exposing animals not usually involved in enzootic transmission of virus. *C. salinarius* previously had been identified as a potential bridge vector of SLE (Mitchell *et al.*, 1980) and eastern equine encephalitis virus (Vaidyanathan *et al.*, 1997) from the epornitic (avian enzootic/epizootic) cycle to humans and/or horses and has been proposed to serve the same function of WNV (Kramer and Bernard, 2001; Kulasekera *et al.*, 2001). Since *Cx. tarsalis* and *Cx. quinequefasciatus* feed on both birds and mammals, a bridge vector is not required. Some other potential vector species, such as *Ochlerotatus japonicus* (Theobald) and *Aedes albopictus* (Skuse) have been implicated in WNV transmission in the United States by the detection of viral RNA in field-collected specimens (Bernard *et al.*, 2001; CDC, unpublished data) and demonstration of high vector competence in the laboratory (Turell *et al.*, 2001). While these introduced species have the potential to pose a serious threat to humans, they do not appear to play a major role in the transmission of WNV at this time. Many of the secondary species may be incidental vectors of little epidemiological significance. Further research is required to determine the importance of mosquitoes other than the *C. pipiens* complex in the ecology and epidemiology of WNV in North America.

Similarly, SLE has been isolated from diverse mosquito species in North America (Day, 2001; Mitchell *et al.*, 1980). The predominant SLE vector is *Culex tarsalis* Coquillett in the West (Levy *et al.*, 1987; Reisen and Reeves, 1990), *C. pipiens* in the eastern and central states to Canada (Reiter *et al.*, 1986; Thorsen *et al.*, 1980), *Culex quinquefasciatus* Say in the central and southern states to Mexico (Savage *et al.*, 1993; Tsai *et al.*, 1988), and *Culex nigripalpus* Theobald in temperate and subtropical Florida (Chamberlain *et al.*, 1964; Day and Curtis, 1993). This same pattern of vector incrimination has been observed with WNV, as described.

Among mosquito-borne nonneurotropic flaviviruses, both DEN and YF are transmitted predominantly by *A. aegypti* (Gubler, 1988; Taylor, 1951). The geographic range of DEN disease has expanded in recent history primarily because of the spread of this mosquito vector (Gratz, 1999; Gubler, 1998), which has accompanied the global increase in human population size (Zanotto *et al.*, 1996) and movement. A variety of other *Aedes* spp. have also been implicated. *A. albopictus* was responsible for dengue fever and dengue hemorrhagic fever epidemics in many Far Eastern and Pacific regions, including Hawaii (Usinger, 1944) (ProMED-mail, 10/6/01, 10/14/01, http://www.promedmail.org/pls/promed/promed.home), Japan (Sabin, 1952), Thailand (Gould *et al.*, 1968), Singapore (Chan *et al.*, 1971), Indonesia (Jumali *et al.*, 1979), Seychelles (Metselar *et al.*, 1980), and southern China (Qiu *et al.*, 1981). More recently, *A. albopictus* has been incriminated as the primary vector in rural parts of Asia (Fan *et al.*, 1989; Gubler, 1988; Monath, 1994; World Health Organization, 1997). The first record of *A. albopictus* naturally infected with DEN in the Americas was during the 1995 outbreak in Reynosa, Mexico (Ibanez-Bernal *et al.*, 1997). Interestingly, virus was isolated from both male and female mosquitoes, implying that vertical transmission may occur naturally. *Ochlerotatus (Gymnometopa) mediovittatus* Coquillett, a forest and peridomestic mosquito found in the Caribbean area, was found to be an efficient vertical and horizontal vector of DEN in the laboratory (Freier and Rosen, 1988).

YF is maintained in a sylvan cycle involving lower primates and canopy-dwelling mosquitoes in tropical Africa and South America. In an outbreak in Gambia in 1978–1979, YF transmission initially was driven by sylvatic vectors, including the *Aedes furcifer-taylori* (Edwards) group of mosquitoes and possibly others, such as *Aedes luteocephalus* (Newstead), *Aedes metallicus* (Edwards), and *Aedes vittatus* (Bigot) (Germain *et al.*, 1980). In western Africa, two distinct patterns of YF emergence may be due to transmission by two different vectors: in forested regions, where *Aedes africanus* (Theobald) is the vector, transmission is limited and isolated as compared to savanna regions, where *A. furcifer* is the major vector. The different behavior of these two vectors and their population dynamics determined the degree of human–vector contact and are responsible for these differing patterns of emergence (Cordellier, 1991). *Haemagogus janthinomys* Dyar is a potential vector of sylvan YF in the Brazilian Amazon (Degallier *et al.*, 1991) and Trinidad (Chadee *et al.*, 1992), transmitting YF between monkeys. Within sylvatic habitats, the same vector populations are responsible for both epizootic and epidemic transmission cycles. Humans become infected when they invade the sylvan environment, and subsequently carry the virus to urban areas, initating an *A. aegypti*-driven urban cycle (Solomon and Mallewa, 2001). *A. albopictus* has been hypothesized to pose a substantial risk as a bridge vector for sylvan YF in Brazil (Gomes *et al.*, 1999; Marques *et al.*, 1998).

II. Flavivirus Infection of and Replication in the Mosquito Vector

Early studies of flaviviruses in mosquitoes used infectious virus assays, immunofluorescence assays (IFA), immunocytochemical studies, and electron microscopy (EM) to delineate the sequential spread of virus in the mosquito vector. More recently, molecular tools such as reverse transcription–polymerase chain reaction (RT–PCR) (Tardieux *et al.*, 1992) and *in situ* RNA–RNA hybridization (Miller *et al.*, 1989a) have been implemented to study the kinetics of viral infection in the mosquito. The accumulated findings of traditional and molecular methods have facilitated a detailed understanding of viral replication in mosquito hosts, although questions still abound. Flavivirus infection in competent mosquitoes proceeds in a stepwise fashion, beginning with acquisition of infectious virus during blood feeding, infection of the posterior portion of the mesenteron, replication in the mesenteron, dissemination from the mesenteron, secondary amplification in fat body cells, infection of the salivary glands, and transmission of virus to a susceptible host during a subsequent blood meal (Fig. 1).

The first stage of flavivirus replication occurs in cells of the mesenteron following an infectious blood meal. This stage of the infectious

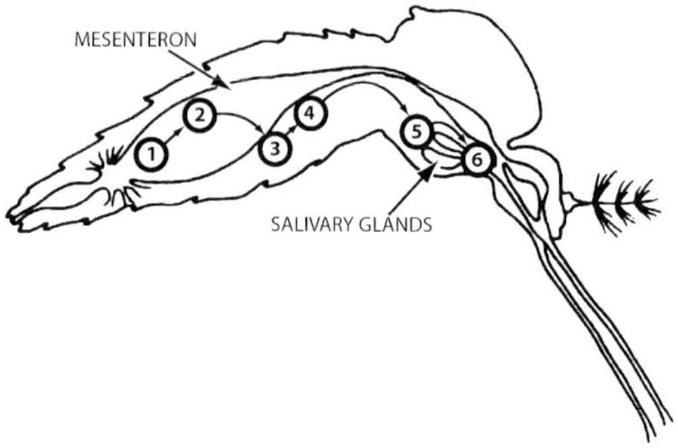

FIG 1. (Modified from Hardy, J. L. in Monath, T. P., 1988). Sequential steps required for a competent female mosquito to transmit a flavivirus after ingestion of an infective blood meal. (1) Infectious blood meal ingested. (2) Virus infects and multiplies in mesenteronal epithelial cells. (3) Virus released (escapes) from mesenteronal epithelial cells. (4) Secondary amplification: Virus infects and amplifies in fat body cells. (5) Virus infects and multiplies in salivary gland epithelial cells and other tissues. (6) Virus released from salivary glands into salivary secretion and is transmitted by feeding.

process was reviewed thoroughly by Hardy *et al.* (1983). Briefly, SLE was demonstrated to penetrate the mesenteronal epithelial cells of *C. pipiens* by binding to cell surface receptors (Marsh *et al.*, 1981), followed by endocytosis (Whitfield *et al.*, 1973). Nascent SLE virions were first observed within cisternae of the endoplasmic reticulum of mesenteronal epithelial cells in *C. pipiens* and subsequently were found in extracellular spaces between the plasma membrane of the cell and the basal lamina surrounding the meseneteron on the hemocoel side (Whitfield *et al.*, 1973). *A. albopictus* infected perorally with DEN-2 (Kuberski, 1979), and *C. tritaeniorhynchus* and *C. pipiens* infected with JEV had evidence of viral antigen in a few cells in the posterior portion of the mesenteron by 4 days postinfection. Within a few days, viral antigen was detected in most cells in the posterior mesenteron. DEN-2 was never found in the anterior portion of the mesenteron, whereas JE was eventually observed throughout (Hardy and Reeves, 1990). Electron microscopic examination of the mesenteron demonstrated SLE infection in $\leq$1 in 5 mesenteronal epithelial cells (Whitfield *et al.*, 1973). However, the distribution of infected cells within the mesenteron and the number of cells infected following oral infection vary with the virus and the mosquito species.

Doi *et al.* (1967, 1970) demonstrated that a second stage of JE multiplication in *C. tritaeniorhynchus* and *C. pipiens* occurred in fat body cells immediately adjacent to the mesenteron. This was also observed in *A. albopictus* infected with YF (Miller *et al.*, 1989a). Third-stage multiplication occurred in the salivary glands and other susceptible tissue. The DEN-2 viral antigen was found in large amounts in neural tissue 6 days after *A. albopictus* ingested virus and before the salivary glands had evidence of viral antigen by IFA (Kuberski, 1979). DEN-2 viral RNA loads were demonstrated to peak in *A. aegypti* on day 6 pi (Armstrong and Rico-Hesse, 2001). Peak infectious titers of YF (3.5 $\log_{10}$ PFU/mosquito) were observed 7 days after *A. albopictus* were fed on viral suspensions containing 7.2 $\log_{10}$ PFU/ml (Miller *et al.*, 1989a) and held at 26.7 °C. Viral antigen was first detected in head tissues 6 days after feeding; salivary glands were infected by 11 days, as many as 5 days after first detection in head tissues. Neither infectious virus nor viral RNA was detected in the reproductive tract. A similar pattern of viral dissemination is described for YF virus in *A. aegypti*, although viral titers were higher (B. R. Miller unpublished data).

III. Vectorial Capacity

A. Definition

The mechanisms of virus transmission by mosquitoes have been described thoroughly (Chamberlain and Sudia, 1961). Intrinsic and extrinsic factors have an impact on the dynamics of virus transmission cycles (virus–vector–vertebrate host interactions) (Hardy *et al.*, 1983). Included among intrinsic factors are mosquito vector competence [susceptibility, extrinsic incubation period (EIP), transmission rate], longevity, and host blood meal preferences. Each of these intrinsic factors is affected by extrinsic factors such as larval density and nutrition, temperature, rainfall, host availability, and host immunity.

A mathematical model that represents the interplay of factors that impact the dynamics of arthropod-borne pathogen transmission was described by MacDonald (1961) and later modified by other researchers. One equation for vectorial capacity, the "basic reproductive rate" (R_0) of a vector-borne pathogen, is represented mathematically by $V = ma^2p^nb/{-\ln(p)}$ (Black and Moore, 1996) where V is vectorial capacity (R_0); m is vector density in relation to the host; a is probability a vector feeds on a host in 1 day (= host preference index × feeding frequency); p is probability of vector surviving through 1 day; n is duration of EIP (in days); b is vector competence, the proportion of vectors ingesting an infective meal that successfully become infective; and $1/(-\ln p)$ is duration of vector's life in days after surviving the extrinsic incubation period.

Various researchers have refined the Black and Moore mathematical model to include additional factors that impact arbovirus transmission cycles. Kay *et al.* (1987) applied a cyclic mosquito-feeding pattern in a model of MVE transmission. A DEN model that describes the EIP as a function of temperature has also been described (Focks *et al.*, 1993a, 1993b, 1995). A further modification of the DEN model for container-breeding vectors included improved mechanisms to estimate water depth and temperature in the immature aquatic environment (Cheng *et al.*, 1998). Two studies analyzing risk factors for DEN infection (Carbajo *et al.*, 2001; Rodriguez-Figueroa *et al.*, 1995) determined that the two most important factors were duration of the virus transmission cycle and survival of the vectors. Esteva and Vargas (1998, 2000) formulated models describing the dynamics of DEN disease incorporating mechanical and vertical transmission of the virus (see Sections IV,A and V,A).

Among other factors that merit consideration in modeling disease risk are (1) viral serotype and strain and (2) probability of importation of virus, which may increase with urbanization and high human population density. Global warming may also increase the probability that a flavivirus introduced into a temperate environment would progress into an epidemic (Focks *et al.*, 1995). It may also increase mosquito activity, thereby increasing viral transmission (Tsai, 1997). Rigorous evaluations taking into account the multiple parameters of vectorial capacity are crucial to estimate risk of infection.

Environmental factors, especially temperature and precipitation, significantly affect arbovirus transmission by impacting vector abundance (Epstein, 2001; Hubalek, 2000), feeding behavior (Van Handel *et al.*, 1994), longevity (Dye, 1992; Smith, 1987), length of the mosquito gonotrophic cycle (Pant and Yasuno, 1973), and EIP of the virus (Reisen *et al.*, 1993; Watts *et al.*, 1987). The complex interactions of factors affected by changes in temperature and precipitation and the implications for flavivirus transmission have been reviewed thoroughly (Gubler *et al.*, 2001; Reeves *et al.*, 1994). These factors, discussed in detail later, have the greatest impact on the transmission of vector-borne diseases, and understanding their interaction is indispensable in designing public health interventions.

B. Vector Competence

Vector competence is a measure of the ability of a mosquito to become infected with, allow replication of, and transmit virus to a susceptible host. Viruses confront a series of "barriers" in the mosquito vector that may limit dissemination and/or transmission of the virus, rendering the mosquito incompetent. These barriers have been well described and include the mesenteronal infection barrier (MIB), which prevents infection of the mesenteron (Chamberlain and Sudia, 1961; Paulson *et al.*, 1989); the mesenteronal escape barrier (MEB), which prevents dissemination of virus from the infected mesenteronal epithelium (Hardy *et al.*, 1983; Kramer *et al.*, 1981); and salivary gland infection and escape barriers (Grimstad *et al.*, 1985; Kramer *et al.*, 1981; Paulson *et al.*, 1989). Although the molecular determinants of these barriers remain poorly understood, their description has significantly enhanced our understanding of vector competence. Less common modes of flavivirus transmission include vertical transmission from the female mosquito to the progeny and venereal transmission by male mosquitoes to females.

Numerous determinants affect vector competence. It is under genetic control of both the vector (Hardy *et al.*, 1983; Wallis *et al.*, 1985) and the virus (Armstrong and Rico-Hesse, 2001; Hardy *et al.*, 1986; Miller and Mitchell, 1986), but is also dependent on extrinsic factors such as temperature of extrinsic incubation, infecting dose, and nutritional status of the adult. These factors are discussed in detail in subsequent sections of this review. Some of these variables are not explicitly included in the vectorial capacity equation, but are captured in the term for vector competence, b.

Vector competence is a linear component of the vectorial capacity equation and thus is a relatively weak contributor to overall vectorial capacity. A poorly competent vector may therefore drive intense transmission of flaviviruses (and other vector-borne pathogens) if other conditions included in the vectorial capacity equation are favorable. Highly focused feeding, for example, compensated for minimal vector competence in an *A. aegypti*-driven urban outbreak of YF in Nigeria (Miller *et al.*, 1989b). Similarly, although the vector competence of *C. pipiens* for WNV is moderate (Turell *et al.*, 2001), its focus on avian hosts has driven epizootic transmission of WNV in the United States and Canada.

C. Vector Abundance

1. Impact on Vectorial Capacity

The abundance of vector mosquitoes, m, is also a linear factor in the vectorial capacity equation. It is determined by behavioral and environmental factors. Some mosquitoes, such as *A. aegypti*, lay eggs peridomestically where their preferred hosts, human beings, are readily available. The abundance of *A. aegypti* is directly proportional to the availability of domestic and peridomestic water-filled oviposition containers (Moore *et al.*, 1978). Interestingly, tremendous variation has been observed in the productivity of different types of containers (Focks and Chadee, 1997; Focks *et al.*, 2001; MacDonald, 1957; Southwood *et al.*, 1972; Teng *et al.*, 1999). In some types of containers (e.g., flower pots), although larvae are present, adult emergence rarely occurs because the water is changed frequently (Focks and Chadee, 1997). Stochastic models that incorporate estimates of water temperature and depth in containers are now being applied to describe the dynamics of DEN transmission more accurately (Hopp and Foley, 2001).

The importance of proximity of breeding sites to susceptible vertebrate hosts was also evident in Australia in 1998 where a JE outbreak was most likely initiated by the close proximity of domestic pigs, human dwellings, and *C. annulirostris* breeding sites (e.g., paddy fields, ditches) (Hanna *et al.*, 1996). Similarly, following the introduction of WNV to New York City in 1999, dry summer conditions brought together humans, dense populations of susceptible American crows, and the predominant vector, *C. pipiens*. Although vector abundance is an important factor in vectorial capacity, like vector competence it acts weakly: a mosquito species that is host specific or long-lived will have more impact on arbovirus transmission than one that is highly abundant but lacks host specificity or longevity.

2. Effect of Temperature

Temperature affects the abundance and feeding behavior of mosquitoes. At hotter temperatures, biting activity may be more frequent because of a relatively rapid metabolism of nutrients (Van Handel *et al.*, 1994). This may, in turn, impact the mosquito reproductive rate and thereby density of the mosquito population. Critical vector densities are required for efficient virus transmission. An integrative transmission model of SLE epidemics in Florida indicated that seasonal variation in mosquito populations did not correlate with outbreaks (Lord and Day, 1999). Similarly, *C. tarsalis* abundance in California, which was greatest during the spring and fall, did not correlate with field-derived measures of virus transmission, mean environmental temperature, humidity, or rainfall (Reisen *et al.*, 1995a).

WNV outbreaks are associated with high populations of mosquitoes (especially *Culex* spp.) frequently caused by flooding and subsequent dry and hot weather (Hubalek, 2000). An explosive increase in the population of *C. pipiens* as a result of poor plumbing and sewage in the basements of apartments was thought to be instrumental in the 1996 outbreak of WNV in Romania following a prolonged drought and excessive heat (Savage *et al.*, 1999; Tsai *et al.*, 1998). The outbreak of WNV in North America following its introduction to New York in 1999 was preceded by a mild winter and a dry hot summer (Epstein, 2001).

3. *Effect of Rainfall*

The effect of rainfall on mosquito abundance is complex. It is generally understood that increases in rainfall promote arbovirus transmission. Devastating floods occurred after heavy rains in Moravia, Czech Republic, leading to an abrupt increase in *C. pipiens* populations and an increase in WNV activity (Hubalek and Halouzka, 1999). The density of *A. aegypti* was correlated positively with rainfall, the relationship being more marked in drier locations of Puerto Rico (Moore *et al.*, 1978). In 2000, following an increase in temperature and rainfall, an epidemic of YF occurred in Brazil (Vasconcelos *et al.*, 2001). It is thought that increased rainfall provides additional suitable habitat for immature mosquitoes. Conversely, drought conditions may increase the intentional storage of drinking water in areas without a stable piped water supply, also leading to an increase in the number of developmental sites for mosquitoes. Similarly, the dry summer conditions in New York City in 1999 led to natural water collections containing high concentrations of organic matter that favored breeding of *C. pipiens* (Epstein, 2001).

The impact of rainfall on the abundance of *C. annulirostris* and *C. australicus*, vectors of MVE, which breed in permanent or semipermanent habitats, was difficult to assess in Murray Valley, Australia, during the dry summer months of 1979–1985 (Russell, 1986a, 1986b). Urban SLE activity in the United States associated with *C. pipiens* complex mosquitoes is frequently associated with periods of excessive rainfall followed by periods of drought (Bowen and Francy, 1980) largely because the breeding habits of these vectors (Day, 2001) favor water high in organic content (Epstein, 2001). The drought conditions foster contact between mosquito populations and avian hosts in urban sites, allowing enzootic transmission to occur. Following late summer rainfall, mosquito abundance increases and infected populations expand to initiate epidemic transmission. Similar climatic conditions fostered the establishment of WNV in New York in 1999. Therefore, rainfall clearly effects mosquito abundance, but the magnitude and direction of this effect appear to vary with the virus, mosquito species, and other environmental determinants.

D. Vector Longevity and Extrinsic Incubation Period

1. Impact on Vectorial Capacity

The most powerful components in the vectorial capacity equation are the terms that represent the longevity of the vector and the EIP of the pathogen, p^n. The EIP determines how long a mosquito must survive after an infective blood meal before it can transmit virus and varies for each virus–vector system (Chamberlain *et al.*, 1954). Vector longevity is raised to the power of the EIP, thus entering the vectorial capacity equation as an exponential term. In combination, then, these two factors are the most powerful contributors to vectorial capacity. A long-lived mosquito contributes the most to the transmission of vector-borne pathogens. Determinants of vector longevity are complex and partially determined by extrinsic factors that are discussed in more detail later.

2. Effect of Temperature

Higher temperatures shorten the EIP of the virus (Chamberlain and Sudia, 1961; Taylor, 1951). Replication of SLE in the mosquito decreases as a linear function of temperature at temperatures greater than 17 °C (Reisen *et al.*, 1993). An increased temperature of extrinsic incubation has been demonstrated to decrease the length of time following ingestion of an infectious blood meal before mosquitoes can transmit most flaviviruses, including SLE (Chamberlain *et al.*, 1959; Hurlbut, 1973), JE (Takahashi, 1976), DEN (Watts *et al.*, 1987), YF (Bates and Roca-Garcia, 1946; Khozinskaya *et al.*, 1985), WNV (Dohm and Turell, 2001; Jupp, 1974), and MVE (Kay *et al.*, 1989c). At 32 °C, duration of the EIP can be as short as 7 days for *A. aegypti* infected with DEN (Watts *et al.*, 1987). Flaviviruses can be transmitted experimentally, however, after lengthy incubation at low temperatures: a Canadian isolate of SLE (Mclean *et al.*, 1978) and a southwest Pacific isolate of DEN-2 (Mclean *et al.*, 1975) were transmitted by *Aedes* spp. after 3 weeks at 13 °C. Kay *et al.* (1989b) demonstrated that the vector competence of *C. annulirostris* for Murray Valley encephalitis was depressed only if the extrinsic incubation temperature was decreased relative to the rearing temperature and found no significant difference in infection or transmission rates in mosquitoes reared and maintained at lower temperatures as compared to those reared and maintained at 27 °C.

Temperature also affects mosquito longevity, an important component of vectorial capacity (Dye, 1992; Smith, 1987). Mortality rates depend on age (Lord and LeFevre, 2001; Muir and Kay, 1998; Su and Mulla, 2001) and increase with increasing temperature. The effect of temperature on mosquito longevity and vectorial capacity, however, is not always straightforward or intuitive. Sheppard *et al.* (1969) found no seasonal variation in life expectancy of *A. aegypti* in Thailand. In Puerto Rico, DEN transmission is most intense from June to December when the temperatures are highest and there is the most rainfall (Rodriguez-Figueroa *et al.*, 1995) but *A. aegypti* survival is shortest at this time (Costero *et al.*, 1999). Nonetheless, because *A. aegypti*, and presumably other mosquito species, seek shelter in cooler resting places during hours of excessive heat, the effect of high temperatures may be ameliorated (Christophers, 1960). Reisen *et al.* (1995a) concluded that SLE transmission progressed efficiently during midsummer because elevated temperatures shortened the extrinsic incubation period without decreasing survivorship markedly. This resulted in an increased proportion of females surviving extrinsic incubation (Reisen *et al.*, 1995a). Warmer temperatures may intensify or extend the transmission season for DEN, SLE (Patz and Reisen, 2001), and other flaviviruses.

E. Global Climate Change

The International Council of Scientific Unions and the Intergovernmental Panel on Climate Change have projected a 2 °C increase in temperature by the end of the twenty-first century (IPCC, 1996). Because of the complex relationship of arboviruses and their mosquito vectors with temperature, such global change is expected to influence arthropod-borne disease transmission (Jetten and Focks, 1997; Reeves *et al.*, 1994), although the relative importance of global climate change remains under discussion (Gubler *et al.*, 2001). Epidemics of DEN (Hales *et al.*, 1996, 1999) and MVE (Nicholls, 1986, 1993) have been associated with the El Niño phase of southern oscillation, supporting the observation that arbovirus transmission cycles are affected by climate, as are other arthropod-borne diseases (Bouma *et al.*, 1997a, 1997b). Global climate change would presumably affect vectorial capacity indirectly by changing the temperature and precipitation patterns in endemic areas (Cayan, 1999). It is crucial to understand the relationship between these factors and mosquito biology in order to elucidate the relationships between climate and disease ecology and ultimately human health (Hopp and Foley, 2001).

IV. Impact of Reproductive Biology on Dynamics of Flavivirus Transmission

A. Feeding Behavior

Critical components of vectorial capacity are the degree to which a vector species focuses on a single blood meal source and biting frequency (MacDonald, 1957). The importance of feeding behavior on virus transmission is exemplified by the highly anthropophilic mosquito *A. aegypti*, the principal vector of DEN (Gubler, 1988) and urban YF (Taylor, 1951). *A. albopictus* is a more competent DEN vector than *A. aegypti*, but is a less selective feeder and therefore generally appears to be of less epidemiological significance. In vectorial capacity, the importance of focused feeding is captured by squaring the term a, indicating that for transmission to continue, two blood meals must be taken from competent vertebrate hosts. A twofold increase in host fidelity thus impacts vectorial capacity much more profoundly than a twofold increase in vector competence or vector abundance.

Various determinants affect host preference, and thereby vectorial capacity. Olfaction plays a major role in mosquito host preference and blood feeding, integral behaviors for disease transmission. Host specificity is determined genetically, as demonstrated by inherited differences within members of the *Anopheles gambiae* complex (Takken, 1996). The mechanisms of *A. aegypti* host preference may be due to a large extent to differences in the amount of lactic acid (Steib *et al.*, 2001) and perhaps ammonia (Geier *et al.*, 1999) among other components of the human skin in sweat. To evaluate host preferences of various mosquito species, studies have applied serologic analyses (Niebylski and Meek, 1992; Robertson *et al.*, 1993; Tempelis, 1975; Tempelis *et al.*, 1965, 1976; Wang, 1975) and comparison of feeding behavior in the field (Loftin *et al.*, 1997; Mwandawiro *et al.*, 2000). More recently, molecular analyses have been employed (Boakye *et al.*, 1999; Lee *et al.*, 2001; Ngo and Kramer, 2003).

Another component of feeding behavior that may impact host fidelity is the temporal pattern in mosquito and human or other vertebrate activity. The temporal patterns of feeding activity for *A. aegypti* depend on locality and subspecies (reviewed in Chadee and Martinez, 2000; Gubler, 1988), but appear to be predominantly diurnal with peak feeding in the early morning and late afternoon, coinciding with peridomestic human activity (Christophers, 1960; Gubler, 1970; Ho *et al.*, 1973). A nocturnal feeding component has also been observed in Africa (McClelland, 1959, 1960) and Trinidad (Chadee and Martinez,

2000) that has significant epidemiological impact, as it extends the period of peak mosquito activity. It has been postulated that artificial light may be contributing to the feeding behavior of *A. aegypti* in these areas.

Nutritional requirements during the gonotrophic cycle of mosquitoes may also impact the intensity of arbovirus transmission. In most mosquito species, adult females feed on plant sugars, supplementing nutrients acquired with blood (Bidlingmayer and Hem, 1973; Reisen *et al.*, 1986). In contrast, *A. aegypti* feed almost exclusively on human blood (Edman *et al.*, 1992; Van Handel *et al.*, 1994) and frequently take multiple partial blood meals during each gonotrophic cycle (Gubler, 1988; MacDonald, 1956; Platt *et al.*, 1997; Scott *et al.*, 1993b; Yasuno and Tonn, 1970). This phenomenon may be seasonal, demonstrating the interrelationship between factors generally viewed as "intrinsic" and "extrinsic" (Costero *et al.*, 1998a; Day *et al.*, 1994; Scott *et al.*, 1993a, 1993b; Van Handel *et al.*, 1994). The incidence of multiple blood meals correlates with the DEN transmission season in rural Thailand (Scott *et al.*, 1993a). This effectively increases the speed of virus spread, as virus may be transmitted each time a mosquito expectorates saliva into a susceptible host during feeding (Putnam and Scott, 1995). *A. aegypti* that were infected perorally transmitted at a greater rate after multiple probing than controls that fed once. This phenomenon was noted earlier by Hurlbut (1966) with SLE and *C. pipiens*. It was hypothesized that the first fraction of saliva contained old secretions with fewer or inactivated virions. Salivation stimulated increased shedding of virus from salivary gland tissues, and these virions were more infectious.

The feeding habits of *A. aegypti* may contribute to the observation of clusters of DEN disease within individual households in Puerto Rico (Morrison *et al.*, 1998). In addition to *A. aegypti* (Scott *et al.*, 1993a), *C. tarsalis* (Anderson and Brust, 1997) and *C. salinarius* (Thapar *et al.*, 1998) have been demonstrated to take multiple blood meals during each gonotrophic cycle. Multiple blood meals translate into more frequent human–mosquito contact, increased virus transmission, increased fecundity, and enhanced survival (Day *et al.*, 1994; Scott *et al.*, 1993a, 1993b) (i.e., enhanced vectorial capacity). This behavior may be critical for the maintenance of DEN viral transmission at low but detectable levels during interepidemic periods. It also confers an evolutionary advantage to the mosquitoes that take multiple human blood meals: the number of offspring is higher and the females reproduce more quickly than those imbibing blood and plant-derived carbohydrates (Costero *et al.*, 1998b; Vaidyanathan *et al.*, 1997).

Some pathogens have been observed to affect blood feeding behavior in the infected vector, increasing the duration of contact between the arthropod and the vertebrate host, for example, LaCrosse-infected *Oclerotatus triseriatus* (Say) (Grimstad *et al.*, 1980) and *Plasmodium gallinaceum*–infected *A. aegypti* (Rossignol *et al.*, 1984). Putnam and Scott (1995) found no evidence of similar manipulation of infected highly colonized *A. aegypti* by passaged DEN, but Platt *et al.* (1997) found that the mean total time required for probing and feeding by infected *A. aegypti* was significantly longer than the time required by uninfected mosquitoes. Platt *et al.* (1997) related the difference in results to the specific mosquito tissues that were infected.

B. Oviposition Behavior

Mosquitoes undergo complete metamorphosis as they develop and are affected by several critical environmental determinants. The type of larval habitat available affects adult nutritional status, body size, survival, and possibly vector competence (Lord, 1998; Sumanochitrapon *et al.*, 1998). Climatic factors such as temperature and rainfall affect development of the immature aquatic stages as much as adult survivorship and abundance (Focks *et al.*, 1993a, 1993b; Macfie, 1920; Rueda *et al.*, 1990) (Fig. 2). Climate and habitat type are related to one another, as are socioeconomic factors, including sanitation and the availability of piped water, that may also affect the incidence of disease. For example, the difference in DEN incidence in Ho Chi Minh City and its periphery in Vietnam was correlated inversely with access to a piped drinking water supply (Mirovsky *et al.*, 1965), indicating that provision of an adequate water supply system precludes the need for artificial containers to catch rainwater and consequently reduces mosquito density and DEN transmission (Focks *et al.*, 1999).

The duration of immature development of *A. aegypti* varied according to the container position (i.e., shaded or exposed) and the availability of food resources, as well as inversely with temperature (Tun-Lin *et al.*, 2000). In southeast Asia, even where *A. aegypti* breeds in artificial containers that are kept predominantly indoors, increased transmission of dengue is still most pronounced during the rainy season. This suggests that peak transmission in southeast Asia may not be influenced greatly by mosquito density, but rather that temperature and humidity during the rainy season are more conducive to survival of the adult mosquito, thus increasing the likelihood that infected mosquitoes will survive the EIP and transmit to new individuals. However, in Brazil, local periods of drought have been observed to promote

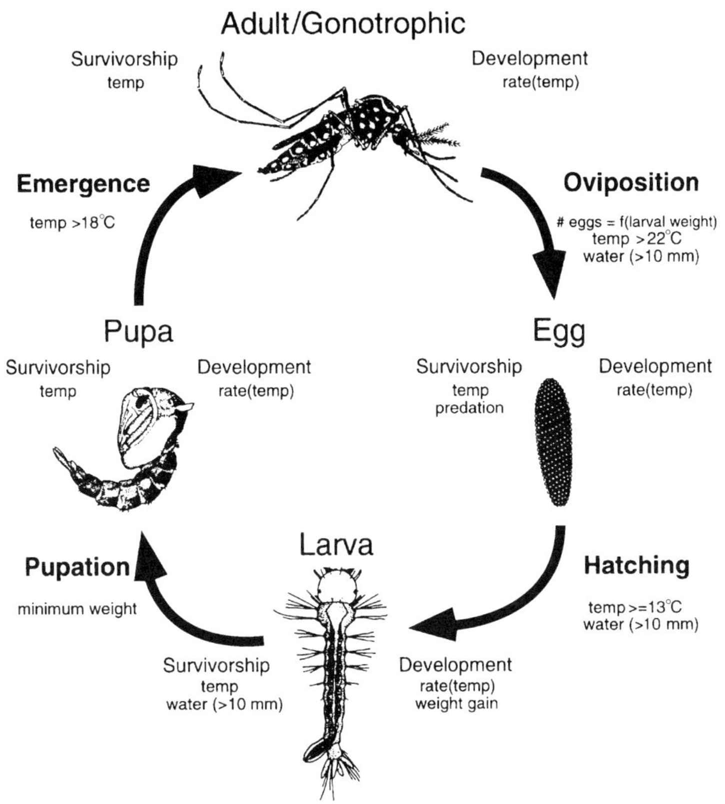

FIG 2. Survivorship and development requirements are life stage dependent. Metamorphosis is dependent on daily cumulative development and other life stage–specific factors. Reprinted with permission from Hopp and Foley (2001).

vector abundance in two ways: residents stored water in which vector mosquitoes could breed and cholera outbreaks due to contaminated water diverted local health workers from routine antivector activities (Pontes *et al.*, 2000). This points out the difficulty of extrapolating from one environment to another.

An increase in the population density of noncontainer-breeding mosquitoes is dependent on the flooding of breeding sites. Vernal increases in *C. tarsalis* abundance typically were associated with flooding of saline marshes along the north shore of the Salton Sea in the Coachella Valley in southern California and were followed 6 to 8 weeks later by the onset of SLE virus activity. Virus then spread to managed marsh and agricultural habitats in a nearby flood plain and, depending on the intensity of amplification, to agricultural and residential areas in the more elevated northwestern portion of the Valley (Reisen *et al.*, 1995b).

C. *Larval Nutrition*

Nutrients available to developing mosquitoes further impact vectorial capacity. Mosquitoes from nutrient-rich environments tend to be larger than those from nutrient-poor environments; larvae that are uncrowded develop into larger adults than those that are crowded. Fecundity is a function of adult weight, thereby affecting population density (Blackmore and Lord, 2000; Nayar and Sauerman, Jr., 1975a, 1975b, 1975c). Size differences did not lead to any statistically significant differences in the vector competence of *C. tritaeniorhynchus* for JE (Takahashi, 1976) and WNV (Baqar, 1980) and of *C. annulirostris* Skuse for MVE (Kay *et al.*, 1989a). These results are in contrast to those observed for *O. triseriatus* (Say) infected with LaCrosse virus, where low nutrition adults were found to be more efficient transmitters of virus than adults reared under normal conditions (Grimstad and Haramis, 1984). However, Nasci (1986, 1991) hypothesized that larger females had greater vector potential due to increased survival time and blood-feeding success. Larval nutrition also influences nutritional reserves such as glycogen and lipids in adult mosquitoes (Chambers and Klowden, 1990) and consequently the rate at which multiple blood meals are taken by adult *A. aegypti*. Nutritionally deprived larvae develop into smaller nutritionally deprived adults that require multiple blood meals. This may lead to increased human–vector contact and increases in vectorial capacity.

V. Role of the Vector in Perpetuation of Virus over Adverse Seasons and Years

A. *Vertical Transmission*

The means by which most flaviviruses perpetuate themselves over adverse seasons and years are poorly understood. In temperate climates where winter occurs, arboviruses must generally persist during periods, up to 9 months, when mosquito activity ceases and transmission stops. In tropical areas, adversity is also related to seasonal changes in rainfall and temperature. On a larger time scale, years frequently occur where conditions are less favorable for arbovirus transmission due to the availability of hosts and breeding sites for vector mosquitoes. Possible mechanisms of viral survival have been reviewed by Reeves (1990) and Rosen (1987b). One potential means of overwintering is through vertical transmission from parent to progeny. Flaviviruses appear to enter the fully formed egg through the micropyle at the time of fertilization (Rosen, 1987a, 1987b). This is a much less efficient mechanism of vertical transmission than true "transovarial" transmission in which the virus infects the germ layer of the developing egg, as occurs with bunyaviruses (Tesh and Shroyer, 1980); however, it does permit the infection of progeny following a single maternal blood meal (Rosen, 1987a). Flaviviruses have been isolated from field-collected larvae and/or male mosquitoes, including JEV from *C. tritaeniorhynchus* (Rosen *et al.*, 1978), WNV from *C. univittatus* (Miller *et al.*, 2000), DEN from *A. aegypti* (Hull *et al.*, 1984; Khin and Than, 1983), and *A. albopictus* (Ibanez-Bernal *et al.*, 1997; Kow *et al.*, 2001) and yellow fever from *A. aegypti* (Fontenille *et al.*, 1997). Laboratory studies have also successfully demonstrated vertical transmission of flaviviruses. YF was transmitted from parent to progeny *Haemogogus equinus* Theobald (Dutary and LeDuc, 1981) and intrathoracically inoculated *Aedes* spp. (Aitken *et al.*, 1979; Beaty *et al.*, 1980). JEV was transmitted vertically by *C. tritaeniorhynchus* (Rosen *et al.*, 1978) and *O. japonicus* (Takashima and Rosen, 1989) and WNV by *C. tritaeniorhynchus, A. albopictus, A. aegypti* (Baqar *et al.*, 1993), and *C. pipiens* (Turell *et al.*, 2001). Kunjin was demonstrated to be transmitted vertically by *A. albopictus* (Tesh, 1980). MVE (Kay and Carley, 1980) and DEN (Rosen *et al.*, 1983) were also transmitted vertically in the laboratory at a low rate by *A. aegypti*. Evidence for higher rates of vertical transmission of DEN, possibly sufficient to assist in the maintenance of DEN in nature, have been found in *A. albopictus* (Bosio *et al.*, 1992; Mitchell and Miller, 1990; Shroyer, 1990).

SLE has been demonstrated in the laboratory to be transmitted vertically by several mosquito species (Hardy *et al.*, 1984), but appeared to depend at least in part on mosquito rearing conditions. Temperature of rearing affected vertical transmission rates of SLE in *Aedes taeniorhynchus* (Wiedemann) and *Aedes epactius* Dyar and Knab, but not *Culex* spp: MIRs were significantly higher in F_1 adults following rearing at 18 °C than they were in F_1 adults reared at 27 °C (Hardy *et al.*, 1980). Virus may be inactivated during the metamorphic changes that occur during larval pupation and/or adult emergence. In addition, use of different strains of SLE appeared to influence the vertical transmission of virus by some mosquito species (e.g.,

A. albopictus) and not others, (e.g. *A. epacticus*). Three geographic strains of *A. albopictus* infected parenterally with DEN-1 and DEN-4 transmitted virus vertically to progeny (Mitchell and Miller, 1990). As had been demonstrated previously, the rate of vertical transmission varied with the DEN strain and serotype (Rosen *et al.*, 1983), with the geographic strain of mosquito (Bosio *et al.*, 1992), and on the mosquito stage assayed. It is therefore difficult to compare transmission rates from different studies. DEN-infected male mosquitoes (via vertical transmission) have been demonstrated to pass the virus to females through venereal transmission (Rosen, 1987c; Tu *et al.*, 1998). Such transmission was enhanced if the females had taken a blood meal 2 to 7 days prior to mating. Allowing F_1 female progeny of *C. pipiens* to take a blood meal prior to being tested for SLE infection did not increase levels of vertical transmission to equal those of California serogroup viruses, which are truly transmitted transovarially (Francy *et al.*, 1981). Thus, the role of vertical transmission in flaviviral persistence is at present unclear. Further work is required to resolve some of the ambiguities in the literature on the role of vertical transmission in flavivirus persistence.

B. Hibernating Adult Mosquitoes

An alternate mechanism of carrying virus through adverse seasons is through the female becoming infected in the fall by feeding on a viremic host or through vertical transmission, surviving winter, and reinitiating the transmission cycle in the spring. Cool temperatures facilitate viral persistence in mosquitoes over extended periods of time. Experiments simulating winter conditions have demonstrated a long survival of flaviviruses in mosquitoes. SLE was demonstrated to survive under natural winter conditions in *C. quinquefasciatus* more than 100 days (Reeves *et al.*, 1958). JE was demonstrated to survive in and be transmitted by the same mosquito species following initial incubation for 5 to 15 days at 20–30 °C, followed by an extended incubation at 8–13 °C, with final incubation at 30 °C (Hurlbut, 1950; LaMotte, 1958, 1963; Shichijo *et al.*, 1972). Similarly, *C. pipiens* infected with WNV and held exclusively at overwintering temperatures of 10 °C during incubation contained no detectable virus. However, when mosquitoes were transferred to 26 °C after being held at 10 °C for 21–42 days, infection and dissemination rates increased with increased incubation at 26 °C (Dohm and Turell, 2001). SLE (Bailey *et al.*, 1978) and WNV (Nasci *et al.*, 2001a) were isolated from pools of *C. pipiens* collected during the winter in Maryland and New York, respectively; however, the mode of infection of the adults is not clear.

C. *Mosquito Dispersal of Virus*

A third possible mechanism of viral perpetuation in which mosquitoes are involved is through reintroduction from an endemic area. The probability of reintroduction of virus via migrant-infected mosquitoes decreases with time and distance from the infectious source (Johnson, 1969). Hot dry climatic conditions may restrict dispersal in some habitats (Muir and Kay, 1998). For this mode of virus appearance after adverse seasons to be successful, the source of infection would have to support substantial populations of infected insects that "invade" neighboring uninfected locations.

Dispersal of mosquitoes can occur through unassisted means as on wind currents or via assisted means inadvertently on boats, in containers, or on planes. Natural and assisted dispersal of *Aedes* has been described (Provost, 1951; Teesdale, 1950) and has been hypothesized for the transport of *A. aegypti* from west Africa to the New World on slave ships in the fifteenth to seventeenth centuries. Dispersal of *C. annulirostris* over 1800–2100 km (Anderson and Eagle, 1953; Miles and Howes, 1953) has been hypothesized to contribute to the dissemination of MVE. The high degree of genetic relatedness between Australian isolates of JE from 1995 to 1998 with isolates from Papua New Guinea in 1997 and 1998 suggests that the New Guinea mainland was the likely source of incursions of JE virus into Australia (Johansen *et al.*, 2000, 2001). It has been hypothesized that JE-infected mosquitoes may have been carried southward on monsoonal depressions (Hanna *et al.*, 1999). Studies conducted by Kay and Farrow (2000) provide circumstantial support that warm weather will permit crepuscular-nocturnal mosquitoes to be carried into the upper air layers and be transported significant distances. Sellers and Maarouf have analyzed weather conditions in North America and concluded that viruses may be transported 1250–1350 km by windborne mosquitoes (Sellers, 1980; Sellers and Maarouf, 1988, 1991, 1993). Viral population genetic studies, however, indicate that this has not been a significant means of perpetuation of SLE in North America: South American strains of SLE are genetically distinct from those in North America (Kramer and Chandler, 2001).

VI. Genetics of Infection

A. *Variation in Natural Populations*

The idea that vector susceptibility to flaviviruses was under genetic control was first proposed by Bruce-Chwatt (1950) to explain variation in YF infection rates in *A. aegypti*. Variation in the susceptibility of field populations and variation in their ability to transmit the flaviviruses YF (Aitken *et al.*, 1977; Beaty and Aitken, 1979; Tabachnick *et al.*, 1985), DEN (Gubler and Rosen, 1976; Gubler *et al.*, 1979; Rosen *et al.*, 1985), JE (Rosen *et al.*, 1985; Weng *et al.*, 1997), WNV (Ahmed *et al.*, 1979; Hayes *et al.*, 1984), and SLE (Meyer *et al.*, 1988) have been well documented.

Most work with flaviviruses has focused on DEN and its two main vectors, *A. albopictus* and *A. aegypti*. Observations documenting variation in the susceptibility of colonized geographic "strains" of *A. aegypti* and *A. albopictus* to infection with the DEN viruses suggested that in these species, susceptibility to infection was controlled genetically (Gubler *et al.*, 1979; Gubler and Rosen, 1976). Studies of colonized populations of *A. albopictus* and *A. aegypti* from Hawaii, Malaysia, Japan, and North America indicated that all were highly susceptible to infection with DEN-1 (70 to 100%), but varied significantly in the proportion that developed disseminated infections (30 to 94%) (Boromisa *et al.*, 1987). Significant differences in the ability of mosquitoes to transmit virus occurred only rarely once virus disseminated from the midgut, but significant differences in the proportion of mosquitoes transmitting virus were noted when the entire test populations were considered.

Additional studies have documented variation in vector competence within a narrower geographic scale. Significant variation was observed in susceptibility to DEN-2 within *A. aegypti* populations collected in and around Ho Chi Minh City, Vietnam (Tran *et al.*, 1999). While susceptibility to DEN-2 was variable among populations collected from the city center, those collected from the periphery were more homogeneous. These differences paralleled the degree of genetic differentiation between populations collected from either region: within the city center, populations were highly differentiated, indicating little gene flow among them, whereas there appeared to be more genetic homogeneity on the periphery. This study confirmed earlier reports that *A. aegypti* populations from highly urbanized cities tend to differ greatly in genetic structure (Tabachnick, 1991). The link between variation in susceptibility to DEN-2 and genetic differentiation was also suggested as an explanation for the pattern of susceptibility observed among populations of *A. aegypti* collected in French Polynesia (Vazeille-Falcoz *et al.*, 1999). On Tahiti, differences in susceptibility are partitioned geographically: Populations collected from the more populated and urbanized western coast of the island were highly variable, whereas those collected from the less populated and less environmentally disrupted eastern coast were homogeneous.

These studies extended earlier evidence that vector susceptibility to flavivirus infection is determined genetically by demonstrating that variation is apparent on a much finer geographic scale. Importantly, they demonstrated that this variation may be related to particular landscape features and/or human land use: relatively inbred mosquito populations may be more variable in susceptibility than populations where gene flow is less restricted

Significant variation in susceptibility and, presumably, vector competence, however, is not uniformly present in vector populations. Variation was lacking in recently colonized *A. aegypti* and *A. albopictus* collected in Madagascar (Vazeille *et al.*, 2001) and in *A. aegypti* in French Guiana (Fouque *et al.*, 2001). In the *A. albopictus* populations from Madagascar, genetic differentiation was detected, implying that the genetic differentiation observed was not sufficient to account for differences in vector competence.

Studies on natural variation in susceptibility to infection have also been described for WNV and its main vectors, *Culex* spp. mosquitoes. Ahmed *et al.* (1979) described low variation among geographic strains of *C. tritaeniorhynchus*. A second study (Hayes *et al.*, 1980) evaluated the vector competence of a colonized Pakistani strain of *C. tritaeniorhynchus* and found ID_{50} values similar to those obtained by Ahmed *et al.* (1979). A further evaluation of the susceptibility of 20 mutant and geographic strains of *C. tritaeniorhynchus* demonstrated that one mutant mosquito strain had significantly reduced susceptibility to WNV infection (Hayes *et al.*, 1984). Infection in the remaining strains was homogeneous. These studies, in contrast to those examining variation in the susceptibility of *Aedes* mosquitoes to DEN, failed to document significant differences in susceptibility or vector competence for WNV among recently colonized *C. tritaeniorhynchus* populations. Genetic studies of WNV vectors, however, have tended to use highly colonized strains of mosquitoes and tend not to include colony controls. It may be that natural variation in important WNV vectors exists in nature, but has not been identified in laboratory experiments to date.

Natural variation among *A. aegypti* populations for YF was documented by Aitken *et al.* (1977). Three geographically disparate strains of *A. aegypti* from Africa, Asia, and the New World were susceptible to infection and able to transmit virus. At low infectious titers, however, a strain from Uganda appeared less susceptible to infection and transmitted virus less frequently than the others. Unfortunately, the number of individual mosquitoes studied was insufficient for statistical significance, but these data again suggested that genetic factors unique to a mosquito population may control susceptibility to infection and vector competence.

Strains of *C. tarsalis* differ in their vector competence for SLE virus. Assays with mosquitoes collected from various locations in California indicated that one strain of *C. tarsalis* was highly susceptible to SLE infection (100% susceptibility) and transmitted virus efficiently (70%), whereas the others were incompetent (less than 70% infection and no transmission) (Meyer *et al.*, 1988).

Variation in susceptibility to, and in some cases vector competence for, flaviviruses has thus been demonstrated amply by numerous investigators since the mid-1970s. Variation has been observed among populations that differ widely in their geographic origin, but also among adjacent populations that are partitioned by the environment into populations that are essentially isolated. These studies established that vector competence is at least partly determined genetically.

B. Selection of Resistant and Susceptible Mosquitoes

The finding of significant variation in the susceptibility, and in some cases the vector competence, of different arthropod vectors of flaviviruses led to experiments aimed at selecting mosquito strains that are more or less susceptible or resistant to particular viruses. The link between variation in susceptibility/competence and genetic differentiation was taken as evidence that such efforts would be fruitful, leading some investigators to frame their experiments within the context of "permanent disease control" (Gubler *et al.*, 1979) or the regulation of wild mosquito populations (Hayes *et al.*, 1980, 1984).

Gubler and Rosen (1976) exploited natural variation in susceptibility to determine the mode of inheritance of susceptibility/resistance to DEN-2 virus in *A. albopictus*. Following at least five generations of selective inbreeding, most strains remained relatively stable in their susceptibility. One strain, however, had significantly reduced susceptibility, decreasing from 73.5 to 12.5% infection from the parental to F_2 generation. The offspring of reciprocal crosses between resistant and susceptible strains were of intermediate susceptibility, and the mode of inheritance of susceptibility was unclear.

Similar experiments conducted with *A. aegypti* sought to exploit natural variation in susceptibility to determine the mode of inheritance of DEN-2 virus susceptibility and resistance (Gubler *et al.*, 1979). In each of two experiments, hybrid offspring inherited the phenotype of the less susceptible parent, suggesting that in *A. aegypti* the gene or genes controlling decreased susceptibility to DEN-2 were dominant.

Studies of the susceptibility of *C. tritaeniorhynchus* to WNV seemingly came to the opposite conclusion regarding the mode of inheritance of susceptibility and resistance. Selective inbreeding of strains of *C. tritaeniorhynchus* with differing susceptibilities to WNV produced a strain of consistently increased susceptibility, but failed to produce a strain of consistently reduced susceptibility to infection (Hayes *et al.*, 1984). Viral titers were greater in more susceptible strains by approximately one-half $\log_{10}$. The F_1 offspring of crosses between susceptible and resistant parents were similar to the more susceptible parent in their infection rate. Infectious titers in each of the reciprocal crosses were intermediate to the parental strains, but one (susceptible female parents) was significantly greater than the other (susceptible male parents). These observations were interpreted as evidence that susceptibility was dominant. Further backcrosses to the parental lines with a mutant carrying recessive alleles on each of the three mosquito chromosomes suggested that genes on chromosomes 1 and 2 play a role in determining increased susceptibility to WN virus infection and that genetic traits blocking midgut infection and those modulating subsequent viral replication may be discrete.

Variation in the susceptibility of isofemale lines of *A. aegypti* to YF was exploited to generate lines of susceptible (~30% infection) and resistant (~10% infection) mosquitoes (Wallis *et al.*, 1985). The difference in susceptibility of these lines was statistically significant and remained fairly constant over a number of generations. Selection over these generations failed to produce lines that differed significantly from the previous generation, leading these investigators to suggest that it may be impossible to produce mosquitoes that are either completely susceptible or refractory to infection with YF. It was concluded that although susceptibility to infection was under some genetic control, the epidemiological significance of the findings was unclear.

Later studies of isofemale *A. albopictus* lines, however, established strains of mosquitoes that were highly susceptible (~90% infection) or highly refractory (~0% infection) not only to YF virus, but also to DEN and other flaviviruses, including Uganda S and Zika viruses (Miller and Mitchell, 1991). Susceptibility and resistance phenotypes were stable across a number of generations, suggesting that a small number of loci control susceptibility. A MEB was responsible for the resistant phenotype, supporting the previous observations of Boromisa *et al.* (1987), who had determined that midgut escape was the most frequent barrier to disseminated DEN-1 virus infections in *A. albopictus*, and the variation in virus titers observed in WNV-infected *Cx tritaeniorhynchus* observed by Hayes *et al.* (1984). Crossing experiments between susceptible and refractory parents produced offspring of intermediate susceptibility, implying codominance (i.e., that alleles at vector competence loci act additively). Results of backcrosses were consistent with a single controlling locus for susceptibility, but suggested that the resistant phenotype was more complex.

C. Quantitative Genetics and Viral Contribution

Because vector susceptibility to infection and presumably vector competence are under the control of multiple genes, and also influenced by environmental factors such as larval nutrition and temperature (Dohm and Turell, 2001; Grimstad and Haramis, 1984; Kramer *et al.*, 1983; Takahashi, 1976), Bosio *et al.* (1998) used quantitative genetics to estimate the relative contribution of genetic and environmental factors to the titer of DEN-2 in the midguts and heads of susceptible and incompetent mosquitoes and their offspring. The mosquito strains differed in the proportion that were infected 14 days following an infectious blood meal, implying the presence of an MIB in the resistant strain. The presence of an MEB was documented by the presence in each strain of individual mosquitoes with midgut infections, but no infectious virus was detected in the head. Once infection occurred, virus titers in each mosquito strain were similar in both the midgut and the head, implying that factors influencing the infection of, and escape from, particular tissues, and not factors influencing the kinetics of viral replication, modulate vector competence of *A. aegypti* for DEN-2. The results supported earlier work (Miller and Mitchell, 1991) suggesting that at least two loci control vector competence in *A. aegypti*. The authors hypothesized that one locus controlled an MIB and another the MEB. Environmental effects, however, accounted for most of the phenotypic variance observed. It is likely that under less well-controlled conditions (i.e., in the field), the environmental contribution to the total phenotypic variance would be even greater. Further studies demonstrated the existence of two quantitative trait loci controlling a MIB in *A. aegypti* (Bosio *et al.*, 2000). These studies mapped these alleles to chromosomes 2 and 3. Midgut escape, however, was correlated only weakly with genotype, perhaps due to the small number of mosquitoes that permitted midgut infection but blocked escape. Overall, the results of these studies supported the hypothesis that variation in dengue susceptibility and dissemination is partly due to quantitative genetic factors, but that a significant proportion of variation is best explained by environmental factors.

Viral genotype may also be a significant determinant of vector competence. Studies by Armstrong and Rico-Hesse (2001), confirming earlier studies (Rosen *et al.*, 1985), have demonstrated that particular strains of DEN-2 infect *A. aegypti* more efficiently than others. DEN-2 strains of the southeast Asian genotype (Rico-Hesse *et al.*, 1997) tend to infect a higher proportion of mosquitoes than American strains. Significantly, these differences in infectivity among viral genotypes were observed solely in low-generation mosquito colonies. The magnitude of difference in mosquito susceptibility between viral genotype decreased after maintaining the mosquito colony for successive generations in the laboratory (P. Armstrong, personal communication), supporting observations that the susceptibility of *A. aegypti* to YF virus changed as mosquitoes were passaged in the laboratory (Lorenz *et al.*, 1984). Additionally, viral passage history affects vector infection and transmission rates (Miller and Mitchell, 1986). Indeed, laboratory studies of vector competence have led to inaccurate assessments of vectorial capacity (Miller *et al.*, 1989b).

VII. Vector Control

A. *Identification of Vector Species*

Laboratories are increasingly applying RT–PCR assays, both standard and real-time fluorogenic (Applied Biosystems, Inc., CA), to their surveillance detection protocols in order to increase speed, sensitivity, and throughput of their system (Bernard *et al.*, 2001; Nasci *et al.*, 2001b; Shi *et al.*, 2001; White *et al.*, 2001). Detection of viral RNA or infectious virus in a pool of field-collected mosquitoes, however, does not indicate that species is a competent vector. The virus must be demonstrated to disseminate in the mosquito and replicate to sufficient titers in the salivary glands to be transmitted during expectoration of the salivary secretion. The use of RT–PCR as a surveillance tool leads to complications in the interpretation of vector status in that infectious virus is not isolated from all of the specimens that are RNA positive (Bernard and Kramer, 2001), and it exacerbates the interpretation of minimal infection rates, which historically have been based on infectious virus. While this technology provides a means of sensitive and rapid detection, the significance of finding RNA in the absence of infectious virus is not clear until vector competence studies are conducted (Bernard and Kramer, 2001).

Reliable mosquito identification is crucial to unraveling the complexity of virus transmission cycles. Morphological characters are extremely difficult to recognize when mosquitoes are older and have lost scales during trapping and transport to the laboratory. The application of PCR to confirm *Culex* species identification (Crabtree *et al.*, 1995), for example, increases the confidence in conclusions drawn regarding epidemiological significance of various species. The molecular identification of species implicated in WNV transmission in the United States indicated problems in morphological identification, which have an impact on control issues (Bernard and Kramer, 2001; Nasci *et al.*, 2001b). Once the target species are identified accurately, in-depth ecological studies are a prerequisite to effective control.

B. *Classical Methods of Control*

Integrated pest management applies a combination of several approaches to intervention, including environmental management and biological, chemical, and mechanical control measures in an ecologically sound manner (Lacey and Lacey, 1990). Efforts to control WNV in the northeastern United States in 2000 and 2001 used such an approach, focusing on source reduction and larviciding to reduce the population of the enzootic vector *C. pipiens* and encouragement of the use of personal protection measures. Spraying for adults is expensive and often ineffective in the urban environment (Newton and Reiter, 1992) where DEN and WNV human vectors are active. Control of rice land mosquitoes, which are vectors of JE, has also focused on IPM, using biological control agents such as larvivorous fish, a nematode, fungus, and bacteria, as well as environmental management (Lacey and Lacey, 1990). Mosquito control experts need to balance the short- and long-term risks of pesticide use, high cost, sustainability, and resistance with the predicted risk of disease (Thier, 2001). In the case of WNV, guidelines were composed to aid in this estimation (Centers for Disease Control and Prevention, 2000b).

C. *Novel Strategies*

Novel control strategies may be necessary as insecticide resistance evolves in some vectors and spreads through their populations (Chareonviriyahpap *et al.*, 1999; Crampton *et al.*, 1994; McClelland,

1972; Rawlins, 1998; Roberts and Andre, 1994; Vaughan *et al.*, 1998). One novel approach that may be exploited for disease control is to block arbovirus transmission by reducing mosquito infection rates through the immunization of vertebrate reservoirs with antimosquito antibodies (Ramasamy *et al.*, 1990). This approach would be most applicable where domestic animals play an important role in the epidemiology of the disease (e.g., pigs and JE) (Nandi *et al.*, 1982).

The vector competence for flaviviruses is at least partly genetically determined (Bosio *et al.*, 1998, 2000; Miller and Mitchell, 1991). Genetic manipulation of vectors may therefore reduce their competence. Efforts to create incompetent "transgenic" mosquitoes have been reviewed extensively (Beerntsen *et al.*, 2000; Gubler, 1982) and promoted vigorously by some as the key to control of many vector-borne infections (James *et al.*, 1999; Kokoza *et al.*, 2000), including flaviviruses (Gubler *et al.*, 1979; Hayes *et al.*, 1984; Jasinskiene *et al.*, 1998; Olson *et al.*, 1996).

A control strategy based on the genetic transformation of vector populations would require a mechanism that reduces vector competence dramatically (effector) and one that carries the effector through wild mosquito populations (drive). Much research in this area has centered on parasitic vector-borne diseases such as malaria and lymphatic filariasis where the natural mechanisms of vector resistance to parasite infection are better understood (reviewed by Beerntsen *et al.*, 2000). Progress toward genetically manipulating mosquitoes for decreased vector competence for arboviruses has been somewhat slower because the mechanisms of mosquito resistance to flavivirus infections are at present relatively obscure.

Most studies aimed at creating flavivirus-refractory genetically modified mosquitoes have not pursued natural mosquito resistance to flaviviruses, which appears to be fairly weak (Bosio *et al.*, 1998, 2000). Rather, they have sought to introduce novel antiviral mechanisms that do not occur naturally in mosquito populations. Studies have shown that DEN viruses may be inhibited by an RNA interference mechanism when DEN sequences are inserted into double subgenomic Sindbis viruses (Adelman *et al.*, 2001). Similar studies have been described for other arboviral agents (Blair *et al.*, 2000; Higgs *et al.*, 1998; Olson *et al.*, 1996), indicating that this approach may be broadly applicable. Interventions based on the genetic manipulation of vector mosquitoes nonetheless face nontrivial technical, theoretical, and political hurdles. It is likely that classical control methods will prevail in the foreseeable future.

VIII. Future Directions

Many issues surrounding the dynamics of flavivirus infection in mosquitoes need to be resolved. While progress has been made in the genetics of mosquito resistance and susceptibility to flaviviruses, the mechanisms of such have not been described. The genetic basis for host preference and other components of vectorial capacity have also not been investigated adequately. The ecological and environmental conditions that favor emergence and perpetuation of mosquito-borne flaviviruses are poorly understood. The effect of changing global climate is more complex than initially acknowledged and requires rigorous study. A better understanding of all these issues will elucidate flavivirus transmission cycles and risk of disease.

References

Adelman, Z. N., Blair, C. D., Carlson, J. O., Beaty, B. J., and Olson, K. E. (2001). Sindbis virus-induced silencing of dengue viruses in mosquitoes. *Insect Mol. Biol.* **10:**265–273.

Ahmed, T., Hayes, C. G., and Baqar, S. (1979). Comparison of vector competence for West Nile virus of colonized populations of *Culex tritaeniorhynchus* from southern Asia and the Far East. *Southeast Asian J. Trop. Med. Public Health* **10:**498–504.

Aitken, T. H., Downs, W. G., and Shope, R. E. (1977). *Aedes aegypti* strain fitness for yellow fever virus transmission. *Am. J. Trop. Med. Hyg.* **26:**985–989.

Aitken, T. H., Tesh, R. B., Beaty, B. J., and Rosen, L. (1979). Transovarial transmission of yellow fever virus by mosquitoes *(Aedes aegypti). Am. J. Trop. Med. Hyg.* **28:**119–121.

Anderson, R. A., and Brust, R. A. (1997). Interrupted blood feeding by *Culex* (Diptera: Culicidae) in relation to individual host tolerance to mosquito attack. *J. Med. Entomol.* **34:**95–101.

Anderson, S. G., and Eagle, M. (1953). Murray Valley encephalitis: The contrasting epidemiological picture in 1951 and 1952. *Med. J. Aust.* **1:**478–481.

Armstrong, P. M., and Rico-Hesse, R. (2001). Differential susceptibility of *Aedes aegypti* to infection by the American and southeast Asian genotypes of dengue type 2 virus. *Vector Borne Zoonotic Dis.* **1:**159–168.

Bailey, C. L., Eldridge, B. F., Hayes, D. E., Watts, D. M., Tammariello, R. F., and Dalrymple, J. M. (1978). Isolation of St. Louis encephalitis virus from overwintering *Culex pipiens* mosquitoes. *Science* **199:**1346–1349.

Baqar, S. (1980). The effect of larval rearing conditions and adult age and the susceptibility of *Culex tritaeniorhynichus* to infection with West Nile virus. *Mosq. News* **40:**165–173.

Baqar, S., Hayes, C. G., Murphy, J. R., and Watts, D. M. (1993). Vertical transmission of West Nile virus by *Culex* and *Aedes* species mosquitoes. *Am. J. Trop. Med. Hyg.* **48:**757–762.

Bates, M., and Roca-Garcia, M. (1946). The development of the virus of yellow fever in *Haemagogus* mosquitoes. *Am. J. Trop. Med. Hyg.* **26:**585–605.

Beaty, B. J., and Aitken, T. H. G. (1979). In vitro transmission of yellow fever virus by geographic strains of *Aedes aegypti*. *Mosq. News* **39:**232–238.

Beaty, B. J., Tesh, R. B., and Aitken, T. H. (1980). Transovarial transmission of yellow fever virus in *Stegomyia* mosquitoes. *Am. J. Trop. Med. Hyg.* **29:**125–132.

Beerntsen, B. T., James, A. A., and Christensen, B. M. (2000). Genetics of mosquito vector competence. *Microbiol. Mol. Biol. Rev.* **64:**115–137.

Bernard, K. A., and Kramer, L. D. (2001). West Nile virus activity in the United States, 2001. *Viral Immunol.* **14:**319–338.

Bernard, K. A., Maffei, J. G., Jones, S. A., Kauffman, E. B., Ebel, G. D., Dupuis, A. P., II, Ngo, K. A., Nicholas, D. C., Young, D. M., Shi, P.-Y., Kulasekera, V. L., Eidson, M., White, D. J., Stone, W. B., and Kramer, L. D. (2001). West Nile virus infection in birds and mosquitoes, New York State, 2000. *Emerg. Infect. Dis.* **7:**679–685.

Bidlingmayer, W. L., and Hem, D. G. (1973). Sugar feeding by Florida mosquitoes. *Mosq. News* **33:**535–538.

Black, W. I., and Moore, C. G. (1996). Population biology as a tool for studying vector-borne diseases. *In* "The Biology of Disease Vectors" (B. J. Beaty and W. C. Marquards, eds.), pp. 393–416. University Press of Colorado, Niwot.

Blackmore, M. S., and Lord, C. C. (2000). The relationship between size and fecundity in *Aedes albopictus*. *J. Vector Ecol.* **25:**212–217.

Blair, C. D., Adelman, Z. N., and Olson, K. E. (2000). Molecular strategies for interrupting arthropod-borne virus transmission by mosquitoes. *Clin. Microbiol. Rev.* **13:**651–661.

Boakye, D. A., Tang, J., Truc, P., Merriweather, A., and Unnasch, T. R. (1999). Identification of bloodmeals in haematophagous Diptera by cytochrome B heteroduplex analysis. *Med. Vet. Entomol.* **13:**282–287.

Boromisa, R. D., Rai, K. S., and Grimstad, P. R. (1987). Variation in the vector competence of geographic strains of *Aedes albopictus* for dengue 1 virus. *J. Am. Mosq. Control Assoc.* **3:**378–386.

Bosio, C. F., Beaty, B. J., and Black, W. C. (1998). Quantitative genetics of vector competence for dengue-2 virus in *Aedes aegypti*. *Am. J. Trop. Med. Hyg.* **59:**965–970.

Bosio, C. F., Fulton, R. E., Salasek, M. L., Beaty, B. J., and Black, W. C. (2000). Quantitative trait loci that control vector competence for dengue-2 virus in the mosquito *Aedes aegypti*. *Genetics* **156:**687–698.

Bosio, C. F., Thomas, R. E., Grimstad, P. R., and Rai, K. S. (1992). Variation in the efficiency of vertical transmission of dengue-1 virus by strains of *Aedes albopictus* (Diptera: Culicidae). *J. Med. Entomol.* **29:**985–989.

Bouma, M. J., Kovats, R. S., Goubet, S. A., Cox, J. S., and Haines, A. (1997a). Global assessment of El Niño's disaster burden. *Lancet* **350:**1435–1438.

Bouma, M. J., Poveda, G., Rojas, W., Chavasse, D., Quinones, M., Cox, J., and Patz, J. (1997b). Predicting high-risk years for malaria in Colombia using parameters of El Niño Southern Oscillation. *Trop. Med. Int. Health* **2:**1122–1127.

Bowen, G. S., and Francy, D. B. (1980). Surveillance. *In* "St. Louis Encephalitis" (T. P. Monath, ed.), pp. 473–499. American Public Health Association, Inc., Washington DC.

Bruce-Chwatt, L. J. (1950). Recent studies on insect vectors of yellow fever and malaria in British West Africa. *J. Trop. Med. Hyg.* **53:**71–79.

Burke, D. S., and Leake, C. J. (1988). Japanese encephalitis. *In* "The Arboviruses: Epidemiology and Ecology" (T. P. Monath, ed.), pp. 63–92. CRC Press, Boca Raton, Fla.

Carbajo, A. E., Schweigmann, N., Curto, S. I., de Garin, A., and Bejaran, R. (2001). Dengue transmission risk maps of Argentina. *Trop. Med. Int. Health* **6:**170–183.

Cayan, D. R. (1999). ENSO and hydrologic extremes in the Western United States. *J. Climate* **12:**2893.

Centers for Disease Control and Prevention (2000a). Update: West Nile Virus activity—Eastern United States, 2000. *Morb. Mortal. Wkly. Rep.* **49:**1044–1047.
Centers for Disease Control and Prevention (2000b). Guidelines for surveillance, prevention, and control of West Nile virus infection—United States. *J. Am. Med. Assoc.* **283:**997–998.
Centers for Disease Control and Prevention (2000c). Surveillance for West Nile virus in overwintering mosquitoes—New York, 2000. *J. Am. Med. Assoc.* **283:**2380–2381.
Centers for Disease Control and Prevention (2000d). West Nile virus activity—New York and New Jersey. *J. Am. Med. Assoc.* **284:**823–824.
Chadee, D. D., and Martinez, R. (2000). Landing periodicity of *Aedes aegypti* with implications for dengue transmission in Trinidad, West Indies. *J. Vector Ecol.* **25:**158–163.
Chadee, D. D., Tikasingh, E. S., and Ganesh, R. (1992). Seasonality, biting cycle and parity of the yellow fever vector mosquito *Haemagogus janthinomys* in Trinidad. *Med. Vet. Entomol.* **6:**143–148.
Chamberlain, R. W., Corristan, E. C., and Sikes, R. K. (1954). Studies on the North American arthropod-borne encephalitides. V. The extrinsic incubation of eastern and western equine encephalitis in mosquitoes. *Am. J. Hyg.* **60:**269–277.
Chamberlain, R. W., and Sudia, W. D. (1961). Mechanism of transmission of viruses by mosquitos. *Annu. Rev. Entomol.* **6:**371–390.
Chamberlain, R. W., Sudia, W. D., Coleman, P. H., and Beadle, L. D. (1964). Vector studies in the St. Louis encephalitis epidemic, Tampa Bay area, Florida, 1962. *Am. J. Trop. Med. Hyg.* **13:**456–461.
Chamberlain, R. W., Sudia, W. D., and Gillett, J. D. (1959). St. Louis encephalitis virus in mosquitoes. *Am. J. Hyg.* **70:**221–236.
Chambers, G. M., and Klowden, M. J. (1990). Correlation of nutritional reserves with a critical weight for pupation in larval *Aedes aegypti* mosquitoes. *J. Am. Mosq. Control Assoc.* **6:**394–399.
Chan, Y. C., Ho, B. C., and Chan, K. L. (1971). *Aedes aegypti* (L.) and *Aedes albopictus* (Skuse) in Singapore City. V. Observations in relation to dengue haemorrhagic fever. *Bull. World Health Organ.* **44:**651–657.
Chareonviriyahpap, T., Aum-aung, B., and Ratanatham, S. (1999). Current insecticide resistance patterns in mosquito vectors in Thailand. *Southeast Asian J. Trop. Med. Public Health* **30:**184–194.
Cheng, S., Kalkstein, L. S., Focks, D. A., and Nnaji, A. (1998). New procedures to estimate water temperatures and water depths for application in climate-dengue modeling. *J. Med. Entomol.* **35:**646–652.
Christophers, S. R. (1960). "*Aedes aegypti* (L.), The Yellow Fever Mosquito: Its Life History, Bionomics and Structure." Cambridge Univ. Press, London.
Cordellier, R. (1991). The epidemiology of yellow fever in Western Africa. *Bull. World Health Organ.* **69:**73–84.
Costero, A., Attardo, G. M., Scott, T. W., and Edman, J. D. (1998a). An experimental study on the detection of fructose in *Aedes aegypti*. *J. Am. Mosq. Control Assoc.* **14:**234–242.
Costero, A., Edman, J. D., Clark, G. G., Kittayapong, P., and Scott, T. W. (1999). Survival of starved *Aedes aegypti* (Diptera: Culicidae) in Puerto Rico and Thailand. *J. Med. Entomol.* **36:**272–276.
Costero, A., Edman, J. D., Clark, G. G., and Scott, T. W. (1998b). Life table study of *Aedes aegypti* (Diptera: Culicidae) in Puerto Rico fed only human blood versus blood plus sugar. *J. Med. Entomol.* **35:**809–813.

Crabtree, M. B., Savage, H. M., and Miller, B. R. (1995). Development of a species-diagnostic polymerase chain reaction assay for the identification of *Culex* vectors of St. Louis encephalitis virus based on interspecies sequence variation in ribosomal DNA spacers. *Am. J. Trop. Med. Hyg.* **53:**105–109.

Crampton, J. M., Warren, A., Lycett, G. J., Hughes, M. A., Comley, I. P., and Eggleston, P. (1994). Genetic manipulation of insect vectors as a strategy for the control of vector-borne disease. *Ann. Trop. Med. Parasitol.* **88:**3–12.

Day, J. F. (2001). Predicting St. Louis encephalitis virus epidemics: Lessons from recent, and not so recent, outbreaks. *Annu. Rev. Entomol.* **46:**111–138.

Day, J. F., and Curtis, G. A. (1993). Annual emergence patterns of *Culex nigripalpus* females before, during and after a widespread St. Louis encephalitis epidemic in south Florida. *J. Am. Mosq. Control Assoc.* **9:**249–255.

Day, J. F., Edman, J. D., and Scott, T. W. (1994). Reproductive fitness and survivorship of *Aedes aegypti* (Diptera: Culicidae) maintained on blood, with field observations from Thailand. *J. Med. Entomol.* **31:**611–617.

Degallier, N., Travassos da Rosa, A. P., Vasconcelos, P. F., Guerreiro, S. C., Travassos da Rosa, J. F., and Herve, J. P. (1991). Estimation of the survival rate, the relative density and the infection rate of a population of *Haemagogus janthinomys* Dyar (Diptera: Culicidae) from which strains of yellow fever were isolated in Brazilian Amazon. *Bull. Soc. Pathol. Exot.* **84:**386–397.

Dhileepan, K. (1996). Mosquito seasonality and arboviral disease incidence in Murray Valley, southeast Australia. *Med. Vet. Entomol.* **10:**375–384.

Dohm, D. J., and Turell, M. (2001). Effect of incubation at overwintering temperatures on the replication of West Nile virus in New York *Culex pipiens* (Diptera: Culicadae). *J. Med. Entomol.* **38:**462–464.

Doi, R. (1970). Studies on the mode of development of Japanese encephalitis virus in some groups of mosquitoes by the fluorescent antibody technique. *Jpn. J. Exp. Med.* **40:**101–115.

Doi, R., Shirasaki, A., and Sasa, M. (1967). The mode of development of Japanese encephalitis virus in the mosquito *Culex tritaeniorhynchus summorosus* as observed by the fluorescent antibody technique. *Jpn. J. Exp. Med.* **37:**227–238.

Dutary, B. E., and LeDuc, J. W. (1981). Transovarial transmission of yellow fever virus by a sylvatic vector, *Haemagogus equinus*. *Trans. R. Soc. Trop. Med. Hyg.* **75:**128.

Dye, C. (1992). The analysis of parasite transmission by bloodsucking insects. *Annu. Rev. Entomol.* **37:**1–19.

Edman, J. D., Strickman, D., Kittayapong, P., and Scott, T. W. (1992). Female *Aedes aegypti* (Diptera: Culicidae) in Thailand rarely feed on sugar. *J. Med. Entomol.* **29:**1035–1038.

Epstein, P. R. (2001). West Nile virus and the climate. *J. Urban Health* **78:**367–371.

Esteva, L., and Vargas, C. (1998). Analysis of a dengue disease transmission model. *Math. Biosci.* **150:**131–151.

Esteva, L., and Vargas, C. (2000). Influence of vertical and mechanical transmission on the dynamics of dengue disease. *Math. Biosci.* **167:**51–64.

Fan, W. F., Yu, S. R., and Cosgriff, T. M. (1989). The reemergence of dengue in China. *Rev. Infect. Dis.* **11**(Suppl. 4)**:**S847–S853.

Focks, D. A., Brenner, R. J., Chadee, D. D., and Trosper, J. H. (1999). The use of spatial analysis in the control and risk assessment of vector-borne diseases. *Am. J. Entomol.* **45:**173–183.

Focks, D. A., and Chadee, D. D. (1997). Pupal survey: An epidemiologically significant surveillance method for *Aedes aegypti*. *Am. J. Trop. Med. Hyg.* **56:**159–167.

Focks, D. A., Daniels, E., Haile, D. G., and Keesling, J. E. (1995). A simulation model of the epidemiology of urban dengue fever: Literature analysis, model development, preliminary validation, and samples of simulation results. *Am. J. Trop. Med. Hyg.* **53:**489–506.

Focks, D. A., Haile, D. G., Daniels, E., and Mount, G. A. (1993a). Dynamic life table model for *Aedes aegypti* (Diptera: Culicidae): Analysis of the literature and model development. *J. Med. Entomol.* **30:**1003–1017.

Focks, D. A., Haile, D. G., Daniels, E., and Mount, G. A. (1993b). Dynamic life table model for *Aedes aegypti* (Diptera: Culicidae): Simulation results and validation. *J. Med. Entomol.* **30:**1018–1028.

Focks, D. A., Lele, S., Hayes, J., Morrison, A. C., Bangs, M. J., Church, C., Scott, T. W., and Clark, G. (2001). A potential dengue control strategy focusing on rare but extremly productive breeding containers. *Am. J. Trop. Med.* **65S:**281–282.

Fontenille, D., Diallo, M., Mondo, M., Ndiaye, M., and Thonnon, J. (1997). First evidence of natural vertical transmission of yellow fever virus in *Aedes aegypti*, its epidemic vector. *Trans. R. Soc. Trop. Med. Hyg.* **91:**533–535.

Fouque, F., Vazeille, M., Mousson, L., Gaborit, P., Carinci, R., Issaly, J., Rodhain, F., and Failloux, A. B. (2001). *Aedes aegypti* in French Guiana: Susceptibility to a dengue virus. *Trop. Med. Int. Health* **6:**76–82.

Francy, D. B., Rush, W. A., Montoya, M., Inglish, D. S., and Bolin, R. A. (1981). Transovarial transmission of St. Louis encephalitis virus by *Culex pipiens* complex mosquitoes. *Am. J. Trop. Med. Hyg.* **30:**699–705.

Freier, J. E., and Rosen, L. (1988). Vertical transmission of dengue viruses by *Aedes mediovittatus*. *Am. J. Trop. Med. Hyg.* **39:**218–222.

Gaunt, M. W., Sall, A. A., de Lamballerie, X., Falconar, A. K., Dzhivanian, T. I., and Gould, E. A. (2001). Phylogenetic relationships of flaviviruses correlate with their epidemiology, disease association and biogeography. *J. Gen. Virol.* **82:**1867–1876.

Geier, M., Bosch, O. J., and Boeckh, J. (1999). Ammonia as an attractive component of host odour for the yellow fever mosquito, *Aedes aegypti*. *Chem. Senses* **24:**647–653.

Germain, M., Francy, D. B., Monath, T. P., Ferrara, L., Bryan, J., Salaun, J. J., Heme, G., Renaudet, J., Adam, C., and Digoutte, J. P. (1980). Yellow fever in the Gambia, 1978–1979: Entomological aspects and epidemiological correlations. *Am. J. Trop. Med. Hyg.* **29:**929–940.

Gomes, A. C., Bitencourt, M. D., Natal, D., Pinto, P. L. S., Mucci, L. F., *et al.* (1999). *Aedes albopictus* em area rural do Brasil e implicacoes na transmissao de febre amarela silvestre. *Saude* **33:**95–97.

Gould, D. J., Yuill, T. M., Moussa, M. A., Simasathien, P., and Rutledge, L. C. (1968). An insular outbreak of dengue hemorrhagic fever. III. Identification of vectors and observations on vector ecology. *Am. J. Trop. Med. Hyg.* **17:**609–618.

Gratz, N. G. (1999). Emerging and resurging vector-borne diseases. *Annu. Rev. Entomol.* **44:**51–75.

Grimstad, P. R., and Haramis, L. D. (1984). *Aedes triseriatus* (Diptera: Culicidae) and La Crosse virus. III. Enhanced oral transmission by nutrition-deprived mosquitoes. *J. Med. Entomol.* **21:**249–256.

Grimstad, P. R., Paulson, S. L., and Craig, G. B., Jr. (1985). Vector competence of *Aedes hendersoni* (Diptera: Culicidae) for La Crosse virus and evidence of a salivary-gland escape barrier. *J. Med. Entomol.* **22:**447–453.

Grimstad, P. R., Ross, Q. E., and Craig, G. B., Jr. (1980). *Aedes triseriatus* (Diptera: Culicidae) and La Crosse virus. II. Modification of mosquito feeding behavior by virus infection. *J. Med. Entomol.* **17:**1–7.

Gubler, D. (1982). Arthropod vector competence: Epidemiological, genetic, and biological considerations. *In* "Recent Developments In the Genetics of Insect Disease Vectors," pp. 343–378. Stipes, Champaign, Ill.
Gubler, D. (1988). Dengue. *In* "The Arboviruses: Epidemiology and Ecology" (T. P. Monath, ed.), pp. 223–260. CRC Press, Boca Raton, Fla.
Gubler, D. J. (1970). The ecology of *Aedes albopictus*. *In* "The Johns Hopkins University CMRT Annual Report," p. 74. Johns Hopkins Univ. Baltimore, Md.
Gubler, D. J. (1998). The global pandemic of dengue-dengue haemorrhagic fever: Current status and prospects for the future. *Ann. Acad. Med. Singapore* **27:**227–234.
Gubler, D. J., Nalem, S., Tan, R., Saipan, H., and Saroso, J. S. (1979). Variation in suseptibility to oral infection with dengue viruses among geographic strains of *Aedes aegypti*. *Am. J. Trop. Med. Hyg.* **28:**1045–1052.
Gubler, D. J., Reiter, P., Ebi, K. L., Yap, W., Nasci, R., and Patz, J. A. (2001). Climate variability and change in the United States: Potential impacts on vector- and rodent-borne diseases. *Environ. Health Perspect.* **109**(Suppl. 2)**:**223–233.
Gubler, D. J., and Rosen, L. (1976). Variation among geographic strains of *Aedes albopictus* in susceptibility to infection with dengue viruses. *Am. J. Trop. Med. Hyg.* **25:**318–325.
Hales, S., Weinstein, P., Souares, Y., and Woodward, A. (1999). El Niño and the dynamics of vectorborne disease transmission. *Environ. Health Perspect.* **107:**99–102.
Hales, S., Weinstein, P., and Woodward, A. (1996). Dengue fever epidemics in the South Pacific: Driven by El Niño Southern Oscillation? *Lancet* **348:**1664–1665.
Hanna, J. N., Ritchie, S. A., Phillips, D. A., Lee, J. M., Hills, S. L., van Den Hurk, A. F., Pyke, A. T., Johansen, C. A., and MacKenzie, J. S. (1999). Japanese encephalitis in North Queensland, Australia, 1998. *Med. J. Aust.* **170:**533–536.
Hanna, J. N., Ritchie, S. A., Phillips, D. A., Shield, J., Bailey, M. C., MacKenzie, J. S., Poidinger, M., McCall, B. J., and Mills, P. J. (1996). An outbreak of Japanese encephalitis in the Torres Strait, Australia, 1995. *Med. J. Aust.* **165:**256–260.
Hardy, J. L. (1988). Susceptibility and resistance of vector mosquitos. *In* "The Arboviruses: Epidemiology and Ecology." (T. P. Monath, ed.), pp. 87–126. CRC Press, Boca Raton, Fla.
Hardy, J. L., Houk, E. J., Kramer, L. D., and Reeves, W. C. (1983). Intrinsic factors affecting vector competence of mosquitoes for arboviruses. *Annu. Rev. Entomol.* **28:**229–262.
Hardy, J. L., Presser, S. B., Meyer, R. P., Reisen, W. K., Kramer, L. D., and Vorndam, A. V. (1986). Comparison of a 1984 Los Angeles strain of SLE virus with earlier California strains of SLE virus: Mouse virulence, chicken viremogenic, RNA oligonucleotide and vector competence characteristics. *Proc. Calif. Mosq. Vector Control Assoc.* **53:**10–15.
Hardy, J. L., and Reeves, W. C. (1990). Experimental studies in infection in vectors. *In* "Epidemiology and Control of Mosquito-Borne Arboviruses in California" (W. C. Reeves, ed.), pp. 145–253. Calif. Mosq. Vect. Cont. Assoc., Sacramento, Calif.
Hardy, J. L., Rosen, L., Kramer, L. D., Presser, S. B., Shroyer, D. A., and Turell, M. J. (1980). Effect of rearing temperature on transovarial transmission of St. Louis encephalitis virus in mosquitoes. *Am. J. Trop. Med. Hyg.* **29:**963–968.
Hardy, J. L., Rosen, L., Reeves, W. C., Scrivani, R. P., and Presser, S. B. (1984). Experimental transovarial transmission of St. Louis encephalitis virus by *Culex* and *Aedes* mosquitoes. *Am. J. Trop. Med. Hyg.* **33:**166–175.
Hayes, C. G. (1989). West Nile fever. *In* "The Arboviruses: Epidemiology and Ecology" (T. P. Monath, ed.), Vol. V, pp. 59–88. CRC Press, Boca Raton, Fla.

Hayes, C. G., Baker, R. H., Baqar, S., and Ahmed, T. (1984). Genetic variation for West Nile virus susceptibility in *Culex tritaeniorhynchus*. *Am. J. Trop. Med. Hyg.* **33:**715–724.

Hayes, C. G., Basit, A., Bagar, S., and Akhter, R. (1980). Vector competence of *Culex tritaeniorhynchus* (Diptera: Culicidae) for West Nile virus. *J. Med. Entomol.* **17:**172–177.

Higgs, S., Rayner, J. O., Olson, K. E., Davis, B. S., Beaty, B. J., and Blair, C. D. (1998). Engineered resistance in *Aedes aegypti* to a West African and a South American strain of yellow fever virus. *Am. J. Trop. Med. Hyg.* **58:**663–670.

Ho, B. C., Chan, Y. C., and Chan, K. L. (1973). Field and laboratory observations on landing and biting periodicities of *Aedes albopictus* (Skuse). *Southeast Asian J. Trop. Med. Public Health* **4:**238–244.

Hopp, M. J., and Foley, J. A. (2001). Global-scale relationships between climate and the dengue fever vector, *Aedes aegypti*. *Climate Change* **48:**441–463.

Hubalek, Z. (2000). European experience with the West Nile virus ecology and epidemiology: Could it be relevant for the New World? *Viral Immunol.* **13:**415–426.

Hubalek, Z., and Halouzka, J. (1999). West Nile fever: A reemerging mosquito-borne viral disease in Europe. *Emerg. Infect. Dis.* **5:**643–650.

Hubalek, Z., Halouzka, J., Juricova, Z., Prikazsky, Z., Zakova, J., and Sebesta, O. (1999). Surveillance of mosquito-borne viruses in Breclav after the flood of 1997. *Epidemiol. Mikrobiol. Imunol.* **48:**91–96.

Hull, B., Tikasingh, E., de Souza, M., and Martinez, R. (1984). Natural transovarial transmission of dengue 4 virus in *Aedes aegypti* in Trinidad. *Am. J. Trop. Med. Hyg.* **33:**1248–1250.

Hurlbut, H. S. (1950). Japanese encephalitis virus survival in and transmitted in winter conditions. *Am. J. Hyg.* **51:**265–268.

Hurlbut, H. S. (1966). Mosquito salivation and virus transmission. *Am. J. Trop. Med. Hyg.* **15:**989–993.

Hurlbut, H. S. (1973). The effect of environmental temperature upon the transmission of St. Louis encephalitis virus by *Culex pipiens quinquefasciatus*. *J. Med. Entomol.* **10:**1–12.

Ibanez-Bernal, S., Briseno, B., Mutebi, J. P., Argot, E., Rodriguez, G., Martinez-Campos, C., Paz, R., de la Fuente-San Roman, P., Tapia-Conyer, R., and Flisser, A. (1997). First record in America of *Aedes albopictus* naturally infected with dengue virus during the 1995 outbreak at Reynosa, Mexico. *Med. Vet. Entomol.* **11:**305–309.

Intergovernmental Panel on Climate Change (IPCC) (1996). Climate change 1995: The science of climate change. *In* "Contribution of Working Group I to the Second Assessment Report of the Intergovernmental Panel on Climate Change." Cambridge Univ. Press, Cambridge, UK.

James, A. A., Beerntsen, B. T., Capurro, M., Coates, C. J., Coleman, J., Jasinskiene, N., and Krettli, A. U. (1999). Controlling malaria transmission with genetically-engineered, Plasmodium-resistant mosquitoes: Milestones in a model system. *Parassitologia* **41:**461–471.

Jasinskiene, N., Coates, C. J., Benedict, M. Q., Cornel, A. J., Rafferty, C. S., James, A. A., and Collins, F. H. (1998). Stable transformation of the yellow fever mosquito, *Aedes aegypti*, with the Hermes element from the housefly. *Proc. Natl. Acad. Sci. USA* **95:**3743–3747.

Jetten, T. H., and Focks, D. A. (1997). Potential changes in the distribution of dengue transmission under climate warming. *Am. J. Trop. Med. Hyg.* **57:**285–297.

Johansen, C. A., van Den Hurk, A. F., Pyke, A. T., Zborowski, P., Phillips, D. A., MacKenzie, J. S., and Ritchie, S. A. (2001). Entomological investigations of an outbreak of Japanese encephalitis virus in the Torres Strait, Australia, in 1998. *J. Med. Entomol.* **38:**581–588.

Johansen, C. A., van Den Hurk, A. F., Ritchie, S. A., Zborowski, P., Nisbet, D. J., Paru, R., Bockarie, M. J., Macdonald, J., Drew, A. C., Khromykh, T. I., and MacKenzie, J. S. (2000). Isolation of Japanese encephalitis virus from mosquitoes (Diptera: Culicidae) collected in the Western Province of Papua New Guinea, 1997–1998. *Am. J. Trop. Med. Hyg.* **62:**631–638.

Johnson, C. G. (1969). "Migration and Dispersal of Insects by Flight." Methuen, London.

Jumali, Sunarto, Gubler, D. J., Nalim, S., Eram, S., and Sulianti Saroso, J. (1979). Epidemic dengue hemorrhagic fever in rural Indonesia. III. Entomological studies. *Am. J. Trop. Med. Hyg.* **28:**717–724.

Jupp, P. G. (1974). Laboratory studies on the transmission of West Nile virus by *Culex (Culex) univittatus* Theobald; factors influencing the transmission rate. *J. Med. Entomol.* **11:**455–458.

Kay, B. H., and Carley, J. G. (1980). Transovarial transmission of Murray Valley encephalitis virus by *Aedes aegypti* (L). *Aust. J. Exp. Biol. Med. Sci.* **58:**501–504.

Kay, B. H., Edman, J. D., Fanning, I. D., and Mottram, P. (1989a). Larval diet and the vector competence of *Culex annulirostris* (Diptera: Culicidae) for Murray Valley encephalitis virus. *J. Med. Entomol.* **26:**487–488.

Kay, B. H., Fanning, I. D., and Mottram, P. (1989b). Rearing temperature influences flavivirus vector competence of mosquitoes. *Med. Vet. Entomol.* **3:**415–422.

Kay, B. H., Fanning, I. D., and Mottram, P. (1989c). The vector competence of *Culex annulirostris, Aedes sagax* and *Aedes alboannulatus* for Murray Valley encephalitis virus at different temperatures. *Med. Vet. Entomol.* **3:**107–112.

Kay, B. H., and Farrow, R. A. (2000). Mosquito (Diptera: Culicidae) dispersal: Implications for the epidemiology of Japanese and Murray Valley encephalitis viruses in Australia. *J. Med. Entomol.* **37:**797–801.

Kay, B. H., Saul, A. J., and McCullagh, A. (1987). A mathematical model for the rural amplification of Murray Valley encephalitis virus in southern Australia. *Am. J. Epidemiol.* **125:**690–705.

Khin, M. M., and Than, K. A. (1983). Transovarial transmission of dengue 2 virus by *Aedes aegypti* in nature. *Am. J. Trop. Med. Hyg.* **32:**590–594.

Khozinskaya, G. A., Chunikhin, S. P., Khozinsky, V. V., and Stefutkina, L. F. (1985). Variability of Powassan virus cultured in tissue explants and organism of *Hyalomma anatolicum* ticks. *Acta Virol.* **29:**305–311.

Kokoza, V., Ahmed, A., Cho, W. L., Jasinskiene, N., James, A. A., and Raikhel, A. (2000). Engineering blood meal-activated systemic immunity in the yellow fever mosquito, *Aedes aegypti*. *Proc. Natl. Acad. Sci. USA* **97:**9144–9149.

Kow, C. Y., Koon, L. L., and Yin, P. F. (2001). Detection of dengue viruses in field caught male *Aedes aegypti* and *Aedes albopictus* (Diptera: Culicidae) in Singapore by type-specific PCR. *J. Med. Entomol.* **38:**475–479.

Kramer, L. D., and Bernard, K. A. (2001). West Nile virus in the Western Hemisphere. *Curr. Opin. Infect. Dis.* **14:**519–525.

Kramer, L. D., and Chandler, L. J. (2001). Phylogenetic analysis of the envelope gene of St. Louis encephalitis virus. *Arch. Virol.* **146:**2341–2355.

Kramer, L. D., Hardy, J. L., and Presser, S. B. (1983). Effect of temperature of extrinsic incubation on the vector competence of *Culex tarsalis* for western equine enephalomyelitis virus. *Am. J. Trop. Med. Hyg.* **32:**1130–1139.

Kramer, L. D., Hardy, J. L., Presser, S. B., and Houk, E. J. (1981). Dissemination barriers for western equine encephalomyelitis virus in *Culex tarsalis* infected after ingestion of low viral doses. *Am. J. Trop. Med. Hyg.* **30:**190–197.

Kuberski, T. (1979). Fluorescent antibody studies on the development of dengue-2 virus in *Aedes albopictus* (Diptera: Culicidae). *J. Med. Entomol.* **16:**343–349.

Kulasekera, V. L., Kramer, L. D., Nasci, R. S., Mostashari, F., Cherry, B., Trock, S. C., Glaser, C., and Miller, J. R. (2001). West Nile virus infection in mosquitoes, birds, horses and humans, Staten Island, New York, 2000. *Emerg. Infect. Dis.* **7:**722–725.

Lacey, L. A., and Lacey, C. M. (1990). The medical importance of riceland mosquitoes and their control using alternatives to chemical insecticides. *J. Am. Mosq. Control Assoc. Suppl.* **2:**1–93.

LaMotte, L. C. (1958). The infection of mosquitoes and bats with Japanese B encephalitis virus, with reference to survival of virus during simulated hibernation. Doctoral thesis, Johns Hopkins University, Baltimore, Md.

LaMotte, L. C. (1963). Effect of low environmental temperature upon Japanese B encephalitis virus multiplication in the mosquito. *Mosq. News* **23:**330–335.

Lee, J. H., Hassan, H. K., Hill, G., Higazi, T. B., Mitchell, C. J., Godsey, M. S., Komar, N., and Unnasch, T. R. (2001). Identification of mosquito avian derived bloodmeals to the species level by PCR-HDA. *Am. J. Trop. Med. Hyg.* **65S:**187.

Levy, C. E., Doll, J. M., and Wright, M. E. (1987). Control of an outbreak of mosquito-borne encephalitis along the Colorado River in 1983. *J. Am. Mosq. Control Assoc.* **3:**100–101.

Loftin, K. M., Byford, R. L., Loftin, M. J., Craig, M. E., and Steiner, R. L. (1997). Host preference of mosquitoes in Bernalillo County, New Mexico. *J. Am. Mosq. Control Assoc.* **13:**71–75.

Lord, C. C. (1998). Density dependence in larval *Aedes albopictus* (Diptera: Culicidae). *J. Med. Entomol.* **35:**825–829.

Lord, C. C., and Day, J. F. (1999). Modeling St. Louis encephalitis in Florida: What factors affect yearly variation in transmission? *Suppl. Am. J. Trop. Med. Hyg.* **61:**396–397.

Lord, C. C., and LeFevre, L. C. (2001). Age dependent mortality in *Culex nigripalpus*. *Am. J. Trop. Med. Hyg.* **65S:**188.

Lorenz, L., Beaty, B. J., Aitken, T. H., Wallis, G. P., and Tabachnick, W. J. (1984). The effect of colonization upon *Aedes aegypti* susceptibility to oral infection with yellow fever virus. *Am. J. Trop. Med. Hyg.* **33:**690–694.

MacDonald, G. (1957). "The Epidemiology and Control of Malaria." Oxford Univ. Press, London.

MacDonald, G. (1961). Epidemiologic models in studies of vector-borne diseases. *Public Health Rep.* **76:**753–764.

Macdonald, W. W. (1956). *Aedes aegypti* in Malaya. I. Larval and adult biology. *Ann. Trop. Med. Parasitol.* **50:**399–414.

Macfie, J. W. S. (1920). Heat and *Stegomyia fasciata*, short exposures to raised temperatures. *Ann. Trop. Med. Parasitol.* **14:**73–82.

Marques, C. A., Marques, G. R. A., and Degallier, N. (1998). Is *Aedes albopictus* only pest mosquito or also a vector of arboviruses in Brazil? *In* "In an Overview of Arbovirology in Brazil and Neighbouring Countries" (A. P. A. Travassos da Rosa,

P. F. C. Vasconcelos, and J. F. S. Travassos da Rosa, eds.), pp. 248–260. Instit. Evandro Chagas, Belem, Brazil.

Marsh, M., Matlin, K., Simons, K., Reggio, H., White, J., Kratenbeck, J., and Helenius, A. (1981). Are lysosomes a site of envelope-virus penetration. *Cold Spring Harb. Symp. Quant. Biol.* **46:**835–843.

Marshall, I. D. (1988). Murray Valley and Kunjin encephalitis. *In* "The Arboviruses: Epidemiology and Ecology" (T. P. Monath, ed.), pp. 151–189. CRC Press, Boca Raton, Fla.

McClelland, G. A. H. (1959). Observations on the mosquito *Aedes (Stegomyia) aegypti* (L.) in East Africa. I. The biting cycle in an outdoor population at Entebbe, Uganda. *Bull. Entomol. Res.* **50:**227–235.

McClelland, G. A. H. (1960). Observations on the mosquito *Aedes (Stegomyia) aegypti* (L.) in East Africa. II The biting cycle in a domestic population on the Kenya coast. *Bull. Entomol. Res.* **51:**687–697.

McClelland, G. A. H. (1972). Some man-made mosquito problems in Africa and prospects for their rational solution. Proc. 5th Tall Timbers Conf. Ecol. Anim. Control, Tallahassee, Fla.

McDonald, G. (1980). Population studies of *Culex annulirostris* Skuse and other mosquitoes (Diptera: Culicidae) at Mildura in the Murray Valley of southern Australia. *J. Aust. Entomol. Soc.* **19:**37–40.

Mclean, D. M., Grass, P. N., Judd, B. D., Stolz, K. J., and Wong, K. K. (1975). Dengue virus transmission by mosquitoes incubated at low temperatures. *Mosq. News* **35:**322–327.

Mclean, D. M., Grass, P. N., Judd, B. D., Stolz, K. J., and Wong, K. K. (1978). Transmission of Northway and St. Louis encephalitis viruses by arctic mosquitoes. *Arch. Virol.* **57:**315–322.

Metselar, D., Grainger, C. R., Oei, K. G., Reynolds, D. G., Pudney, M., Leake, C. J., Tukei, P. M., D'Offay, R. M., and Simpson, D. I. H. (1980). An outbreak of type 2 dengue fever in the Seychelles, probably transmitted by *Aedes albopictus* (Skuse). *Bull. World Health Organ.* **58:**937–943.

Meyer, R. P., Hardy, J. L., Presser, S. B., and Reisen, W. K. (1988). Preliminary evaluation of the vector competence of some southern California mosquitos to western equine encephalomyelitis (WEE) and St. Louis encephalitis (SLE) viruses. *In* "Proceedings and Papers of the 56th Annual Conference of the California Mosquito and Vector Control Association," pp. 42–48.

Miles, J. A. R., and Howes, D. W. (1953). Observations on virus encephalitis in South Australia. *Med. J. Aust.* **1:**7–12.

Miller, B. R., and Mitchell, C. J. (1986). Passage of yellow fever virus: Its effect on infection and transmission rates in *Aedes aegypti*. *Am. J. Trop. Med. Hyg.* **35:**1302–1309.

Miller, B. R., and Mitchell, C. J. (1991). Genetic selection of a flavivirus-refractory strain of the yellow fever mosquito *Aedes aegypti*. *Am. J. Trop. Med. Hyg.* **45:**399–407.

Miller, B. R., Mitchell, C. J., and Ballinger, M. E. (1989a). Replication, tissue tropisms and transmission of yellow fever virus in *Aedes albopictus*. *Trans. R. Soc. Trop. Med. Hyg.* **83:**252–255.

Miller, B. R., Monath, T. P., Tabachnick, W. J., and Ezike, V. I. (1989b). Epidemic yellow fever caused by an incompetent mosquito vector. *Trop. Med. Parasitol.* **40:**396–399.

Miller, B. R., Nasci, R. S., Godsey, M. S., Savage, H. M., Lutwama, J. J., Lanciotti, R. S., and Peters, C. J. (2000). First field evidence for natural vertical transmission of West

Nile virus in *Culex univittatus* complex mosquitoes from Rift Valley Province, Kenya. *Am. J. Trop. Med. Hyg.* **62:**240–246.

Mirovsky, J., Vymola, F., and Hoang, T. T. (1965). The 1960 summer dengue epidemic in the Democratic Republic of Vietnam. I. Epidemiology observations. *J. Hyg. Epidemiol. Microbiol. Immunol.* **IX:**356–363.

Mitchell, C. J., Francy, D. B., and Monath, T. P. (1980). Arthropod vectors. *In* "St. Louis Encephalitis" (T. P. Monath, ed.), pp. 313–379. American Public Health Association, Washington, DC.

Mitchell, C. J., and Miller, B. R. (1990). Vertical transmission of dengue viruses by strains of *Aedes albopictus* recently introduced into Brazil. *J. Am. Mosq. Control Assoc.* **6:**251–253.

Monath, T. P. (1988). Yellow fever. *In* "The Arboviruses: Epidemiology and Ecology" (T. P. Monath, ed.), pp. 139–231. CRC Press, Boca Raton, Fla.

Monath, T. P. (1994). Dengue: The risk to developed and developing countries. *Proc. Natl. Acad. Sci. USA* **91:**2395–2400.

Moore, C. G., Cline, B. L., Ruiz-Tiben, E., Lee, D., Romney-Joseph, H., and Rivera-Correa, E. (1978). *Aedes aegypti* in Puerto Rico: Environmental determinants of larval abundance and relation to dengue virus transmission. *Am. J. Trop. Med. Hyg.* **27:**1225–1231.

Morrison, A. C., Getis, A., Santiago, M., Rigau-Perez, J. G., and Reiter, P. (1998). Exploratory space-time analysis of reported dengue cases during an outbreak in Florida, Puerto Rico, 1991–1992. *Am. J. Trop. Med. Hyg.* **58:**287–298.

Muir, L. E., and Kay, B. H. (1998). *Aedes aegypti* survival and dispersal estimated by mark–release–recapture in northern Australia. *Am. J. Trop. Med. Hyg.* **58:**277–282.

Mwandawiro, C., Boots, M., Tuno, N., Suwonkerd, W., Tsuda, Y., and Takagi, M. (2000). Heterogeneity in the host preference of Japanese encephalitis vectors in Chiang Mai, northern Thailand. *Trans. R. Soc. Trop. Med. Hyg.* **94:**238–242.

Nandi, A. K., Mukherjee, K. K., Chakravarti, S. K., and Chakraborty, M. S. (1982). Activity of Japanese encephalitis virus among certain domestic animals in West Bengal. *Indian J. Med. Res.* **76:**499–503.

Nasci, R. S. (1986). The size of emerging and host-seeking *Aedes aegypti* and the relation of size to blood-feeding success in the field. *J. Am. Mosq. Control Assoc.* **2:**61–62.

Nasci, R. S. (1991). Influence of larval and adult nutrition on biting persistence in *Aedes aegypti* (Diptera: Culicidae). *J. Med. Entomol.* **28:**522–526.

Nasci, R. S., Savage, H. M., White, D., Miller, J. R., Cropp, B. C., Godsey, M. S., Kerst, A. J., Bennett, P., Gottfried, K., and Lanciotti, R. S. (2001a). West Nile virus in overwintering *Culex* mosquitoes, New York City, 2000. *Emerg. Infect. Dis.* **7:**742–744.

Nasci, R. S., White, D. J., Stirling, H., Oliver, J., Daniels, T. J., Falco, R. C., Campbell, S., Crans, W. J., Savage, H. M., Lanciotti, R. S., Moore, C. G., Godsey, M. S., Gottfried, K. L., and Mitchell, C. J. (2001b). West Nile virus isolates from mosquitoes in New York and New Jersey, 1999. *Emerg. Infect. Dis.* **7:**626–630.

Nayar, J. K., and Sauerman, D. M., Jr. (1975a). The effects of nutrition on survival and fecundity in Florida mosquitoes. I. Utilization of sugar for survival. *J. Med. Entomol.* **12:**92–98.

Nayar, J. K., and Sauerman, D. M., Jr. (1975b). The effects of nutrition on survival and fecundity in Florida mosquitoes. II. Utilization of a blood meal for survival. *J. Med. Entomol.* **12:**99–103.

Nayar, J. K., and Sauerman, D. M., Jr. (1975c). The effects of nutrition on survival and fecundity in Florida mosquitoes. III. Utilization of blood and sugar for fecundity. *J. Med. Entomol.* **12:**220–225.

Newton, E. A., and Reiter, P. (1992). A model of the transmission of dengue fever with an evaluation of the impact of ultra-low volume (ULV) insecticide applications on dengue epidemics. *Am. J. Trop. Med. Hyg.* **47:**709–720.

Ngo, K. A., and Kramer, L. D. (2003). Identification of mosquito bloodmeals using polymerase chain reaction (PCR) with order-specific primers. *J. Med. Entomol.* **40:** 215–222.

Nicholls, N. (1986). A method for predicting Murray Valley encephalitis in southeast Australia using the Southern Oscillation. *Aust. J. Exp. Biol. Med. Sci.* **64**(Pt. 6): 587–594.

Nicholls, N. (1993). El Niño-Southern oscillation and vector-borne disease. *Lancet* **342:**1284–1285.

Niebylski, M. L., and Meek, C. L. (1992). Blood-feeding of *Culex* mosquitoes in an urban environment. *J. Am. Mosq. Control Assoc.* **8:**173–177.

Olson, K. E., Higgs, S., Gaines, P. J., Powers, A. M., Davis, B. S., Kamrud, K. I., Carlson, J. O., Blair, C. D., and Beaty, B. J. (1996). Genetically engineered resistance to dengue-2 virus transmission in mosquitoes. *Science* **272:**884–886.

Pant, C. P., and Yasuno, M. (1973). Field studies on the gonotrophic cycle of *Aedes aegypti* in Bangkok, Thailand. *J. Med. Entomol.* **10:**219–223.

Patz, J. A., and Reisen, W. K. (2001). Immunology, climate change and vector-borne diseases. *Trends Immunol.* **22:**171–172.

Paulson, S. L., Grimstad, P. R., and Craig, G. B., Jr. (1989). Midgut and salivary gland barriers to La Crosse virus dissemination in mosquitoes of the *Aedes triseriatus* group. *Med. Vet. Entomol.* **3:**113–123.

Platt, K. B., Linthicum, K. J., Myint, K. S., Innis, B. L., Lerdthusnee, K., and Vaughn, D. W. (1997). Impact of dengue virus infection on feeding behavior of *Aedes aegypti*. *Am. J. Trop. Med. Hyg.* **57:**119–125.

Pontes, R. J., Freeman, J., Oliveira-Lima, J. W., Hodgson, J. C., and Spielman, A. (2000). Vector densities that potentiate dengue outbreaks in a Brazilian city. *Am. J. Trop. Med. Hyg.* **62:**378–383.

Provost, M. W. (1951). The occurrence of salt marsh mosquitoes in the interior of Florida. *Florida Entomol.* **34:**48–53.

Putnam, J. L., and Scott, T. W. (1995). Blood-feeding behavior of dengue-2 virus-infected *Aedes aegypti*. *Am. J. Trop. Med. Hyg.* **52:**225–227.

Qiu, F., Zhang, H., Shao, L., Li, X., Luo, H., and Yu, Y. (1981). Studies on the rapid detection of dengue virus antigen by immunofluorescence and radioimmunoassay. *Chin. Med. J.* **94:**653–658.

Ramasamy, M. S., Sands, M., Kay, B. H., Fanning, I. D., Lawrence, G. W., and Ramasamy, R. (1990). Anti-mosquito antibodies reduce the susceptibility of *Aedes aegypti* to arbovirus infection. *Med. Vet. Entomol.* **4:**49–55.

Rawlins, S. C. (1998). Spatial distribution of insecticide resistance in Caribbean populations of *Aedes aegypti* and its significance. *Rev. Panam. Salud Publica.* **4:**243–251.

Reeves, W. C. (1990). Overwintering of arboviruses. *In* "Epidemiology and Control of Mosquito-Borne Arboviruses in California, 1943–1987" (W. C. Reeves, ed.), pp. 357–382. California Mosquito and Vector Association, Inc., Sacramento, Calif.

Reeves, W. C., Bellamy, R. E., and Scrivani, R. P. (1958). Relationships of mosquito vectors to winter survival of encephalitis viruses. I. Under natural conditions. *Am. J. Hyg.* **67:**78–89.

Reeves, W. C., Hardy, J. L., Reisen, W. K., and Milby, M. M. (1994). Potential effect of global warming on mosquito-borne arboviruses. *J. Med. Entomol.* **31:**323–332.

Reisen, W. K., Lothrop, H. D., and Hardy, J. L. (1995a). Bionomics of *Culex tarsalis* (Diptera: Culicidae) in relation to arbovirus transmission in southeastern California. *J. Med. Entomol.* **32:**316–327.

Reisen, W. K., Lothrop, H. D., Presser, S. B., Milby, M. M., Hardy, J. L., Wargo, M. J., and Emmons, R. W. (1995b). Landscape ecology of arboviruses in southern California: Temporal and spatial patterns of vector and virus activity in Coachella Valley, 1990–1992. *J. Med. Entomol.* **32:**255–266.

Reisen, W. K., Meyer, R. P., and Milby, M. M. (1986). Patterns of fructose feeding by *Culex tarsalis* (Diptera: Culicidae). *J. Med. Entomol.* **23:**366–373.

Reisen, W. K., Meyer, R. P., Presser, S. B., and Hardy, J. L. (1993). Effect of temperature on the transmission of western equine encephalomyelitis and St. Louis encephalitis viruses by *Culex tarsalis* (Diptera: Culicadae). *J. Med. Entomol.* **30:**151–160.

Reisen, W. K., and Reeves, W. C. (1990). Bionomics and ecology of *Culex tarsalis* and other potential mosquito vector species. *In* "Epidemiology and Control of Mosquito-Borne Arboviruses in California, 1943–1987" (W. C. Reeves, ed.), pp. 254–329. California Mosquito and Vector Control Association, Inc., Sacramento, Calif.

Reiter, P., Jakob, W. L., Francy, D. B., and Mullenix, J. B. (1986). Evaluation of the CDC gravid trap for the surveillance of St. Louis encephalitis vectors in Memphis, Tennessee. *J. Am. Mosq. Control Assoc.* **2:**209–211.

Rico-Hesse, R., Harrison, L. M., Salas, R. A., Tovar, D., Nisalak, A., Ramos, C., Boshell, J., de Mesa, M. T., Nogueira, R. M., and da Rosa, A. T. (1997). Origins of dengue type 2 viruses associated with increased pathogenicity in the Americas. *Virology* **230:**244–251.

Ritchie, S. A., Phillips, D., Broom, A., Mackenzie, J., Poidinger, M., and van den, H. A. (1997). Isolation of Japanese encephalitis virus from *Culex annulirostris* in Australia. *Am. J. Trop. Med. Hyg.* **56:**80–84.

Roberts, D. R., and Andre, R. G. (1994). Insecticide resistance issues in vector-borne disease control. *Am. J. Trop. Med. Hyg.* **50:**21–34.

Robertson, L. C., Prior, S., Apperson, C. S., and Irby, W. S. (1993). Bionomics of *Anopheles quadrimaculatus* and *Culex erraticus* (Diptera: Culicidae) in the Falls Lake basin, North Carolina: Seasonal changes in abundance and gonotrophic status, and host-feeding patterns. *J. Med. Entomol.* **30:**689–698.

Rodriguez-Figueroa, L., Rigau-Perez, J. G., Suarez, E. L., and Reiter, P. (1995). Risk factors for dengue infection during an outbreak in Yanes, Puerto Rico in 1991. *Am. J. Trop. Med. Hyg.* **52:**496–502.

Rosen, L. (1987a). Mechanism of vertical transmission of the dengue virus in mosquitoes. *C. R. Acad. Sci. III* **304:**347–350.

Rosen, L. (1987b). Overwintering mechanisms of mosquito-borne arboviruses in temperate climates. *Am. J. Trop. Med. Hyg.* **37:**69S–76S.

Rosen, L. (1987c). Sexual transmission of dengue viruses by *Aedes albopictus*. *Am. J. Trop. Med. Hyg.* **37:**398–402.

Rosen, L., Roseboom, L. E., Gubler, D. J., Lien, J. C., and Chaniotis, B. N. (1985). Comparative susceptibility of mosquito species and strains to oral and parenteral infection with dengue and Japanese encephalitis viruses. *Am. J. Trop. Med. Hyg.* **34:**603–615.

Rosen, L., Shroyer, D. A., Tesh, R. B., Freier, J. E., and Lien, J. C. (1983). Transovarial transmission of dengue viruses by mosquitoes: *Aedes albopictus* and *Aedes aegypti*. *Am. J. Trop. Med. Hyg.* **32:**1108–1119.

Rosen, L., Tesh, R. B., Lien, J. C., and Cross, J. H. (1978). Transovarial transmission of Japanese encephalitis virus by mosquitoes. *Science* **199:**909–911.

Rossignol, P. A., Ribeiro, J. M., and Spielman, A. (1984). Increased intradermal probing time in sporozoite-infected mosquitoes. *Am. J. Trop. Med. Hyg.* **33:**17–20.

Rueda, L. M., Patel, K. J., Axtell, R. C., and Stinner, R. E. (1990). Temperature-dependent development and survival rates of *Culex quinquefasciatus* and *Aedes aegypti* (Diptera: Culicidae). *J. Med. Entomol.* **27:**892–898.

Russell, R. C. (1986a). Population age composition and female longevity of the arbovirus vector *Culex annulirostris* (Skuse) near Echuca, Victoria, in the Murray Valley of southeastern Australia 1979–1985. *Aust. J. Exp. Biol. Med. Sci.* **64**(Pt. 6)**:**595–606.

Russell, R. C. (1986b). Seasonal activity and abundance of the arbovirus vector *Culex annulirostris* (Skuse) near Echuca, Victoria, in the Murray Valley of southeastern Australia 1979–1985. *Aust. J. Exp. Biol. Med. Sci.* **64**(Pt. 1)**:**97–103.

Russell, R. C. (1995). Arboviruses and their vectors in Australia: An update on the ecology and epidemiology of some mosquito-borne arboviruses. *Rev. Med. Vet. Entomol.* **83:**141–158.

Sabin, A. B. (1952). Research on dengue during World War II. *Am. J. Trop. Med. Hyg.* **1:**30–50.

Savage, H. M., Ceianu, C., Nicolescu, G., Karabatsos, N., Lanciotti, R., Vladimirescu, A., Laiv, L., Ungureanu, A., Romanca, C., and Tsai, T. F. (1999). Entomologic and avian investigations of an epidemic of West Nile fever in Romania in 1996, with serologic and molecular characterization of a virus isolate from mosquitoes [published erratum appears in *Am. J. Trop. Med. Hyg.* **62**(1)**:**162 (2000)]. *Am. J. Trop. Med. Hyg.* **61:**600–611.

Savage, H. M., Smith, G. C., Moore, C. G., Mitchell, C. J., Townsend, M., and Marfin, A. A. (1993). Entomologic investigations of an epidemic of St. Louis encephalitis in Pine Bluff, Arkansas, 1991. *Am. J. Trop. Med. Hyg.* **49:**38–45.

Scherer, W. F., Buescher, E. L., Southam, C. M., Flemings, M. B., and Noguchi, A. (1959). Ecologic studies of Japanese encephalitis virus in Japan. VIII. Survey for infection of wild rodents. *Am. J. Trop. Med. Hyg.* **8:**716.

Scott, T. W., Chow, E., Strickman, D., Kittayapong, P., Wirtz, R. A., Lorenz, L. H., and Edman, J. D. (1993a). Blood-feeding patterns of *Aedes aegypti* (Diptera: Culicidae) collected in a rural Thai village. *J. Med. Entomol.* **30:**922–927.

Scott, T. W., Clark, G. G., Lorenz, L. H., Amerasinghe, P. H., Reiter, P., and Edman, J. D. (1993b). Detection of multiple blood feeding in *Aedes aegypti* (Diptera: Culicidae) during a single gonotrophic cycle using a histologic technique. *J. Med. Entomol.* **30:**94–99.

Sellers, R. F. (1980). Weather, host and vector: Their interplay in the spread of insect-borne animal virus diseases. *J. Hyg. (Lond.)* **85:**65–102.

Sellers, R. F., and Maarouf, A. R. (1988). Impact of climate on western equine encephalitis in Manitoba, Minnesota and North Dakota, 1980–1983. *Epidemiol. Infect.* **101:**511–535.

Sellers, R. F., and Maarouf, A. R. (1991). Possible introduction of epizootic hemorrhagic disease of deer virus (serotype 2) and bluetongue virus (serotype 11) into British Columbia in 1987 and 1988 by infected Culicoides carried on the wind. *Can. J. Vet. Res.* **55:**367–370.

Sellers, R. F., and Maarouf, A. R. (1993). Weather factors in the prediction of western equine encephalitis epidemics in Manitoba. *Epidemiol. Infect.* **111:**373–390.

Sheppard, P. M., Macdonald, W. W., Tonn, R. J., and Grabs, B. (1969). The dynamics of an adult population of *Aedes aegypti* in relation to dengue hemorragic fever in Bangkok. *J. Anim. Ecol.* **38:**661–697.

Shi, P.-Y., Kauffman, E. B., Ren, P., Felton, A., Tai, J. H., Dupuis, A. P., II, Jones, S. A., Ngo, K. A., Nicholas, D. C., Maffei, J. G., Ebel, G. D., Bernard, K. A., and Kramer, L. D. (2001). High throughput detection of West Nile virus RNA. *J. Clin. Microbiol.* **39:**1264–1271.

Shichijo, A., Mifune, L., Hayashi, K., Wada, Y., Oda, T., and Omori, N. (1972). Experimental infection of *Culex tritaeniorhynchus summorosus* mosquitoes reared in biotron with Japanese encephalitis virus. *Trop. Med.* **14:**218–229.

Shroyer, D. A. (1990). Vertical maintenance of dengue-1 virus in sequential generations of *Aedes albopictus*. *J. Am. Mosq. Control Assoc.* **6:**312–314.

Smith, C. E. (1987). Factors influencing the transmission of western equine encephalomyelitis virus between its vertebrate maintenance hosts and from them to humans. *Am. J. Trop. Med. Hyg.* **37:**33S–39S.

Solomon, T., and Mallewa, M. (2001). Dengue and other emerging flaviviruses. *J. Infect.* **42:**104–115.

Southwood, T. R., Murdie, G., Yasuno, M., Tonn, R. J., and Reader, P. M. (1972). Studies on the life budget of *Aedes aegypti* in Wat Samphaya, Bangkok, Thailand. *Bull. World Health Organ.* **46:**211–226.

Steib, B. M., Geier, M., and Boeckh, J. (2001). The effect of lactic acid on odour-related host preference of yellow fever mosquitoes. *Chem. Senses* **26:**523–528.

Su, T., and Mulla, M. S. (2001). Effects of temperature on development, mortality, mating and blood feeding behavior of *Culiseta incidens* (Diptera: Culicidae). *J. Vector Ecol.* **26:**83–92.

Sumanochitrapon, W., Strickman, D., Sithiprasasna, R., Kittayapong, P., and Innis, B. L. (1998). Effect of size and geographic origin of *Aedes aegypti* on oral infection with dengue-2 virus. *Am. J. Trop. Med. Hyg.* **58:**283–286.

Tabachnick, W. J. (1991). The evolutionary relationships among arboviruses and the evolutionary relationships of their vectors provides a method for understanding vector-host interactions. *J. Med. Entomol.* **28:**297–298.

Tabachnick, W. J., Wallis, G. P., Aitken, T. H., Miller, B. R., Amato, G. D., Lorenz, L., Powell, J. R., and Beaty, B. J. (1985). Oral infection of *Aedes aegypti* with yellow fever virus: Geographic variation and genetic considerations. *Am. J. Trop. Med. Hyg.* **34:**1219–1224.

Takahashi, M. (1976). The effects of environmental and physiological conditions of *Culex tritaeniorhynchus* on the pattern of transmission of Japanese encephalitis virus. *J. Med. Entomol.* **13:**275–284.

Takashima, I., and Rosen, L. (1989). Horizontal and vertical transmission of Japanese encephalitis virus by *Aedes japonicus* (Diptera: Culicidae). *J. Med. Entomol.* **26:**454–458.

Takken, W. (1996). Synthesis and future challenges: The response of mosquitoes to host odours. *Ciba Found. Symp.* **200:**302–312.

Tardieux, I., Poupel, O., Rodhain, F., and Lapchin, L. (1992). Oral susceptibility of *Aedes albopictus* to dengue type 2 virus: A study of infection kinetics, using the polymerase chain reaction for viral detection. *Med. Vet. Entomol.* **6:**311–317.

Taylor, R. M. (1951). Epidemiology. *In* "Yellow Fever" (G. K. Strode, ed.), pp. 431–538. McGraw-Hill, New York.

Teesdale, C. (1950). An apparent invasion of *Aedes (Banksinella) lineatopennis* and *Aedes (b.) albicosta* into Mombasa Island. *Proc. R. Soc. Lond. Ser.* **25:**99–102.

Tempelis, C. H. (1975). Host-feeding patterns of mosquitoes, with a review of advances in analysis of blood meals by serology. *J. Med. Entomol.* **11:**635–653.

Tempelis, C. H., Reeves, W. C., Bellamy, R. E., and Lofy, M. F. (1965). A three-year study of the feeding habits of *Culex tarsalis* in Kern County, California. *Am. J. Trop. Med. Hyg.* **14:**170–177.

Tempelis, C. H., Reeves, W. C., and Nelson, R. L. (1976). Species identification of blood meals from *Culex tarsalis* that had fed on passeriform birds. *Am. J. Trop. Med. Hyg.* **25:**744–746.

Teng, H. J., Wu, Y. L., and Lin, T. H. (1999). Mosquito fauna in water-holding containers with emphasis on dengue vectors (Diptera: Culicidae) in Chungho, Taipei County, Taiwan. *J. Med. Entomol.* **36:**468–472.

Tesh, R. B. (1980). Experimental studies on the transovarial transmission of Kunjin and San Angelo viruses in mosquitoes. *Am. J. Trop. Med. Hyg.* **29:**657–666.

Tesh, R. B., and Shroyer, D. A. (1980). The mechanism of arbovirus transovarial transmission in mosquitoes: San Angelo virus in *Aedes albopictus. Am. J. Trop. Med. Hyg.* **29:**1394–1404.

Thapar, B. R., Sharma, S. N., Dasgupta, R. K., Kaul, S. M., Bali, A., Chhabra, K., and Lal, S. (1998). Blood meal identification by using Microdot ELISA in vector mosquitoes. *J. Commun. Dis.* **30:**283–287.

Thier, A. (2001). Balancing the risks: Vector control and pesticide use in response to emerging illness. *J. Urban Health* **78:**372–381.

Thorsen, J., Artsob, H., Spence, L., Surgeoner, G., Helson, B., and Wright, R. (1980). Virus isolations from mosquitoes in southern Ontario, 1976 and 1977. *Can. J. Microbiol.* **26:**436–440.

Tran, K. T., Vazeille-Falcoz, M., Mousson, L., Tran, H. H., Rodhain, F., Ngugen, T. H., and Failloux, A. B. (1999). *Aedes aegypti* in Ho Chi Minh City (Viet Nam): Susceptibility to dengue 2 virus and genetic differentiation. *Trans. R. Soc. Trop. Med. Hyg.* **93:**581–586.

Traore-Lamizana, M., Zeller, H. G., Mondo, M., Hervy, J. P., Adam, F., and Digoutte, J. P. (1994). Isolations of West Nile and Bagaza viruses from mosquitoes (Diptera: Culicidae) in central Senegal (Ferlo). *J. Med. Entomol.* **31:**934–938.

Tsai, T. F. (1997). Factors in the changing epidemiology of Japanese encephalitis and West Nile fever. *In* "Factors in the Emergence of Arbovirus Diseases" (J. F. Saluzzo and B. Dodet, eds.), pp. 179–189. Elsevier, Paris.

Tsai, T. F., Canfield, M. A., Reed, C. M., Flannery, V. L., Sullivan, K. H., Reeve, G. R., Bailey, R. E., and Poland, J. D. (1988). Epidemiological aspects of a St. Louis encephalitis outbreak in Harris County, Texas, 1986. *J. Infect. Dis.* **157:**351–356.

Tsai, T. F., Popovici, F., Cernescu, C., Campbell, G. L., and Nedelcu, N. I. (1998). West Nile encephalitis epidemic in southeastern Romania. *Lancet* **352:**767–771.

Tsai, T. T., and Mitchell, C. J. (1989). St. Louis encephalitis. *In* "The Arboviruses: Epidemiology and Ecology" (T. P. Monath, ed.), pp. 113–144. CRC Press, Boca Raton, Fla.

Tu, W. C., Chen, C. C., and Hou, R. F. (1998). Ultrastructural studies on the reproductive system of male *Aedes aegypti* (Diptera: Culicidae) infected with dengue 2 virus. *J. Med. Entomol.* **35:**71–76.

Tun-Lin, W., Burkot, T. R., and Kay, B. H. (2000). Effects of temperature and larval diet on development rates and survival of the dengue vector *Aedes aegypti* in north Queensland, Australia. *Med. Vet. Entomol.* **14:**31–37.

Turell, M. J., O'Guinn, M. L., Dohm, D. J., and Jones, J. W. (2001). Vector competence of North American mosquitoes (Diptera: Culicidae) for West Nile virus. *J. Med. Entomol.* **38:**130–134.

Usinger, R. L. (1944). Entomological phases of the recent dengue epidemic in Honolulu. *Public Health Rep.* **59:**423–430.

Vaidyanathan, R., Edman, J. D., Cooper, L. A., and Scott, T. W. (1997). Vector competence of mosquitoes (Diptera: Culicidae) from Massachusetts for a sympatric isolate of eastern equine encephalomyelitis virus. *J. Med. Entomol.* **34:**346–352.

Van Handel, E., Edman, J. D., Day, J. F., Scott, T. W., Clark, G. G., Reiter, P., and Lynn, H. C. (1994). Plant sugar, glycogen, and lipid assay of *Aedes aegypti* collected in urban Puerto Rico and rural Florida. *J. Am. Mosq. Control Assoc.* **10:**149–153.

Vasconcelos, P. F., Costa, Z. G., Travassos da Rosa, E. S., Luna, E., Rodrigues, S. G., Barros, V. L., Dias, J. P., Monteiro, H. A., Oliva, O. F., Vasconcelos, H. B., Oliveira, R. C., Sousa, M. R., Barbosa, D. S., Cruz, A. C., Martins, E. C., and Travassos da Rosa, J. F. (2001). Epidemic of jungle yellow fever in Brazil, 2000: Implications of climatic alterations in disease spread. *J. Med. Virol.* **65:**598–604.

Vaughan, A., Chadee, D. D., and French-Constant, R. (1998). Biochemical monitoring of organophosphorus and carbamate insecticide resistance in *Aedes aegypti* mosquitoes from Trinidad. *Med. Vet. Entomol.* **12:**318–321.

Vazeille-Falcoz, M., Mousson, L., Rodhain, F., Chungue, E., and Failloux, A. B. (1999). Variation in oral susceptibility to dengue type 2 virus of populations of *Aedes aegypti* from the islands of Tahiti and Moorea, French Polynesia. *Am. J. Trop. Med. Hyg.* **60:**292–299.

Vazeille, M., Mousson, L., Rakatoarivony, I., Villeret, R., Rodhain, F., Duchemin, J. B., and Failloux, A. B. (2001). Population genetic structure and competence as a vector for dengue type 2 virus of *Aedes aegypti* and *Aedes albopictus* from Madagascar. *Am. J. Trop. Med. Hyg.* **65:**491–497.

Wallis, G. P., Aitken, T. H., Beaty, B. J., Lorenz, L., Amato, G. D., and Tabachnick, W. J. (1985). Selection for susceptibility and refractoriness of *Aedes aegypti* to oral infection with yellow fever virus. *Am. J. Trop. Med. Hyg.* **34:**1225–1231.

Wang, L. Y. (1975). Host preference of mosquito vectors of Japanese encephalitis. *Zhonghua Min Guo. Wei Sheng Wu Xue. Za Zhi.* **8:**274–279.

Watts, D. M., Burke, D. S., Harrison, B. A., Whitmire, R. E., and Nisalak, A. (1987). Effect of temperature on the vector efficiency of *Aedes aegypti* for dengue 2 virus. *Am. J. Trop. Med. Hyg.* **36:**143–152.

Weng, M. H., Lien, J. C., Wang, Y. M., Wu, H. L., and Chin, C. (1997). Susceptibility of three laboratory strains of *Aedes albopictus* (Diptera: Culicidae) to Japanese encephalitis virus from Taiwan. *J. Med. Entomol.* **34:**745–747.

White, D. J., Kramer, L. D., Backenson, P. B., Lukacik, G., Johnson, G., Oliver, J., Howard, J. J., Means, R. G., Eidson, M., Gotham, I., Kulasekera, V., and Campbell, S. (2001). Mosquito surveillance and polymerase chain reaction detection of West Nile virus, New York State. *Emerg. Infect. Dis.* **7:**643–649.

Whitfield, S. G., Murphy, F. A., and Sudia, W. D. (1973). St. Louis encephalitis virus: An ultrastructural study of infection in a mosquito vector. *Virology* **56:**70–87.

World Health Organization (1997). Vector surveillance and control. *In* "Dengue Haemorrhagic Fever: Diagnosis, Treatment, Prevent and Control," Vol. 2, pp. 48–59.

Yasuno, M., and Tonn, R. J. (1970). A study of biting habits of *Aedes aegypti* in Bangkok, Thailand. *Bull. World Health Organ.* **43:**319–325.

Zanotto, P. M., Gould, E. A., Gao, G. F., Harvey, P. H., and Holmes, E. C. (1996). Population dynamics of flaviviruses revealed by molecular phylogenies. *Proc. Natl. Acad. Sci. USA* **93:**548–553.

ADVANCES IN VIRUS RESEARCH, VOL 60

DYNAMICS OF INFECTION IN TICK VECTORS AND AT THE TICK–HOST INTERFACE

P. A. Nuttall* and M. Labuda†

*CEH Institute of Virology and Environmental Microbiology
Oxford, OX1 3SR
United Kingdom
†Institute of Zoology
Slovak Academy of Sciences
Dubravska cesta 9
83164 Bratislava, Slovakia

Tick-borne flaviviruses are common, widespread, and successfully adapted to their mode of transmission. Most tick vectors of flaviviruses are ixodid species. These ticks are characterized by a comparatively long life cycle, lasting several years, during which the infecting virus may be maintained from one developmental stage of the tick to the next. Hence ticks act as highly efficient reservoirs of flaviviruses. Many tick-borne flaviviruses are transmitted vertically, from adult to offspring, although the frequency is too low to maintain the viruses

0065-3527/03 $35.00

solely in the tick population. Instead, the survival of tick-borne flaviviruses is dependent on horizontal transmission, both from an infected tick to a susceptible vertebrate host and from an infected vertebrate to uninfected ticks feeding on the animal. The dynamics of transmission and infection have traditionally been considered in isolation: in the tick, following virus uptake in the infected blood meal, infection of the midgut, passage through the hemocoel to the salivary glands, and transmission via the saliva; and in the vertebrate host, virus delivery into the skin at the site of tick feeding, infection of the draining lymph nodes, and dissemination to target organs. However, there is now compelling evidence of a complex interaction between the tick vector and its vertebrate host that affects virus transmission profoundly. The feeding site in the skin is a battleground in which the hemostatic, inflammatory, and immune responses of the host are countered by antihemostatic, anti-inflammatory, and immunomodulatory molecules (mostly proteins and peptides) secreted in tick saliva. Here we speculate that exploitation of the tick pharmacopeia, rather than development of viremia, is the key step in successful tick-borne flavivirus transmission.

I. Introduction to Ticks

Ticks are not insects. Related to spiders and scorpions, they are arachnids belonging to the order Acari (mites and ticks). There are some 850 species of ticks. Except for one species (*Nuttalliella namaqua*), they are classified into two families: the Ixodidae (ixodid or hard ticks) and the Argasidae (argasid or soft ticks). The common names reflect the presence of a tough, sclerotized plate (scutum) on the dorsal body surface of ixodid ticks that is absent from the leathery cuticle of argasid ticks.

Ticks transmit the greatest variety of pathogens of any bloodsucking arthropod, and they are second only to mosquitoes in their medical importance as disease vectors. Examples of tick-borne pathogens are found among viruses, bacteria (including *Ehrlichia* and *Rickettsia* species), fungi, protozoa, and nematodes (Service, 2001; Sonenshine and Mather, 1994). One of the reasons why ticks are such good vectors of pathogens is their feeding behavior. Generally, each parasitic stage (larva, nymph, and adult female) takes a blood meal before developing into the next stage, or producing eggs. For three-host ixodid ticks (typical of flavivirus vectors), the life cycle is completed by feeding on three different individual animals, which may be small mammals and birds for the immature stages and larger mammals such as ungulates for the adult stages (Fig. 1). When an immature-stage tick feeds on an infected host, the infecting virus is acquired by the feeding (recipient) tick. If the tick is a competent vector of the particular virus, the virus will replicate in the tick and subsequently be transmitted to a new host when the molted tick takes its next blood meal. Thus the recipient (uninfected) tick becomes a donor (infected) tick. Furthermore, the infection is retained by the donor tick and hence can be transmitted again to another new host the next time the newly molted tick takes a blood meal. Alternatively, the virus may be transmitted from the adult tick, through the eggs, to the larvae.

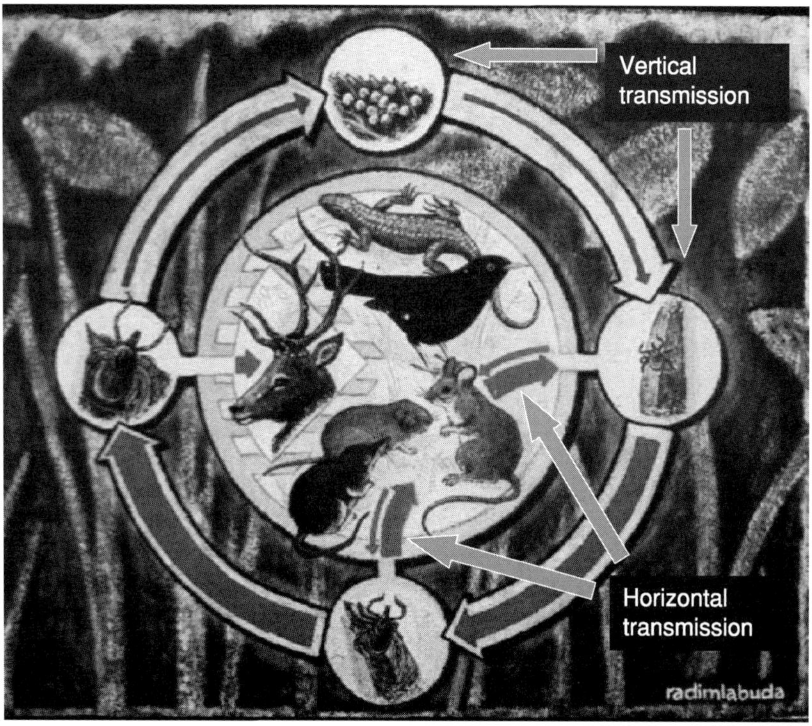

FIG 1. Schematic transmission cycle of TBE virus in Europe. Vertical transmission indicates virus transfers from one tick generation to the next by passage from the infected female to the eggs and the emergent larvae. Horizontal transmission, the principal transmission mechanism, indicates transmission from infected tick to uninfected host and vice versa. Width of the arrows within the transmission cycle indicates force of transmission. (See Color Insert.)

Transmission between tick and host is referred to as horizontal transmission, whereas transmission from one tick generation to the next is called vertical transmission (Fig. 1). The ability of each tick life stage to survive for relatively long periods (weeks to months and even years) without a blood meal greatly enhances the role of ticks as reservoirs of pathogens. New concepts regarding the transmission dynamics and survival of tick-borne pathogens have emerged from studies of tick-borne viruses, including the flaviviruses, tick-borne encephalitis (TBE) virus, and louping ill virus.

II. Tick-Borne Flaviviruses

Viruses that replicate in blood-feeding arthropods, and are transmitted to vertebrate hosts by arthropods, are known as arboviruses. They constitute the largest biological group of vertebrate viruses. Their considerable number (more than 530) suggests that transmission via an arthropod vector offers distinct benefits for viruses. Approximately 50% of the viruses isolated from field-collected arthropods and listed in the International Catalogue of Arboviruses (Karabatsos, 1985) are from mosquitoes and 25% are from ticks. The rest are mostly from sand flies and biting midges. Most of the 70-odd flaviviruses are mosquito borne, but at least 14 are transmitted by ticks, generally ixodid species (e.g., certain species of *Ixodes*, *Haemaphysalis*, and *Dermacentor*) (Table I). Some tick-borne flaviviruses are associated with disease in vertebrates (Table II).

III. Anatomy and Dynamics of Infection in Ticks

Ticks possess several unusual features that contribute to their remarkable success and vector potential, features that distinguish them from other arachnids and from insects. These include their characteristics of blood feeding, digestive processes, oviposition, and longevity.

A. Blood Feeding and Infection in Ticks

Ticks are hemorrhagic pool feeders, sawing through the epidermis and tiny blood vessels of the dermis with their chelicerae and sucking the fluids that are exuded into the wound (Sonenshine, 1991). Ixodid ticks feed for periods ranging from a few days to several weeks. To concentrate the dilute blood meal, ticks use their salivary glands to

TABLE I
TICK VECTOR SPECIES

Flavivirus	Primary tick vector species	Secondary tick vector species
TBE (European)	*Ixodes ricinus*	*Ixodes* spp., *Haemaphysalis* spp.; all ixodid spp.?
TBE (Far Eastern)	*I. persulcatus*	*Ixodes* spp., *Haemaphysalis* spp.; all ixodid spp.?
Louping ill	*I. ricinus*	None
Langat	*I. granulatus*	Unknown
Omsk hemorrhagic fever	*Dermacentor reticulatus*, *I. apronophorus*	*D. marginalis*, *I. persulcatus*
Kyasanur forest disease	*Haemaphysalis spinigera*, *H. turturis*	*Haemaphysalis* spp., *I. ceylonensis*, *I. petauristae*
Powassan	*I. cookei*, *I. persulcatus*	*I. marxi*, *I. scapularis*, *Dermacentor andersoni*
Karshi	*Ornithodoros papillipes*	Unknown
Royal Farm	*Argas hermanni*	Unknown
Gadgets Gully	*I. uriae*	Unknown
Kadam	*Rhipicephalus pravus*	Unknown
Tyuleniy	*I. uriae*	Unknown
Meaban	*O. maritimus*	Unknown
Saumaraez Reef	*I. eudyptidis*, *O. capensis*	Unknown

periodically secrete excess water from the blood meal back into the host. As a result, ixodid ticks secrete large volumes of saliva (Bowman *et al.*, 1997; Sauer *et al.*, 1995). All developmental stages (except males) consume many times their original body weight in blood and other fluids; mated females often imbibe more than 100 times their unfed body weight. The duration of feeding, secretion of copious volumes of saliva, and intake of such large blood meals contribute to the effectiveness of ticks as virus vectors.

Tick-borne flavivirus are transmitted to vertebrate hosts in the saliva secreted by their feeding tick vectors (Nosek *et al.*, 1972; Nuttall *et al.*, 1994). Hence when an uninfected tick feeds on an infected vertebrate host, the goal of the tick-borne flavivirus, when taken up in the blood meal, is to get to the salivary glands of the tick in order to be transmitted to a new host when the tick next feeds. The first challenge to survival of the imbibed virus is the process of blood meal digestion.

One of the most important and unusual features ticks possess, and which clearly distinguishes them from the other arthropod vectors, is their digestive process. Except for hemolysis of the blood cells in the

TABLE II

TICK-BORNE FLAVIVIRUSES ASSOCIATED WITH DISEASE

Virus	Susceptible vertebrate host	Clinical sign	Distribution
Kyasanur forest disease[a]	Humans, monkeys	Hemorrhagic disease	Asia (India)
Louping ill	Sheep, grouse, humans	Encephalitis	Europe (Britain, Ireland)
Omsk hemorrhagic fever	Humans, muskrats	Hemorrhagic disease	Asia (Western Siberia)
Powassan[b]	Humans	Encephalitis	North America, Asia
Tick-borne encephalitis	Humans	Encephalitis	Asia, Europe

[a] Alkhurma virus, a relative of Kyasanur forest disease virus, is associated with severe hemorrhagic fever in humans in Saudi Arabia (Charrel *et al.*, 2001).

[b] The pathogenic potential of deer tick virus, a relative of Powassan virus, is unknown (Kuno *et al.*, 2001a).

midgut lumen, digestion in ticks is entirely intracellular, a process known as heterophagy. All digestion is accomplished within epithelial cells of the midgut. Tick saliva apparently lacks hemolytic enzymes, and the erythrocytes and other cellular elements are ingested unchanged. Intracellular digestion of the blood meal is an unusual phenomenon characteristic of ticks and possibly some other Acari and is not believed to occur in other arthropods.

Blood taken into the midgut of the tick remains largely undigested for long periods. Without the rapid infusion of digestive enzymes into the lumen of the gut, the contents of the blood meal (including virions and virus-infected host cells) remain within a nonhostile environment, allowing time for penetration of the virus into tick body tissues. In addition, the absence of proteolytic enzymes in the midgut lumen favors virus entry as infectious extracellular virions and via infected host cells in which virus replication may continue for some time, increasing the infectious dose. Indeed, conditions within the midgut are unlikely to trigger the conformational changes of viral surface glycoproteins, which are the first step in the entry of virions into cells (Heinz and Allison, 2001). The initial conditions of virus infection differ markedly for ticks and insects and may be the principal reason tick-borne flaviviruses are rarely, if ever, transmitted by insects.

The undigested blood meal remains as a food reserve and, except during oviposition or ecdysis, is consumed gradually. Having ingested and stored this rich nutrient source, ticks are able to survive extended periods of starvation, yet another feature that greatly enhances their importance as arbovirus vectors (Section III,G).

Replication in tick cells commences after the entry of virions into cells of the tick midgut wall. Studies with Powassan virus in *Dermacentor andersoni* demonstrated that midgut epithelial cells are the first site of infection. Rajcani *et al.* (1976) found TBE virus also in the esophagus and subesophageal ganglion in *D. marginatus* larvae and in columnal epidermal cells of *D. reticulatus* nymphs. The susceptibility of arthropod midgut cells to virus infection is one of the most important determinants of vector competence, which is the combined effect of all the physiological and ecological factors of vector, host, pathogen, and environment (McKelvey *et al.*, 1981). Experiments comparing different methods of infecting ticks have demonstrated the presence of a "gut barrier" to virus infection (Nuttall *et al.*, 1994). The presence of such a barrier indicates that a specific interaction exists between virus (imbibed in the blood meal) and midgut cells, but the nature of the gut barrier has not been determined.

Despite saliva being the medium for virus transmission from infected tick to vertebrate host, only a few studies report virus detection in tick saliva, or salivary glands (Nosek *et al.*, 1972). Most evidence of a salivary gland route of transmission has been based on observations of infection of vertebrate hosts on which infected ticks have engorged. Host infection has usually been detected by virus isolation from the blood and/or by seroconversion (e.g., Kozuch *et al.*, 1990).

B. Transstadial Survival

As a result of the feeding behavior of ticks and of their life cycles, a tick-borne virus must persist from one instar to the next to be transmitted to a vertebrate host (Fig. 1). This means that the "extrinsic incubation period," which is so important in determining the transmission dynamics of insect-borne viruses (Turell, 1988), is not significant for virus transmission by ixodid ticks because it is unlikely to exceed the comparatively long molting period. The histolytic enzymes and tissue replacement associated with molting provide a potentially hostile environment for infecting viruses (Balashov, 1983). Several authors have suggested that the dynamics of viral replication within the tick reflect these events: a fall in virus titer, followed by an increase in titer as the virus infects and replicates in replacement tick tissues (Rehacek, 1965). However, the replication of Langat virus in *I. ricinus* was not obviously correlated with any particular stage of the molting period (Varma and Smith, 1972).

C. Oviposition and Transovarial Transmission

Mating occurs on the host in all but *Ixodes* species. Females then feed to repletion, drop from their hosts, and commence oviposition in a sheltered microenvironment. The ability of ticks to produce large numbers of eggs is important in their population dynamics and hence in the natural history of tick-borne flaviviruses. Few terrestrial organisms equal ticks in their productivity. Females of most ixodid ticks produce thousands of eggs. An *I. ricinus* or *I. persulcatus* female is capable of laying 350 to 5000 eggs, depending mostly on the amount of consumed blood (Korenberg and Kovalevskii, 1994).

One method of arbovirus persistence in the vector population is via vertical transmission, in which virus from the infected parent, usually the female, is transmitted via the egg to the succeeding generation (Fig. 1). Although evidence of vertical transmission has been recorded for tick-borne flaviviruses, the levels of vertical transmission and filial infection in nature generally appear low (Nuttall and Labuda, 1994). Certainly, the high levels of vertical transmission of certain insect-borne viruses associated with stabilized infections of their vectors have not been recorded for any tick-borne flavivirus. However, vertical transmission of insect-borne viruses can have adverse effects on the vector population, although flaviviruses appear to be benign (Burke and Monath, 2001). If tick-borne flaviviruses rely on their vectors for persistence, then any deleterious effects of vertical transmission on ticks may outweigh the advantages to the virus. The balance between costs and benefits of vertical transmission, together with the gains from cofeeding and nonviremic transmission (see Section IV), may explain why vertical transmission is common among tick-borne flaviviruses but occurs at a low level. Moreover, low levels of TBE virus infection in a population of larval ticks (detectable only by polymerase chain reaction) were shown to be amplified by "nonviremic" cofeeding to yield a significant number of infected nymphal ticks (Labuda *et al.*, 1993a). Because tick larvae quest and feed in clusters, there are many opportunities for amplification of vertically acquired TBE virus in the vector population through nonviremic transmission between cofeeding larvae. As a result of such amplification, vertical transmission might be the difference between survival and extinction of TBE virus in certain foci of infection (Section IV).

The anatomical basis of infection of the egg is poorly understood. Differences in vertical transmission rates in female ticks infected parenterally by inoculation, as opposed to *per os*, led Rehacek (1962) to postulate the existence of an "ovary barrier" to infection. For TBE virus, the transovarial and filial infection rates observed in experimental studies were relatively low and were related to the time when the tick fed and the level of viremia in the vertebrate host at the time of feeding (Rehacek, 1962). In theory, tick-borne flaviviruses could be maintained solely in the vector population by maternal inheritance. Presumably, the costs to the virus of this strategy of maintenance are too high, otherwise the virus would no longer be an arbovirus capable of infecting a vertebrate host.

D. Vector Species

Identifying tick species that act as vectors of flaviviruses is one of the key tasks in explaining the transmission cycle and survival strategy of these viruses in nature. In principle, two main approaches to identifying the tick vector have been used: (1) analysis of infection rates in field-collected ticks and (2) data from laboratory transmission experiments. Experiments have been performed using many tick species, often applying methods that are not applicable to the natural situation.

Demonstration of virus replication in ticks after parenteral inoculation of virus into the tick body cavity provides the first basic information about the ability of tick cells to replicate a particular flavivirus. It is very artificial because the natural infection barriers are bypassed and usually large doses of virus are used. However, negative results exclude the tick species as a potential vector of a particular virus. From these and other experiments, the ability to replicate TBE virus appears to be a general feature of ixodid ticks. There is sufficient evidence that artificially infected ticks of several species are capable of supporting both virus replication and efficient transmission of TBE virus to a vertebrate host when feeding. Such evidence for susceptible tick species outside the range of TBE virus distribution does not help explain the ecology of TBE virus in a natural focus; however, it allows the use of "exotic" species (such as *Rhipicephalus appendiculatus*) as experimental laboratory vectors of the virus (e.g., Labuda *et al.*, 1993b).

Evidence of primary tick vectors (Table I) is based mostly on virus isolations from field-collected ticks. For example, since the expedition of Zilber and colleagues (Zilber and Soloviev, 1946), repeated field studies have identified *I. persulcatus* as the principal vector of TBE virus (Far Eastern subtype). Similarly, following the first isolation of TBE virus from *I. ricinus* ticks in former Czechoslovakia (Rampas and Gallia, 1949), large numbers of isolates have been obtained from *I. ricinus* in various geographic regions of Europe, reflecting the primary role of *I. ricinus* as a vector of the European TBE viral subtype. For a more extensive review of primary and secondary tick vectors of tick-borne flaviviruses, see Nuttall and Labuda (1994).

E. Vector Efficiency

Ticks feeding on an infected host take up virus with their blood meal. The virus enters the tick gut lumen as extracellular virus particles (virions) and/or as infected host cells (Section III,A). A general finding from several studies is that the higher the blood titer of virus, the greater the proportion of ticks that becomes infected (Nuttall *et al.*, 1994). The proportion of infected ticks is species dependent, and the ability to acquire infection is an integral part of the vectorial efficiency of a given species.

Vector efficiency is the major determinant of vector status. It is reflected in the infection threshold and the transmission rate.

Infection threshold is defined as the lowest concentration of virus capable of causing an infection in approximately 1 to 5% of the vector population (Chamberlain *et al.*, 1954). Thus the lower the infection threshold, the smaller the viral dose required to infect the vector and hence the greater the probability of infection of the vector in nature. Comparatively high infection thresholds of $\geq 4.5 \log_{10} LD_{50}$ ml blood were recorded for TBE virus and either *I. persulcatus* feeding on guinea pigs or *I. ricinus* feeding on chicks, whereas infection thresholds were at least 100-fold less on natural rodent hosts (Nuttall and Labuda, 1994). The concept of infection thresholds based on the levels of viremia was questioned in studies of TBE virus infection of *I. persulcatus* that were fed on nonviremic rabbits (Galimov *et al.*, 1989). Similar studies using *I. persulcatus*, *I. ricinus*, *Dermacentor marginatus*, *D. reticulatus*, and *R. appendiculatus* demonstrated TBE virus transmission between infected and uninfected ticks feeding either in contact with each other or separated physically on nonviremic guinea pigs (Alekseev and Chunikhin, 1990; Labuda *et al.*, 1993b) and natural rodent hosts (Labuda *et al.*, 1993d). The phenomenon has been called nonviremia transmission (Section IV) and appears to be mediated by bioactive molecules secreted in tick saliva (Section VIII).

In addition to infection threshold, the transmission rate is also important in determining vector efficiency. Transmission of TBE virus by ticks to the host appears to be an efficient process when the vertebrate host is susceptible to infection. Even *H. inermis* nymphs feeding for only about 90 min successfully transmitted TBE virus to *Micromys minutus* (Nosek *et al.*, 1981), *Sorex araneus* (Kozuch *et al.*, 1967), and *Capreolus capreolus* (Nosek *et al.*, 1967b).

Despite the vector potential of many ixodid ticks for TBE virus, only *I. ricinus* (European subtype of TBE virus) and *I. persulcatus* (Far Eastern subtype) are recognized as primary vectors of the virus. The reasons for this have not been fully defined. Obviously, virus and tick must be sympatric in their distribution. However, this is not the complete story because many competent tick vectors are found within the geographical range of TBE virus but virus transmission cycles depend (almost) exclusively on *I. ricinus* or *I. persulcatus*.

F. Tick Host Preferences

In addition to vector efficiency, two other parameters are significant in tick-borne flavivirus transmission, namely vertebrate host preferences of vector tick species and susceptibility of these vertebrate hosts to infection. For example, immature stages of *I. ricinus* and *I. persulcatus* often feed on rodents and birds, whereas adults feed on goats and other ungulates. Several rodent species are highly competent in supporting the tick-borne transmission of TBE virus, whereas avian species are not (Nuttall and Labuda, 1994). The reasons for this difference are unknown. Perhaps more intriguing is the observation that, in experimental studies, goats did not support tick-borne TBE virus transmission (Labuda *et al.*, 2002; Nosek *et al.*, 1967a) and yet goats' milk is a common source of infection in humans. The largest outbreak of TBE attributed to drinking goats' milk was registered in southern Slovakia in 1951 (Blaskovic, 1954). Sporadic cases and family outbreaks of alimentary TBE virus infections are observed in Slovakia almost annually (Kohl *et al.*, 1996).

G. Longevity

The tick life span is measured in years rather than the days or weeks characteristic of mosquito-borne flaviviruses. Specific life stages readily survive from one year to the next. This exceptional longevity not only perpetuates the ticks, it also perpetuates the viruses they carry, making it considerably more difficult to control these infections. As a result, ticks are not only vectors but also efficient reservoirs of flaviviruses in nature. For example, TBE virus persisted in *I. ricinus* ticks for at least 9 months in experimentally infected starving females maintained at either ambient temperature or 4 °C (Benda, 1958a) and survived in fed larvae for 102 days maintained under natural overwintering conditions (Rehacek, 1960). Moreover, TBE virus is transmitted from one developmental stage to the next and, in a small percentage of infections, also transovarially (Benda, 1958a, 1958b; Rehacek, 1962; Sections III,B and III,C), thus expanding the virus–vector association into many months if not years. Each stage of *I. ricinus* takes approximately 1 year to develop to the next, hence the life cycle takes 3 years to complete, although it may vary from 2 to 6 years throughout the geographical range (Nuttall and Labuda, 1994). Tick-borne flaviviruses probably spend at least 95% of their evolutionary lifespan in ticks rather than in vertebrate hosts (Nuttall *et al.*, 1991). This helps explain the comparatively slow rate of evolution of tick-borne compared with mosquito-borne flaviviruses and the apparent cline of TBE complex viruses across the northern hemisphere (Zanotto *et al.*, 1995, 1996).

IV. Nonviremic Transmission

By definition, arthropod vectors become infected with arboviruses (such as tick-borne flaviviruses) by feeding on animals that have a substantial viremia (i.e., levels of virus in their blood that equal or exceed the threshold level for infection of the vector) (Karabatsos, 1985; WHO, 1985, 1986). Threshold levels are often determined experimentally by feeding vectors on animals infected by needle and syringe inoculation of the virus. A corollary of the arbovirus definition is that epidemiologically significant hosts of arboviruses are identified by their ability to sustain readily detectable levels of viremia when infected by a particular arbovirus. Again, experiments are generally undertaken using hosts infected by needle and syringe inoculation. Thus, for example, sheep were identified as maintenance hosts for louping ill virus (Reid, 1984).

More recent studies of arbovirus transmission suggest that the definition of an arbovirus may be fundamentally flawed, at least for tick-borne arboviruses. Indeed, results of experiments mimicking natural conditions of tick-borne virus transmission indicate that viremia may be nothing more than a "red herring"—a byproduct of infection rather than a prerequisite for transmission (Section X).

The first evident contradiction of the role of viremia in arbovirus transmission was reported in studies involving Thogoto virus. This unusual arbovirus is classified in the same family (Orthomyxoviridae) as the influenza viruses (Kuno *et al.*, 2001b). However, it is undoubtedly an arbovirus, transmitted by *Rhipicephalus appendiculatus*, the African brown ear tick (Davies *et al.*, 1986). Hamsters are highly susceptible to Thogoto virus, developing high levels of viremia; death occurs 5 to 6 days postinfection. In contrast, guinea pigs infected with the virus show no clinical signs and typically no viremia. Nevertheless, when Thogoto virus-infected and uninfected ticks were allowed to feed together on guinea pigs, most of the uninfected ticks became infected, even though they were separated physically from the infected ticks, and the guinea pigs showed no detectable viremia. Indeed, more uninfected nymphal ticks became infected by cofeeding with infected adult ticks on nonviremic guinea pigs than by cofeeding on highly viremic hamsters (Jones *et al.*, 1987). The minimum overlap in the cofeeding period was approximately 3 days; when fed together for a shorter duration, the number of nymphs that became infected was reduced (Jones and Nuttall, 1989b). Transmission between cofeeding ticks was inhibited when the guinea pig hosts were immunized previously against Thogoto virus (Jones and Nuttall, 1989a) and was reduced when the animals were immune to tick infestation (Jones and Nuttall, 1990). Although these studies with Thogoto virus challenged the emphasis placed on viremia in arbovirus transmission, they were undertaken with an atypical arbovirus and an artificial laboratory model. The generality of the phenomenon needs to be tested with other tick-borne viruses and natural hosts.

As a more typical arbovirus, TBE virus is an ideal candidate for investigating the concept of nonviremic transmission, defined as virus transmission between competent infected and uninfected vectors feeding together on a vertebrate host that has an undetectable or subthreshold level of viremia. Initial studies reported nonviremic transmission between TBE virus-infected and uninfected ticks feeding together on nonviremic guinea pigs (Alekseev and Chunikhin, 1990; Alekseev *et al.*, 1991; Labuda *et al.*, 1993b). In nature, the principal European vector of TBE virus, *I. ricinus* (the wood or sheep tick), feeds on a wide range of species: immature stages commonly infest small mammals, birds, and medium-sized mammals such as squirrels and hares, whereas adults prefer larger mammals (e.g., deer, sheep, and goats). To determine whether nonviremic transmission occurs with natural host species, field mice (*Apodemus flavicollis* and *A. agrarius*), bank voles (*Clethrionomys glareolus*), pine voles (*Pitymys subterraneus*), hedgehogs (*Erinaceus europaeus*), and pheasants (*Phasianus colchicus*) were captured in the wild (Table III). Individual animals that had no evidence of prior TBE virus infection (i.e., no detectable neutralizing antibodies to the virus) were infested with *I. ricinus* ticks retained in two neoprene chambers attached to the back of the animals. In chamber 1 were placed two infected adult female ticks, two uninfected males, and 20 uninfected nymphs; chamber 2 contained 20 uninfected nymphs only. After a feeding period of 4 days, the animals were killed and ticks, blood, and target organs (spleen, lymph nodes, and brain) were assayed for virus (Table III). Only two engorged nymphs from hedgehogs, and none of the nymphs from pheasants, were infected, indicating that these two species did not support TBE virus transmission. In contrast, pine voles were highly susceptible to virus infection. Three of the six pine voles died before the ticks completed engorgement, and virus was isolated from their brains; the remaining individuals showed comparatively high levels of virus in their spleens, lymph nodes, brain, and blood. Although 71% of the nymphs from pine voles were infected, only 14% of the uninfected nymphs, placed on the three surviving pine voles, fed successfully. In contrast, field mice showed comparatively low levels of virus infection but produced the greatest yield of infected ticks compared with pine

TABLE III

TRANSMISSION OF TICK-BORNE ENCEPHALITIS VIRUS BY COFEEDING OF *Ixodes ricinus* TICKS FED ON DIFFERENT WILD-CAUGHT TICK HOST SPECIES[a]

Vertebrate host (number)	Nymphs in chamber 1: No. infected/No. fed	Nymphs in chamber 1: % infected	Nymphs in chamber 2: No. infected/No. fed	Nymphs in chamber 2: % infected	Viremia	Spleen and lymph nodes
A. flavicollis (6)	76/95	80	46/90	51	3/6	3/43 (7%)
A. agrarius (2)	18/24	75	7/8	88	2/2	1/13 (8%)
C. glareolus (8)	15/46	33	21/84	25	8/8	23/48 (48%)
P. subterraneus (6)	11/12	92	1/5	20	3/3 (+ 3 died)	12/19 (63%)
E. europaeus (2)	2/21	10	0/27	0	0/2	0%
P. colchicus (5)	0/38	0	0/59	0	Trace	0%

[a] Data summarized from Labuda *et al.* (1993d).

voles and bank voles. In fact, three of six *A. flavicollis* field mice had no detectable viremia and yet 58% (47/81) of ticks that fed on these individuals became infected. Similarly designed experiments, but with virus-immune hosts, provide further evidence of nonviremic transmission (Section V). Cofeeding studies have also demonstrated nonviremic transmission of louping ill virus. This virus is mostly confined to upland regions of the United Kingdom and Ireland where it is transmitted by *I. ricinus* to a range of mammalian and avian species, although sheep were thought to be the key host in virus survival (Reid, 1984). Mountain hares (*Lepus timidus*) develop subthreshold levels of viremia following syringe inoculation with louping ill virus and consequently were discounted from playing a significant epidemiological role. However, when louping ill virus-infected and uninfected ticks were allowed to cofeed on wild-caught hares, 47% uninfected nymphs became infected, although feeding on apparently nonviremic animals (Jones *et al.*, 1997). The results of cofeeding TBE and louping ill virus-infected and uninfected ticks on wild vertebrate hosts provide compelling evidence that nonviremic transmission occurs in nature.

Nonviremic transmission takes advantage of the fact that ticks, like most parasites, show a typical negative binomial distribution on their hosts. Thus, at any one time point, a small number of individual animals are heavily infested with ticks, whereas the majority of the host population are uninfested or support low numbers of ticks (Randolph *et al.*, 1996, 1999). The multi-infested hosts appear to be ideally suited to nonviremic transmission, especially as ticks are gregarious feeders (Wang *et al.*, 2001). For example, >90% immature *I. ricinus* feed together on the ears of rodents or around the bill of birds (Randolph *et al.*, 1996). Feeding aggregation reduces the distance between cofeeding infected and uninfected ticks, thereby facilitating nonviremic transmission (Labuda *et al.*, 1996, 1997).

As an additional consequence of the overdispersed distribution of ticks on their hosts, nonviremic transmission may even play an important role in virus survival when transovarial transmission occurs. Many tick-borne flaviviruses are transmitted vertically from one generation (Section III,C) to the next, but the numbers of infected adults that transmit the virus via the eggs to the larvae and the numbers of transovarially infected larvae are generally <5% (reviewed by Nuttall and Labuda, 1994; Nuttall *et al.*, 1994). However, larvae show a highly nonrandom distribution on their hosts as individuals from an egg batch quest together. Even if only a few larvae from an egg batch are infected transovarially, the number of infections may be increased as a result of nonviremic transmission between cofeeding larvae (Labuda *et al.*, 1993a). By this means, a low prevalence of transovarial infections may be amplified to yield much higher numbers of nymphal infections and thereby make a significant contribution to virus survival and R_0 (Section V).

Generally, natural hosts of arboviruses had been identified by their ability to develop substantial levels of viremia following syringe inoculation of the arbovirus under investigation. The concept of nonviremic transmission provides an alternative approach to such studies that may help answer some of the existing conundrums about how tick-borne flaviviruses survive in nature (Section V). It remains to be determined whether this new approach can be extrapolated to insect-borne flaviviruses.

V. Population Biology of Tick-Borne Flaviviruses

Viruses are absolute parasites; they survive by infecting and replicating within a cell and then passing on to and infecting a new susceptible cell. The population biology of viruses (and all other parasites) is reflected in the basic reproductive number of the infection (R_0), where R_0 of a virus is defined as the average number of secondary infections produced when one infected individual is introduced into a host population in which every individual is susceptible (Anderson and May, 1991). To survive, R_0 must be at least unity ($R_0 = 1$). The feeding behavior of ixodid species dictates that one infected tick gives rise to only one new infection at the level of the organism, such as an infected nymph ($\equiv$ one infection) feeding on and transmitting the infection to a susceptible vertebrate host ($\equiv$ one new infection). If, for example, three infected nymphs feed together on the same susceptible host, the three infections still only generate one new infection (the vertebrate host on which they all fed). In contrast, one infected vertebrate host can give rise to many new infections if competent tick vectors feed on that host during its infective period. Thus the vertebrate host has the potential to amplify the infection ($R_0 > 1$), thereby playing an important role in virus survival. There are several anecdotal reports of increases in TBE virus infection prevalence associated with increases in rodent populations and even conjecture that global warming may be a contributing factor.

Nonviremic transmission has important implications for the population biology of tick-borne flaviviruses. First, it adds to the host population some species that may have been dismissed because they do not develop patent viremia (e.g., hares (*L. timidus*) infected with louping ill virus) (Hudson *et al.*, 1995; Jones *et al.*, 1997). Second, it provides an alternative mode of transmission that may contribute significantly to virus survival. This was demonstrated theoretically by estimating relative R_0 values for viremic compared with nonviremic transmission of TBE virus in rodent populations (Randolph *et al.*, 1996). Estimates of how many infected ticks arose from a susceptible host gave relative measures for R_0 of 0.98 for viremic transmission and 1.65 for systemic transmission. This quantitative comparison indicates that nonviremic transmission yields a >50% increase in the amplification of TBE virus than does viremic transmission.

Although tick-borne flaviviruses are generally only transmitted once by an infected stage of the tick vector (interstadial transmission), an exception occurs when a host dies before the infected tick has completed feeding. The partially fed, infected tick will then urgently seek an alternative host on which to complete engorgement. In completing engorgement on a second host, the infected tick may transmit the infection a second time (intrastadial transmission). Experimentally, an infected female adult tick was shown to transmit a tick-borne virus infection three times following repeated interrupted feeding (Wang and Nuttall, 2001). A comparable situation may apply in nature to tick-borne flavivirus infections that cause high mortality in vertebrate hosts, for example, TBE virus in pine voles (*Pitymys subterraneus*), louping-ill virus transmitted by *I. ricinus* to red grouse (*Lagopus scoticus*), Kyasanur forest disease transmitted by *Haemaphysalis spinigera* to monkeys, and Omsk hemorrhagic fever virus in muskrats. Indeed, interrupted feeding followed by intrastadial transmission may account for the high incidence of Kyasanur fever in humans that had been in contact with dead monkeys in India (Banerjee, 1988; Sreenivasan *et al.*, 1979). It also highlights the risks of handling fresh tick-infested carcasses. The frequency of interrupted feeding in ticks and the contribution of intrastadial transmission to R_0 are unknown. Possibly, intrastadial transmission provides a mechanism to support the amplification and persistence of highly virulent tick-borne flaviviruses. Optimum virulence depends on the relationship between pathogen transmissibility and its effect on host mortality (Lenski and May, 1994). Susceptible host populations may, before they die out, support intrastadial transmission (via interrupted feeding), thus providing the feedbacks between ecological and evolutionary processes that favor high virulence.

Despite the importance of vertebrate hosts as amplifiers of tick-borne flavivirus infections, their contribution to R_0 is considered to be limited by immunity. However, this limitation may not be as great as originally thought (see Section VI).

VI. Tick-Borne Transmission on Immune Hosts

A key epidemiological question is the role of immune hosts in natural tick-borne flaviviral transmission cycles. Hosts may develop immunity to the virus and/or to tick infestation.

To investigate the effect of virus immunity on transmission, natural rodent hosts of TBE virus, field mice and bank voles, were immunized with the virus (Labuda *et al.*, 1997). Animals were immunized by either subcutaneous syringe inoculation of virus (the typical experimental approach) or by infective tick bite (the natural route of infection). Unlike laboratory mice, field mice and bank voles survive TBE virus infection. Following the development of neutralizing antibodies, immune and control (nonimmune) animals were infested with infected donor and uninfected recipient *I. ricinus* ticks. Uninfected (recipient) nymphs were allowed to feed for 3 days and were then assayed for infectious virus to determine the ability of their virus-immune hosts to support "cofeeding transmission" from infected to uninfected ticks. Both species supported TBE virus transmission, although a greater number of ticks became infected on virus-immune field mice compared with bank voles (Table IV). On the former, 98/328 (30%) ticks feeding in chambers 1 + 2 acquired TBE virus compared with 33/187 (18%) on bank voles.

The efficiency of virus transmission depended on the location of recipient nymphs compared with infected donor ticks. Localized transmission in chamber 1, in which uninfected nymphs cofed with two infected female ticks (donors), yielded a higher proportion of infected recipient nymphs than disseminated transmission from infected donor ticks to ticks feeding in chamber 2, separated physically from donor ticks (45% compared with 10% for field mice and 26% compared with 1.5% for bank voles, respectively). The same was true for nonimmune hosts (Table IV). Clustering of ticks within a retaining chamber mimics the natural situation in which feeding ticks aggregate together in localized sites, such as on the ears of mammals or around the beaks of birds (Randolph *et al.*, 1996). Hence the natural feeding aggregation of ticks promotes tick-borne flavivirus transmission on immune hosts, as well as on nonimmune hosts as noted previously (Sections IV and V).

Despite the presence of neutralizing antibodies, 89% (24/27) of immune animals indicated in Table IV supported virus transmission

TABLE IV

TRANSMISSION OF TICK-BORNE ENCEPHALITIS VIRUS BY COFEEDING OF *Ixodes ricinus* TICKS ON VIRUS-IMMUNE FIELD MICE (*Apodemus flavicollis*) AND BANK VOLES (*Clethrionomys glareolus*)[a]

Host species, No. used (No. with viremia)	Method of immunization	Chamber 1	Chamber 2
A. flavicollis, 6 (0)	Tick feeding	24% (16/68)	3% (2/61)
C. glareolus, 6 (0)	Tick feeding	29% (14/49)	2% (1/43)
A. flavicollis, 8 (5)	sc inoculation	58% (67/115)	15% (13/84)
C. glareolus, 7 (2)	sc inoculation	25% (18/73)	0% (0/22)
A. flavicollis, 4 (4)	No immunization	72% (44/61)	38% (26/68)
C. glareolus, 6 (6)	No immunization	42% (28/67)	22% (8/37)

[a] Data summarized from Labuda *et al.* (1997).

between infected and uninfected cofeeding ticks. Furthermore, 17/20 of these immunized animals, which supported cofeeding transmission, had no detectable viremia.

Vertebrate hosts immune to virus infection are generally considered dead-end hosts. As a result, immunity has a negative effect on the basic reproductive number, R_0, of the virus because the size of the host population is effectively reduced (Section V). In studies on TBE virus (Dumina, 1958) and louping ill virus (Alexander and Neitz, 1935), recipient ticks did not acquire virus when cofeeding with infected ticks on immune hosts. However, neither of these studies apparently examined localized cofeeding transmission, analogous to conditions in chamber 1 (Table IV), which exploit the aggregated distribution of feeding ticks. When three hares (*L. timidus*) immune to louping ill virus were challenged with infected and uninfected ticks that fed together, 2/46 (4%) uninfected ticks became infected with louping ill virus (Jones *et al.*, 1997). Although this was a significant reduction in transmission efficiency compared with nonimmune hosts, the result confirms that immunity to the virus does not necessarily mean that the host is a dead end for the virus.

A 5-year survey of small mammals trapped in western Slovakia revealed a 15% neutralizing antibody prevalence for TBE virus. The antibody prevalence varied seasonally and according to species (e.g., in March–May, 25% subadult and in June–August, 25% adult field mice were positive, whereas in September–November, 18% subadult and 11% adult were positive) (Kozuch *et al.*, 1990). Considering the low relative R_0 values of TBE virus (Section V), it is unlikely that TBE virus would survive in nature if immune hosts are a dead end. However, if immune hosts can indeed contribute to the susceptible host population, the effective host density is much higher than previously thought, especially if these immune hosts repeatedly support virus transmission and thereby contribute significantly to the numbers of newly infected ticks. As discussed by Korenberg (1976), antibodies to TBE virus disappear relatively quickly in the absence of reexposure to the virus in natural hosts and, consequently, such animals probably play a crucial role in supporting TBE virus survival in nature.

Although TBE virus immunization of natural rodent hosts resulted in significantly fewer infected ticks when compared with controls, tick bite (the natural mode of immunization) was more effective than immunization via syringe inoculation in suppressing transmission (Table IV). The apparent difference in the "transmission-blocking" capacity of tick bite versus syringe inoculation suggests that tick-borne transmission induces a stronger immune response (possibly by stimulating cell-mediated immunity). Alternatively, immunity to tick infestation inhibited virus transmission. The latter point may explain the greater efficiency of field mice in supporting TBE virus transmission between cofeeding ticks (Table III), also observed in other studies (Labuda *et al.*, 1993d, 1994). Bank voles, but not field mice, develop a resistance response to tick feeding that is believed to be immune mediated (Dizij and Kurtenbach, 1995; Randolph and Nuttall, 1994). All the animals used in the experiments were captured in the field and consequently would have been exposed to ticks in nature. Further studies are needed to clarify the effect of host immunity to ticks on tick-borne flavivirus transmission.

VII. Role of Skin in Tick-Borne Transmission

Early events during tick-borne delivery of a flavivirus to a susceptible vertebrate host are poorly understood. Most studies have involved needle and syringe inoculation of relatively high doses of the virus, often by unnatural routes (e.g., intracerebral inoculation of day-old mice) and with highly susceptible laboratory animals. Following subcutaneous administration, Málková (1968) showed that TBE virus quickly leaves the site of inoculation and passes within minutes into the regional lymphatic system. The results indicated that the draining lymph nodes are the first and decisive place of virus replication in the previremic phase of infection, on which is dependent the development of viremia. However, a very different picture emerged when wild and laboratory rodents were exposed to cofeeding infected and uninfected *I. ricinus* ticks.

To study the early events of TBE virus replication in the skin of natural hosts and its importance for virus transmission, yellow-necked field mice (*A. flavicollis*) and bank voles (*C. glareolus*) were infested with infected and uninfected ticks (Labuda *et al.*, 1996). On each animal, infected adult and uninfected nymphal ticks were placed in retaining chamber 1 and uninfected nymphs were placed in chamber 2 (Fig. 2). Infection was monitored in skin sample site A excised from a tick-infested area within chamber 1 and skin sample B taken from within chamber 2 and two uninfested skin sites (C and D). Infection levels in the skin biopsies and in blood were compared with the efficiency of virus transmission to uninfected nymphal ticks. In neither species was viremia detectable at day 1 of cofeeding, and yet a relatively high percentage of nymphs feeding with infected adult ticks in chamber 1 had become infected (Table V). By day 2, most of the animals were viremic, and uninfected ticks in chamber 2, approximately 1 cm distant from infected and uninfected ticks in chamber 1, had become infected. Tick infection dynamics paralleled the detection of infectious virus in the skin sites of tick feeding (site A followed by site B).

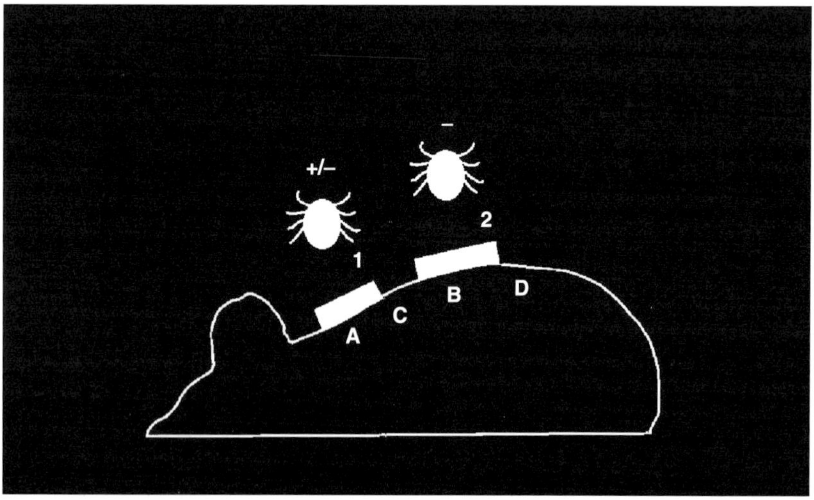

FIG 2. Role of skin in TBE virus transmission. Infected (+) and uninfected (–) ticks placed in chambers 1 and 2. Skin biopsies taken from within the feeding chambers (A and B) and two noninfested sites (C and D).

TABLE V
IMPORTANCE OF SKIN INFECTION IN TRANSMISSION DYNAMICS WHEN NATURAL HOSTS ARE EXPOSED TO COFEEDING UNINFECTED AND TBE VIRUS-INFECTED TICKS[a]

Host species (number)	Days of cofeeding	Log_{10} viremia	% infected ticks		Virus detected in skin site			
			Chamber 1	Chamber 2	A	B	C	D
Apodemus flavicollis (6)	1	< 1	44	9	+	−	−	−
	2	1.4	43	35	+	+	−	(+)
	3	1.4	74	34	+	+	−	−
Clethrionomys glareolus (12)	1	< 1	33	0	+	−	−	−
	2	1.7	47	40	+	+	−	−
	3	2.5	37	18	+	+	−	−

[a] Adapted from Labuda *et al.* (1996). (+) indicates that infection was not detected in all animals.

However, a striking result was the general absence of patent infection in skin sites (C and D) not infested by ticks. Clearly, virus transmission from infected to uninfected cofeeding ticks was correlated with virus replication in the skin rather than with virus levels in peripheral blood. Just as significantly, uninfected ticks appeared to attract the virus to their site of feeding. The greater susceptibility to TBE virus infection of field mice supports previous observations that field mice are more efficient than bank voles in supporting virus transmission between infected and uninfected cofeeding ticks (Labuda *et al.*, 1993d, 1996).

Infection dynamics at the cellular level were examined with laboratory mice for which there are cell markers (Labuda *et al.*, 1996). Balb/c and C57B1/6 mice were infested with TBE virus-infected and uninfected ticks. Instead of assaying skin samples for infectious virus, whole skin explants (20 mm^2) excised from sites of tick infestation (site A within chamber 1 and site B within chamber 2) and from untreated site C were cultured by floating the explants on medium for 24 h at 37 °C. Cells migrating out from the explants were collected for immunocytochemistry, and the culture medium was assayed for infectious virus.

Numerous leukocytes (10^4–10^5 cells/explant) were observed to emigrate from the skin explants where ticks were feeding (A and B). Cells migrating from the uninfected posterior site C were 10- to 100-fold fewer in number. From skin excised on day 1 of tick feeding and cultured for 24 h, most of the emigrating cells were neutrophils and

MHC class II-positive Langerhans cells made up about 5–10% of the total nonadherent population. By 2 and 3 days of feeding, emigrating Langerhans cells had increased to 17–32% (2–8 $\times$ 10^3 cells/culture) of the total nonadherent cell population. Using two-color immunocytochemistry, viral antigen was observed in MHC class II-negative cells (predominantly neutrophils) but not in nonadherent MHC class II-positive cells at day 1 of tick feeding. However, both MHC class II-positive Langerhans cells and MHC class II-negative neutrophils migrating from explants taken at days 2 and 3 of tick feeding were positive. In Langerhans cells, viral antigen was detectable in a distinct cytoplasmic compartment, possibly Golgi or endoplasmic reticulum, and some of the cells appeared to form syncytia. In addition to virus in Langerhans cells and neutrophils, migratory monocyte/macrophages were shown to produce infectious virus.

As observed with natural hosts of TBE virus (Table V), virus infection in the skin matched the presence of infected (site A) and uninfected ticks (site B), with comparatively little virus detected in uninfested site C except for the more sensitive C57B1/6 mice at day 3 of infestation (Table VI). Again, this suggests that the virus is attracted to skin sites of tick feeding. However, when uninfested skin site A* was examined, positioned on the neck of mice, the infection dynamics appeared similar to those for infested site B. The contrasting observations suggest that virus dissemination is associated with the draining lymphatic system.

TABLE VI

INFECTION DYNAMICS IN SKIN EXPLANTS OF LABORATORY MICE EXPOSED TO COFEEDING UNINFECTED AND TBE VIRUS-INFECTED TICKS[a]

Mouse strain (number)	Days of cofeeding	Log_{10} viremia	Virus detected in skin site			
			A	B	C	A*
Balb/c (11)	1	<1	+	(+)	–	(+)
	2	1.9	+	+	–	(+)
	3	3.6	+	+	–	+
C5781/6 (11)	1	1.6	+	(+)	–	(+)
	2	3.0	+	+	(+)	+
	3	3.2	+	+	+	+

[a] Adapted from Labuda *et al.* (1996). (+) indicates that infection was not detected in all animals.

A* see text.

Assays of the amount of infectious virus shed into the culture medium from the explants indicated considerable replication capacity in the skin, particularly where infected ticks were feeding. Indeed, the levels of virus detected in the blood (viremia) may well have originated from virus replication in the skin. Nevertheless, the prevalence of viral antigen in Langerhans cells suggests that these migratory cells play a key role in TBE virus transmission. A model of virus dissemination from infected to cofeeding uninfected ticks has been proposed based on the infection of Langerhans cells and lymphocyte trafficking (Nuttall, 1999).

In contrast to infection by tick bite, when laboratory mice infested with uninfected ticks were infected by intradermal syringe inoculation of virus (0.1 ml 6.0 $\log_{10}$ pfu/ml TBE virus), comparatively few ticks became infected. Relatively few cells (10^2–10^3 cells/explant) were observed to emigrate into culture medium when skin explants were prepared from the site of syringe inoculation.

These results illustrate the important role that the skin site of tick feeding plays in both virus transmission from infected (donor) ticks and virus acquisition by uninfected (recipient) cofeeding ticks. They also illustrate the misleading picture that may result from studies relying on artificial methods of arbovirus inoculation.

VIII. Saliva-Activated Transmission

Results described in the preceding sections clearly show the different picture generated by experiments mimicking natural conditions of tick-borne virus transmission in which infected and uninfected ticks feed together on the same individual host compared with the more conventional method of infecting animals by needle and syringe inoculation. The first evidence that this difference may be due to tick salivary gland products was revealed by syringe inoculation experiments of Thogoto virus (see Section IV) mixed with an extract prepared from the salivary glands of uninfected feeding ticks. Remarkably, 10 times more nymphs became infected compared to the numbers of nymphs infected by feeding on guinea pigs inoculated with virus alone. As with nonviremic virus transmission between cofeeding infected and uninfected ticks (Section IV), none of the inoculated animals showed a detectable viremia (Jones *et al.*, 1989). The enhancement of virus transmission was only observed when the virus inoculum was mixed with salivary gland extracts (SGE) of feeding ticks and was not observed with salivary glands from unfed ticks or with any other tick organ.

To investigate the involvement of salivary gland products in the transmission of TBE virus, SGE was prepared from partially fed uninfected female *I. ricinus*, *D. reticulatus*, and *R. appendiculatus* ticks. Guinea pigs were infested with uninfected *R. appendiculatus* nymphs and inoculated with a mixture of TBE virus and SGE or virus alone. The number of ticks that became infected from feeding on animals inoculated with TBE virus and SGE was fourfold greater than the number infected by feeding on animals inoculated with virus alone or virus plus SGE from unfed *I. ricinus* ticks (Labuda *et al.*, 1993c). This was the first clear evidence that TBE virus transmission is enhanced by factors associated with the salivary glands of feeding ticks and that these factors may facilitate efficient transmission of TBE virus between infected and uninfected ticks.

Because the enhancing factor in the salivary glands of feeding ticks appears to act within the skin of the vertebrate host rather than directly on the virus, this factor is most likely secreted in tick saliva. This hypothesis was tested with saliva collected from uninfected adult female *Amblyomma variegatum* removed from uninfected guinea pigs at different days of feeding (Jones *et al.*, 1992b). After collecting saliva from this large tick species, each tick was dissected and the uninfected salivary glands were removed and prepared as SGE. Each saliva and equivalent SGE was mixed separately with Thogoto virus and inoculated into different tick-infested guinea pigs. The enhancing activity of saliva and SGE showed similar dynamics; there was a gradual increase in the number of recipient ticks that became infected, reaching a maximum with saliva or SGE collected from uninfected ticks that had fed for 6 days, and then followed by a decline. The obvious similarity in activity profiles is as yet the best evidence that the virus transmission-enhancing factor is synthesized in the salivary glands during tick feeding and secreted into the skin-feeding lesion in tick saliva. The phenomenon was named saliva-activated transmission (SAT) (Nuttall and Jones, 1991).

Promotion of virus transmission (SAT) by tick saliva or SGE has been demonstrated experimentally with only one flavivirus (TBE virus) to date, although the phenomenon has also been demonstrated with Thogoto virus, a tick-borne member of the Orthomyxoviridae (Alekseev *et al.*, 1991; Jones *et al.*, 1989, 1992a; Labuda *et al.*, 1993c). Interestingly, SAT has only been demonstrated with arthropod species that are competent vectors. Thus SGE of *I. ricinus* does not show SAT with Thogoto virus (for which *I. ricinus* is not a competent vector), but SAT occurs with TBE virus for which *I. ricinus* is the principal European vector species (Jones *et al.*, 1992a; Labuda *et al.*, 1993c).

This implies that the mechanism underlying SAT differs for different vector–virus associations.

The SAT factor is produced by uninfected ticks. Furthermore, enhancement of virus transmission has no obvious benefits for the tick (except perhaps in the case of high virulence; see Labuda *et al.*, 1993d). These two observations indicate that the SAT factor is independent of virus infection. Most likely, the function of the SAT factor is to modulate the skin site of tick attachment and thereby facilitate feeding. Tick-borne viruses that demonstrate SAT appear to have coevolved with their vectors and vertebrate hosts to exploit the unique environment of the vector–host interface.

Although there is no reported evidence of SAT for mosquito-borne flaviviruses, the phenomenon has been implicated in the transmission of other insect-borne arboviruses. These include La Crosse virus (Osorio *et al.*, 1996), Cache Valley virus (Edwards *et al.*, 1998), and vesicular stomatitis virus (Limesand *et al.*, 2000; Mead *et al.*, 2000).

IX. Host Modulation by Tick Saliva

What is the general mechanism underlying SAT of tick-borne viruses, and might it apply to insect-borne flaviviruses?

Ticks have been described as "smart pharmacologists, skilfully manipulating their hosts immune system and associated inflammatory and anti-hemostatic reactions to have what they want, a blood meal" (Ribeiro, 1995). In fact, increasing evidence shows that the saliva of all blood-sucking arthropods contains a wealth of pharmacologically active components that inhibit hemostasis and inflammation and modulate immune responses (Wikel, 1996a). Some of these antihemostatic and immunomodulatory effects may be exploited by vector-borne pathogens.

The saliva of ixodid ticks is probably the richest cocktail of bioactive ingredients among hematophagous arthropods (Table VII). This is partly due to the numerous roles of tick saliva: in water balance, attachment (via cement), ion transport, secretion of various enzymes, enzyme inhibitors and toxins, hemostatic and inflammatory/immune modulation, and excretion of host immunoglobulins (Kaufman, 1989; Neitz and Vermeulen, 1987; Ribeiro, 1987b; Sauer and Bowman, 2000; Wang and Nuttall, 1999). The complex nature of tick saliva reflects the protracted feeding process of ixodid species. Full engorgement may take up to 2 weeks or more to achieve and involve processing several hundred milligrams of host blood (Bowman *et al.*, 1997;

TABLE VII

TICK PHARMACOPEIA: PHARMACOLOGICAL ACTIVITIES OF SALIVARY GLAND EXTRACTS OR SALIVA OF *Ixodes* TICK VECTORS OF FLAVIVIRUSES

Activity	Target	Reference
Antihemostatic	ADP	Ribeiro *et al.* (1985)
	Prostaglandin receptor	Ribeiro *et al.* (1985)
	Prostacyclin receptor	Ribeiro *et al.* (1988)
	Thrombin	Hoffmann *et al.* (1991)
Anti-inflammatory	Anaphylatoxins	Ribeiro and Spielman (1986)
	Bradykinin	Ribeiro *et al.* (1985)
	Histamine	*G.C. Paesen*, unpublished data
Anti-immune	Alternative complement system	Ribeiro (1987a); Lawrie *et al.* (1999); Valenzuela *et al.* (2000)
	Neutrophils	Ribeiro *et al.* (1990)
	Splenic T lymphocytes	Urioste *et al.* (1994)
	Interleukin 2	Urioste *et al.* (1994)
	Interleukin 8	Hajnicka *et al.* (2001)
	Macrophages, nitric oxide	Urioste *et al.* (1994)
	Interferon α	Hajnicka *et al.* (2000)
	Immunoglobulin G	Wang and Nuttall (1999)

Kaufman, 1989;). Acquisition of a blood meal proceeds through several stages that include sawing through the epidermis, deposition of a cement cone to secure the mouthparts of the tick, and formation of a feeding pool below the mouthparts comprising interstitial fluids and cellular infiltrate. Most of the blood meal is imbibed during the latter half of the feeding period—the rapid engorgement phase.

The complex sequence of events during tick feeding is mirrored by changes in the protein profile of tick salivary glands as functional demands on tick saliva change during feeding (McSwain *et al.*, 1992). Maintaining such a long period of attachment on the host imposes enormous selection pressure on the tick to counteract the host responses of hemostasis, inflammation, and specific and nonspecific immune-mediated rejection mechanisms. Indeed, the sophisticated salivary cocktail of ticks may be considered the product of at least 100 million years of coevolutionary adaptation in which ticks have responded to countless host-imposed selection pressures.

One of the first antihemostatic molecules identified in tick saliva was prostaglandin E_2 (PGE_2) (Ribeiro *et al.*, 1985). Prostaglandins play a role in vasodilation and antiplatelet aggregation, as well as immunomodulation. They are produced from arachidonic acid taken up in the blood meal by feeding ticks. Concentrations of this autacoid hormone are high (50 to 469 ng/ml) compared with concentrations found in vertebrate inflammatory exudates (0.5 to 20 ng/ml). Many of the immunomodulatory effects of tick salivary gland products are thought to be mediated by PGE_2 (Bowman *et al.*, 1996; Champagne, 1994). These include suppression of T-cell functions demonstrated by the inhibition of interleukin 1 (IL-1), IL-2, tumor necrosis factor α (TNF-α), interferon γ (IFN-γ) activity, and lymphocyte proliferative responses to concanavalin A (Con A) (Ramachandra and Wikel, 1992). However, other observations suggest that the role of PGE_2 may have been overstated, particularly when the lability of this molecule is taken into consideration (Urioste *et al.*, 1994).

Peptides and proteins also appear to play an important role in the pharmacology of tick saliva. Of these, several immunomodulatory molecules are reviewed as SAT factor candidates in promoting tick-borne flavivirus transmission.

A. *Complement Inhibitors*

Although complement provides a means of controlling flavivirus infections through both antibody-independent and antibody-mediated mechanisms, its relative importance (particularly at the initial skin site of infection) is poorly understood. Nevertheless, the ability of tick vectors to control complement is likely to benefit tick-borne flaviviruses. Both ixodid and argasid ticks control the complement cascade. Inhibitory activity of *Ixodes* salivary gland products is directed against the alternative complement cascade and shows host specificity. The cosmopolitan *I. ricinus* demonstrates anticomplement activity to a range of mammalian hosts, whereas *I. hexagonus* has particularly high levels of anticomplement activity against hedgehog serum (hedgehogs are its preferred host) (Lawrie *et al.*, 1999). Unlike most bioactive tick saliva proteins, significant levels of anticomplement activity were demonstrated in SGE from both unfed and feeding *I. ricinus* ticks (Lawrie *et al.*, 1999). This reflects the fact that activation of the alternative complement cascade, through the acute-phase inflammatory response, is one of the first events following tick attachment and presents a constant threat to the tick (and tick-borne flaviviruses) throughout blood feeding. The anticomplement activity of *I. scapularis* is provided by a novel 18.5-kDa protein that appears to act in a similar fashion to factor H, interfering with the binding of factor B to C3b (Valenzuela *et al.*, 2000). It remains to be determined whether a similar complement control protein is produced by other *Ixodes* species, particularly the more important flavivirus vectors, *I. persulcatus* and *I. ricinus*.

B. Immunoglobulin-Binding Proteins

The observation that adult female *R. appendiculatus* excrete intact IgG in their saliva led to the discovery of a family of immunoglobulin-binding proteins (IGBPs) in the hemolymph and salivary glands of ixodid ticks (Wang and Nuttall, 1994, 1999). Although IGBPs have not been reported for argasid tick species, circumstantial evidence suggests they may occur (Minoura *et al.*, 1985). Three IGBPs have been cloned and sequenced; they are unique and unrelated to immunoglobulin receptors. The prevalence and abundance of IGBPs indicate that they play an important role in blood feeding. Feeding experiments using laboratory animals immunized with a recombinant form of a secreted IGBP revealed a role for the protein in facilitating female tick blood feeding, possibly through local immunosuppression of the host (Wang *et al.*, 1998). The mechanism is unknown but might involve saturation of Fc receptors (on Langerhans cells, neutrophils, eosinophils, basophils, and dermal macrophages) by the tick-excreted IgG (Wang and Nuttall, 1999). IGBP-mediated immunosuppression within the feeding lesion should help any tick-borne flaviviruses entering the skin site. On virus-immune hosts, saturation of Fc receptors at the feeding site may interfere with the protection normally mediated by the Fc portion of the flavivirus-specific antibody (Schlesinger *et al.*, 1993). An alternative role for IGBPs also has benefits for tick-borne flaviviruses in the provision of a tick immunoglobulin excretion system. This system, comprising different IGBPs in the hemolymph, salivary glands, and saliva, enables ticks to ferry potentially damaging antibodies safely through the hemocoel to their salivary glands from where they are excreted (Wang and Nuttall, 1999). If such an IgG-transporting system exists, it may benefit tick-borne flaviviruses by protecting them (and possibly the infected tick) from virus-specific antibodies taken up in the blood meal. The results of interrupted feeding experiments, in which ticks were fed on virus-infected hosts followed by virus-immune hosts and vice versa, are consistent with the presence of an antibody-protective mechanism within ticks (Jones and Nuttall, 1989a).

C. Natural Killer Cell Suppressors

Natural killer (NK) cells are a component of the innate immune system and play an important role in killing cells that express viral antigens on their surface (Biron, 1997). In severe flavivirus infections, NK cells may be involved in the lysis of infected neurons (Liu *et al.*, 1989); their role in the skin during the first stages of infection is unknown. Cytokines secreted by activated NK cells provide an important communication between innate and acquired immunity (Lanier, 2000). Therefore, it is perhaps not surprising that tick salivary gland extracts suppress NK cell activity (Kopecky and Kuthejlova, 1998; Kubes *et al.*, 1994). The activity is directed against the first step in NK cell-mediated cytotoxicity, namely formation of effector/target cell conjugates (Kubes *et al.*, 2002). Although anti-NK cell activity has been detected in salivary gland extracts of several ixodid tick species, it was not demonstrated for *I. ricinus* using human effector cells. However, a suppressive effect of salivary gland extracts derived from partially fed *I. ricinus* females has been described in a mouse model. Further studies are needed to determine whether SGE-mediated effects on NK cell activity vary according to the host source of NK cells. Thus it is unclear whether anti-NK cell activity facilitates TBE virus transmission by its most common European vector.

D. Cytokine Inhibitors

Cytokines are the chemical mediators of inflammation and immunity. A number of anticytokine activities have been described for tick salivary gland products (Kopecky *et al.*, 1999; Wikel, 1996b). Potentially, one of the most significant for SAT is suppression of the antiviral action of type 1 interferon (IFN-α/β). This activity was demonstrated *in vitro* using vesicular stomatitis virus, an insect-borne virus used commonly in interferon research. Suppression was most apparent when cells were treated with salivary gland extracts prior to virus infection, suggesting that the effect is directed against the IFN-α/β receptor rather than a direct interaction with IFN-α/β molecules (Hajnicka *et al.*, 2000). Type 1 interferons are elicited during the early stages of viral infections. Although TBE virus has been shown to induce interferon in chick cell cultures (Vilcek, 1960), few (if any) studies have been undertaken on interferon induction at the feeding site of infected ticks. Low viral doses are delivered repeatedly into the skin, possibly over a period of days. Such conditions are ideal for eliciting a local interferon-mediated antiviral response. Under these conditions, a tick inhibitor of IFN-α/β may be crucial for successful virus transmission. Indeed, it would be difficult to conceive of how a single infected tick could infect a host without the virus exploiting the pharmacy of the tick.

A group of cytokines that are particular enemies of viruses are the chemokines. These chemotactic proteins orchestrate the movement of inflammatory and immune cells to sites of injury and infection. Saliva and salivary gland extracts from several ixodid tick species (including *I. ricinus*) contain molecules that bind interleukin 8 (IL-8) and inhibit IL-8-induced chemotaxis of neutrophils (Hajnicka *et al.*, 2001). This recent discovery offers yet another potential SAT mechanism for flaviviruses to exploit.

E. *Other Activities*

In addition to the immunomodulatory properties of tick salivary gland products described in the preceding sections, many other reported activities are likely to facilitate flavivirus transmission. For example, inhibition by *I. scapularis* saliva of splenic T-cell proliferation in response to Con A or phytohemagglutinin (PHA) stimulation, reduction in IL-2 secretion by T cells, and suppression of nitric oxide production by lipopolysaccharide (LPS)-treated macrophages were attributed to a protein of $\geq$5 kDa (Urioste *et al.*, 1994). The SAT factor of *R. appendiculatus*, promoting transmission of certain tick-borne viruses, appears to be a protein(s) of approximately 30 kDa (Jones *et al.*, 1990; unpublished data). However, its primary function in tick feeding has not been identified. The SAT factor of *I. ricinus*, promoting transmission of TBE virus, is even less clearly defined. Recent studies showing protection of mice immunized with a tick salivary gland protein against lethal challenge with tick-transmitted TBE virus confirm the importance of the tick-host response in flavivirus infections (M. Labuda *et al.*, unpublished data). Clearly, a great deal more research needs to be done, particularly in identifying the bioactive ingredients and in determining how their functions in tick feeding affect tick-borne virus transmission. Such studies will help explain the importance of mimicking natural conditions of vector-borne pathogen transmission and the crucial first steps of transmission that determine the outcome of the infection, whether it be disease or a benign infection. They may also lead to the development of novel transmission-blocking vaccines.

X. The "Red Herring" Hypothesis

Studies on tick-borne virus transmission have revealed the important role that tick-feeding behavior and the bioactive ingredients of tick saliva play in virus transmission. How might these factors combine to facilitate the transmission of a virus from a cofeeding-infected donor tick to an uninfected recipient tick via a nonviremic host?

At the site of tick feeding there is a profound cellular response with marked infiltration and emigration of leukocytes (Mbow *et al.*, 1994). For example, epidermal dendritic cells (Langerhans cells) attracted to the feeding site internalize and process tick antigens, express the relevant peptide–MHC complexes, and migrate to the draining lymph nodes where they deliver costimulatory signals for T-cell activation and initiation of primary immune responses (Austyn, 1992; Brossard and Wikel, 1997; Larsen *et al.*, 1990). Interestingly, Langerhans cells collected from cultured explants taken from the skin site of TBE virus-infected tick feeding were infected and formed syncytia (Labuda *et al.*, 1996). In contrast, Langerhans cells isolated from skin sites on which uninfected ticks had not been feeding, when inoculated with virus *in vitro*, showed no evidence of syncytium formation. Possibly, the immunomodulatory effects of tick saliva (Section IX) facilitate virus infection of target cells at the site of tick feeding. Targets such as Langerhans cells may then act to shuttle the virus to the draining lymph nodes where the infection may spread to lymphocytes with which Langerhans cells interdigitate as part of the priming of the immune response. Such lymphocytes, primed to migrate from the lymph nodes to the skin, will be attracted to sites where uninfected (and infected) ticks are feeding. This hypothesis is consistent with the observation that TBE virus infection is localized in skin sites where ticks feed and not in adjoining uninfested sites (Labuda *et al.*, 1996). In summary, an alternative model of tick-borne flavivirus transmission (see Nuttall, 1999) is one involving (1) tick-induced immunomodulation at the skin site of tick feeding; (2) infection of Langerhans cells, which shuttle the virus to the draining lymph nodes; (3) infection and priming of lymphocytes in lymph nodes; (4) lymphocyte trafficking that shuttles the virus to the skin site of uninfected tick feeding; and finally (5) virus acquisition by uninfected cofeeding ticks. Because of the exquisite efficiency of virus transmission inherent in this model, viremia is a redundant feature of the infection—a "red herring."

References

Alekseev, A. N., and Chunikhin, S. P. (1990). Exchange of tick-borne encephalitis virus between Ixodidae simultaneously feeding on animals with subthreshold levels of viremia. *Med. Parazitol. Parazit. Bolezni* **2**:48–50.

Alekseev, A. N., Chunikhin, S. P., Rukhkyan, M. Y., and Stefutkina, L. F. (1991). Possible role of Ixodidae salivary gland substrate as an adjuvant enhancing arbovirus transmission. *Med. Parazitol. Parazit. Bolezni* **1**:28–31.

Alexander, R. A., and Neitz, W. O. (1935). The transmission of louping ill by ticks (*Rhipicephalus appendiculatus*). *Onderstepoort J. Vet. Sc. Anim. Ind.* **5**:15–33.

Anderson, R. M., and May, R. M. (1991). "Infectious Diseases of Humans: Dynamics and Control," p. 757. Oxford Univ. Press, Oxford.

Austyn, J. M. (1992). Antigen uptake and presentation by dendritic leukocytes. *Sem. Immunol.* **4**:227–236.

Balashov, Y. S. (1983). "An Atlas of Ixodid Tick Ultrastructure," p. 289. Entomological Society of America.

Banerjee, K. (1988). Kyasanur forest disease. *In* "The Arboviruses; Epidemiology and Ecology" (T. P. Monath, ed.), pp. 93–116. CRC Press, Boca Raton, Fla.

Benda, R. (1958a). The common tick, *Ixodes ricinus* L., as a reservoir and vector of tick-borne encephalitis. I. Survival of the virus (strain B3) during the development of the tick under laboratory conditions. *J. Hyg. Epidemiol. Microbiol. Immunol.* **2**:314–330.

Benda, R. (1958b). The common tick, *Ixodes ricinus* L., as a reservoir and vector of tick-borne encephalitis. II. Experimental transmission of encephalitis to laboratory animals by ticks at various stages of development. *J. Hyg. Epidemiol. Microbiol. Immunol.* **2**:331–344.

Biron, C. (1997). Activation and function of natural killer cell responses during viral infections. *Curr. Op. Immunol.* **9**:24–34.

Blaskovic, D. (1954). "An Epidemic of Encephalitis in the Natural Infective Focus in Roznava," p. 314. Bratislava Slovenska Akademia Vied, Bratislava.

Bowman, A. S., Dillwith, J. W., and Sauer, J. R. (1996). Tick salivary gland prostaglandins: Presence, origin and significance. *Parasitol. Today* **12**:388–396.

Bowman, A. S., Coons, L. B., Needham, G. R., and Sauer, J. R. (1997). Tick saliva: Recent advances and implications for vector competence. *Med. Vet. Ent.* **11**:277–285.

Brossard, M., and Wikel, S. K. (1997). Immunology of interactions between ticks and hosts. *Med. Vet. Ent.* **11**:270–276.

Burke, D. S., and Monath, T. P. (2001). Flaviviruses. *In* "Fields Virology" (D. M. Knipe and P. M. Howley, eds.), 4th Ed., pp. 1043–1125. Lippincott Williams and Wilkins, Philadelphia.

Chamberlain, R. W., Sikes, R. K., Nelson, D. B., and Sudia, W. D. (1954). Studies on the North American arthropod-borne encephalitides. *Am. J. Hyg.* **60**:278–285.

Champagne, D. (1994). The role of salivary vasodilators in bloodfeeding and parasite transmission. *J. Parasitol.* **10**:430–433.

Charrel, R. N., Zaki, A. M., Attoui, H., Fakeeh, M., Billoir, F., Yousef, A. I., de Chesse, R., De Micco, P., Gould, E. A., and de Lamballerie, X. (2001). Complete coding sequence of the Alkhurma virus, a tick-borne flavivirus causing severe hemorrhagic fever in Saudi Arabia. *Biochem. Biophys. Res. Commun.* **287**:455–461.

Davies, C. R., Jones, L. D., and Nuttall, P. A. (1986). Experimental studies on the transmission cycle of Thogoto virus, a candidate orthomyxovirus, in *Rhipicephalus appendiculatus* ticks. *Am. J. Trop. Med. Hyg.* **35**:1256–1262.

Dizij, A., and Kurtenbach, K. (1995). *Clethrionomys glareous*, but not *Apodemus flavicollis*, acquires resistance to *Ixodes ricinus L*, the main European vector of *Borrelia burgdorferi*. *Parasite Immunol.* **17:**177–183.

Dumina, A. L. (1958). Experimental study of the extent to which the tick *Ixodes persulcatus* becomes infected with Russian spring-summer encephalitis virus as a result of sucking the blood of immune animals. *Vopr. Virusol.* **3:**156–159.

Edwards, J. F., Higgs, S., and Beaty, B. J. (1998). Mosquito feeding-induced enhancement of Cache Valley Virus (Bunyaviridae) infection in mice. *J. Med. Entomol.* **35:**261–265.

Galimov, V. R., Galimova, E. Z., Katin, A. A., and Kolchanova, L. P. (1989). The transmission of TBE virus to adult taiga ticks when virusemia in their hosts is absent. *In* "Proceedings of the XII All-Union Conference on Natural Focality of Diseases," pp. 43–44. Novosibirsk, Russia.

Hajnicka, V., Kocakova, P., Slavikova, M., Slovak, M., Gasperik, J., Fuchsberger, N., and Nuttall, P. A. (2001). Anti-interleukin-8 activity of tick salivary gland extracts. *Parasite Immunol.* **23:**483–489.

Hajnicka, V., Kocakova, P., Slovak, M., Labuda, M., Fuchsberger, N., and Nuttall, P. A. (2000). Inhibition of the antiviral action of interferon by tick salivary gland extract. *Parasite Immunol.* **22:**201–206.

Heinz, F. X., and Allison, S. L. (2001). The machinery for flavivirus fusion with host cell membranes. *Curr. Opin. Microbiol.* **4:**450–455.

Hoffmann, A., Walsmann, P., Riesener, G., Paintz, M., and Markwardt, F. (1991). Isolation and characterization of a thrombin inhibitor from the tick *Ixodes ricinus*. *Pharmazie* **46:**209–212.

Hudson, P. J., Norman, R., Laurenson, M. K., Newborn, D., Gaunt, M., Jones, L., Reid, H., Gould, E., Bowers, R., and Dobson, A. (1995). Persistence and transmission of tick-borne viruses: *Ixodes ricinus* and louping-ill virus in red gouse populations. *Parasitology* **111**(Suppl.):S49–S58.

Jones, L. D., and Nuttall, P. A. (1989a). The effect of virus-immune hosts on Thogoto virus infection of the tick, *Rhipicephalus appendiculatus*. *Virus Res.* **14:**129–140.

Jones, L. D., and Nuttall, P. A. (1989b). Non-viremic transmission of Thogoto virus: Influence of time and distance. *Trans. Roy. Soc. Trop. Med. Hyg.* **83:**712–714.

Jones, L. D., and Nuttall, P. A. (1990). The effect of host resistance to tick infestation on the transmission of Thogoto virus by ticks. *J. Gen. Virol.* **71:**1039–1043.

Jones, L. D., Davies, C. R., Steele, G. M., and Nuttall, P. A. (1987). A novel mode of arbovirus transmission involving a nonviremic host. *Science* **237:**775–777.

Jones, L. D., Gaunt, M., Hails, R. S., Laurenson, K., Hudson, P. J., Reid, H., Henbest, P., and Gould, E. A. (1997). Transmission of louping-ill virus between infected and uninfected ticks co-feeding on mountain hares. *Med. Vet. Entomol.* **11:**172–176.

Jones, L. D., Hodgson, E., and Nuttall, P. A. (1989). Enhancement of virus transmission by tick salivary glands. *J. Gen. Virol.* **70:**1895–1898.

Jones, L. D., Hodgson, E., and Nuttall, P. A. (1990). Characterization of tick salivary gland factor(s) that enhance Thogoto virus transmission. *Arch. Virol.* **Suppl. 1:**227–234.

Jones, L. D., Hodgson, E., Williams, T., Higgs, S., and Nuttall, P. A. (1992a). Saliva activated transmission (SAT) of Thogoto virus: Relationship with vector potential of different hematophagous arthropods. *Med. Vet. Entomol.* **6:**261–265.

Jones, L. D., Kaufman, W. R., and Nuttall, P. A. (1992b). Modification of the skin feeding site by tick saliva mediates virus transmission. *Experientia* **48:**779–782.

Karabatsos, N. (1985). "International Catalogue of Arthropod-Borne Viruses," 3rd Ed. American Society for Tropical Medicine and Hygiene, San Antonio, Tex.

Kaufman, W. (1989). Tick-host interaction: A synthesis of current concepts. *Parasitol. Today* **5**:47–56.

Kohl, I., Kozuch, O., Eleckova, E., Labuda, M., and Zaludko, J. (1996). Family outbreak of alimentary tick-borne encephalitis in Slovakia associated with a natural focus of infection. *Eur. J. Epidemol.* **12**:373–375.

Kopecky, J., Grubhoffer, L., Kovar, V., Jindrak, L., and Vokurkova, D. (1999). A putative host cell receptor for tick-borne encephalitis virus identified by anti-idiotypic antibodies and virus affinoblotting. *Intervirology* **42**:9–16.

Kopecky, J., and Kuthejlova, M. (1998). Suppressive effect of *Ixodes ricinus* salivary gland extract on mechanisms of natural immunity in vitro. *Parasite Immunol.* **20**:169–174.

Korenberg, E. I. (1976). Some contemporary aspects of natural focality and epidemiology of tick-borne encephalitis. *Folia Parasitol.* **23**:357–366.

Korenberg, E. I., and Kovalevskii, Y. V. (1994). A model for relationships among the tick-borne encephalitis virus, its main vectors, and hosts. *In* "Advances in Disease Vector Research," pp. 65–92. Springer-Verlag, New York.

Kozuch, O., Gresikova, M., Nosek, J., Lichard, M., and Sekeyova, M. (1967). The role of small rodents and hedgehogs in a natural focus of tick-borne encephalitis. *Bull. World Health Organ.* **36**(Suppl. 1):61–66.

Kozuch, O., Labuda, M., Lysy, J., Weismann, P., and Krippel, E. (1990). Longitudinal study of natural foci of Central European encephalitis virus in west Slovakia. *Acta Virol.* **34**:537–544.

Kubes, M., Fuchsberger, N., Labuda, M., Zuffova, E., and Nuttall, P. A. (1994). Salivary gland extracts of partially fed *Dermacentor reticulatus* ticks decrease natural killer cell activity *in vitro*. *Immunology* **82**:113–116.

Kubes, M., Kocáková, P., Slovák, M., Sláviková, M., Fuchsberger, N., and Nuttall, P. A. (2002). Heterogeneity in the effect of different ixodid tick species on human natural killer cell activity. *Parasite Immunol.* **24**:23–28.

Kuno, G., Artsob, N., Karabatsos, N., Tsuchiya, K. R., and Chang, G. J. (2001a). Genomic sequencing of deer tick virus and phylogeny of Powassan-related viruses in North America. *Am. J. Trop. Med. Hyg.* **65**:671–676.

Kuno, G., Chang, G. J., Tsuchiya, K. R., and Miller, B. R. (2001b). Phylogeny of Thogoto virus. *Virus Genes* **23**:211–214.

Labuda, M., Austyn, J., Zuffova, E., Kozuch, O., Fuchsberger, N., Lysy, J., and Nuttall, P. (1996). Importance of localized skin infection in tick-borne encephalitis virus transmission. *Virology* **219**:357–366.

Labuda, M., Danielova, V., Jones, L. D., and Nuttall, P. A. (1993a). Amplification of tick-borne encephalitis virus infection during co-feeding of ticks. *Med. Vet. Entomol.* **7**:339–342.

Labuda, M., Eleckova, E., Lickova, M., and Sabo, A. (2002). Tick-borne encephalitis virus foci in Slovakia *Int. J. Med. Microbiol.* **291**(Suppl. 33):43–47.

Labuda, M., Jones, L. D., Williams, T., Danielova, D., and Nuttall, P. A. (1993b). Efficient transmission of tick-borne encephalitis virus between cofeeding ticks. *J. Med. Entomol.* **30**:295–299.

Labuda, M., Jones, L. D., Williams, T., and Nuttall, P. A. (1993c). Enhancement of tick-borne encephalitis virus transmission by tick salivary gland extracts. *Med. Vet. Entomol.* **7**:193–196.

Labuda, M., Kozuch, O., Zuffova, E., Eleckova, E., Hails, R. S., and Nuttall, P. A. (1997). Tick-borne encephalitis virus transmission between ticks co-feeding on specific immune natural rodent hosts. *Virology* **235**:138–143.

Labuda, M., Nuttall, P. A., Austyn, J., Zuffova, E., and Kozuch, O. (1994). Cellular basis of tick-borne encephalitis virus "non-viremic" transmission: Involvement of Langerhans cells? *In* "Acarology IX," (R. Mitchell, D. J. Horn, G. R. Needham, and W. C. Welbourn, eds.), pp. 461–463. Ohio Biological Survey, Columbus, OH.

Labuda, M., Nuttall, P. A., Kozuch, O., Eleckova, E., Williams, T., Zuffova, E., and Sabo, A. (1993d). Non-viremic transmission of tick-borne encephalitis virus: A mechanism for arbovirus survival in nature. *Experientia* **49:**802–805.

Lanier, L. S. (2000). The origin and functions of natural killer cells. *Clin. Immunol.* **95:**S14–S18.

Larsen, C. P., Steinman, R. M., Witmer-Pack, M., Hankins, D. F., Morris, P. J., and Austyn, J. M. (1990). Migration and maturation of Langerhans cells in skin transplants and explants. *J. Exp. Med.* **172:**1483–1493.

Lawrie, C. H., Randolph, S. E., and Nuttall, P. A. (1999). *Ixodes* ticks: Serum species sensitivity of anti-complement activity. *Exp. Parasitol.* **93:**207–214.

Lenski, R. E., and May, R. M. (1994). The evolution of virulence in parasites and pathogens: Reconciliation between two competing hypotheses. *J. Theor. Biol.* **169:**253–265.

Limesand, K. H., Higgs, S., Pearson, L. D., and Beaty, B. J. (2000). Potentiation of vesicular stomatitis New Jersey virus infection in mice by mosquito saliva. *Parasite Immunol.* **22:**461–467.

Liu, Y., Blanden, R. V., and Mullbacher, A. (1989). Identification of cytolytic lymphocytes in West Nile virus-infected murine central nervous system. *J. Gen. Virol.* **70:**565–573.

Málková, D. (1968). The significance of the skin and the regional lymph nodes in the penetration and multiplication of tick-borne encephalitis virus after subcutaneous infection of mice. *Acta Virol.* **12:**222–228.

Mbow, M. L., Rutti, B., and Brossard, M. (1994). Infiltration of $CD4^+$ $CD8^+$ T cells, and expression of ICAM-1, Ia antigens, IL-1a and TNF-α in the skin lesion of BALB/c mice undergoing repeated infestations with nymphal *Ixodes ricinus* ticks. *Immunology* **82:**596–602.

McKelvey, J. J., Jr., Eldridge, B. F., and Maramorosch, K. (1981). "Vectors of Disease Agents." Praeger, New York.

McSwain, J. L., Essenberg, R. C., and Sauer, J. R. (1992). Oral secretion elicited by effectors of signal transduction pathways in the salivary glands of *Amblyomma americanum* (Acari: Ixididae). *J. Med. Entomol.* **29:**41–48.

Mead, D. G., Ramberg, F. B., Besselsen, D. G., and Mare, C. J. (2000). Transmission of vesicular stomatitis virus from infected to uninfected black flies co-feeding on non-viremic deer mice. *Science* **287:**485–487.

Minoura, H., Chinzei, Y., and Kitamura, S. (1985). *Ornithodoros moubata*: Host immunoglobulin G in tick hemolymph. *Exp. Parasitol.* **60:**355–363.

Neitz, A. W. H., and Vermeulen, N. M. J. (1987). Biochemical studies on the salivary glands and hemolymph of *Amblyomma hebraeum*. *Onderstepoort J. Vet. Res.* **54:**443–450.

Nosek, J., Ciampor, F., Kozuch, O., and Rajcani, J. (1972). Localization of tick-borne encephalitis virus in alveolar cells of salivary glands of *Dermacentor marginatus* and *Haemaphysalis inermis* ticks. *Acta Virol.* **16:**493–497.

Nosek, J., Kozuch, O., Ernek, E., and Lichard, M. (1967a). The importance of goats in the maintenance tick-borne encephalitis virus in nature. *Acta Virol.* **11:**470–472.

Nosek, J., Kozuch, O., Ernek, E., and Lichard, M. (1967b). Uebertragung des Zeckenencephalitis-Virus (TBE) durch die Weibchen von *Ixodes ricinus* und die Nymphen

von *Haemaphysalis inermis* auf die Rehkitze (*Capreolus capreolus*). *Zentralbl. Bakteriol. I. Orig.* **203:**162–166.
Nosek, J., Kozuch, O., and Lysy, J. (1981). The survival of tick-borne encephalitis (TBE) virus in nymphs of *Haemaphysalis inermis* ticks and its transmission to pigmy mouse (*Micromys minutus*). *Cahiers O.R.S.T.O.M. Ser. Entomol. Med. Parasitol.* **19:**67–69.
Nuttall, P. A. (1999). Pathogen-tick-host interactions: *Borrelia burgdorferi* and TBE virus. *Zentralbl. Bakteriol.* **289:**492–505.
Nuttall, P. A., and Jones, L. D. (1991). Non-viremic tick-borne virus transmission: Mechanism and significance. *In* "Modern Acarology," (F. Dusbábek and V. Bukva, eds.), Vol. 2, pp. 3–6. Academia, Prague and SPB Academic Publishing bv.
Nuttall, P. A., Jones, L. D., and Davies, C. R. (1991). Advances in disease vector research: The role of arthropod vectors in arbovirus evolution. *Adv. Dis. Vector Res.* **8:**16–45.
Nuttall, P. A., Jones, L. D., Labuda, M., and Kaufman, W. R. (1994). Adaptations of arboviruses to ticks. *J. Med. Entomol.* **31:**1–9.
Nuttall, P. A., and Labuda, M. (1994). Tick-borne encephalitis subgroup. *In* "Ecological Dynamics of Tick-Borne Zoonoses" (D. E. Sonenshine and T. N. Mather, eds.), Oxford Univ. Press, Oxford, UK.
Osorio, J. E., Godsey, M. S., Defoliart, G. R., and Yuill, T. M. (1996). La Crosse viremias in white-tailed deer and chipmunks exposed by injection or mosquito bite. *Am. J. Trop. Med. Hyg.* **54:**338–342.
Rajcani, J., Nosek, J., Kozuch, O., and Waltinger, H. (1976). Reaction of the host to the tick bite. II. Distribution of tick-borne encephalitis virus in sucking ticks. *Zentralblt. Bakteriol. Parsiten. Abt. I. Orig. A* **236:**1–9.
Ramachandra, R. N., and Wikel, S. K. (1992). Modulation of host-immune responses to ticks (Acari: Ixodidae): Effect of salivary gland extracts on host macrophages and lymphocyte cytokine production. *J. Med. Entomol.* **29:**818–826.
Rampas, J., and Gallia, F. (1949). Isolation of encephalitis virus from *Ixodes ricinus* ticks. *Casopis Lekaru Ceskoslovenskych* **88:**1179–1180.
Randolph, S. E. (1994). Density-dependent acquired resistance to ticks in natural hosts, independent of concurrent infection with *Babesia microti*. *J. Parasitol.* **108:**413–419.
Randolph, S. E., Gern, L., and Nuttall, P. A. (1996). Co-feeding ticks: Epidemiological significance for tick-borne pathogen transmission. *Parasitol. Today* **12:**472–479.
Randolph, S. E., Miklisova, D., Lysy, J., Rogers, D. J., and Labuda, M. (1999). Incidence from coincidence: Patterns of tick infestations on rodents facilitate transmission of tick-borne encephalitis virus. *Parasitology* **118:**177–186.
Randolph, S. E., and Nuttall, P. A. (1994). Nearly right or precisely wrong? Natural vs laboratory studies of vector-borne diseases. *Parasitol. Today* **10:**458–462.
Rehacek, J. (1960). Experimental hibernations of the tick-borne encephalitis virus in engorged larvae of the tick *Ixodes ricinus*. *Acta Virol.* **4:**106–109.
Rehacek, F. (1962). Transovarial transmission of tick-borne encephalitis virus by ticks. *Acta Virol.* **6:**220–226.
Rehacek, F. (1965). Development of animal viruses and rickettsiae in ticks and mites. *Annu. Rev. Entomol.* **10:**1–24.
Reid, H. (1984). Epidemiology of louping-ill. *In* "Vectors in Virus Biology" (M. A. Mayo and K. A. Harrap, eds.), pp. 161–178. Academic Press, London.
Ribeiro, J., and Spielman, A. (1986). *Ixodes dammini*: Salivary anaphylatoxin inactivating activity. *Exp. Parasitol.* **62:**292–297.
Ribeiro, J. M. C. (1987a). *Ixodes dammini*: Salivary anti-complement activity. *Exp. Parasitol.* **64:**347–353.

Ribeiro, J. M. C. (1987b). Role of saliva in blood-feeding by arthropods. *Annu. Rev. Entomol.* **32:**463–478.

Ribeiro, J. M. C. (1995). How ticks make a living. *Parasitol. Today* **11:**91–93.

Ribeiro, J. M. C., Makoul, G. T., Levine, J., Robinson, D. R., and Spielman, A. (1985). Antihemostatic, antiinflammatory and immunosuppressive properties of the saliva of a tick, *Ixodes dammini. J. Exp. Med.* **161:**332–344.

Ribeiro, J. M. C., Makoul, G. T., and Robinson, D. R. (1988). *Ixodes dammini*: Evidence for salivary prostacyclin secretion. *J. Parasitol.* **74:**1068–1069.

Ribeiro, J. M. C., Weis, J. J., and Telford, S. R. (1990). Saliva of the tick *Ixodes dammini* inhibits neutrophil function. *Exp. Parasitol.* **70:**382–388.

Sauer, J. R., and Bowman, A. S. (2000). Prostaglandins in tick salivary gland physiology. *In* "XXI International Congress of Entomology," Session 1/0006. Iguassu Falls, Brazil.

Sauer, J. R., McSwain, J. L., Bowman, A. S., and Essenberg, R. C. (1995). Tick salivary gland physiology. *Annu. Rev. Entomol.* **40:**245–267.

Schlesinger, J. J., Foltzer, M., and , S. C. (1993). The Fc portion of antibody to yellow fever virus NS1 is a determinant of protection against YF encephalitis in mice. *Virology* **192:**132–141.

Service, M. W. (2001). "The Encyclopedia of Arthropod-Transmitted Infections," p. 579. CABI, Wallingford, UK.

Sonenshine, D. E. (1991). "Biology of Ticks," 1st Ed. Oxford Univ. Press, New York.

Sonenshine, D. E., and Mather, T. N. (1994). "Ecological Dynamics of Tick-Borne Zoonoses," p. 447. Oxford Univ. Press, New York.

Sreenivasan, M. A., Bhat, H. R., and Rajagopalan, P. K. (1979). Studies on the transmission of Kyasanur forest disease by partly fed ixodid ticks. *Ind. J. Med. Res.* **69:**708–713.

Turell, M. J. (1988). Horizontal and vertical transmission of viruses by insect and tick vectors. *In* "The Arboviruses: Ecology and Epidemiology" (T. P. Monath, ed.), pp. 127–152. CRC Press, Boca Raton, Fla.

Urioste, S., Hall, L. R., Telford, S. R., III, and Titus, G. R. (1994). Saliva of the Lyme disease vector, *Ixodes dammini*, blocks cells activation by a nonprostaglandin E2-dependent mechanism. *J. Exp. Med.* **180:**1077–1085.

Valenzuela, J. G., Charlab, R., Mather, T. N., and Ribeiro, J. M. C. (2000). Purification, cloning, and expression of a novel salivary anticomplement protein from the tick, *Ixodes scapularis*. *J. Biol. Chem.* **275:**18717–18723.

Varma, M. G. R., and Smith, C. E. G. (1972). Multiplication of Langat virus in the tick *Ixodes ricinus*. *Acta Virol.* **16:**159–167.

Vilcek, J. (1960). An interferon-like substance released from tick-borne encephalitis virus infected chick embryo fibroblast cells. *Nature* **187:**73–74.

Wang, H., Hails, R. S., Cui, W. W., and Nuttall, P. A. (2001). Feeding aggregation of the *tick Rhipicephalus appendicul*atus (Ixodidae): benefits and costs in the contest with host responses. *Parasitology* **123:**447–453.

Wang, H., and Nuttall, P. A. (1994). Excretion of host immunoglobulin in tick saliva and detection of IgG-binding proteins in tick hemolymph and salivary glands. *Parasitology* **109:**525–530.

Wang, H., and Nuttall, P. A. (1999). Immunoglobulin binding proteins in ticks: New target for vaccine development against a blood-feeding parasite. *Cell. Mol. Life Sci.* **56:**286–295.

Wang, H., and Nuttall, P. A. (2001). Intra-stadial tick-borne Thogoto virus (Orthomyxoviridae) transmission: Accelerated arbovirus transmission triggered by host death. *Parasitology* **122:**439–446.

Wang, H., Paesen, G. C., Nuttall, P. A., and Barbour, A. G. (1998). Male ticks help their mates to feed. *Nature* **391:**753–754.

Wikel, S. K. (1996a). "The Immunology of Host-Ectoparasitic Arthropod Relationships," p. 331. CAB International, Wallingford, UK.

Wikel, S. K. (1996b). Tick modulation of host cytokines. *Exp. Parasitol.* **84:**304–309.

World Health Organization (1985). "Arthropod-Borne and Rodent-Borne Viral Diseases." World Health Organization, Technical Report Series No. 719, Geneva.

World Health Organization (1986). "Tick-Borne Encephalitis and Hemorrhagic Fever with Renal Syndrome." Report on a WHO Meeting, Baden, 3–5 October, 1983. *EURO Reports and Studies* **104**.

Zanotto, P. M., Gao, G. F., Gritsun, T., Marin, M. S., Jlang, W. R., Venugopal, K., Reid, H. W., and Gould, E. A. (1995). An arbovirus cline across the Northern Hemisphere. *Virology* **210:**152–159.

Zanotto, P. M. D. A., Gould, E. A., Gao, G. F., and Harvey, P. H. (1996). Population dynamics of flaviviruses revealed by molecular phylogenies. *Proc. Nat. Acad. Sci. USA* **93:**548–553.

Zilber, L. A., and Soloviev, V. D. (1946). Far Eastern tick-borne spring-summer encephalitis. *Amer. Rev. Sov. Med.* (Special Supp.)**:** 80 pp.

ADVANCES IN VIRUS RESEARCH, VOL 60

PATHOGENESIS OF FLAVIVIRUS ENCEPHALITIS

Thomas J. Chambers* and Michael S. Diamond†

*Department of Molecular Microbiology and Immunology
St. Louis University Health Sciences Center, School of Medicine
St. Louis, Missouri 63104
†Department of Molecular Microbiology, Medicine, Pathology, and Immunology
Washington University School of Medicine
St. Louis, Missouri 63110

I. Introduction

Within the flavivirus family, viruses that cause natural infections of the central nervous system (CNS) principally include members of the Japanese encephalitis virus (JEV) serogroup and the tick-borne encephalitis virus (TBEV) serocomplex. Neuroinvasion follows infection of host organisms in the periphery by bites of chronically infected mosquito and tick vectors. Syndromes that result from CNS infection in humans range from mild aseptic meningitis to acute encephalitis of variable morbidity and mortality and are often complicated by neurologic sequelae among survivors. The pathogenesis of these diseases involves complex interactions of viruses, which differ in neurovirulence potential, and a number of host factors, which govern susceptibility to infection and the capacity to mount effective antiviral immune responses both in the periphery and within the CNS. Animal models

0065-3527/03 $35.00

have been instrumental for providing insight into how virus-specific and host factors influence the course of disease. Rodent models have been used in classic experiments on pathogenesis and continue to be relied upon for studies of viral neurovirulence determinants and immune system requirements for a successful antiviral response, particularly because of readily available knockout strains. Nonhuman primates have also been useful for some studies of peripheral immune responses to encephalitic viruses and for quantitating the disease burden caused by replication of virus in the CNS (monkey neurovirulence testing) and have also been applied as models for immunization and challenge. This review summarizes progress in the field of flavivirus neuropathogenesis since previous reviews on this topic (Monath, 1986; Monath and Heinz, 1996). Mosquito-borne and tick-borne viruses are considered together, although it is important to note that there are differences in the pathogenesis of these two groups of viruses.

II. Host Factors

It is generally acknowledged that the ratio of apparent to inapparent infections with flaviviruses is quite low (on the order of 1:100 to 1:300), implying that a number of host factors are involved in protection against CNS disease. The most well known of these are age, genetic factors, and preexisting flavivirus immunity. The effect of age has been recognized in both clinical and experimental studies of flavivirus encephalitis (Eldadah *et al.*, 1967a; Grossberg and Scherer, 1966; Luby *et al.*, 1967; O'Leary *et al.*, 1942; Powell and Kappus, 1978; Weiner *et al.*, 1970), in some cases accompanied by effects of gender (Andersen and Hanson, 1974). In terms of human infections, clinical disease with JEV is primarily a pediatric entity, suggesting that certain features of the developing nervous system predispose to pathogenesis of encephalitis. Experimental evidence in support of this is based on the known propensity of arboviruses to infect and spread more rapidly through nervous tissue of young rodents; however, resistance to fatal infection occurs abruptly soon after the weanling stage. This resistance occurs in conjunction with neural ontogeny and is associated with restricted replication of virus in neurons as a function of their degree of differentiation, with low levels of virus being observed in mature neurons in some models (Hase *et al.*, 1993; Ogata *et al.*, 1991). However, this phenomenon is very dependent on the intrinsic level of viral neurovirulence, being most apparent with strains of lower virulence (Eldadah *et al.*, 1967a; Ogata *et al.*, 1991; Oliver *et al.*, 1997), and is also affected by the dose and route of inoculation (Eldadah *et al.*, 1967a; Fitzgeorge and Bradish, 1980). Changes in the expression of cellular receptors and/or intracellular factors required for efficient replication have been invoked as explanations for the age-related resistance. In support of these hypotheses, it is known that the replication block can be overcome by the neuroadaptation of flaviviruses, which results in genetic changes in the envelope region, as well as in nonstructural regions of the genome (Chambers and Nickells, 2001; McMinn, 1997; Schlesinger *et al.*, 1996).

Studies with the mouse model of Sindbis virus encephalitis suggest that the differential expression of neural genes in young versus older mice is the determinant of age-related resistance. Candidate genes involved in this process include apoptotic regulators, interferon-responsive genes, and other classes that are regulated developmentally (Labrada *et al.*, 2002; see Section IX). These factors are presumed to operate by activating innate antiviral effector systems in neurons and promoting survival of these cells after viral infection.

Despite these experimental data, age-dependent resistance to encephalitic viruses in areas endemic for human disease is influenced by the effects of immunization and recurrent inapparent infections with homologous and heterologous viruses, such that the cumulative immune responses in adults may confer protection or ameliorate the severity of disease (Kurane, 2002; Solomon and Vaughn, 2002). However, susceptibility of young children (below the age of 9 months) to yellow fever virus (YFV) 17D vaccine-associated encephalitis is well documented (Freestone, 1994), and resistance to this adverse event in older individuals is clearly not based on immunity from recurrent subclinical infections. This supports the phenomenon of age-related susceptibility to encephalitis. A predisposition of elderly individuals to develop encephalitis from West Nile Virus (WNV) and St. Louis encephalitis (SLE) virus has also been observed. In this case, lack of previous cross-reactive immunity may be an important factor, but comorbid illnesses may contribute to the risk of complications, and the decline in immune function with advanced age is also likely to play a role (Grubeck-Loebenstein and Wick, 2002). In support of this, immunocompromised individuals encountering flavivirus infections appear to be less able to mount effective immune responses, as in reports of HIV-infected subjects with complicated CNS disease (Neogi *et al.*, 1998; Okhuysen *et al.*, 1993; Szilak and Minamoto, 2000; Wasay *et al.*, 2000) and in transplant recipients sustaining WNV infection (Iwamoto *et al.*, 2003). YFV 17D vaccine appears to be tolerated in

HIV infection, provided immunosupression is not severe, although this isuue deserves further investigation (Kengsakul *et al.*, 2002; Receveur *et al.*, 2000).

Coexistent infection with other infectious agents has been suggested as a modifying factor for flavivirus encephalitis. The association between cysticercosis and JEV infection was investigated in a controlled study that did not provide evidence for predisposition to viral encephalitis (Azad *et al.*, 2003). Association of TBEV encephalitis with borrelia infection has been described (Korenberg *et al.*, 2001), but probably reflects coexistence of these pathogens in the tick vector. Autopsy studies have suggested that herpes simplex infection may predispose to JEV encephalitis by altering the integrity of the blood–brain barrier (Hayashi and Arita, 1977). Experimentally, infection of mice with *Toxocara canis* or *Trichinella spiralis* predisposes to JEV encephalitis as a result of T-cell immunosuppression (Cypess *et al.*, 1973; Gupta and Pavri, 1987; Lubiniecki *et al.*, 1974; Pavri *et al.*, 1975), suggesting the possibility of increased disease severity in endemic areas where parasitic pathogens and encephalitic viruses coexist.

The susceptibility of mice to flavivirus encephalitis is controlled genetically and is associated with host factors that map to chromosome 5 at the oligoadenylate synthetase (OAS) gene cluster (Mashimo *et al.*, 2002; Perelygin *et al.*, 2002; Sangster *et al.*, 1994). Although OAS is involved in the activation of RNase L, the mechanism of resistance associated with the gene is not known and could conceivably relate to other potential functions of OAS proteins in cellular responses to viral injury (Samuel, 2002). The importance of the OAS gene in human susceptibility to these viruses requires investigation, as differences exist between murine and human gene clusters. Evidence shows that 2′, 5′-OAS is activated in response to peripheral flavivirus infection in humans (Bonnevie-Nielsen *et al.*, 1989, 1995), but studies to correlate this with the effectiveness of the antiviral responses are not yet available.

The effect of different class I HLA-A and B alleles on the immune response to flavivirus infection has been investigated for dengue viruses, where associations have been found with increased and decreased susceptibility to dengue hemorrhagic fever (Loke *et al.*, 2001; Stephens *et al.*, 2002), suggesting a relationship between the extent of T-cell activation and severe disease. It is unclear whether similar results will be found for encephalitic viruses, as these associations occurred in the context of secondary infections and presumably affect the strength of the memory rather than the primary T-cell response.

Interactions between flavivirus untranslated regions (UTRs) and intracellular proteins indicate another potential level of host-mediated control over virus infection (Brinton, 2000). Multiple proteins have been reported to bind to the 3′ UTR of the positive strand of various flaviviruses, including EF-1α and Mov34, which bind to the positive strand 3′ UTR of WNV and JEV, respectively (Blackwell and Brinton, 1995, 1997; Ta and Vrati, 2000), and as many as eight cellular proteins in the case of dengue (De Nova-Ocampo *et al.*, 2002). Several proteins have also been reported to bind to the 3′ UTR of the negative strand of dengue virus (Yocupicio-Monroy *et al.*, 2003) and four proteins in the case of West Nile virus (Li *et al.*, 2002). These proteins have been suggested to participate in flavivirus replication by influencing viral transcription and/or translation in conjunction with host intracellular membranes. There is some evidence that interactions of the viral RNA with such proteins are in fact important for virus replication (Li *et al.*, 2002). It is not known whether compartmentalization of their interactions in different cell types is a determinant of tissue tropism in infected hosts.

III. Arthropod Factors Affecting Pathogenesis

It has become evident from studies on arthropod vectors that certain components of their saliva influence pathogenesis in vertebrate hosts. For mosquitoes, flaviviruses are deposited principally in the extravascular tissue during probing, as virus that is injected intravascularly is reingested rapidly during the blood meal (Burke and Monath, 2001; Turell and Spielman, 1992; Turell *et al.*, 1995). Instead of a rapid dissemination of virus via the bloodstream, flaviviruses may then undergo some replication locally in subcutaneous tissues accompanied by spread to regional lymph nodes through lymphatic channels. Dendritic cells in the skin are likely to serve as a vehicle for the transport of virus to lymphoid tissues.

Components of the insect saliva modulate the earliest steps in flavivirus infection by altering the local host immune response. Feeding by *Aedes aegypti* or *Culex pipiens* mosquitoes or administration of sialokinin-I, a mosquito salivary protein, downregulates interferon-γ (IFN-γ) production and upregulates the T_H2 cytokines interleukin (IL)-4 and IL-10 (Zeidner *et al.*, 1999). Salivary gland extracts from *Dermacentor* and *Ixodes* ticks decrease natural killer activity (Kubes *et al.*, 1994, 2002), suppress the antiviral actions of interferons in cell culture (Hajnicka *et al.*, 2000), and enhance the transmission of TBEV in rodents (Labuda *et al.*, 1993). Overall, these findings suggest that insect factors facilitate flavivirus transmission by interfering with aspects of both innate and adaptive responses. Whether these effects are required to establish systemic infection and neuroinvasion is not known.

IV. Extraneural Infection

Flavivirus neuropathogenesis involves both *neuroinvasiveness* (capacity to enter the CNS) and *neurovirulence* (replication within the CNS) (Monath, 1986), both of which can be manipulated experimentally. In rodent models, neurovirulence is an inherent property of most of these viruses, and the quantity of virus needed to cause infection in the CNS is usually quite small. In classic studies of arbovirus pathogenesis (Albrecht, 1968; Huang and Wong, 1963), distinctions were made among neuropathogenic phenotypes on the basis of replication efficiency and pathogenic potential in the peripheral tissues versus the CNS, with various phenotypes being distinguished (high peripheral susceptibility, low neurotropism; low neuroinvasiveness, high neurotropism; and high neuroinvasiveness, high neurotropism). These phenotypes were related to different clinical outcomes, which ranged from inapparent infection to acute encephalitis of varying severity, and are influenced in the rodent model by host age and species (Kozuch *et al.*, 1981). These concepts have held up over time and have been supplemented by additional data concerning viral determinants of virulence and host innate and adaptive immune mechansisms. A main principle that applies is the relationship between peripheral virus burden and the propensity to cause neuroinvasion. Viruses with a low capacity to replicate in the periphery generally can be classified as low in neuroinvasive potential, regardless of their intrinsic level of neurovirulence. A relationship between systemic virus burden and viremia is also apparent (Albrecht, 1968), with the potential of the virus to generate viremia being a correlate of neuroinvasion as it applies to most naturally acquired encephalitic infections. Aerosol-acquired infections are probably an exception, but some of these may also cause systemic infection with viremia after gaining access to the lower respiratory tract. Mucosal infection of the alimentary tract has also been implicated both experimentally and in naturally acquired cases of TBEV encephalitis (Gresikova *et al.*, 1975; Odeola and Oduye, 1977). Furthermore, data from a number of studies indicate that such factors as the time of onset, magnitude, and duration of the viremia, as well as the integrity of the host innate immune system, also influence the risk of entry into the CNS, prior to the onset of the virus-specific immune response (see Section VIII). Type I interferons and macrophages in particular have been identified as important factors in limiting infection and clearing systemic virus. Thus this process is largely a balance between the replicative efficiency of the virus and the effectiveness of early host defenses in clearing viremia.

V. Cellular Receptors for Flaviviruses

The cellular receptors that mediate attachment and entry of flaviviruses have been only partially characterized, and to date there has been no definitive identification of the molecules required for these processes in either peripheral or CNS tissues. Although several groups have used biochemical approaches to identify candidate protein receptors (Kopecky *et al.*, 1999; Martinez-Barragan and del Angel, 2001; Ramos-Castaneda *et al.*, 1997) on mammalian cells, their physiologic relevance remains unclear, as there is heterogeneity in the proteins that bind flaviviruses in different cell types. Most recently, CD209 or DC-SIGN (the dendritic cell-specific intercellular adhesion molecule 3–grabbing nonintegrin) has been proposed as a cell surface ligand for dengue virus (Navarro-Sanchez *et al.*, 2003; Tassaneetritep *et al.*, 2003). Additional studies are required to evaluate the *in vivo* significance of DC-SIGN as an attachment or entry ligand and whether this is a common determinant of tropism for other flaviviruses. One possibility to reconcile these observations is that multiple independent cellular receptor molecules are utilized either during the spread of flaviviruses within the host or during entry into the CNS. A number of early studies have described binding of flaviviruses to mouse brain substances (reviewed in Albrecht, 1968), suggesting that this tissue is enriched for a unique receptor activity, which may enhance tropism for this organ, particularly in developing brain (Kimura-Kuroda *et al.*, 1992). In this regard, it has been shown that mouse and monkey brain membrane-receptor preparations preferentially bind neurovirulent strains of flaviviruses, but not attenuated variants (Ni and Barrett, 1998; Ni *et al.*, 2000). Receptor variability *in vivo* may be a general mechanism for promoting wide tissue tropisms of arthropod-borne viruses, which require cycling in both arthropod and vertebrate hosts. Some data suggest that different cell surface proteins may be utilized for the entry of insect versus vertebrate cells (Martinez-Barragan and del Angel, 2001; Munoz *et al.*, 1998). Tissue tropism in mosquitos has also been observed to correlate with expression of a specific receptor molecule (Yazi-Mendoza *et al.*, 2002).

Heparan sulfate has been proposed as a flavivirus receptor based on studies showing the dependence of dengue virus infectivity on binding of the E protein to heparan sulfate on target cells (Chen *et al.*, 1997; Hilgard and Stockert, 2000). Subsequent reports have demonstrated that infectivity of TBEV, yellow fever virus (YFV), JEV, and Murray Valley encephalitis (MVE) viruses is affected by cell surface interactions with glycosaminoglycans that are proposed to mediate initial low-affinity binding to the cell surface (Germi *et al.*, 2002; Lee and Lobigs, 2000; Mandl *et al.*, 2001; Su *et al.*, 2001), but the role of heparin as an authentic receptor for virulent flaviviruses remains uncertain. For instance, the serial passage of JEV and MVE viruses in cell culture results in selection for viruses that exhibit increased binding to heparin but decreased virulence *in vivo* (Lee and Lobigs, 2002). Similar observations have been made with alphaviruses (Bernard *et al.*, 2000; Klimstra *et al.*, 1998), where heparin binding was associated with cell culture adaptation of primary virus isolates and attenuation of viral virulence. Thus enhanced binding to glycosaminoglycans is a marker for attenuation of JEV and MVE viruses in the mouse model and correlates with rapid clearance of the glycosaminoglycan-binding variants from the circulation compared to more pathogenic strains (Lee and Lobigs, 2002). The mechanism responsible for this process has not been defined. The relationship of this observation to classic studies on neuroinvasion is also unclear, as virulent strains were originally characterized by their ability to undergo rapid uptake from the circulation, presumably as a result of highly efficient binding and entry to target cells (Albrecht, 1968). Wild-type and glycosaminoglycan-binding variants may differ, however, with respect to their entry into cells that are permissive for replication or are involved in virus clearance (Lee and Lobigs, 2002).

Antibody-dependent enhancement (ADE) of infection has been reported for some encephalitic flaviviruses. ADE is believed to occur through FcγR I (CD64) and FcγR II (CD32), although a second type of ADE that requires complement has also been described (Cardosa *et al.*, 1983, 1986). In tissue culture, ADE occurs with several flaviviruses in cells of myeloid lineage (Brandriss and Schlesinger, 1984; Brandt *et al.*, 1982; Cardosa *et al.*, 1983, 1986; Diamond *et al.*, 2000b; Halstead and O'Rourke, 1977; Halstead *et al.*, 1980, 1984; Schlesinger and Brandriss, 1983). Most of these data are relevant to the pathogenesis of dengue infection, and the significance for encephalitic viruses is less certain. However, there is some experimental evidence that ADE may be involved (Hawkes, 1964), and this notion is consistent with the concepts described for dengue viruses based on the wide antibody cross-reactivity among flaviviruses with respect to the E protein. Neutralizing homologous or cross-reactive antiflavivirus antibodies can enhance neurovirulence and mortality associated with YFV and JEV infection (Gould and Buckley, 1989; Gould *et al.*, 1987; Lobigs *et al.*, 2003b). However, in some cases, the pathogenesis was associated with complement-mediated cytolysis and not with enhancement of infection *in vivo* (Gould *et al.*, 1987). The strongest data in support of ADE in dengue infections are epidemiological in nature. In this regard, there is not abundant evidence to support a phenomenon of enhancement of JEV or other encephalitic viruses by preexisting cross-reactive antibodies in natural infections. There is evidence of cross-protection in experimental models, particularly among JEV serogroup members, and for amelioration of JEV encephalitis by prior immunity to related viruses, including dengue (Kurane, 2002; Solomon and Vaughn, 2002). Experimentally, the protective effect is presumably antibody mediated and is most apparent following infection with live virus or transfer of serum from animals infected with virus (Broom *et al.*, 2000; Tesh *et al.*, 2002). Antibody reponses elicited by inactivated virus do not exhibit much cross-protection in animals and may lead to enhanced infections (Broom *et al.*, 2000; Lobigs *et al.*, 2003b). Consistent with these observations, recipients of inactivated JEV vaccine or live-attenuated dengue vaccine did not generate neutralizing activity in sera against WNV (Kanesa-Thasan *et al.*, 2002). Severe forms of TBEV encephalitis have been observed after passive immunization with hyperimmune globulin (Waldvogel *et al.*, 1996), but it is not clear if this represents an enhancement phenomenon as opposed to either failure of antibody to penetrate the CNS or suppression of peripheral or CNS immune responses by high-titer immunoglobulin. The entire issue is somewhat limited by the fact that the encephalitic viruses may not necessarily target cells with abundance of Fc receptors, such as monocyte–macrophages, which are generally considered more important for the pathogenesis of dengue viruses.

VI. Cellular Tropism of Encephalitic Flaviviruses

In cell culture, flaviviruses readily infect a variety of cell types, including epithelial, endothelial, and fibroblasts (Avirutnan *et al.*, 1998; Bielefeldt-Ohmann, 1998; Diamond *et al.*, 2000b; Kurane *et al.*, 1992), but the relationship of these findings to *in vivo* replication is uncertain. After peripheral inoculation, flaviviruses probably do not replicate extensively in the skin, but are spread from local lymph nodes by immature dendritic or Langerhans cells, which are permissive for infection (Byrne *et al.*, 2001; Johnston *et al.*, 1996, 2000; Libraty *et al.*, 2001; McMinn *et al.*, 1996; Wu *et al.*, 2000). Within 1 day of infection, epidermal Langerhans cells that express viral antigens migrate from the skin to the draining lymph node (Byrne *et al.*, 2001; Wu *et al.*, 2000) while expressing maturation markers such as B7-1, B7-2, class II MHC molecules, CD11b, and CD83 (Ho *et al.*, 2001). These cells produce tumor necrosis factor (TNF)-α and IFN-α (Ho *et al.*, 2001; Libraty *et al.*, 2001) and become more resistant to flavivirus infection (Wu *et al.*, 2000). Thus infected dendritic cells probably serve to promote antigen presentation in the lymph node and also participate in the spread of infection to lymphoid compartments. The consequences of DC infection, whether apoptosis (as for alphaviruses) or persistent infection, as in the case of Kunjin virus replicons (Varnavski and Khromykh, 1999), and their effects on subsequent shaping of the immune response remain important areas for further investigation. For instance, data from other viral models indicate that there is a quantitative requirement for activated dendritic cells in order to induce T-cell responses (Ludewig *et al.*, 1998). Survival versus death of these cells as a result of virus infection may have an important impact on this requirement. After replication in lymphoid tissue, encephalitic viruses are believed to exit via efferent lymphatics (Malkova and Frankova, 1959) and gain access to the circulation, whereby systemic infection is established.

Although a tropism of encephalitic viruses for lymphoid tissues has been observed, the identities of the cell types in other compartments that support replication to the levels needed to generate a viremia sufficient to cause neuroinvasion have not been determined definitively. Replication in various peripheral tissues occurs, but vascular endothelial cells have not necessarily been implicated as important sites of replication (Albrecht, 1968 and references therein). However, it should be noted that dengue, JEV, and probably other flaviviruses can enter and, in some cases, establish infection in endothelial cells (Dropulic and Masters, 1990; Liou and Hsu, 1998) and modulate their activation state and cytokine production (Anderson *et al.*, 1997; Avirutnan *et al.*, 1998; Bosch *et al.*, 2002; Huang *et al.*, 2000). *In vitro* studies with endothelial cells are complicated by the fact that variable responses can be observed depending on the cell types and assay systems; however, given the potential role of these cells in immune activation, further studies on the effects of flavivirus infection on cytokine and chemokine production by these cells are needed.

In the CNS, neurons are the primary targets for encephalitic flaviviruses (Eldadah *et al.*, 1967b; Hase *et al.*, 1987; Iwasaki *et al.*, 1986; Kimura-Kuroda *et al.*, 1992; Wang *et al.*, 1997, 1998; Weiner *et al.*, 1970; Xiao *et al.*, 2001). Although viral replication and antigen production have also been observed in cultured oligodendrocytes (Jordan *et al.*, 2000) and astrocytes (Chen *et al.*, 2000; Liu *et al.*, 1988; Suri and Banerjee, 1995), the significance of these reports is difficult to judge, as there is scant evidence for infection of glial cells *in vivo*. Neurotropism and neurovirulence are governed to a large extent by determinants in the viral E protein, as indicated by an abundance of genetic data indicating that mutations in this protein modulate these phenotypes (McMinn, 1997; Ni and Barrett, 1998; Ni *et al.*, 2000), presumably through their effects on receptor targeting and postreceptor events involved in virus entry. In the absence of an experimental system to manipulate the virus receptor on neurons, one cannot conclude whether high neurotropism of these viruses is dependent solely on the binding of the E protein. Comparison of sequence data from virulent and attenuated strains of encephalitic viruses also suggests that the nonstructural region and 3′ UTRs contain determinants that influence pathogenesis. It is important in these types of studies to differentiate effects of genetic mutations on overall replication fitness of the virus versus specific effects in terms of interactions of viral proteins and RNA structures with host factors that affect pathogenesis uniquely. Thus the molecular basis for neurotropism is not understood adequately, and further studies using genetic clones of well-characterized viruses in animal models should help address this issue. Furthermore, the use of primate models to investigate the issue is needed greatly, as the observations in rodent models may not have direct correlates to human infections.

VII. Immune Responses to Flaviviruses and Their Role in Pathogenesis

Susceptibility to flavivirus encephalitis implies a failure at some stage of the immune response that theoretically may be defined in either qualitative or quantitative terms. There is substantial clinical and experimental evidence for a correlation between protection against encephalitic disease and the presence of virus-specific antibodies, but the molecular and cellular basis for the development of this response has not been defined thoroughly. In studies that have shown protection by antibodies, the roles of other immune system components in the process have not often been assessed. Furthermore, there is increasing evidence that flaviviruses have evolved mechanisms to manipulate the effector functions of both innate and adaptive immune responses. The magnitude and importance of these responses probably vary from one experimental model to another and account for differences observed in studies that have examined the immune system in the context of a either a primary or a memory response.

A. Innate Immunity

1. Interferons

In vitro and *in vivo* studies have demonstrated that interferon-dependent responses are relevant to protection against flavivirus infections. Dendritic cells in the skin may be the first cells to produce type I interferons (IFN-α or -β) in response to flavivirus infection and initiate this antiviral response (Libraty *et al.*, 2001). These interferons inhibit flavivirus infection by preventing translation and replication of infectious viral RNA at least partially through an RNase L-, Mx1-, and Protein Kinase R (PKR)-independent mechanism (Anderson and Rahal, 2002; Diamond and Harris, 2001; Diamond *et al.*, 2000a). These studies have been supported by experiments in immunodeficient and therapeutic mouse models of disease. Pretreatment of mice with IFN-α or its inducers prevents or ameliorates flavivirus infections (Brooks and Phillpotts, 1999; Charlier *et al.*, 2002; Harrington *et al.*, 1977; Leyssen *et al.*, 2003b; Lucas *et al.*, 2003), and mice that are deficient in type I IFNs or their receptor have increased susceptibility to flaviviruses (Johnson and Roehrig, 1999; Lobigs *et al.*, 2003a). The role of type II IFN (IFN-γ) in protection versus immunopathogenesis of flavivirus encephalitis is less clear, as this cytokine has a multitude of effects on the host response to these viruses. In part, this includes induction of proinflammatory and antiviral molecules, including nitric oxide (Lin *et al.*, 1997), and enhancement of the phagocytic activity of monocytes/macrophages through increased Fc receptor expression (Rothman and Ennis, 1999). Data from various models support the importance of IFN-γ production in the context of a T_H1 virus-specific immune response for the control of infection with encephalitic viruses (Johnson and Roehrig, 1999; Liu and Chambers, 2001; Lobigs *et al.*, 2003a). Distinctions should be made, however, concerning the effects of IFN-γ in the periphery and in the context of the CNS immune response (see Section X).

Flaviviruses appear to be capable of attenuating some of the IFN-dependent antiviral effector mechanisms. Treatment of cells or animals with IFN-α as few as 4 h after infection with dengue or SLE viruses resulted in almost a complete loss of antiviral activity (Brooks and Phillpotts, 1999; Diamond *et al.*, 2000a). Similarly, IFN-α treatment of patients with documented JEV encephalitis had no significant effect on outcome (Solomon *et al.*, 2003), despite anecdotal reports of its benefit (Harinasuta *et al.*, 1985). The mechanisms by which the antiviral effect is avoided remain uncharacterized, but data suggest that it may act at a very early step in infection. Future studies should also help determine if the role of type I IFNs in these infections lies only in innate intracellular effector defenses or whether there are effects on the quality and magnitude of the subsequent cell-mediated immune response through their immunoregulatory effects.

2. Macrophages

Activation of macrophages and modulation of their effector functions are integral parts of flavivirus pathogenesis (Rothman and Ennis, 1999; Spain-Santana *et al.*, 2001). In addition to their role in nonspecific defense, macrophages are targets of infection by some flaviviruses and have the potential to contribute to pathogenesis through antibody-dependent enhancement of infection mediated by Fc and complement receptors (Cardosa *et al.*, 1986; Gollins and Porterfield, 1984; Hawkes, 1964; Peiris *et al.*, 1981). The preponderance of data supports the protective role of macrophages in control of infection by means of cytokine production and antigen presentation to B and T cells (Kulkarni *et al.*, 1991a; Marianneau *et al.*, 1999). Classic studies have shown that abrogation of phagocytic activity of macrophages results in higher viremia, neuroinvasion, and more severe encephalitis (Ben-Nathan *et al.*, 1996a; Khozinsky *et al.*, 1985; Monath, 1971; Zisman *et al.*, 1971). Some of the protective effect provided appears to be mediated by the stimulation of inducible nitric oxide synthetase (NOS-2) to produce nitric oxide (NO) and other reactive oxygen intermediates such as peroxynitrites (Saxena *et al.*, 2000). Pretreatment of macrophages with agents that induce NO synthesis have been shown to inhibit JEV infection *in vitro* (Lin *et al.*, 1997). Moreover, treatment of mice with a NOS-2 inhibitor increased mortality after JEV infection (Lin *et al.*, 1997). However, other studies suggest that the inflammatory actions of NO and other reactive oxygen intermediates may, in some cases, contribute to flavivirus pathogenesis. The *in vivo* administration of a competitive inhibitor of NOS-2 improved survival in mice infected with TBEV (Kreil and Eibl, 1995, 1996). Activation of macrophages in response to flavivirus infection promotes not only production of NO, but also release of TNF-α, IL-1β, IL-8, and other mediators of acute inflammation that may contribute to tissue damage where macrophages accumulate (Atrasheuskaya *et al.*, 2003; Bosch *et al.*, 2002; Raghupathy *et al.*, 1998; Rothman and Ennis, 1999). These various studies indicate that the behavior of macrophages is fundamental to the pathogenesis of flavivirus disease, but the contribution to virus clearance versus deleterious effects driven by IFN-γ and other proinflammatory stimuli depends on regulation of their activity, and the properties of this innate defense may therefore vary from one context to another.

3. *Natural Killer Cells*

Natural killer (NK) lymphocytes lyse infected cells by releasing cytotoxic granules that contain perforin and granzymes or by binding to apoptosis-inducing receptors on target cells (Orange *et al.*, 2002). NK cell activation is finely regulated through a balance of activating (Ly49D, Ly49H, and NKG2D) and inhibitory receptors [killer cell immunoglobulin-like receptors (KIR), immunoglobulin-like inhibitory receptors (ILT), and CD94-NKG2A] (Smith, *et al.*, 2001). A decrease in expression of class I MHC molecules on a cell may prompt NK cell activation by attenuating the inhibitory signals. Thus NK cell target recognition occurs after ligation of activating receptors and repression of inhibitory receptors on the cell surface. NK responses have been analyzed in various experimental models of flavivirus infection, but as noted (Hill *et al.*, 1993), the characterization of the responding cells has been limited. Studies of NK cell activity during WNV infection have revealed blunted cytolytic activity against virus-infected cells, associated with upregulation of MHC antigen and ICAM-1 expression on the target cells by interferon-independent mechanisms (Müllbacher *et al*, 1989). However, NK cell-dependent lysis of dengue virus-infected target cells by both natural killer and antibody-dependent cell-mediated cytotoxicity has been observed (Kurane *et al.*, 1984). Infection of mice with Langat, WNV, and TBEV transiently activated and then suppressed NK cell activity (Vargin and Semenov, 1986). Despite these conflicting observations, the bulk of evidence currently suggests that flaviviruses have evolved a mechanism to evade NK cell responses through an augmentation of cell surface class I MHC expression (King and Kesson, 1988; King *et al.*, 1989; Liu *et al.*, 1988, 1989b), driven by a TAP-dependent process (Momburg *et al.*, 2001; Müllbacher and Lobigs, 1995; Lobigs *et al.*, 1999) and NF-κB-dependent transcriptional activation of MHC class I genes (Kesson and King, 2001). Thus, flaviviruses may overcome susceptibility to NK cell-mediated lysis at the expense of increased class I MHC expression and later recognition by virus-specific cytotoxic lymphocytes (CTLs). Consistent with this hypothesis, splenocytes from WNV-immunized mice had poor NK cell lytic activity (Momburg *et al.*, 2001), and mice that are genetically deficient in NK cells demonstrate no increased morbidity or mortality compared to wild-type controls in response to WNV infection (M. Engle, W. Yokoyama, and M. Diamond, unpublished results). Some residual NK cell function may still be important during flavivirus pathogenesis, as suggested by studies with perforin and Fas knockout mice, which are partially protected from encephalitic disease, through events that operate at the level of neuroinvasion. This may involve the cytotoxic activities of NK cells and/or CTLs (Lincon Luna *et al.*, 2002). In addition, the lack of a substantial effect of IFN-α in the therapy of flavivirus encephalitis is also consistent with inhibition of the NK response, as type I interferons are normally potent activators of these cells.

4. *Natural Antibody*

Natural antibodies are primarily of the IgM class, although activity of IgG has also been described. They are secreted constitutively by $CD5^+$ B-1 cells without specific stimulation, have widely variable binding avidities, and represent an initial nonspecific defense against pathogens (Baumgarth *et al.*, 2000; Casali and Notkins, 1989; Ochsenbein *et al.*, 1999a) through direct neutralization of some bacteria and viruses in the circulation (Gobet *et al.*, 1988; Ochsenbein *et al.*, 1999a), enhancement of phagocytosis (Navin *et al.*, 1989), and complement activation (Baumgarth *et al.*, 2000). Although the role of natural antibody in flavivirus infection remains unexplored, mice that genetically lack secreted IgM, (sIgM −/−), but in which cell surface IgM and IgG responses are intact have increased mortality in certain viral infections, involving a defect in the antiviral IgG responses (Baumgarth *et al.*, 2000; Boes *et al.*, 1998). Such mice are also very susceptible to infection with WNV (Fig. 1; M. Engle and M. Diamond, unpublished data). This observation, along with other data, suggests an important role for natural antibody and complement during the early antiviral defense against flaviviruses, although a virus-specific IgM response is likely to be more important (see Section VII,A,5).

5. *Complement*

The complement system is an important innate defense for limiting infection by fungal, bacterial, and viral pathogens. Complement inhibits viruses by several mechanisms (Volanakis, 2002), including lysis

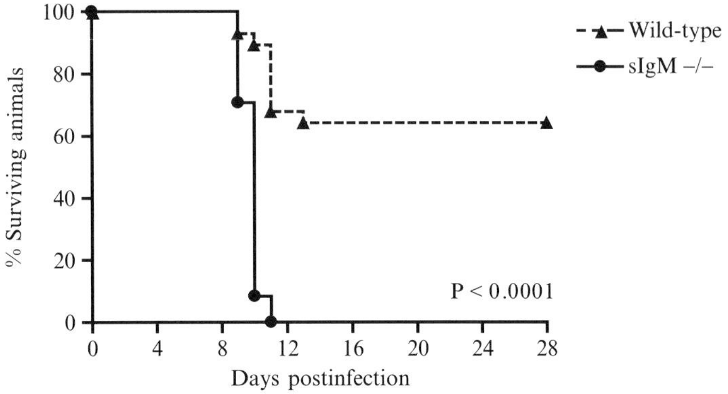

FIG 1. Soluble IgM-deficient mice are highly sensitive to peripheral infection with WNV. Mice were infected by subcutaneous inoculation of the footpad with 100 plaque-forming units of WNV NY 99 and monitored for mortality from CNS disease.

of enveloped viral particles and virus-infected cells by the C5–C9 membrane attack complex; recruitment and activation of monocytes and granulocytes by C3a and C5a; clearance of virus from circulation after opsonization by proteolytic fragments of C3, C3b, and C3bi, followed by uptake into cells that express complement receptor; and C3-facilitated uptake of antigen and presentation by macrophages and dendritic cells (Ochsenbein and Zinkernagel, 2000) during priming of T and B lymphocytes (Da Costa *et al.*, 1999; Kopf *et al.*, 2002; Ochsenbein *et al.*, 1999b). Preliminary studies indicate that complement plays an essential role in limiting WNV infection. Mice that are genetically deficient in C3 uniformly succumb to infection even at low inoculating doses (Fig. 2; E. Mehlhop, M. Engle, and M. Diamond, unpublished data). Additional studies must be performed to determine which individual mechanisms are most critical for this complement-mediated control. A deficiency of C3 could exacerbate WNV infection because of depressed C5–C9 lytic or C3 opsonic activity that results in a failure to clear virus from the circulation. Alternatively, C3 may play important roles in linking the innate and adaptive immune responses (Barrington *et al.*, 2001; Carroll, 1998; Ochsenbein and Zinkernagel, 2000) against WNV. C3 is required for normal IgG production and T-cell priming against influenza and herpes viruses (Da Costa *et al.*, 1999; Fischer *et al.*, 1996; Kopf *et al.*, 2002; Ochsenbein *et al.*, 1999), and a deficiency in C3 decreases opsonization and viral antigen presentation, leading to deficits in the adaptive B-and T-cell responses (Ochsenbein and Zinkernagel, 2000). Although the lytic and proinflammatory activity of complement

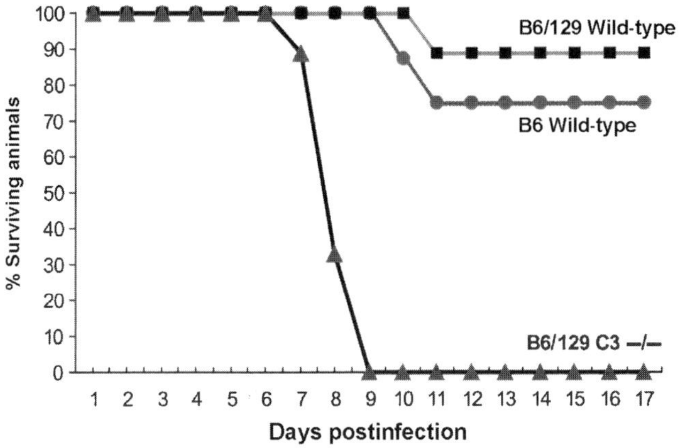

FIG 2. C3-deficient mice are highly sensitive to peripheral infection with WNV. Mice were infected as described in Fig. 1 and monitored for mortality from CNS disease.

may contribute to the defense against WNV, preliminary data indicate that a deficiency in either C3 compromises the antiviral B-cell immune response, as mice that lack C3 have markedly depressed titers of WNV-specific IgG (E. Mehlhop, M. Engle, and M. Diamond, unpublished data).

B. Adaptive Immunity

The roles of both humoral and cellular immunity during the pathogenesis of flavivirus disease have been studied in various models involving active and passive immunization in normal and immunodeficient animals, and understanding of the immunologic basis for protection against acute encephalitis is beginning to advance. B cells, $CD4^+$ cells, and $CD8^+$ T cells have all been implicated in contributing to protection, although as stated earlier, their relative importance seems to vary depending on the context of the experimental model under investigation.

1. B Cells and Antibody

Antibody responses to the E and NS1 proteins involve many epitopes, and both neutralizing and nonneutralizing antibodies (including those against NS1) can prevent fatal encephalitis, as demonstrated in many passive transfer and active immunization experiments in experimental animal models (Brandriss *et al.*, 1986; Diamond *et al.*, 2003; Gould *et al.*, 1987; Henchal *et al.*, 1988; Kimura-Kuroda and

Yasui, 1988; Putnak and Schlesinger, 1990; Schlesinger *et al.*, 1985, 1987; Zhang *et al.*, 1989). A protective effect can occur even after establishment of infection in the CNS. Mechanisms of antibody-mediated protection during flavivirus infection include direct neutralization of receptor binding (Crill and Roehrig, 2001), blocking of virus uncoating (Gollins and Porterfield, 1984), and Fc receptor-dependent virus clearance via the reticuloendothelial system. Most neutralizing antibodies recognize the structural E protein, and antibody epitopes appear to be broadly distributed over its surface (Heinz, 1986; Roehrig *et al.*, 1989); however, these do not all represent potent sites for neutralization. A subset of neutralization epitopes is found on the prM protein (Colombage *et al.*, 1998; Falconar, 1999; Pincus *et al.*, 1992). The presence of nonneutralizing, yet protective antibodies against NS1 is also well documented (Cane and Gould, 1988; Després *et al.*, 1991; Falgout *et al.*, 1990; Henchal *et al.*, 1988; Putnak and Schlesinger, 1990; Schlesinger *et al.*, 1986, 1987). Antibodies to NS1 are proposed to mediate the lysis of virus-infected cells, which express this protein on their surface, by complement-mediated lysis and/or antibody-dependent cellular cytotoxicity (Hill *et al.*, 1993; Kurane *et al.*, 1984; Schlesinger *et al.*, 1990). These humoral responses are believed to be important components of the protective immune response to flavivirus infection, but antibodies to NS1 are not often measured in experimental models of infection or immunization and challenge.

Although the neutralizing antibody responses are considered correlates of protection (Markoff, 2000), this process is probably also a function of additional innate and adaptive immune effector systems whose roles in the control of infection are less easily demonstrated. In this regard, challenge in the context of passively administered virus-specific antibodies is not necessarily associated with sterilizing immunity, indicating that antiviral defenses other than antibodies are involved in protection (Kreil *et al.*, 1998a, 1998b). Definitive experiments to determine the extent of immune activation in this setting are likely to expand our understanding of the correlates of protection and the immunological basis for a successful antiviral response.

Mice that lack B cells are very vulnerable to flavivirus infections and encephalitis (Diamond *et al.*, 2003; Liu and Chambers, 2001), purportedly as a consequence of lacking antibodies. However, these models must be explored further to determine whether T-cell responses are otherwise fully activated and effective, as it is possible that B cells could influence $CD4^+$ and $CD8^+$ T-cell responses through antigen presentation or other immunoregulatory events, as has been observed in other neurotropic viral infections (Bergmann *et al.*, 2001).

Factors that drive B-cell activation and maturation during flavivirus infections are not well understood. Multiple T helper epitopes have been identified in the E protein of JEV and MVE viruses, some of which are dominant, broadly reactive among JE serogroup viruses, and prime for the neutralizing antibody (Kutubuddin *et al.*, 1991; Mathews *et al.*, 1991, 1992; Roehrig *et al.*, 1992). Immunization with plasmid DNA encoding the E region is more effective than viral E antigen itself, presumably due to the induction of T-cell responses to the E protein (Chen *et al.*, 1999). Immunization of HIV-infected individuals with inactivated flavivirus vaccines has suggested that $CD4^+$ T cells may not be critical for the induction of protective antibodies (Panasiuk *et al.*, 2003). However, others have observed that antibody responses to inactivated or live viral vaccines are weak in these subjects (Rojanasuphot *et al.*, 1998; Sibailly *et al.*, 1997). Immunization of CD4 knockout or class II knockout mice with YFV markedly diminished or abrogated the neutralizing antibody response, respectively (Chambers and Liang, unpublished data), suggesting that there is dependence on functional $CD4^+$ T cells to generate long-lasting B-cell memory against flaviviruses. Differences in immunization schedules and numbers of residual $CD4^+$ lineage T cells in these various reports may explain the discrepancies. Studies on the role of dendritic cells in early B-cell activation and the nature of the toll-like receptor signals induced in response to viral antigens on these cells may provide new insights into the determinants that establish and drive the memory response to flavivirus antigens.

An IgM response to flaviviruses is a feature of most clinical and experimental infections (Martin *et al.*, 2002) and has been reported to be a correlate of protection during clinical JEV encephalitis in some studies (Burke *et al.*, 1985a, 1985b; Ravi *et al.*, 1993). Flavivirus infections typically elicit IgM responses that can often persist for prolonged periods (Edelman *et al.*, 1976; Monath, 1971; Roehrig *et al.*, 2003). However, experimentally, the initial IgM response to encephalitic viruses may possess variable neutralization activity and protective capacity (Diamond *et al.*, 2003; Hofmann *et al.*, 1978; Ishii *et al.*, 1968); in some studies, the complement-fixing activity was limited compared to that provided by the ensuing IgG response (Ishii *et al.*, 1968; Lee and Scherer, 1961). This contrasts with what has been observed with YF 17D vaccination (Monath, 1971) and may reflect differences in early B-cell activation among these infections or differences between humans and mice with respect to the process. However, the role of a vigorous IgM response may be in providing temporary neutralizing activity, while more importantly activating complement-dependent pathways involved in programming virus-specific B- and T-cell responses. Studies with the WNV model (see later) suggest a critical early function of both IgM and complement in the control of extraneural infection. The neutralizing activity of sera rises in conjunction with the appearance of IgG, and this response includes a variety of biological activities, including hemagglutination inhibition, complement fixation, and virus neutralization. The memory or "antigenic sin" response elicited by cross-reactive antigens involves all of these responses, but is weakest for the neutralization response (Innis, 1997), suggesting that there is some hierarchical pattern of B-cell epitopes or some selectivity in recognizing the most cross-reactive antigens. This phenomenon is probably evolutionarily adaptive for recurrent infections with heterologous flaviviruses, but is capitalized upon by dengue viruses during the pathogenesis of dengue hemorrhagic fever and shock syndrome. In any case, the anamnestic antibody response has been demonstrated to be critical to a defense against encephalitic viruses in the context of immunization and challenge models, which are surrogates for the efficacy of vaccines (Konishi *et al.*, 1999; Lee and Sherer, 1961; Pan *et al.*, 2001). It nevertheless remains unclear at the present time the extent to which antibody responses alone contribute to the control of acute infections.

2. T Cells

T-cell responses to flavivirus proteins have been best studied for members of the DEN and JEV serogroups. Both $CD4^+$ and $CD8^+$ T lymphocyte responses involve broad flavivirus cross-reactivity, although this varies significantly among different MHC haplotypes (Hill *et al.*, 1992; Kulkarni *et al.*, 1992; Kurane *et al.*, 1991; Uren *et al.*, 1987; reviewed in Hill *et al.*, 1993). Multiple epitopes for T-cell responses have been identified on both structural and nonstructural proteins; however, genetic factors restrict the number of targets and fine specificities differ considerably. Determinants for class I responses are more frequent within the viral nonstructural region, particularly the NS3 protein, which contains dominant epitopes in both humans and mice. In contrast, viral structural protein antigens elicit class II responses more consistently. The role of virus-specific $CD4^+$ T cells in flavivirus encephalitis is not well understood, although experimental models indicate a requirement for such cells in protection against acute disease (see Section X). Some of the cytotoxic T-cell response to dengue and JE viruses is contained in the $CD4^+$ compartment and is probably mediated by Fas/FasL interactions (Aihara *et al.*, 2000; Gagnon *et al.*, 1999), which has implications for possible immunopathogenic responses both in the periphery and in the CNS.

Flavivirus-specific $CD8^+$ T cells have multiple effector functions, including cytotoxic activity and production of IFN-γ (Douglas *et al.*, 1994; Kesson *et al.*, 1987; Kulkarni *et al.*, 1991b; Kurane *et al.*, 1989, 1991, 1995; Liu *et al.*, 1989a; Murali-Krishna *et al.*, 1996; Takada *et al.*, 2000), suggesting that polarization of the immune response toward a T_H1 phenotype is involved in control of these viruses. Although the importance of type I interferons in protection against these viruses would suggest its involvement in promoting this T-cell response, as has been described for other viruses (Cousens *et al.*, 1999), the role of IL-12 in driving this process has received only limited study (Chen *et al.*, 2001; Phillpotts *et al.*, 2003), and its importance may vary from one context to another (Dalod *et al.*, 2002). Cellular immunity clearly contributes to the control of virus infection in experimental animal models, but this varies depending on the context examined and with respect to the virulence of the challenge virus. T-cell-deficient mice fail to generate protective immunity after a sublethal challenge with YFV strains (Bradish *et al.*, 1980). Moreover, animals that are treated with drugs that impair T-cell function develop a rapidly progressive flavivirus encephalitis (Camenga *et al.*, 1974; Cole and Nathanson, 1968; Nathanson and Cole, 1970). The adoptive transfer of immune spleen cells can protect against encephalitis, but the lymphocyte subpopulations that mediate this protection have not been very well characterized in classic studies (Bradish *et al.*, 1980; Camenga *et al.*, 1974; Jacoby *et al.*, 1980).

More recent studies have, to some degree, clarified the role of $CD8^+$ T cells in these infections. Quantitation of the $CD8^+$ T-cell response to YFV in experimental mice (van der Most *et al.*, 2002) reveals activation by immunodominant epitopes and is supported by the observation that $CD8^+$ knockout mice exhibit a defect in the clearance of infectious YFV from the CNS (Liu and Chambers, 2001; T. J. Chambers, unpublished data) and also have increased mortality after WNV infection (B. Shrestha and M. Diamond, manuscript in preparation). Human recipients of YFV 17D vaccine exhibit an increase in $CD8^+$ T cells as well, and epitopes have been mapped to multiple proteins (Co *et al.*, 2002). While data from most of these models would indicate that CTL responses are primarily protective *in vivo*, their potential for immunopathogenic effects requires further investigation. Some studies with MVE virus (Licon Luna *et al.*, 2002) and dengue virus (Rothman and Ennis, 1999) suggest that the cytotoxicity of CTL may contribute to the disease pathogenesis. Studies with knockout mice continue to provide novel information on the role of T cells in flavivirus pathogenesis and immunity. However, an important limitation of these experiments is that the effect of gene knockout on immunologic development and function is not known. Better attempts to assess the entire range of properties of the immune response in these types of experimental models are needed, as gene knockouts could have multiple effects beyond simply loss of the targeted function.

Human studies on immune responses to JEV or JEV structural protein antigens encoded in recombinant vaccinia virus revealed proliferation and induction of cytolytic activity in the $CD8^+$ T-cell compartment (Konishi *et al.*, 1995, 1998a), similar to what has been observed in mice (Konishi *et al.*, 1997, 1998b). The significance of these responses is uncertain, as another study found no correlation between T-cell proliferation and either the antibody response or the clinical outcome (Desai *et al.*, 1995a). Inactivated JEV vaccine induced $CD4^+$, class II-restricted T cells with cytolytic activity (Aihara *et al.*, 2000), suggesting that it is not capable of inducing high levels of $CD8^+$ CTLs; in mice, this vaccine elicts a T_H2 immune response (Ramakrishna *et al.*, 2003). As noted earlier, in the context of clinical infections with flaviviruses, the integrity of cellular immune function appears to be important (Iwamoto *et al.*, 2003; Neogi *et al.*, 1998; Okhuysen *et al.*, 1993; Szilak and Minamoto, 2000). However, better understanding of the effector properties of T cells and their role in protection in humans and in experimental animals is needed.

VIII. Neuroinvasion

Neuroinvasiveness is a critical step in the pathogenesis of flavivirus encephalitis and is affected by both viral and host factors. In terms of viral factors, characterization of various virulent and attenuated strains of JEV, TBEV, YFV, and WNV has revealed that viral determinants of neuroinvasiveness map principally the E protein (reviewed in McMinn, 1997). The mechanisms associated with these genetic determinants have not been completely determined, but are believed to relate to increased viral infectivity toward important target cells through enhanced binding and penetration. Entry into the CNS has been proposed to involve a number of potential processes, none of which has been definitively demonstrated *in vivo*. The proposed pathways include (1) transport across the cerebrovascular endothelium, or infection of these and other cells constituting the blood–brain barrier (Dropulic and Masters, 1990; Liou and Hsu, 1998); (2) access to the CNS after loss of blood–brain barrier integrity (Kobiler *et al.*, 1989; Lustig *et al.*, 1992); and (3) entry through the olfactory epithelium

(McMinn *et al.*, 1996; Monath *et al.*, 1983). There are regions of the CNS where lack of a blood–brain barrier may constitute sites of vulnerability to infection in the presence of viremia (i.e., the choroid plexus and the circumventricular organs). Some of these may be supported by nonspecific defenses that are sufficient to withstand a low level of viremia, as the choroid is a site of IFN-α expression and OAS induction (Asada-Kubota *et al.*, 1997; Khan *et al.*, 1989). Entry of virus into the CNS by passage across the small cerebrovascular vessels is consistent with the accumulation of perivascular infiltrates of inflammatory cells, which is a hallmark of flavivirus encephalitis (Johnson *et al.*, 1985). However, access through the olfactory bulb is believed to occur either after infection by the aerosol route or intranasally (Hambleton *et al.*, 1983; Myint *et al.*, 1999; Nir *et al.*, 1965; Raengsakulrach *et al.*, 1999) or in the context of hematogenous dissemination of virus (McMinn, 1996; Monath, 1986; Monath *et al.*, 1983). The olfactory bulb is especially vulnerable to infection because of the exposure of its nerve terminals within the olfactory mucosa, and this route is exploited by other neurotropic viruses (Fazakerley, 2002).

Disruption of the blood–brain barrier facilitates the entry of noninvasive flaviviruses into the CNS (Lustig *et al.*, 1992), suggesting that neuroinvasion may be influenced by host factors that alter the permeability of this barrier as a result of systemic infection (Chaturvedi *et al.*, 1991; Kaiser and Holzmann, 2000; Mathur and Chaturvedi, 1992), including IFN-γ, TNF-α, and possibly effector functions of CTLs and NK cells (Lincon Luna *et al.*, 2002). Neuroinvasion is also influenced by physical stress and other agents, including inhalational anesthetics (Ben-Nathan *et al.*, 1989, 1992, 1996b, 2000). Immunosuppression by corticosterone and other endogenous immunomodulators is an important factor, as involution of lymphoid tissue has been observed and probably facilitates the generation of viremia by attenuated viral strains. Perturbation of the blood–brain barrier may also be involved in this process (Ben-Nathan *et al.*, 2000). Early viremia and sustained viremia are correlated with neuroinvasion in animal models, consistent with the belief that replication to high titers in peripheral tissues is an important property of invasive strains, at least in immunologically normal hosts (Albrecht, 1968; Huang and Wong, 1963; Monath, 1986). However, the timing of peak viremia is probably critical, as levels sufficient to cause neuroinvasion are dissipated concurrent with the appearance of adaptive immune responses in the periphery (Diamond *et al.*, 2003; Halevy *et al.*, 1994). This is also evident from studies of immunodeficient mice lacking T- and B-cell responses, which exhibit neuroinvasion in the presence of viremia (Chambers and

Nickells, 2001; Charlier *et al.*, 2002; Diamond *et al.*, 2003; Johnson and Roehrig, 1999; Halevy *et al.*, 1994; Lin *et al.*, 1998). At present it appears that the mechanisms responsible for entry into the CNS may vary, depending on a specific virus and the level of host immunocompetence. Continued research with the use of knockout strains harboring single and multiple defects in these redundant immune system defenses will improve the understanding of how this process occurs.

The mechanisms of flavivirus spread within the CNS are not established. Studies with alphaviruses in the mouse model suggest that this occurs in a circuit-specific manner among olfactory neurons that are undergoing developmental synaptogenesis (Oliver and Fazakerley, 1998). Spread of MVE virus in immature mice exhibited a pattern consistent with these observations, but occurred throughout the CNS (McMinn *et al.*, 1996). An unusual accumulation of WNV particles in myelin lamellae was observed in spinal cord cultures (Shahar *et al.*, 1990), suggesting the possibility of virus transport by mechanisms not only involving the axoplasm.

IX. Neuropathology

Flaviviruses, particularly members of the JEV and TBEV serogroups, cause viral encephalitis in many vertebrate species, a dead-end transmission pathway believed to reflect an evolutionarily conserved capacity of these viruses to grow in the CNS of arthropods and vertebrates (Monath, 1986). In contrast to the noncytopathic nature of infection in arthropods, a spectrum of acute and chronic CNS pathologic changes occur in vertebrates and have been documented extensively (Chu *et al.*, 1999; Dominguez and Baruch, 1963; Hase *et al.*, 1993; Levenbook *et al.*, 1987; Manuelidis, 1956; Nathanson *et al.*, 1966; Pogodina *et al.*, 1983; Reyes *et al.*, 1981; Vince and Grevic, 1969; Zlotnik *et al.*, 1971, 1976). Virus can be demonstrated within neurons throughout the brain and spinal cord, but infection of other cell types has been less well characterized. In humans, neuroinvasive flaviviruses cause an acute, often fatal encephalomyelitis (Monath, 1986) associated with characteristic inflammatory changes and often targeted to specific regions (Johnson *et al.*, 1985; Miyake, 1964; Suzuki and Phillips, 1966; Zimmerman, 1946). The pathological changes have been characterized in many experimental animal models, as well as in fatal human cases. These viruses can evoke inflammatory infiltrates extending from the meningeal layers into the brain substance, with features typical of other viral encephalitides, including leptomeningitis, perivascular lymphocytic accumulation, parenchymal infiltrates, and microglial nodules associated with neuronophagia in regions of viral infection (Figs. 3, 4, and 5). The neuropathology can, in some cases, include the destruction of vascular structures with focal hemorrhage, suggesting a vasculitis. Loss of regional blood flow, as well as disruption of the blood–brain barrier, has been described in clinical cases of TBEV (Gunther *et al.*, 1998; Kaiser, 2002; Kaiser and Holzmann, 2000). The nature and degree of the inflammatory disease depend on many factors, including the virulence of the virus, the route of infection, and the age and immunocompetence of the host. Varying levels of CNS inflammation without frank evidence of neuronal damage have been described in some models, including human autopsy cases, suggesting that viral infection can induce lethal pathophysiology prior to or in the absence of recruitment of the peripheral immune response (reviewed in Monath, 1986). Although cytopathic effects have been observed primarily in virus-infected neurons, other noninfected cells can also exhibit pathologic changes, presumably through bystander injury (see later). A number of studies have documented the distribution of the neuropathology and the clinical manifestations that characterize flavivirus infection of the CNS. For instance, infection in cortical

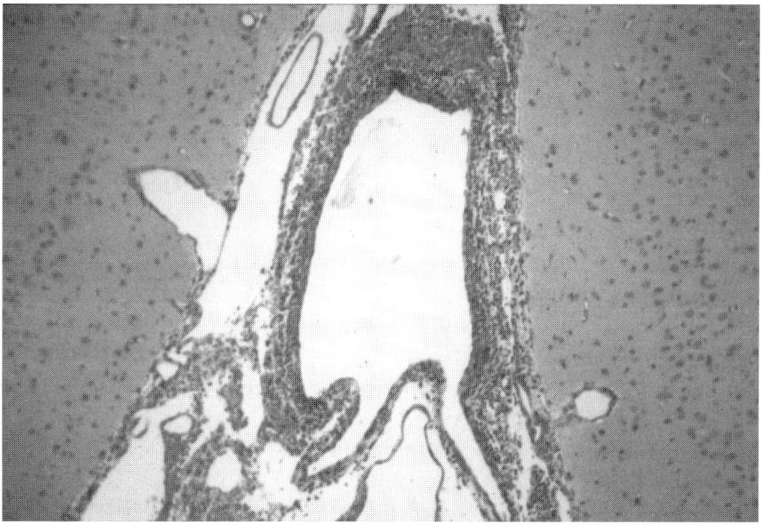

FIG 3. Yellow fever virus meningoencephalitis in the rhesus monkey showing leptomeningeal accumulation of acute inflammatory cells. Courtesy of the United States Army Medical Research Institute of Infectious Diseases (USAMRIID). (See Color Insert.)

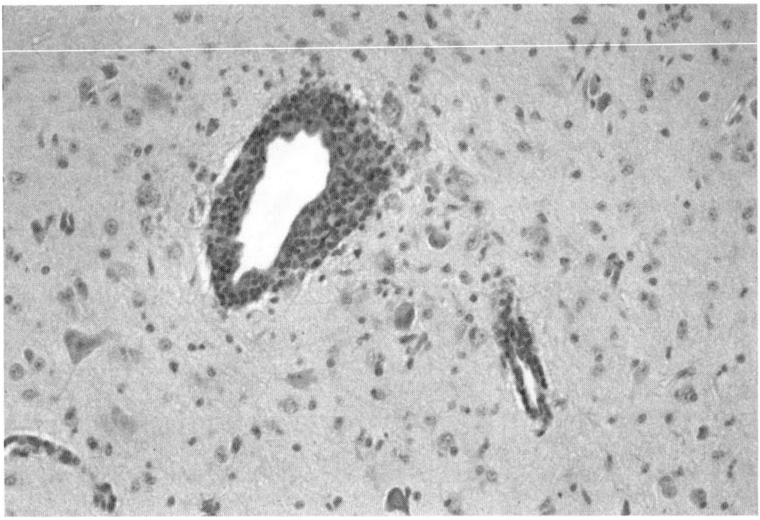

FIG 4. Yellow fever encephalitis in the rhesus monkey showing focus of perivascular infiltrate with mononuclear cells in the cerebral cortex. Courtesy of USAMRIID. (See Color Insert.)

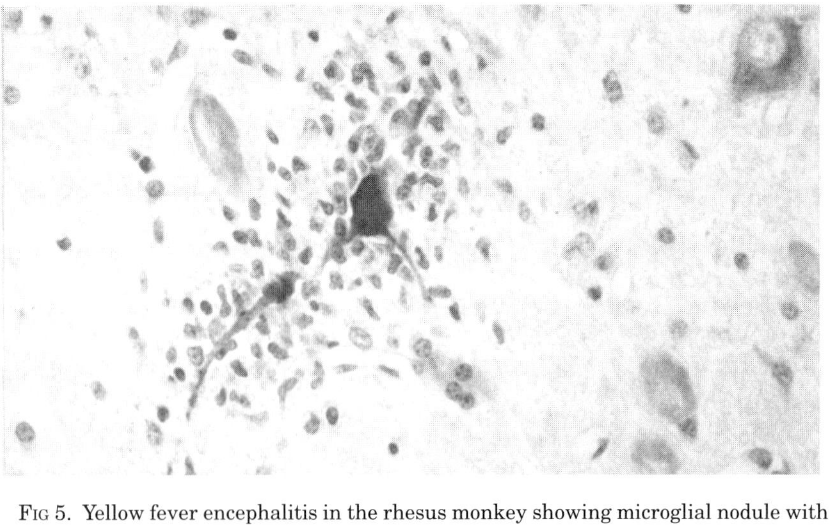

FIG 5. Yellow fever encephalitis in the rhesus monkey showing microglial nodule with neuronophagia of a cortical neuron stained for viral antigen. Courtesy of USAMRIID. (See Color Insert.)

regions typically gives rise to depressed consciousness and seizures, but the involvement of subcortical regions, including the midbrain and brainstem, as well as thalamic and basal ganglial involvement, can give rise to a variety of movement disorders (Asher, 1975; Kalita and Misra, 2000; Misra and Kalita, 1997a; Murgod *et al.*, 2001; Ogata *et al.*, 2000; Pradhan *et al.*, 1999). WNV encephalitis in the United States and JEV encephalitis have both been characterized by a poliomyelitis-like syndrome, suggesting infection of lower motor neurons in association with flaccid paralysis (Glass *et al.*, 2002; Leis *et al.*, 2002; Misra and Kalita, 1997b; Solomon *et al.*, 1998). Lower motor neuron disease is also typical of TBEV infection. Factors governing the differences in neuronal susceptibility to flaviviruses are not known, but may be similar to those that operate in the case of other neurotropic arboviruses, such as Sindbis, where the neuronal response to viral injury may be variable (Griffin and Hardwick, 1999). Neuronal death associated with flavivirus infection has classically been ascribed to degenerative necrosis. Pathologic changes that accompany this process include vacuolization and proliferation of intracellular membranes, which produces a characteristic ultrastructural appearance (Fig. 6; Murphy *et al.*, 1968). Whether there is a greater propensity to cause necrosis versus apoptosis requires further study.

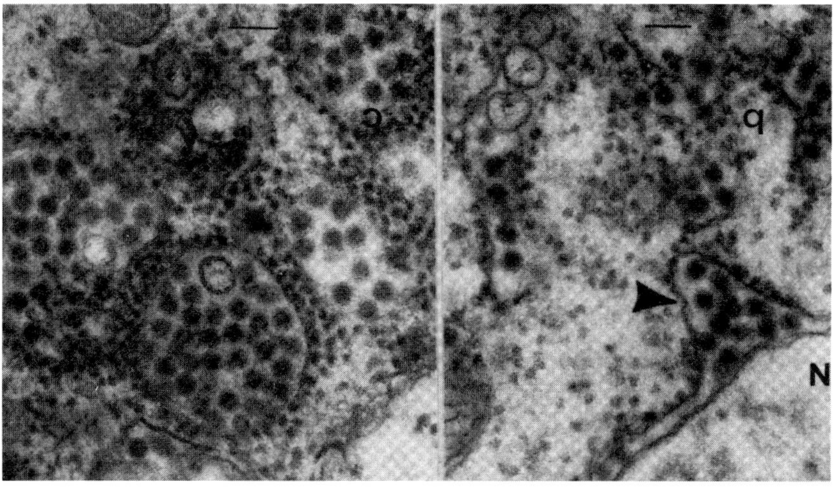

FIG 6. Electron micrograph of a mouse CNS neuron infected with SLE virus showing characteristic cytoplasmic pathology but integrity of the nuclear (N) envelope. Arrow indicates virions within inner and outer nuclear membranes. From Murphy *et al.* (1968), with permission.

Several flaviviruses have been shown to induce apoptosis, both *in vitro* (Isaeva *et al.*, 1998; Liao *et al.*, 1998; Parquet *et al.*, 2001; Prikhod'ko *et al.*, 2002) and *in vivo*, in the rodent CNS (Andrews *et al.*, 1999; Després *et al.*, 1998; Duarte dos Santos *et al.*, 2000; Isaeva *et al.*, 1998; Xiao *et al.*, 2001; Fig. 7). In this regard, reports describing primarily degenerative pathology at the light or electronmicroscopic

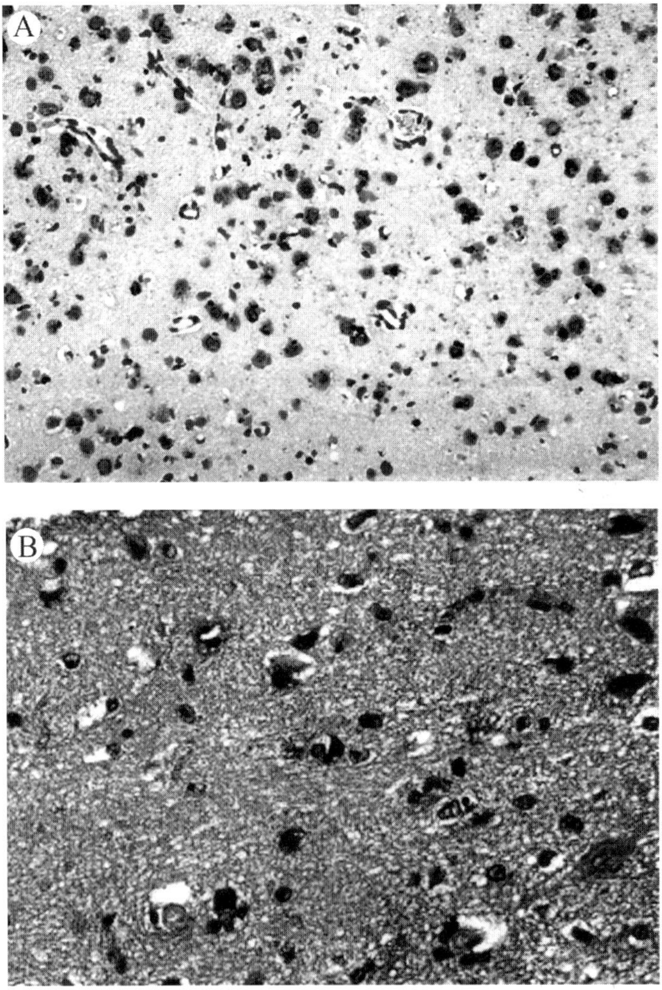

FIG 7. WNV infection in the Syrian golden hamster. (A) Viral antigen-positive neurons in the cerebral cortex. (B) TUNEL-positive apoptotic neurons in the cortex. Courtesy of Dr. Shu-Yan Xiao. From Xiao *et al.*, 2001. (See Color Insert.)

level, typically involving cytoplasmic changes such as chromatolysis, swelling, and dissolution of Nissl substance, also mentioned nuclear pathology, including pyknosis, disruption of the nuclear envelope, and alterations in chromatin (Dominguez and Baruch, 1963; Manuelidis, 1956; Mathews *et al.*, 2000). These findings are of interest given reports that neuronotropic viruses such as Sindbis virus may not induce typical apoptotic morphology in neurons (Griffin and Hardwick, 1999; Havert *et al.*, 2000; Kerr *et al.*, 2002; Sammin *et al.*, 1999) and that cell death may occur through both apoptotic and necrotic mechanisms (Havert *et al.*, 2000; Nargi-Aizenman and Griffin, 2001; Sammin *et al.*, 1999), depending on the integrity of the apoptotic pathways of a given population of neurons, their profile of apoptotic modulators, and the presence of excitotoxic stimuli. The most detailed *in vivo* characterization of neuronal apoptosis by flaviviruses has been reported for dengue. Neuroadapted dengue virus induces apoptosis in infected neurons as well as in noninfected cells (Després *et al.*, 1998), suggesting that indirect mechanisms of cellular injury occur in areas of heavy virus burden. It is important to note that these findings were demonstrated in very young mice, whose neurons are highly susceptible to apoptotic stimuli, and may not reflect the response of more mature cells in older mice. However, studies with WNV in the adult hamster model provide evidence that highly neurovirulent strains are potent inducers of apoptosis (Xiao *et al.*, 2001; Fig. 7). A relationship between viral virulence and the extent of apoptosis has not been clearly established for flaviviruses; however, the general capacity of neuroadapted strains to produce high virus burdens in association with cytopathology suggests that the situation is likely to resemble that of alphaviruses and other neuronotropic viruses (Lewis *et al.*, 1996; Oberhaus *et al.*, 1997; Theerasurakarn and Ubul, 1998). Although the molecular details of this process are not fully known, the expression of some flavivirus proteins appears to directly induce apoptotic cell death of neurons, including the WNV capsid protein (Yang *et al.*, 2002), the Langat virus NS3 protein, which causes apoptosis through the activation of caspases 3, 8, and 9 (Prikhod'ko *et al.*, 2002), and the E proteins of neuroadapted dengue-2 virus (Duarte dos Santos *et al.*, 2000) and JEV, which appear to stimulate an ER unfolded protein response (Su *et al.*, 2002) and a component of oxidative stress (Raung *et al.*, 2001). However, there are many other potential mechanisms for provoking apoptosis by neuronotropic RNA viruses, including signaling through interferon-α-dependent pathways, phospholipase A_2 activation, activation of NF-κB and p53-regulated genes (Fazakerley, 2001), and activation of apoptosis during viral entry (Jan *et al.*, 2000). Some of these processes may also be involved in the pathogenesis of encephalitic flaviviruses.

The role of apoptotic regulators bcl-2, bax, bcl-X, and related gene products in modulating neuronal death has been only partially characterized for flaviviruses. Forced expression of bcl-2-related genes promotes the survival of neuronal cell lines infected with JEV and dengue virus and facilitates viral persistence, primarily by restricting virus-induced cytopathic effects and not viral replication (Liao *et al.*, 1997, 1998). These outcomes vary in neuronal versus nonneuronal cell lines, and differential effects of bcl-2 and bcl-X appear to operate. Such findings suggest that the effects of apoptosis modulators on flaviviruses are similar to those observed with alphaviruses, where it is known that the expression of these proteins can vary in their effects from one cell type to another. Conclusions about the role of these proteins in different *in vivo* models therefore require careful evaluation (Griffin and Hardwick, 1999; Levine, 2002). Furthermore, because neuronotropic viruses can induce both necrosis and apoptosis, neuronal death may require assessment by several criteria. Neuronal injury as a result of bystander effects may also be a factor during flavivirus neuropathogenesis given that microglial activation and elaboration of inflammatory mediators, including IL-1β and TNF-α, occur in the CNS during these infections (Andrews *et al.*, 1999; Liu and Chambers, 2001; Ravi *et al.*, 1997) and may accompany the production of nitric oxide and peroxynitrite, which can cause neurotoxicity. Other potential mechanisms include excitatory cell death due to the activation of NMDA receptors, which has been implicated in the pathogenesis of Sindbis virus and HIV (Nargi-Aizenman and Griffin, 2001). Thus, although it is likely that the neurovirulence phenotype of flaviviruses is linked to the extent of neuronal cell death caused during the encephalitis, there appear to be multiple independent mechanisms by which neuronotropic viruses cause cell death. This process can be affected by the region of the brain affected, the degree of neuronal maturity, the factors that regulate cell death signaling receptors and their pathways, and levels of apoptosis modulators and other innate responses of virus-infected neurons (Fazakerly, 2001; Griffin and Hardwick, 1999; Levine, 2002; Liang *et al.*, 1998). In some cases, changes in the expression of neurotrophins may also be involved in the CNS response to viral injury (Zocher *et al.*, 2000); however, the relevance of this phenomenon to other types of viral encephalitis has yet to be widely investigated. The process is also subject to additional influence by the properties of the immune response recruited into the CNS, including $CD4^+$ and $CD8^+$ T cells, which may be involved in cytotoxicity toward virus-infected and, in the case of $CD4^+$ cells, perhaps even noninfected cells (Després *et al.*, 1998; Gagnon *et al.*, 1999), through Fas-dependent mechanisms under certain circumstances (Medana *et al.*, 2000). The effects of nonspecific inflammation, such as release of toxic substances from neutrophils (Andrews *et al.*, 1999), may also contribute to cellular injury.

X. The Central Nervous System Immune Response

Flavivirus infections induce a CNS inflammatory response of variable intensity. Data from most experimental models suggest that this inflammation is a requirement for protection from lethal infection with neurovirulent strains. The characteristics of the inflammation in experimental models have been shown to be affected by numerous factors, which include the endogenous CNS response, as well as the adequacy of the peripheral immune response and its timely recruitment into the CNS. In some cases of encephalitis, relatively scant inflammatory disease has been noted. This has also been observed experimentally as, for example, with TBEV under conditions where viral neuroinvasion and CNS involvement progress rapidly (Vince and Grevic, 1969). Immunosuppression also dramatically reduces the intensity of the CNS inflammation (Hirsch and Murphy, 1967; Leyssen *et al.*, 2003a). However, acute inflammation may become severe in response to a heavy antigen load and has been implicated in immunopathologic reactions in the CNS (reviewed in Monath, 1986).

A. Innate Responses

Viral infections of the CNS commonly result in the induction of innate responses, which include activation and proliferation of microglia and activation of astrocytes and cerebrovascular endothelium, with ensuing production of chemokines and proinflammatory cytokines (Benveniste, 1997). A consequence of this activation is the conditioning of cells in the CNS parenchyma and the blood–brain barrier to accommodate and modulate the influx of activated lymphocytes from the periphery by the upregulation of adhesion molecules and class I and II antigens. Activation of innate responses within the CNS during flavivirus encephalitis has been suggested by several human and experimental animal model studies in which the expression of chemokine and cytokine genes or their proteins has been analyzed. The production of IL-8 and macrophage inhibitory factor (MIF) has been detected in cerebrospinal fluid (CSF) or brain tissue during early stages of encephalitis with JEV and MVE viruses (Andrews *et al.*, 1999; Singh *et al.*, 2000; Suzuki *et al.*, 2000), with levels of IL-8 correlating directly with the number of neutrophils in the CSF. TNF-α and IL-1β are also elicited in response to flavivirus infection of the CNS (Liu and Chambers, 2001; Ravi *et al.*, 1997; Suzuki *et al.*, 2000), presumably representing intrinsic responses of microglia and astrocytes to the acute injury (Benveniste, 1997). Thus the early stage of the encephalitis involves the endogenous expression of mediators that results in the recruitment of nonspecific acute inflammatory cells with the potential for the production of neurotoxic substances, such as reactive oxygen intermediates, and probably facilitates further stimuli that intensify the inflammation. For instance, IL-1β and TNF-α can also mediate the release of IL-8 from astrocytes (Aloisi *et al.*, 1992). The induction of the nonspecific acute inflammation may be a deleterious process because levels of IL-8 are predictive of fatal disease (Suzuki *et al.*, 2000) and treatment of infected mice with inhibitors of NOS-2 in the acute stage of encephalitis lessened mortality in association with reduced inflammation (Andrews *et al.*, 1999). Because viral infections of the CNS induce the expression of chemokines, which have been implicated in inflammatory cell recruitment, (Liu and Lane, 2001; Liu *et al.*, 2000, 2001), the intense level of inflammation observed in flavivirus encephalitis may be driven by the induction of one or more chemokines and their receptors. The expression of monocyte and T-cell chemokines has not been reported in these infections; however, studies with other models suggest that MCP-1, IP-10, RANTES, and other chemokines are involved in the trafficking of leukocytes into areas of virus infection in the CNS. In conjunction with the effects of TNF-α and IFN-γ, both of which can lead to loss of integrity of the blood–brain barrier, this collection of stimuli may be sufficient to drive the commitment phase of the inflammatory response, during which the unrestricted entry of T lymphocytes then proceeds (Hickey, 1999).

The intrinsic defensive response of the CNS to viral injury also includes the induction of other classes of genes likely to influence the antiviral activity of this compartment through direct effects and by shaping the virus-specific immune response to viral injury (Johnston *et al.*, 2001; Labrada *et al.*, 2002). These include IFN-α and IFN-regulated genes such as ISG12 (Labrada *et al.*, 2002). IFN-α itself does not seem to be strongly upregulated in acute encephalitis. However, IFN-α has been reported in the CSF and brain tissue of human cases of encephalitis with JEV serogroup viruses and, in such cases, appeared to represent a marker of severe infection with a fatal outcome (Burke and

Morill, 1987; Leport *et al.*, 1984; Luby *et al.*, 1969). The role of type I interferons in the CNS response to injury is complex, and differences in the activities of IRFs and IGSFs toward target response elements in neurons and glia can affect the range of genes involved in the response, including MHC antigens, the OAS1b protein, and antiapoptotic factors (Baron-Delage *et al.*, 2000; Hirsch *et al.*, 1986; Lucas *et al.*, 2003; Massa *et al.*, 1999; Njenga *et al.*, 1997). Studies of viral infection in IFN-α and IFN-α-regulated gene knockout mice should help clarify the role of this defense system in directly controlling the viral infection of neurons versus effects on promoting the activity of the immune response recruited from the periphery. The importance of IL-12 in the innate CNS response to flavivirus encephalitis has not been investigated extensively; however, the outcome of CNS disease appears to vary among different viruses when this cytokine is used as an experimental therapy (Phillpotts *et al.*, 2003). Because of the importance of IL-12 in antiviral defense against other CNS viral infections, through its ability to stimulate NOS (Reiss *et al.*, 2002), more studies are needed to determine if it is implicated in protection against flaviviruses.

B. Virus-Specific Responses

Compared to information available on other neurotropic viruses, there has been relatively little work done to characterize the properties of virus-specific T cells recruited into the CNS in response to flavivirus infection (Scheider-Schaulies *et al.*, 1997). This response is subject to a number of influences, including previous immunologic experience with related viruses, the level of immunocompetence, immunogenetic host factors, and the virulence of the infecting virus. Available information comes from a limited number of human and animal model experiments, which partially characterized lymphocytes or soluble markers of activated T cells and cytokines in the CNS (Burke *et al.*, 1985b; Carson *et al.*, 2003; Gunther *et al.*, 1996; Iwasaki *et al.*, 1993; Johnson *et al.*, 1985, 1986; Kuno *et al.*, 1993; Sampson *et al.*, 2000), and other studies that have evaluated the requirements for these cells in the control of infection of this compartment in mouse models (Liu and Chambers, 2001; Liu *et al.*, 1989a; Murali-Krishna *et al.*, 1996; van der Most *et al.*, 2000, 2003). T cells bearing both $CD4^+$ and $CD8^+$ surface markers have been visualized in perivascular infiltrates, in CSF, and in brain parenchyma during human flavivirus encephalitis (Johnson *et al.*, 1985, 1986; Sampson *et al.*, 2000). These cells include a larger proportion of $CD4^+$ to $CD8^+$ T cells and moderate numbers of B cells and macrophages. However, the composition of cells differed regionally in the CNS, with macrophages and T cells more abundant in brain parenchyma and B cells more common within perivascular infiltrates. Factors governing the distribution of these cells and their effector functions are not well understood. Data from other experimental models suggest that $CD4^+$ T cells are important in directing the recruitment of lymphocytes (Hickey, 1999) and in maintaining the effector function of $CD8^+$ T cells within the brain parenchyma (Stohlman *et al.*, 1998). $CD8^+$ T cells have clearly been isolated from the brains of flavivirus-infected mice and, in some cases, demonstrated to have cytolytic activity (Liu *et al.*, 1989a); however, the contributions of such cells to both virus clearance and cellular injury within this compartment have not been defined. This remains a fundamental question, as the role of $CD4^+$ and $CD8^+$ T cells in different viral infections of the CNS and different experimental paradigms can vary considerably (Schneider-Schaulies *et al.*, 1997). For instance, virus clearance from mouse brains acutely infected with JEV has been reported by virus-specific $CD8^+$ T cells adoptively transferred into the CNS; however, $CD4^+$ T cells were also required (Murali-Krishna *et al.*, 1996). The relationship of this adoptive response to the natural immune response recruited from the periphery remains unclear, as the route of transfer may not reflect the normal pathway of lymphocyte trafficking. A requirement for CNS-associated $CD4^+$ and $CD8^+$ T cells was also observed in an immunization/challenge model of dengue virus (van der Most *et al.*, 2000). However, in the context of a memory response to YFV in the mouse model, $CD4^+$ T cells and B cells were required for the control of viral infection, whereas $CD8^+$ T cells were not required, although CD8-deficient mice exhibited a defect in virus clearance (Liu and Chambers, 2001; T. J. Chambers, unpublished data). Part of the effector activity of $CD8^+$ T cells may be mediated by the production of IFN-γ, which has a range of effects on the immunological properties of the CNS, including the upregulation of class I and II antigens on glial cells and injured neurons, activation of microglial, priming of astrocytes for cytokine production, and increasing permeability of the blood–brain barrier, as well as antiviral activity in the brain parenchyma (Benveniste, 1997; Kundig *et al.*, 1993; Popko *et al.*, 1997). IFN-γ knockout mice are defective in the clearance of YFV from the CNS and exhibit decreased inflammatory cell recruitment to this compartment (Liu and Chambers, 2001), indicating an important role of this cytokine in flavivirus encephalitis. IFN-γ may have a primary role in these processes because elimination of virus-infected neurons by CTLs is very tightly constrained by the absence of constitutive expression of MHC class I in these cells and the lack of susceptibility to perforin-mediated lysis (Medana *et al.*, 2000; Neumann *et al.*,

1995). Although killing can occur through Fas–FasL interactions (Medana *et al.*, 2000), induction of FasL expression by neurons can also confer protection against CTL attack under certain conditions (Medana *et al.*, 2001) and even induce apoptosis in infiltrating lymphocytes (Flugel *et al.*, 2000). Virus-specific $CD8^+$ T cells have been shown to persist in the brains of mice recovering from dengue encephalitis for prolonged periods and differentiate into effector–memory cells that lose CTL activity while still producing IFN-γ (van der Most *et al.*, 2003). Thus it appears that cellular immune-mediated mechanisms of neuronal cell death are subject to tight regulation in the presence of acute viral injury.

The role of antibody responses in the control of flavivirus encephalitis has been investigated in many classic studies, and it has been demonstrated repeatedly that immune serum can arrest viral infection in the CNS (Roehrig *et al.*, 2001). Mechanisms responsible for this antibody-mediated control of infection remain unclear; however, studies with YFV encephalitis indicate that the Fc region and the IgG subclass are critical for protection in the mouse model (Schlesinger *et al.*, 1993, 1995). These data suggest that direct effector functions associated with IgG, including interactions with cells bearing Fc receptors, such as microglia and recruited macrophages, are involved in this process. It is also conceivable that antibody–dependent cellular cytotoxicity directed against cell surface NS1 or even complement-mediated lysis also contributes to the protective effect. The antibody-mediated control of neuronal infection is a well-established mechanism for alphaviruses and coronaviruses, in which case the role of antibody in preventing reactivation of viral infection has been demonstrated (Griffin *et al.*, 1997; Levine *et al.*, 1991, 1992; Lin *et al.*, 1999). In the case of Sindbis virus, there is inhibition of viral release and eventual sequestering of viral RNA within neurons without clearance. This process occurs preferentially at cortical sites of infection and obligates the local retention of virus-specific B cells (Tyor *et al.*, 1992) in contrast to infection in the spinal cord where the cytokine-mediated (IFN-γ) elimination of viral RNA apparently predominates (Binder and Griffin, 2001). Antibody-mediated mechanisms may also operate in flavivirus encephalitis, as reactivation or recrudescence of CNS disease has been documented in some experimental animal systems and in clinical cases (Section XII). Factors responsible for the prolonged survival of B cells in the CNS are not known, particularly whether these cells represent a resident population of memory cells or require replenishment from peripheral sites (Tschen *et al.*, 2002).

The beneficial role of antibody responses within the CNS is also supported by clinical studies of JEV and TBEV encephalitis (Burke *et al.*,

1985a, 1985c; Gunther *et al.*, 1997; Han *et al.*, 1988; Hoffman *et al.*, 1978; Kaiser and Holzmann, 2000; Potula *et al.*, 2003; Ravi *et al.*, 1993). High levels of IgM correlate with an improved outcome in some cases, presumably reflecting the recruitment of virus-specific B cells into the CNS, as was reported in JEV cases (Burke *et al.*, 1985a; Gunther *et al.*, 1997), and the presence of IgG1 in particular is associated with the control of infection (Thakare *et al.*, 1991), although IgM can be elevated in severe cases with a fatal outcome (Desai *et al.*, 1994a). The functional activities provided by IgM in the CNS that result in clinical benefit remain unclear, as experimentally, IgM may have limited neutralization activity against encephalitic viruses at least early in infection (Diamond *et al.*, 2003; Hoffman *et al.*, 1978; Ishii *et al.*, 1968). The contribution of complement-fixing activity could be important, as discussed earlier, in view of the fact that complement proteins may be upregulated in the CNS as a result of viral infection (Johnston *et al.*, 2001).

The inflammatory response to flavivirus infection of the CNS may, in some cases, cause deleterious effects on neuronal function and survival. Immune complexes and autoantibodies to neurofilaments and myelin basic protein in the CSF and serum have been reported in severe JEV and TBEV infections (Desai *et al.*, 1994a, 1994b; Fokina *et al.*, 1991; Thakare *et al.*, 1988) and may reflect an immunopathogenic process rather than nonspecific reactions to viral injury, particularly because of their association with poor outcome. Mechanisms leading to resolution of the acute inflammation in CNS viral infections remain undefined (Bradl and Flugel, 2002). At least three potential factors could be involved, including the expression of IL-4, IL-10, and TGF-β, which have been observed in the CNS of Sindbis virus-infected mice (Wesselingh *et al.*, 1994) and are known to have immunomodulatory activities that reduce CNS inflammation; induction of apoptosis of infiltrating T cells; or possibly entry of NKT cells, which have been associated with the suppression of inflammatory responses. Further studies are needed to determine whether these or other factors are required to downregulate potentially harmful immune responses in cases where viral disease is eventually controlled.

XI. Neuropathogenesis: West Nile Virus as a Model

Similar to JEV, infections with WNV can be characterized as protean, involving an extraordinary host range, and an abundance of pathologic and virologic data has been obtained from experiments with

WNV in birds (Kramer and Bernard, 2001; Steele *et al.*, 2000), horses (Bunning *et al.*, 2002), and humans (Asnis *et al.*, 2000; Hubalek and Halouzka, 1999). WNV has emerged as dimorphic in its clinical disease, with the apparent shift from typical WN fever to disease of greater severity, including more frequent cases of acute encephalitis in conjunction with the emergence of New World lineage I strains (Asnis *et al.*, 2000; Cernescu *et al.*, 1997). Studies with virulent strains such as New York 1999 WNV have demonstrated the neuropathogenic potential inherent in this virus. It is therefore important to note that lineage differences (type I versus type II) and perhaps genetic variation within lineages may influence the development of CNS disease.

After peripheral inoculation in the mouse model, WNV is believed to infect Langerhans dendritic cells (Johnston *et al.*, 2000), which migrate to draining lymph nodes, and within 12 to 24 h of infection, viral replication is observed in secondary lymphoid tissue (Diamond *et al.*, 2003; McMinn *et al.*, 1996). Infectious virus is detected in serum within 24 to 48 h of infection (Diamond *et al.*, 2003; Kramer and Bernard, 2001). The course of the early infection is slightly different in wild-type (WT) versus B-cell-deficient mice, as the peak of viremia occurs later in the latter case (day 2 versus days 4 to 6; Fig. 8). Shortly afterward, in WT mice, infectious virus is detected in visceral organs such as the spleen, kidney, and heart but not the liver (Diamond *et al.*, 2003; Kramer and Bernard, 2001; Weiner *et al.*, 1970), which may reflect restricted tropism or a high level of reticuloendothelial clearance in this organ. The levels of infectious virus in visceral tissues and serum peaked by day 4 after infection and thereafter diminished (Diamond *et al.*, 2003; Kramer and Bernard, 2001; Xiao *et al.*, 2001), concurrent with a rise in the titer of neutralizing antibodies (Fig. 9). In B-cell-deficient mice, replication in peripheral lymphoid tissue and visceral organs follows kinetics similar to those of WT mice; however, virus is not cleared from these sites. These data indicate the profound susceptibility of mice to WNV in the absence of antibody-producing B cells.

Virus can first be detected in the CNS by 4 days after peripheral subcutaneous infection in both WT and B-cell-deficient mice. Infectious virus is detected simultaneously in multiple sites in the brain, as well as in the inferior and superior spinal cord, suggesting a hematogenous route of dissemination and/or rapid spread within the CNS (Diamond *et al.*, 2003). However, the route of peripheral inoculation influences the rate of spread to the brain and spinal cord. Infection via an intraperitoneal or intravenous route results in the spread of infectious virus to the brain within 2 days of infection (Kramer and Bernard, 2001), with these animals succumbing to infection several days earlier than

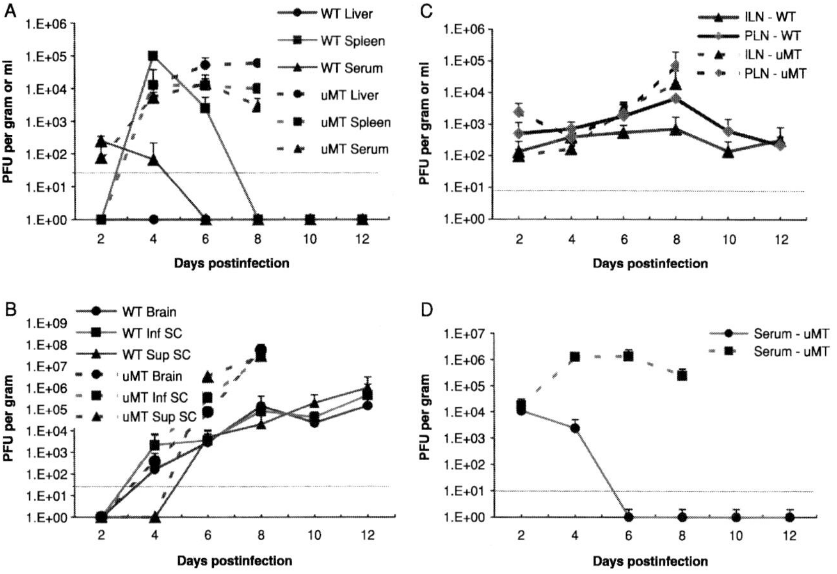

FIG 8. Wild-type or B-cell-deficient (uMT) mice were inoculated with WNV in the footpad, and the virus content in serum, peripheral tissues, and brain was measured serially using plaque assay or quantitative polymerase chain reaction. From Diamond *et al.* (2003), with permission.

those infected by a subcutaneous route. These differences probably reflect the fact that the interaction of virus with peripheral tissues results in the engagement of nonspecific and virus-specific defenses that have an impact on the course of disease. Models of infection that bypass the physiologic route of inoculation must be considered with this limitation in mind, as the balance of viral replication and dissemination versus immune activation can be skewed greatly by intraperitoneal or intravenous injection. Regardless of the route of administration, the course of disease after the onset of CNS infection is rapid and leads to fatal encephalitis beginning by days 9 to 10 postinfection, associated with high burdens of brain-associated virus.

Neutralizing activity associated with the IgM class is detectable by day 4 postinfection in the mouse model (Fig. 9), and this antibody response can confer partial protection against virus challenge in naïve mice (Diamond *et al.*, 2003). Neutralizing IgG is detectable by day 8 and reaches levels of activity that are 10-fold higher than that of IgM by day 12 postinfection. The importance of secreted IgM in protection against disease is indicated from these data, as well as data presented

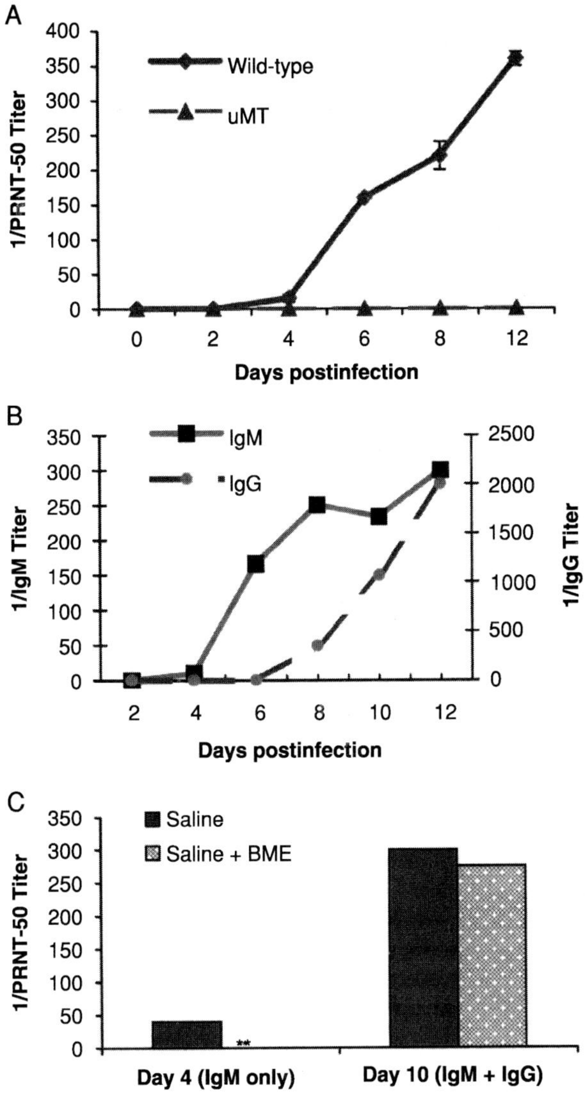

FIG 9. Neutralizing IgM and IgG antibody responses in acute WNV encephalitis in the mouse model. From Diamond *et al.* (2003), with permission.

earlier (Fig. 1). The relative importance of the mechanisms involved, including neutralizing activity per se or complement-assisted priming of the B-cell response to facilitate IgG production, is currently under investigation.

XII. Virus Persistence

Persistent infection *in vitro* and *in vivo* has been described in a number of experimental and clinical settings involving flaviviruses. The phenomenon of persistent infection in cell cultures of vertebrate and arthropod origin is well documented (Brinton, 1982; Chen *et al.*, 1996; Jarman *et al.*, 1968; Igarashi, 1979; Katz and Goldblum, 1968; Lancaster *et al.*, 1998; Loginova *et al.*, 1980; Poidinger *et al.*, 1991; Randolph and Hardy, 1988a, 1988b; Schmaljohn and Blair, 1977, 1979; Shah and Gadkari, 1987; Vlaycheva and Chambers, 2002; Zhang *et al.*, 1993), and the findings are analogous to many other models of viral infections *in vitro* where this process can be observed. Flavivirus persistence *in vitro* typically arises following a cytocidal infection, with survival of a residual population of cells that harbor low levels of replicating virus for long periods of time. The majority of cells in such cultures usually express viral antigen, but only a minority actually are productive of infectious virus. The cultures are generally resistant to superinfection with homologous but not heterologous virus, although in some cases, superinfection can drive virus production in quiescent antigen-positive cells (Schmaljohn and Blair, 1979). The persistence is not dependent on the expression of IFN-α, rather there is evidence for the involvement of host cell antiapoptotic pathways (see later). The infection is not necessarily deleterious for host cells, but in some cases, reduced growth efficiency occurs. Viruses detected in persistently infected cultures frequently undergo phenotypic alterations, including a reduction in plaque size, temperature sensitivity (Randolph and Hardy, 1988b; Shah and Gadkari, 1987), host-range restriction (Randolph and Hardy, 1988b), and loss of neurovirulence for mice (Igarashi, 1979). The genetic basis for the phenotypic change of viral variants has, in some cases, been demonstrated (Vlaycheva and Chambers, 2002). However, these infections are also associated with the emergence of defective viral particles or defective RNAs (Brinton, 1982; Debnath *et al.*, 1991; Lancaster *et al.*, 1998) that form the basis for superinfection resistance, and perhaps in part for the attenuation of mouse neurovirulence in flavivirus–resistant strains of mice. Some persistently infected cultures exhibit alterations in the composition of viral proteins or produce truncated forms of NS1 (Chen *et al.*, 1996), which presumably are believed to reflect deletions in the viral genome associated with the generation of defective interfering viruses. The genome of defective MVE virus in Vero cells was shown to lack a large segment encompassing the prM-E region and a portion of the NS1 region, which leads to the production of truncated NS1 (Blitvich *et al.*,

1999; Lancaster *et al.*, 1998). This finding is entirely consistent with the fact that replicons such as those engineered for Kunjin virus are designed to delete the corresponding region of the genome and are able to establish persistent replication in cell culture without any typical viral cytopathic effects (Varnavski and Khromykh, 1999). At least part of the explanation for the loss of cytocidal activity therefore may result from the deletion of structural proteins, which may provoke an ER overload response and/or trigger apoptosis (Duarte dos Santos *et al.*, 2000; Prikhod'ko *et al.*, 2001; Su *et al.*, 2002). The production of infectious virus from persistently infected cultures can be reduced by treatment with a neutralizing antibody (Randolph and Hardy, 1988a), but it is not clear whether this shifts the cells to preferentially harbor defective viral genomes or whether these are actually eliminated.

The expression of apoptotic modulators is an important factor influencing the establishment and maintenance of persistent infection. In a model of JEV infection, the persistence in some cell lines was facilitated by the expression of bcl-2 (Liao *et al.*, 1997), indicating that the mechanism is analogous to that of alphaviruses, where lytic infection can be converted to persistence in the presence of antiapoptotic proteins (Levine *et al.*, 1993). However, the mechanisms that confer resistance to JEV-induced apoptosis must differ in some way from those of Sindbis and other viruses, as bcl-2 could fully protect N18 neuroblastoma cells against Sindbis, but not JEV. It is likely that the difference involves divergent pathways for activating the apoptotic process (Jan *et al.*, 2000; Su *et al.*, 2002). It is also possible that some cell lines represent variants that are defective in their apoptotic pathways, allowing virus to adapt in the absence of apoptotic events. At present, there is no evidence that flaviviruses encode proteins with antiapoptotic properties that might influence the process of the persistence.

The relevance of data on persistently infected cells *in vitro* to the issue of persistent infection in animal models and in apparent human cases is not straightforward. Although evidence for such clinical entities has been reported, the situation is complicated by the fact that there may be overlap between infections that have protracted convalescence and those that have frank neurological sequelae because flavivirus encephalitis is associated with neurologic complications that confer long-term disability among those who survive acute infection (Baruah *et al.*, 2002; Finley and Riggs, 1980; Greve *et al.*, 2002; Haglund and Gunther, 2003; Huy *et al.*, 1994; Kumar *et al.*, 1993; Richter *et al.*, 1961; Vaneeva, 1969). In some cases, there is radiographic and pathologic evidence of permanent neurologic injury, including reduced cerebral blood flow, areas of abnormal signal density, and cellular dropout and gliosis

(Gunther *et al.*, 1998; Ishii *et al.*, 1977; Kumar *et al.*, 1997; Shoji *et al.*, 1990). Second, the possibility that a recrudescent latent infection really represents reinfection in a host with suboptimal immunity is a difficult question to evaluate. Also, the criteria for defining persistence are somewhat arbitrary. The presence of prolonged expression of IgM has been invoked as a basis for concluding persistence, but only infrequently has the presence of infectious virus been documented in situations where the duration of CNS disease exceeds that which is expected in acute uncomplicated flavivirus encephalitis (2 to 3 weeks).

Mechanisms of flavivirus persistence *in vivo* may theoretically also involve the formation of DI particles, which reduce infectious viral load, but like their *in vitro* counterparts, may not be sufficiently cytopathic to cause neuronal cell death. Such defective viruses may be capable of stimulating immune responses and thus be detected serologically. Experimentally, there is certainly evidence that DI particles can inhibit the production of infectious virus in the CNS (Atkinson *et al.*, 1986; Barrett and Dimmock, 1984; Smith, 1981). The failure to detect infectious virus in both clinical and experimental situations where prolonged evidence of viral activity in the CNS was suspected is consistent with this hypothesis (Pogodina *et al.*, 1983; Ravi *et al.*, 1993). In a study of viral RNA in the brains of mice with the *Flv* resistance allele, a reduction in genome-length RNA rather than appearance of DI RNAs was observed (Urosevic *et al.*, 1997), even though flavivirus–resistance in mice is associated with production of D1 virus (Smith, 1981). Thus, data implicating the role of DI particles in *in vitro* persistence may not correspond directly to an *in vivo* process.

Encephalitic flaviviruses that have been implicated in the occurrence of persistent infection of the CNS include members of the JEV serogroup and TBEV serogroup (Iliyenko *et al.*, 1974; Ogawa *et al.*, 1973; Pogodina *et al.*, 1983; Slavin *et al.*, 1943; Zlotnik *et al.*, 1971), with the latter having a particular propensity to cause chronic infections. In humans, there have been reports of various types of persistent infection. These include (1) reactivation of latent disease, as in children with JEV encephalitis who experienced recurrent infection months after the primary infection and, in some cases, gave positive virus isolations (Sharma *et al.*, 1991); (2) chronic progressive disease with cognitive and motor disturbances that resembled subacute sclerosing panencephalitis, years after a primary infection with Russian spring–summer encephalitis (RSSE) virus (Ogawa *et al.*, 1973); (3) prolonged infection in primary cases of acute encephalitis, with elevation of CSF IgM for as long as 10 months and, in some cases, with virus isolated from spinal fluid (Ravi *et al.*, 1993); and (4) evidence of prolonged circulation of virus-infected cells without clear-cut CNS disease (Southam *et al.*, 1958). Protracted cases of TBEV encephalitis, with either cognitive dysfunction or spinal nerve paralysis, occur frequently (Haglund and Gunther, 2003), although the relationship of these outcomes to persistent viral infection is not known. These different types of clinical infections have been mimicked by various experimental models, and it has been documented that virus can be recovered after a protracted phase of infection. For instance, strains of WNV with differing neurovirulence properties were capable of causing prolonged encephalitis of variable severity, ranging up to 5 months in duration in monkeys (Pogodina *et al.*, 1981, 1983). Eventually the viruses were cleared or were rendered replication defective and detectable only by the presence of viral antigen. Some viruses recovered from brains of these animals had undergone attenuation of mouse neurovirulence, indicating a selection for genetic variants and/or defective interfering particles. Mechanisms preventing the efficient clearance of virus were not determined, but did not involve defects in the production of neutralizing antibodies. Immunosuppression, either by cytotoxic drugs or even that associated with pregnancy, has been demonstrated to cause prolonged flavivirus disease (Mathur *et al.*, 1986; Zlotnik *et al.*, 1971), and reactivation of infection can be elicited during or after the primary infection by immune depletion (Mathur *et al.*, 1986). In some cases, as in JEV infection of mice, viral latency was established in T lymphocytes, which could be subsequently activated by immunosuppression to produce infectious virus (Mathur *et al.*, 1989). These latter findings are consistent with reports of recurrent JEV infection observed in children, but the stimulus for reactivation in such cases is not known (Sharma *et al.*, 1991).

In the CNS, experimental persistent infections have been related to immune response factors, as well as the ability of neurons to survive viral infection by mechanisms involving apoptotic modulators. Immunological tolerance is also a potential factor that has been associated with other viruses such as LCMV that establish persistent CNS infection, but this does not seem to play any obvious role in flavivirus persistence. Failure to clear brain-associated virus may result from a limited effectiveness of the innate CNS defenses and the peripheral immune responses that are activated in response to flavivirus infections. In addition, elimination of viral RNA may require $CD8^+$ T-cell functions [as described for alphaviruses (Binder and Griffin, 2001)], which are severely restricted due to a lack of class I expression on neurons. The generally resistant state of differentiated neurons toward apoptotic stimuli is also likely to be an important factor (Griffin and Hardwick, 1999).

Together, these processes may facilitate the persistence of viral RNA and perhaps foster the evolution of DI particles and noncytopathic-attenuated viral variants. The host immune response, together with properties of the neuronal cellular environment, influences the likelihood of virus perisistence and may both contribute to differences in phenotypes of persisting viruses. For instance, WNV variants that emerged from persistent infection in monkeys tended to acquire attenuation phenotypes (Pogodina *et al.*, 1983), whereas it has been observed in persistent infection of mice with Sindbis virus that neurovirulent mutants arise (Levine and Griffin, 1992, 1993). Selection pressures imposed by neutralizing antibodies that are produced intraparenchymally by virus-specific B cells may facilitate the emergence of viral variants in some cases, but tissue-specific adaptations are also involved. At present, it is reasonable to believe that the phenomenon of flavivirus persistence in the CNS is a function of many variables, which include the heterogeneity of the neuronal response to injury, encompassing innate interferon-regulated antiviral defenses (Johnston *et al.*, 2001) and the competence for and propensity towards apoptosis. Interaction of virus–infected cells with virus-specific T cells possessing cytokine-mediated effector functions that can eliminate viral RNA and B cells that can provide antibodies capable of downregulating viral replication influences this process in a region-specific manner. The balance among these factors results in a spectrum of outcomes that may range from either clearance to merely suppression of viral infection, with variable consequences for long-term neurologic function.

References

Aihara, H., Takasaki, T., Toyosaki-Maeda, T., Suzuki, R., Okuno, Y., and Kurane, I. (2000). T-cell activation and induction of antibodies and memory T cells by immunization with inactivated Japanese encephalitis vaccine. *Viral Immunol.* **13:**179–186.

Albrecht, P. (1968). Pathogenesis of neurotropic arbovirus infections. *Curr. Top. Microbiol. Immunol.* **43:**44–91.

Aloisi, F., Care, A., Borsellino, G., Gallo, P., Rosa, S., Bassani, A., Cabibbo, A., Testa, U., Levi, G., and Peschle, C. (1992). Production of hemolymphopoietic cytokines (IL-6, IL-8, colony-stimulating factors) by normal human astrocytes in response to IL-1 beta and tumor necrosis factor-alpha. *J. Immunol.* **149:**2358–2366.

Andersen, A., A., and Hanson, R. P. (1974). Influence of sex and age on natural resistance to St. Louis encephalitis virus infection in mice. *Infect. Immun.* **9:**1123–1125.

Anderson, J. F., and Rahal, J. J. (2002). Efficacy of interferon alpha-2b and ribavirin against West Nile virus in vitro. *Emerg. Infect. Dis.* **8:**107–108.

Anderson, R., Wang, S., Osiowy, C., and Issekutz, A. C. (1997). Activation of endothelial cells via antibody-enhanced dengue virus infection of peripheral blood monocytes. *J. Virol.* **71:**4226–4232.

Andrews, D. M., Matthews, V. B., Sammels, L. M., Carrello, A. C., and McMinn, P. C. (1999). The severity of Murray Valley encephalitis in mice is linked to neutrophil infiltration and inducible nitric oxide synthase activity in the central nervous system. *J. Virol.* **73:**8781–8790.

Asada-Kubota, M., Ueda, T., Nakashima, T., Kobayashi, M., Shimada, M., Takeda, K., Hamada, K., Maekawa, S., and Sokawa, Y. (1997). Localization of 2′,5′-oligoadenylate synthetase and the enhancement of its activity with recombinant interferon-alpha A/D in the mouse brain. *Anat. Embryol. (Berl.)* **195:**251–257.

Asher, D. M. (1975). Movement disorders in rhesus monkeys after infection with tick-borne encephalitis virus. *Adv. Neurol.* **10:**277–289.

Asnis, D. S., Conetta, R., Teixeira, A. A., Waldman, G., and Sampson, B. A. (2000). The West Nile Virus outbreak of 1999 in New York: The Flushing Hospital experience. *Clin. Infect. Dis.* **30:**413–418.

Atkinson, T., Barrett, A. D. T., Mackenzie, A., and Dimmock, N. J. (1986). Persistence of virulent Semliki Forest virus in mouse brain following co-inoculation with defective interfering particles. *J. Gen. Virol.* **67:**1189–1194.

Atrasheuskaya, A. V., Fredeking, T. M., and Ignatyev, G. M. (2003). Changes in immune parameters and their correction in human cases of tick-borne encephalitis. *Clin. Exp. Immunol.* **131:**148–154.

Avirutnan, P., Malasit, P., Seliger, B., Bhakdi, S., and Husmann, M. (1998). Dengue virus infection of human endothelial cells leads to chemokine production, complement activation, and apoptosis. *J. Immunol.* **161:**6338–6346.

Azad, R., Gupta, R. K., Kumar, S., Pandey, C. M., Prasad, K. N., Husain, N., and Husain, M. (2003). Is neurocysticercosis a risk factor in coexistent intracranial disease? An MRI-based study. *J. Neurol. Neurosurg. Psychiatry* **74:**359–361.

Baron-Delage, S., Abadie, A., Echaniz-Laguna, A., Melki, J., and Beretta, L. (2000). Interferons and IRF-1 induce expression of the survival motor neuron (SMN) genes. *Mol. Med.* **6:**957–968.

Barrett, A. D. T., and Dimmock, N. J. (1984). Modulation of Semliki Forest virus-induced infection of mice by defective interfering virus. *J. Infect. Dis.* **150:**98–103.

Barrington, R., Zhang, M., Fischer, M., and Carroll, M. C. (2001). The role of complement in inflammation and adaptive immunity. *Immunol. Rev.* **180:**5–15.

Baruah, H. C., Biswas, D., Patgiri, D., and Mahanta, J. (2002). Clinical outcome and neurological sequelae in serologically confirmed cases of Japanese encephalitis patients in Assam, India. *Indian Pediatr.* **39:**1143–1148.

Baumgarth, N., Herman, O. C., Jager, G. C., Brown, L. E., Herzenberg, L. A., and Chen, J. (2000). B-1 and B-2 cell-derived immunoglobulin M antibodies are nonredundant components of the protective response to influenza virus infection. *J. Exp. Med.* **192:**271–280.

Ben-Nathan, D., Huitinga, I., Lustig, S., van Rooijen, N., and Kobiler, D. (1996a). West Nile virus neuroinvasion and encephalitis induced by macrophage depletion in mice. *Arch. Virol.* **141:**459–469.

Ben-Nathan, D., Kobiler, D., Rzotkiewicz, S., Lustig, S., and Katz, Y. (2000). CNS penetration by noninvasive viruses following inhalational anesthetics. *Ann. N. Y. Acad. Sci.* **917:**944–950.

Ben-Nathan, D., Lustig, S., and Feuerstein, G. (1989). The influence of cold or isolation stress on neuroinvasiveness and virulence of an attenuated variant of West Nile virus. *Arch. Virol.* **109:**1–10.

Ben-Nathan, D., Lustig, S., and Kobiler, D. (1996b). Cold stress-induced neuroinvasiveness of attenuated arboviruses is not solely mediated by corticosterone. *Arch. Virol.* **141:**1221–1229.

Ben-Nathan, D., Lustig, S., Kobiler, D., Danenberg, H. D., Lupu, E., and Feuerstein, G. (1992). Dehydroepiandrosterone protects mice inoculated with West Nile virus and exposed to cold stress. *J. Med. Virol.* **38:**159–166.

Benveniste, E. N. (1997). Cytokine expression in the nervous system. *In* "Immunology of the Nervous System" (R. W. Keane and W. F. Hickey, eds.), pp. 419–459. Oxford Univ. Press, New York.

Bergmann, C. C., Ramakrishna, C., Kornacki, M., and Stohlman, S. A. (2001). Impaired T cell immunity in B cell-deficient mice following viral central nervous system infection. *J. Immunol.* **167:**1575–1583.

Bernard, K. A., Klimstra, W. B., and Johnston, R. E. (2000). Mutations in the E2 glycoprotein of Venezuelan equine encephalitis virus confer heparan sulfate interaction, low morbidity, and rapid clearance from blood of mice. *Virology* **276:**93–103.

Bielefeldt-Ohmann, H. (1998). Analysis of antibody-independent binding of dengue viruses and dengue virus envelope protein to human myelomonocytic cells and B lymphocytes. *Virus Res.* **57:**63–79.

Binder, G., and Griffin, D. E. (2001). Interferon-gamma-mediated site-specific clearance of alphavirus from CNS neurons. *Science* **293:**303–306.

Blackwell, J. L., and Brinton, M. A. (1995). BHK cell proteins that bind to the 3′ stem-loop structure of the West Nile virus genome RNA. *J. Virol.* **69:**5650–5658.

Blackwell, J. L., and Brinton, M. A. (1997). Translation elongation factor-1 alpha interacts with the 3′ stem-loop region of West Nile virus genomic RNA. *J. Virol.* **71:**6433–6444.

Blitvich, B. J., Scanlon, D., Shiell, B. J., Mackenzie, J. S., and Hall, R. A. (1999). Identification and analysis of truncated and elongated species of the flavivirus NS1 protein. *Virus Res.* **60:**67–79.

Boes, M., Esau, C., Fischer, M. B., Schmidt, T., Carroll, M., and Chen, J. (1998). Enhanced B-1 cell development, but impaired IgG antibody responses in mice deficient in secreted IgM. *J. Immunol.* **160:**4776–4787.

Bonnevie-Nielsen, V., Heron, I., Monath, T. P., and Calisher, C. H. (1995). Lymphocytic 2′,5′-oligoadenylate synthetase activity increases prior to the appearance of neutralizing antibodies and immunoglobulin M and immunoglobulin G antibodies after primary and secondary immunization with yellow fever vaccine. *Clin. Diag. Lab. Immunol.* **2:**302–306.

Bonnevie-Nielsen, V., Larsen, M. L., Frifelt, J. J., Michelsen, B., and Lernmark, A. (1989). Association of IDDM and attenuated response of 2′,5′-oligoadenylate synthetase to yellow fever vaccine. *Diabetes* **38:**1636–1642.

Bosch, I., Xhaja, K., Estevez, L., Raines, G., Melichar, H., Warke, R. V., Fournier, M. V., Ennis, F. A., and Rothman, A. L. (2002). Increased production of interleukin-8 in primary human monocytes and in human epithelial and endothelial cell lines after dengue virus challenge. *J. Virol.* **76:**5588–5597.

Bradish, C. J., Fitzgeorge, R., and Titmuss, D. (1980). The responses of normal and athymic mice to infections by togaviruses: Strain differentiation in active and adoptive immunization. *J. Gen. Virol.* **46:**255–265.

Bradl, M., and Flugel, A. (2002). The role of T cells in brain pathology. *Curr. Top. Microbiol. Immunol.* **265:**141–162.

Brandriss, M. W., and Schlesinger, J. J. (1984). Antibody-mediated infection of P388D1 cells with 17D yellow fever virus: Effects of chloroquine and cytochalasin B. *J. Gen. Virol.* **65:**791–794.

Brandriss, M. W., Schlesinger, J. J., Walsh, E. E., and Briselli, M. (1986). Lethal 17D yellow fever encephalitis in mice. I. Passive protection by monoclonal antibodies to the envelope proteins of 17D yellow fever and dengue 2 viruses. *J. Gen. Virol.* **67**:229–234.

Brandt, W. E., McCown, J. M., Gentry, M. K., and Russell, P. K. (1982). Infection enhancement of dengue type 2 virus in the U-937 human monocyte cell line by antibodies to flavivirus cross-reactive determinants. *Infect. Immun.* **36**:1036–1041.

Brinton, M. A. (1982). Characterization of West Nile virus persistent infection in genetically resistant and susceptible mouse cells. I. Generation of defective nonplaquing virus particles. *Virology* **116**:84–94.

Brinton, M. A. (2000). Host factors involved in West Nile virus replication. *Ann. N. Y. Acad. Sci.* **951**:207–219.

Brooks, T. J., and Phillpotts, R. J. (1999). Interferon-alpha protects mice against lethal infection with St. Louis encephalitis virus delivered by the aerosol and subcutaneous routes. *Antiviral Res.* **41**:57–64.

Broom, A. K., Wallace, M. J., Mackenzie, J. S., Smith, D. W., and Hall, R. A. (2000). Immunisation with gamma globulin to Murray Valley encephalitis virus and with an inactivated Japanese encephalitis virus vaccine as prophylaxis against Australian encephalitis: Evaluation in a mouse model. *J. Med. Virol.* **61**:259–265.

Bunning, M. L., Bowen, R. A., Cropp, C. B., Sullivan, K. G., Davis, B. S., Komar, N., Godsey, M. S., Baker, D., Hettler, D. L., Holmes, D. A., Biggerstaff, B. J., and Mitchell, C. J. (2002). Experimental infection of horses with West Nile virus. *Emerg. Infect. Dis.* **8**:380–386.

Burke, D. S., Lorsomrudee, W., Leake, C. J., Hoke, C. H., Nisalak, A., Chongswasdi, V., and Laorakpongse, T. (1985a). Fatal outcome in Japanese encephalitis. *Am. J. Trop. Med. Hyg.* **34**:1203–1210.

Burke, D. S., Nisalak, A., Lorsomrudee, W., Ussery, M. A., and Laorpongse, T. (1985b). Virus-specific antibody-producing cells in blood and cerebrospinal fluid in acute Japanese encephalitis. *J. Med. Virol.* **17**:283–292.

Burke, D. S., Nisalak, A., Usery, M. A., Laorakpongse, T., and Chantavibul, S. (1985c). Kinetics of IgM and IgG responses to Japanese encephalitis virus in human serum and cerebrospinal fluid. *J. Infect. Dis.* **151**:1093–1099.

Burke, D. S., and Morill, J. C. (1987). Levels of interferon in the plasma and cerebrospinal fluid of patients with acute Japanese encephalitis. *J. Infect. Dis.* **155**:797–799.

Burke, D. S., and Monath, T. P. (2001). Flaviviruses. *In* "Fields Virology" (D. M. Knipe and P. M. Howley, eds.), 4th Ed., Vol. 1, pp. 1043–1125. Lippincott Williams & Wilkins, Philadelphia.

Byrne, S. N., Halliday, G. M., Johnston, L. J., and King, N. J. (2001). Interleukin-1beta but not tumor necrosis factor is involved in West Nile virus-induced Langerhans cell migration from the skin in C57BL/6 mice. *J. Invest. Dermatol.* **117**:702–709.

Camenga, D. L., Nathanson, N., and Cole, G. A. (1974). Cyclophosphamide-potentiated West Nile viral encephalitis: Relative influence of cellular and humoral factors. *J. Infect. Dis.* **130**:634–641.

Cane, P. A., and Gould, E. A. (1988). Reduction of yellow fever virus mouse neurovirulence by immunization with a bacterially synthesized nonstructural protein (NS1) fragment. *J. Gen. Virol.* **69**:1241–1246.

Cardosa, M. J., Porterfield, J. S., and Gordon, S. (1983). Complement receptor mediates enhanced flavivirus replication in macrophages. *J. Exp. Med.* **158**:258–263.

Cardosa, M. J., Gordon, S., Hirsch, S., Springer, T. A., and Porterfield, J. S. (1986). Interaction of West Nile virus with primary murine macrophages: Role of cell activation and receptors for antibody and complement. *J. Virol.* **57**:952–959.

Carroll, M. C. (1998). The role of complement and complement receptors in induction and regulation of immunity. *Annu. Rev. Immunol.* **16:**545–568.

Carson, P. J., Steidler, T., Patron, R., Tate, J. M., Tight, R., and Smego, R. A. Jr. (2003). Plasma cell pleiocytosis in cerebrospinal fluid in patients with West Nile virus encephalitis. *Clin. Infect. Dis.* **37:**e12–e15.

Casali, P., and Notkins, A. L. (1989). $CD5^{+}$ B lymphocytes, polyreactive antibodies and the human B-cell repertoire. *Immunol. Today* **10:**364–368.

Cernescu, C., Ruta, S. M., Tardei, G., Grancea, C., Moldoveanu, L., Spulbar, E., and Tsai, T. (1997). A high number of severe neurologic clinical forms during an epidemic of West Nile virus infection. *Rom. J. Virol.* **48:**13–25.

Chambers, T. J., and Nickells, M. (2001). Neuroadapted yellow fever virus 17D: Genetic and biological characterization of a highly mouse-neurovirulent virus and its infectious molecular clone. *J. Virol.* **75:**10912–10922.

Charlier, N., Leyssen, P., Paeshuyse, J., Drosten, C., Schmitz, H., Van Lommel, A., De Clercq, E., and Neyts, J. (2002). Infection of SCID mice with Montana Myotis leukoencephalitis virus as a model for flavivirus encephalitis. *J. Gen. Virol.* **83:**1887–1896.

Chaturvedi, U. C., Dhawan, R., Khanna, M., and Mathur, A. (1991). Breakdown of the blood-brain barrier during dengue virus infection of mice. *J. Gen. Virol.* **72:**859–866.

Chen, C. J., Liao, S. L., Kuo, M. D., and Wang, Y. M. (2000). Astrocytic alteration induced by Japanese encephalitis virus infection. *Neuroreport* **11:**1933–1937.

Chen, H. W., Pan, C. H., Huan, H. W., Liau, M. Y., Chiang, J. R., and Tao, M. H. (2001). Suppression of immune response and protective immunity to a Japanese encephalitis virus DNA vaccine by coadministration of an IL-12-expressing plasmid. *J. Immunol.* **166:**7419–7426.

Chen, H. W., Pan, C. H., Liau, M. Y., Jou, R., Tsai, C. J., Wu, H. J., Lin, Y. L., and Tao, M. H. (1999). Screening of protective antigens of Japanese encephalitis virus by DNA immunization: A comparative study with conventional viral vaccines. *J. Virol.* **73:**10137–10145.

Chen, L. K., Liao, C. L., Lin, C. G., Lai, S. C., Liu, C. I., Ma, S. H., Huang, Y. Y., and Lin, Y. L. (1996). Persistence of Japanese encephalitis virus is associated with abnormal expression of the nonstructural protein NS1 in host cells. *Virology* **217:**220–229.

Chen, Y., Maguire, T., Hileman, R. E., Fromm, J. R., Esko, J. D., Linhardt, R. J., and Marks, R. M. (1997). Dengue virus infectivity depends on envelope protein binding to target cell heparan sulfate. *Nature Med.* **3:**866–871.

Chu, C. T., Howell, D. N., Morenlander, J. C., Hulette, C. M., McLendon, R. E., and Miller, S. E. (1999). Electron microscopic diagrosis of human flavivirus encephalitis: use of confocal microscopy as an aid. *Am. J. Surg. Path.* **23:**1217–1226.

Co, M. D., Terajima, M., Cruz, J., Ennis, F. A., and Rothman, A. L. (2002). Human cytotoxic T lymphocyte responses to live attenuated 17D yellow fever vaccine: Identification of HLA-B35-restricted CTL epitopes on nonstructural proteins NS1, NS2b, NS3, and the structural protein E. *Virology* **293:**151–163.

Cole, G. A., and Nathanson, N. (1968). Potentiation of experimental arbovirus encephalitis by immunosuppressive doses of cyclophosphamide. *Nature* **220:**399–401.

Colombage, G., Hall, R., Pavy, M., and Lobigs, M. (1998). DNA-based and alphavirus-vectored immunisation with prM and E proteins elicits long-lived and protective immunity against the flavivirus, Murray Valley encephalitis virus. *Virology* **250:**151–163.

Cousens, L. P., Peterson, R., Hsu, S., Dorner, A., Altman, J. D., Ahmed, R., and Biron, C. A. (1999). Two roads diverged: Interferon alpha/beta- and interleukin 12-mediated

pathways in promoting T cell interferon gamma responses during viral infection. *J. Exp. Med.* **189:**1315–1328.
Crill, W. D., and Roehrig, J. T. (2001). Monoclonal antibodies that bind to domain III of dengue virus E glycoprotein are the most efficient blockers of virus adsorption to Vero cells. *J Virol.* **75:**7769–7773.
Cypess, R. H., Lubiniecki, A. S., and Hammon, W. M. (1973). Immunosuppression and increased susceptibility to Japanese B encephalitis virus in *Trichinella spiralis*-infected mice. *Proc. Soc. Exp. Biol. Med.* **143:**469–473.
Da Costa, X. J., Brockman, M. A., Alicot, E., Ma, M., Fischer, M. B., Zhou, X., Knipe, D. M., and Carroll, M. C. (1999). Humoral response to herpes simplex virus is complement-dependent. *Proc. Natl. Acad. Sci. USA* **96:**2708–11272.
Dalod, M., Salazar-Mather, T. P., Malmgaard, L., Lewis, C., Asselin-Paturel, C., Briere, F., Trinchieri, G., and Biron, C. A. (2002). Interferon alpha/beta and interleukin 12 responses to viral infections: Pathways regulating dendritic cell cytokine expression in vivo. *J. Exp. Med.* **195:**517–528.
Debnath, N. C., Tiernery, R., Sil, B. K., Wills, M. R., and Barrett, A. D. (1991). In vitro homotypic and heterotypic interference by defective interfering particles of West Nile virus. *J. Gen. Virol.* **72:**2705–2711.
De Nova-Ocampo, M., Villegas-Sepulveda, N., and del Angel, RM. (2002). Translation elongation factor-1alpha, La, and PTB interact with the 3′ untranslated region of dengue 4 virus RNA. *Virology* **295:**337–347.
Desai, A., Ravi, V., Chandramuki, A., and Gourie-Devi, M. (1995a). Proliferative response of human peripheral blood mononuclear cells to Japanese encephalitis virus. *Microbiol. Immunol.* **39:**269–273.
Desai, A., Ravi, V., Chandramuki, A., and Gourie-Devi, M. (1994a). Detection of immune complexes in the CSF of Japanese encephalitis patients: Correlation of findings with outcome. *Intervirology* **37:**352–355.
Desai, A., Ravi, V., Guru, S. C., Shankar, S. K., Kaliaperumal, V. G., Chandramuki, A., and Gourie-Devi, M. (1994b). Detection of autoantibodies to neural antigens in the CSF of Japanese encephalitis patients and correlation of findings with the outcome. *J. Neurol. Sci.* **122:**109–116.
Despres, P., Dietrich, J., Girard, M., and Bouloy, M. (1991). Recombinant baculoviruses expressing yellow fever virus E and NS1 proteins elicit protective immunity in mice. *J. Gen. Virol.* **72:**2811–2816.
Despres, P., Frenkiel, M.-P., Ceccaldi, P.-E., Dos Santos, C. D., and Deubel, V. (1998). Apoptosis in the mouse central nervous system in response to infection with mouse-neurovirulent dengue viruses. *J. Virol.* **72:**823.
Diamond, M., Roberts, T., Edgil, D., Lu, B., Ernst, J., and Harris, E. (2000a). Modulation of dengue virus infection in human cells by alpha, beta, and gamma interferons. *J. Virol.* **74:**4957–4966.
Diamond, M. S., Edgil, D., Roberts, T. G., Lu, B., and Harris, E. (2000b). Infection of human cells by dengue virus is modulated by different cell types and viral strains. *J. Virol.* **74:**7814–7823.
Diamond, M. S., and Harris, E. (2001). Interferon inhibits dengue virus infection by preventing translation of viral RNA through a PKR-independent mechanism. *Virology* **289:**297–311.
Diamond, M. S., Shrestha, B., Marri, A., Mahan, D., and Engle, M. (2003). B cells and antibody play critical roles in the immediate defense of disseminated infection by West Nile encephalitis virus. *J. Virol.* **77:**2578–2586.

Dominguez, C., and Baruch, E. (1963). Histopathology of the central nervous system in Swiss mice intracerebrally inoculated with 17-D strain of yellow fever virus. *Am. J. Trop Med. Hyg.* **12:**815–819.

Dropulic, B., and Masters, C. L. (1990). Entry of neurotropic arboviruses into the central nervous system: An in vitro study using mouse brain endothelium. *J. Infect. Dis.* **161:**685–691.

Douglas, M. W., Kesson, A. M., and King, N. J. (1994). CTL recognition of West Nile virus-infected fibroblasts is cell cycle dependent and is associated with virus-induced increases in class I MHC antigen expression. *Immunology* **82:**561–570.

Duarte dos Santos, C. N., Frenkiel, M.-P., Courageot, M.-P., Rocja, C. F. S., Vazeille-Falcoz, M.-C., Wien, M. W., Rey, F. A., Deubel, V., and Despres, P. (2000). Determinants in the envelope E protein and viral RNA helicase NS3 that influence the induction of apoptosis in response to infection with dengue type 1 virus. *Virology* **274:**292–308.

Edelman, R., Schneider, R. J., Vejjajiva, A., Pornpibul, R., and Voodhikul, P. (1976). Persistence of virus-specific IgM and clinical recovery after Japanese encephalitis. *Am. J. Trop. Med. Hyg.* **25:**733–738.

Eldadah, A. H., Nathanson, N., and Sarsitis, R. (1967a). Pathogenesis of West Nile virus encephalitis in mice and rats. I. Influence of age and species on mortality and infection. *Am. J. Epidemiol.* **86:**765–775.

Eldadah, A. H., and Nathanson, N. (1967b). Pathogenesis of West Nile virus encephalitis in mice and rats. II. Virus multiplication, evolution of immunofluorescence, and development of histological lesions in the brain. *Am. J. Epidemiol.* **86:**776–790.

Falconar, A. K. (1999). Identification of an epitope on the dengue virus membrane (M) protein defined by cross-protective monoclonal antibodies: Design of an improved epitope sequence based on common determinants present in both envelope (E and M) proteins. *Arch. Virol.* **144:**2313–2330.

Falgout, B., Bray, M., Schlesinger, J. J., and Lai, C. J. (1990). Immunization of mice with recombinant vaccinia virus expressing authentic dengue virus nonstructural protein NS1 protects against lethal dengue virus encephalitis. *J. Virol.* **64**(9)**:**4356–4363.

Fazakerley, J. (2001). Programmed cell death in virus infections of the nervous system. *Curr. Top. Micro. Immunol.* **253:**95–119.

Fazakerley, J. (2002). Pathogenesis of Semliki Forest virus encephalitis. *J. Neurovirol.* **8**(S2)**:**66–74.

Finley, K., and Riggs, N. (1980). Convalescence and sequellae. *In* "St. Louis Encephalitis" (T. P. Monath, ed.), pp. 535–550. APHA, Washington, DC.

Fischer, M. B., Ma, M., Goerg, S., Zhou, X., Xia, J., Finco, O., Han, S., Kelsoe, G., Howard, R. G., Rothstein, T. L., Kremmer, E., Rosen, F. S., and Carroll, M. C. (1996). Regulation of the B cell response to T-dependent antigens by classical pathway complement. *J. Immunol.* **157:**549–556.

Fitzgeorge, R., and Bradish, C. J. (1980). The in vivo differentiation of strains of yellow fever virus in mice. *J. Gen. Virol.* **46:**1–13.

Flugel, A., Schwaiger, F. W., Neumann, H., Medana, I., Willem, M., Wekerle, H., Kreutzberg, G. W., and Graeber, M. B. (2000). Neuronal FasL induces cell death of encephalitogenic T lymphocytes. *Brain Pathol.* **10:**353–364.

Fokina, G. I., Roikhel, V. M., Magaznik, S. S., Volkova, L. I., Frolova, T. V., and Pogodina, V. V. (1991). Development of antibodies to axonal neurofilaments in the progression of chronic tick-borne encephalitis. *Acta Virol.* **35:**458–463.

Freestone, D. S. (1994). Yellow fever vaccine. *In* "Vaccines" (S. A. Plotkin and E. M. Mortimer, eds.), 2nd Ed., pp. 741–779. Saunders, Philadelphia.

Gagnon, S. J., Ennis, F. A., and Rothman, A. L. (1999). Bystander target cell lysis and cytokine production by dengue virus-specific human $CD4^+$ cytotoxic T-lymphocyte clones. *J. Virol.* **73:**3623–3629.

Germi, R., Crance, J. M., Garin, D., Guimet, J., Lortat-Jacob, H., Ruigrok, R. W., Zarski, J. P., and Drouet, E. (2002). Heparan sulfate-mediated binding of infectious dengue virus type 2 and yellow fever virus. *Virology* **292:**162–168.

Glass, J. D., Samuels, O., and Rich, M. M. (2002). Poliomyelitis due to West Nile virus. *N. Engl. J. Med.* **347:**1280–1281.

Gobet, R., Cerny, A., Ruedi, E., Hengartner, H., and Zinkernagel, R. M. (1988). The role of antibodies in natural and acquired resistance of mice to vesicular stomatitis virus. *Exp. Cell. Biol.* **56:**175–180.

Gollins, S. W., and Porterfield, J. S. (1984). Flavivirus infection enhancement in macrophages: Radioactive and biological studies on the effect of antibody on viral fate. *J. Gen. Virol.* **65:**1261–1272.

Gould, E. A., and Buckley, A. (1989). Antibody-dependent enhancement of yellow fever and Japanese encephalitis virus neurovirulence. *J. Gen. Virol.* **70:**1605–1608.

Gould, E. A., Buckley, A., Groeger, B. K., Cane, P. A., and Doenhoff, M. (1987). Immune enhancement of yellow fever virus neurovirulence for mice: Studies of mechanisms involved. *J. Gen. Virol.* **68:**3105–3112.

Gresikova, M., Sekeyova, M., Stupalova, S., and Necas, S. (1975). Sheep milk-borne epidemic of tick-borne encephalitis in Slovakia. *Intervirology* **5:**57–61.

Greve, K. W., Houston, R. J., Adams, D., Stanford, M. S., Bianchini, K. J., Clancy, A., and Rabito, F. J., Jr. (2002). The neurobehavioural consequences of St. Louis encephalitis infection. *Brain Injury* **16:**917–927.

Griffin, D. E., and Hardwick, J. M. (1999). Perspective: Virus infections and the death of neurons. *Trends Microbiol.* **7:**155–160.

Griffin, D. E., Levine, B., Tyor, W., Ubol, S., and Desprès, P. (1997). The role of antibody in recovery from alphavirus encephalitis. *Immunol. Rev.* **159:**155–161.

Grossberg, S. E., and Scherer, W. F. (1966). The effect of host age, virus dose and route of inoculation on inapparent infection in mice with Japanese encephalitis virus. *Proc. Soc. Exp. Biol. Med.* **123:**118–124.

Grubeck-Loebenstein, B., and Wick, G. (2002). The aging of the immune system. *Adv. Immunol.* **80:**243–284.

Gunther, G., Haglund, M., Lindquist, L., Skoldenberg, B., and Forsgren, M. (1996). Intrathecal production of neopterin and beta 2 microglobulin in tick-borne encephalitis (TBE) compared to meningoencephalitis of other etiology. *Scand. J. Infect. Dis.* **28:**131–138.

Gunther, G., Haglund, M., Lindquist, L., Skoldenberg, B., and Forsgren, M. (1997). Intrathecal IgM, IgA and IgG antibody response in tick-borne encephalitis: Long-term follow-up related to clinical course and outcome. *Clin. Diagn. Virol.* **8:**17–29. Erratum in *Clin. Diagn. Virol.* **8:**167–168.

Gunther, G., Haglund, M., Mesko, L., Bremmer, S., Lindquist, L., Forsgren, M., Skoldenberg, B., and Rudberg, U. (1998). Regional cerebral blood flow scintigraphy in tick-borne encephalitis and other aseptic meningoencephalitis. *J. Nuclear Med.* **39:**2055–2061.

Gupta, A. K., and Pavri, K. M. (1987). Alteration in immune response of mice with dual infection of *Toxocara canis* and Japanese encephalitis virus. *Trans. R. Soc. Trop. Med. Hyg.* **81:**835–840.

Haglund, M., and Gunther, G. (2003). Tick-borne encephalitis: Pathogenesis, clinical course and long-term follow-up. *Vaccine* **21**(Suppl. 1):S11–S18.

Hajnicka, V., Kocakova, P., Slovak, M., Labuda, M., Fuchsberger, N., and Nuttall, P. A. (2000). Inhibition of the antiviral action of interferon by tick salivary gland extract. *Parasite Immunol.* **22:**201–206.

Halevy, M., Akov, Y., Ben-Nathan, D., Kobiler, D., Lachmi, B., and Lustig, S. (1994). Loss of active neuroinvasiveness in attenuated strains of West Nile virus: Pathogenicity in immunocompetent and SCID mice. *Arch. Virol.* **137:**355–370.

Halstead, S. B., and O'Rourke, E. J. (1977). Antibody-enhanced dengue virus infection in primate leukocytes. *Nature* **265:**739–741.

Halstead, S. B., Porterfield, J. S., and O'Rourke, E. J. (1980). Enhancement of dengue virus infection in monocytes by flavivirus antisera. *Am. J. Trop. Med. Hyg.* **29:**638–642.

Halstead, S. B., Venkateshan, C. N., Gentry, M. K., and Larsen, L. K. (1984). Heterogeneity of infection enhancement of dengue 2 strains by monoclonal antibodies. *J. Immunol.* **132:**1529–1532.

Hambleton, P., Stephenson, J. R., Baskerville, A., and Wiblin, C. N. (1983). Pathogenesis and immune response of vaccinated and unvaccinated rhesus monkeys to tick-borne encephalitis virus. *Infect. Imm.* **40:**995–1003.

Han, X. Y., Ren, Q. W., and Tsai, T. F. (1988). Serum and cerebrospinal fluid immunoglobulins M, A, and G in Japanese encephalitis. *J. Clin. Micro.* **26:**976–978.

Harinasuta, C., Nimmanitya, S., and Tisyakorn, U. (1985). The effect of interferon alpha on two cases of Japanese encephalitis in Thailand. *Southeast Asia J. Trop. Med. Pub. Health* **16:**332–336.

Harrington, D. G., Hilmas, D. E., Elwell, M. R., Whitmire, R. E., and Stephen, E. L. (1977). Intranasal infection of monkeys with Japanese encephalitis virus: Clinical response and treatment with a nuclease-resistant derivative of poly(I)-poly(C). *Am. J. Trop. Med. Hyg.* **26:**1191–1198.

Hase, T., Dubois, D. R., Summers, P. L., Downs, M. B., and Ussery, M. A. (1993). Comparison of replication rates and pathogenicities between the SA14 parent and SA14-14-2 vaccine strains of Japanese encephalitis virus in mouse brain neurons. *Arch. Virol.* **130:**131–143.

Havert, M. B., Schofield, B., Griffin, D. E., and Irani, D. N. (2000). Activation of divergent neuronal cell death pathways in different target cell populations during neuroadapted Sindbis virus infection in mice. *J. Virol.* **74:**5352–5356.

Hawkes, R. A. (1964). Enhancement of the infectivity of arboviruses by specific antisera produced in domestic fowls. *Aust. J. Exp. Biol. Med. Sci.* **42:**465–482.

Hayashi, K., and Arita, T. (1977). Experimental double infection of Japanese encephalitis virus and herpes simplex virus in mouse brain. *Jpn. J. Exp. Med.* **47:**9–13.

Heinz, F. X. (1986). Epitope mapping of flavivirus glycoproteins. *Adv. Vir. Res.* **31:**103–168.

Henchal, E. A., Henchal, L. S., and Schlesinger, J. J. (1988). Synergistic interactions of anti-NS1 monoclonal antibodies protect passively immunized mice from lethal challenge with dengue 2 virus. *J. Gen. Virol.* **69:**2101–2117.

Hickey, W. F. (1999). Leukocyte traffic in the central nervous system: The participants and their roles. *Semin. Immunol.* **11:**125–137.

Hilgard, P., and Stockert, R. (2000). Heparan sulfate proteoglycans initiate dengue virus infection of hepatocytes. *Hepatology* **32:**1069–1077.

Hill, A. B., Lobigs, M., Blanden, R. V., Kulkarni, A., and Müllbacher, A. (1993). The cellular immune response to flaviviruses. *In* "Viruses and the Cellular Immune Response" (D. Brian Thomas, ed.), pp. 363–388. Dekker, New York.

Hill, A. B., Müllbacher, A., Parrish, C., Coia, G., *et al.* (1992). Broad cross-reactivity with marked fine-specificity in the cytotoxic T cell response to flaviviruses. *J. Gen. Virol.* **73:**1115–1123.

Hirsch, M. R., Cailla, H., Wietzerbin, J., and Goridis, C. (1986). Interferon-alpha, -beta and -gamma induce (2′-5′) oligoadenylate synthetase in cultured mouse brain cells. *Neurosci. Lett.* **65:**139–144.

Hirsch, M. S., and Murphy, F. A. (1967). Effects of anti-thymocyte serum on 17D yellow fever infection in adult mice. *Nature* **216:**179–180.

Ho, L. J., Wang, J. J., Shaio, M. F., Kao, C. L., Chang, D. M., Han, S. W., and Lai, J. H. (2001). Infection of human dendritic cells by dengue virus causes cell maturation and cytokine production. *J. Immunol.* **166:**1499–1506.

Hofmann, H., Frisch-Niggemeyer, W., and Kunz, C. (1978). Protection of mice against tick-borne encephalitis by different classes of immunoglobulin. *Infection* **6:**154–157.

Huang, C. H., and Wong, C. (1963). Relation of the peripheral multiplication of Japanese B encephalitis virus to the pathogenesis of the infection in mice. *Acta Virol.* **7:**322–330.

Huang, Y. H., Lei, H. Y., Liu, H. S., Lin, Y. S., Liu, C. C., and Yeh, T. M. (2000). Dengue virus infects human endothelial cells and induces IL-6 and IL-8 production. *Am. J. Trop. Med. Hyg.* **63:**71–75.

Hubalek, Z., and Halouzka, J. (1999). West Nile fever—a reemerging mosquito-borne viral disease in Europe. *Emerg. Inf. Dis.* **5:**643–650.

Huy, B. V., Tu, H. C., Luan, T. V., and Lindquist, R. (1994). Early mental and neurological sequellae after Japanese B encephalitis. *Southeast Asia J. Trop. Med. Public Health* **25:**549–553.

Igarashi, A. (1979). Characteristics of *Aedes albopictus* cells persistently infected with dengue viruses. *Nature* **280:**690–691.

Iliyenko, V. I., Komandenko, N. I., Paltonov, V. G., Prozorova, I. N., and Panov, A. G. (1974). Investigation of the pathogenesis of chronic forms of tick-borne encephalitis. *Acta Virol.* **18:**341–346.

Innis, B. L. (1997). Antibody responses to dengue virus infection. *In* "Dengue and Dengue Hemorrhagic Fever" (D. J. Gubler and G. Kuno, eds.), pp. 221–243. CAB International, Wallingford, UK.

Isaeva, M. P., Leonova, G. N., Kozhemiako, V. B., Borisevich, V. G., Maistrovskaia, O. S., and Rasskazov, V. A. (1998). Apoptosis as a mechanism for the cytopathic action of tick-borne encephalitis virus. *Vopr. Virusol.* **43:**182–186.

Ishii, K., Matsunaga, Y., and Kono, R. (1968). Immunoglobulins produced in response to Japanese encephalitis virus infections of man. *J. Immunol.* **101:**770–775.

Ishii, T., Matsushita, M., and Hamada, S. (1977). Characteristic residual neuropathological features of Japanese B encephalitis. *Acta. Neuropathol. (Berl.)* **38:**181–186.

Iwamoto, M., Jernigan, D. B., Guasch, A., Trepka, M. J., Blackmore, C. G., Hellinger, W. C., Pham, S. M., Zaki, S., Lanciotti, R. S., Lance-Parker, S. E., DiazGranados, C. A., Winquist, A. G., Perlino, C. A., Wiersma, S., Hillyer, K. L., Goodman, J. L., Marfin, A. A., Chamberland, M. E., Petersen, L. R., and West Nile Virus in Transplant Recipients Investigation Team. (2003). Transmission of West Nile virus from an organ donor to four transplant recipients. *N. Engl. J. Med.* **348:**2196–2203.

Iwasaki, Y., Zhao, J.-X., Yamamoto, Y., and Konno, H. (1986). Immunohistochemical demonstration of viral antigens in Japanese encephalitis. *Acta Neuropathol.* **70:**79–81.

Iwasaki, Y., Sako, K., Tsunoda, I., and Ohara, Y. (1993). Phenotypes of mononuclear cell infiltrates in human central nervous system. *Acta. Neuropathol. (Berl.)* **85:**653–657.

Jacoby, R. O., Bhatt, P. N., and Schwartz, A. (1980). Protection of mice from lethal flavivirus encephalitis by adoptive transfer of splenic cells from donors infected with live virus. *J. Infect. Dis.* **141:**617–624.

Jan, J. T., Chatterjee, S., and Griffin, D. E. (2000). Sindbis virus entry into cells triggers apoptosis by activating sphingomyelinase, leading to the release of ceramide. *J. Virol.* **74:**6425–6432.

Jarman, R. V., Morgan, P. N., and Duffy, C. E. (1968). Persistence of West Nile virus in L-929 mouse fibroblasts. *Proc. Soc. Exp. Med. Biol.* **129:**633–637.

Johnson, A. J., and Roehrig, J. T. (1999). New mouse model for dengue virus vaccine testing. *J. Virol.* **73:**83–86.

Johnson, R. T., Burke, D. S., Elwell, M., Leake, C. J., Nisalak, *et al.* (1985). Japanese encephalitis: Immunocytochemical studies of viral antigen and inflammatory cells in fatal cases. *Ann. Neurol.* **18:**567–573.

Johnson, R. T., Intralawan, P., and Puapanwatton, S. (1986). Japanese encephalitis: Identification of inflammatory cells in cerebrospinal fluid. *Ann. Neurol.* **20:**691–695.

Johnston, L. J., Halliday, G. M., and King, N. J. (1996). Phenotypic changes in Langerhans' cells after infection with arboviruses: A role in the immune response to epidermally acquired viral infection. *J. Virol.* **70:**4761–4766.

Johnston, L. J., Halliday, G. M., and King, N. J. (2000). Langerhans cells migrate to local lymph nodes following cutaneous infection with an arbovirus. *J. Invest Dermatol.* **114:**560–568.

Johnston, C., Jiang, W., Chu, T., and Levine, B. (2001). Identification of genes involved in the host response to neurovirulent alphavirus infection. *J. Virol.* **75:**10431–10445.

Jordan, I., Briese, T., Fischer, N., Lau, J. Y., and Lipkin, W. I. (2000). Ribavirin inhibits West Nile virus replication and cytopathic effect in neural cells. *J. Infect. Dis.* **182:**1214–1217.

Kaiser, R., and Holzmann, H. (2000). Laboratory findings in tick-borne encephalitis: Correlation with clinical outcome. *Infection* **28:**78–84.

Kaiser, R. (2002). Tick-borne encephalitis (TBE) in Germany and clinical course of the disease. *Int. J. Med. Microbiol.* **291**(Suppl. 33)**:**58–61.

Kalita, J., and Misra, U. K. (2000). Markedly severe dystonia in Japanese encephalitis. *Mov. Disord.* **15:**1168–1172.

Kanesa-Thasan, N., Putnak, J. R., Mangiafico, J. A., Saluzzo, J. E., and Ludwig, G. V. (2002). Short report: Absence of protective neutralizing antibodies to West Nile virus in subjects following vaccination with Japanese encephalitis or dengue vaccines. *Am. J. Trop. Med. Hyg.* **66:**115–116.

Katz, E., and Goldblum, N. (1968). Establishment, steady state, and cure of a chronic infection of LLC cells with West Nile virus. *Arch. Ges. Virus Forsch.* **25:**69–82.

Kengsakul, K., Sathirapongsasuti, K., and Punyagupta, S. (2002). Fatal myeloencephalitis following yellow fever vaccination in a case with HIV infection. *J. Med. Assoc. Thai.* **85:**131–134.

Kerr, D. A., Larsen, T., Cook, S. H., Fanjiang, Y. R., Choi, E., Griffin, D. E., Hardwick, J. M., and Irani, D. N. (2002). BCL-2 and BAX protect adult mice from lethal Sindbis virus infection but do not protect spinal cord motor neurons or prevent paralysis. *J. Virol.* **76:**10393–10400.

Kesson, A. M., Blanden, R. V., and Müllbacher, A. (1987). The primary in vivo murine cytotoxic T cell response to the flavivirus, West Nile. *J. Gen. Virol.* **68:**2001–2006.

Kesson, A. M., and King, N. J. (2001). Transcriptional regulation of major histocompatibility complex class I by flavivirus West Nile is dependent on NF-kappaB activation. *J. Infect. Dis.* **184:**947–954.

Khan, N. U., Pulford, K. A., Farquharson, M. A., Howatson, A., Stewart, C., Jackson, R., McNicol, A. M., and Foulis, A. K. (1989). The distribution of immunoreactive interferon-alpha in normal human tissues. *Immunology* **66:**201–206.

Khozinsky, V. V., Semenov, B. F., Gresikova, M., Chunikhin, S. P., Sekeyova, M., and Kozuch, O. (1985). Role of macrophages in the pathogenesis of experimental tick-borne encephalitis in mice. *Acta Virol.* **29:**194–202.

Kimura-Kuroda, J., Ichikawa, M., Ogata, A., Nagashima, K., and Yasui, K. (1992). Specific tropism of Japanese encephalitis virus for developing neurons in primary rat brain culture. *Arch. Virol.* **130:**477–484.

Kimura-Kuroda, J., and Yasui, K. (1988). Protection of mice against Japanese encephalitis virus by passive administration with monoclonal antibodies. *J. Immunol.* **141:**3606–3610.

King, N. J., and Kesson, A. M. (1988). Interferon-independent increases in class I major histocompatibility complex antigen expression follow flavivirus infection. *J. Gen. Virol.* **69:**2535–2543.

King, N. J., Maxwell, L. E., and Kesson, A. M. (1989). Induction of class I major histocompatibility complex antigen expression by West Nile virus on gamma interferon-refractory early murine trophoblast cells. *Proc. Natl. Acad. Sci. USA* **86:**911–915.

Klimstra, W. B., Ryman, K. D., and Johnston, R. E. (1998). Adaptation of Sindbis virus to BHK cells selects for use of heparan sulfate as an attachment receptor. *J. Virol.* **72:**7357–7366.

Kobiler, D., Lustig, S., Gozes, Y., Ben-Nathan, D., and Akov, Y. (1989). Sodium dodecylsulphate induces a breach in the blood-brain barrier and enables a West Nile virus variant to penetrate into mouse brain. *Brain Res.* **496:**314–316.

Konishi, E., Kurane, I., Mason, P. W., Innis, B. L., and Ennis, F. A. (1995). Proliferative responses of human peripheral blood T lymphocytes against Japanese encephalitis antigens. *Am. J. Trop. Med. Hyg.* **53:**278–283.

Konishi, E., Kurane, I., Mason, P. W., Shope, R. E., and Ennis, F. A. (1997). Poxvirus-based Japanese encephalitis vaccine candidates induce JE virus-specific $CD8^+$ cytotoxic lymphocytes in mice. *Virology* **227:**353–360.

Konishi, E., Kurane, I., Mason, P. W., Shope, R. E., Kanesa-thasan, N., Smucny, J., Hoke, C. H., and Ennis, F. A. (1998a). Induction of Japanese encephalitis virus cytotoxic T lymphocytes in humans by poxvirus-based DNA vaccines. *Vaccine* **8:**842–849.

Konishi, E., Yamaoka, M., Win, K. S., Kurane, I., and Mason, P. W. (1998b). Induction of protective immunity against Japanese encephalitis in mice by immunization with a plasmid encoding JE virus premembrane and envelope genes. *J. Virol.* **72:**4925–4939.

Konishi, E., Yamaoka, M., Win, K.-S., Kurane, I., Takada, K., and Mason, P. W. (1999). The anamnestic neutralizing antibody response is critical for protection of mice from challenge following vaccination with a plasmid encoding the Japanese encephalitis virus premembrane and envelope genes. *J. Virol.* **73:**5527–5534.

Kopecky, J., Grubhoffer, L., Kovar, V., Jindrak, L., and Vokurkova, D. (1999). A putative host cell receptor for tick-borne encephalitis virus identified by anti-idiotypic antibodies and virus affinoblotting. *Intervirology* **42:**9–16.

Kopf, M., Abel, B., Gallimore, A., Carroll, M., and Bachmann, M. F. (2002). Complement component C3 promotes T-cell priming and lung migration to control acute influenza virus infection. *Nature Med.* **8:**373–378.

Korenberg, E. I., Gorban, L. Y., Kovalevskii, Y. V., Frizen, V. I., and Karavanov, A. S. (2001). Risk for human tick-borne encephalitis, borrelioses, and double infection in the pre-Ural region of Russia. *Emerg. Infect. Dis.* **7:**459–462.

Kozuch, O., Chunikhin, S. P., Gresikova, M., Nosek, J., Kurenkov, V. B., and Lysy, J. (1981). Experimental characteristics of viraemia caused by two strains of tick-borne encephalitis virus in small rodents. *Acta Virol.* **25:**219–224.

Kramer, L. D., and Bernard, K. A. (2001). West Nile virus infection in birds and mammals. *Ann. N. Y. Acad. Sci.* **951:**84–93.

Kreil, T. R., and Eibl, M. M. (1995). Viral infection of macrophages profoundly alters requirements for induction of nitric oxide synthesis. *Virology* **212:**174–178.

Kreil, T. R., and Eibl, M. M. (1996). Nitric oxide and viral infection: NO antiviral activity against a flavivirus in vitro, and evidence for contribution to pathogenesis in experimental infection in vivo. *Virology* **219:**304–306.

Kreil, T. R., Maier, E., Fraiss, S., Attakpah, E., Burger, I., Mannhalter, J. W., and Eibl, M. M. (1998a). Vaccination against tick-borne encephalitis virus, a flavivirus, prevents disease but not infection, although viremia is undetectable. *Vaccine* **16:**1083–1086.

Kreil, T. R., Maier, E., Fraiss, S., and Eibl, M. M. (1998b). Neutralizing antibodies protect against lethal flavivirus challenge but allow for the development of active humoral immunity to a nonstructural virus protein. *J. Virol.* **72:**3076–3081.

Kubes, M., Fuchsberger, N., Labuda, M., Zuffova, E., and Nuttall, P. A. (1994). Salivary gland extracts of partially fed *Dermacentor reticulatus* ticks decrease natural killer cell activity in vitro. *Immunology* **82:**113–116.

Kubes, M., Kocakova, P., Slovak, M., Slavikova, M., Fuchsberger, N., and Nuttall, P. A. (2002). Heterogeneity in the effect of different ixodid tick species on human natural killer cell activity. *Parasite Immunol.* **24:**23–28.

Kulkarni, A. B., Müllbacher, A., and Blanden, R. V. (1991a). Functional analysis of macrophages, B cells and splenic dendritic cells as antigen-presenting cells in West Nile virus-specific murine T lymphocyte proliferation. *Immunol. Cell. Biol.* **69:**71–80.

Kulkarni, A. B., Müllbacher, A., and Blanden, R. V. (1991b). In vitro T-cell proliferative response to the flavivirus, West Nile. *Viral Immunol.* **4:**73–82.

Kulkarni, A. B., Müllbacher, A., Parrish, C., *et al.* (1992). Analysis of murine major histocompatibility complex class II-restricted T-cell responses to the Kunjin virus using vaccinia virus expression. *J. Virol.* **66:**3583–3592.

Kumar, R., Mathur, A., Singh, K. B., *et al.* (1993). Clinical sequellae of Japanese encephalitis in children. *Ind. J. Med. Res.* **97:**9–13.

Kumar, S., Misra, U. K., Kalita, J., Salwani, V., Gupta, R. K., and Gujral, R. (1997). MRI in Japanese encephalitis. *Nueroradiology* **39:**180–184.

Kundig, T. M., Hengartner, H., and Zinkernagel, R. M. (1993). T cell-dependent IFN-gamma exerts an antiviral effect in the central nervous system but not in peripheral solid organs. *J. Immunol.* **150:**2316–2321.

Kuno, G., Hayes, C. G., and Chen, W. J. (1993). Cytokine concentrations in cerebrospinal fluid in flavivirus infections. *Southeast Asian J. Trop. Med. Public Health* **24:**781–782.

Kurane, I. (2002). Immune responses to Japanese encephalitis virus. *In* "Japanese Encephalitis and West Nile Viruses," (J. S. Mackenzie, A. D. T. Barrett, and V. Deubel, eds.), pp. 91–104. Springer-Verlag, Berlin.

Kurane, I., Hebblewaite, D., Brandt, W. E., and Ennis, F. A. (1984). Lysis of dengue virus-infected cells by natural cell-mediated cytotoxicity and antibody-dependent cell-mediated cytotoxicity. *J. Virol.* **52:**223–230.

Kurane, I., Innis, B. L., Nisalak, A., Hoke, C., Nimmannitya, S., Meager, A., and Ennis, F. A. (1989). Human T cell responses to dengue virus antigens: Proliferative responses and interferon gamma production. *J. Clin. Invest.* **83:**506–513.

Kurane, I., Brinton, M. A., Samson, A. L., and Ennis, F. A. (1991). Dengue virus-specific, human $CD4^+$ $CD8^-$ cytotoxic T-cell clones: Multiple patterns of virus cross-reactivity recognized by NS3-specific T-cell clones. *J. Virol.* **65:**1823–1828.

Kurane, I., Innis, B. L., Hoke, C. H., Jr., Eckels, K. H., Meager, A., Janus, J., and Ennis, F. A. (1995). T cell activation in vivo by dengue virus infection. *J. Clin. Lab. Immunol.* **46:**35–40.

Kurane, I., Janus, J., and Ennis, F. A. (1992). Dengue virus infection of human skin fibroblasts in vitro production of IFN-beta, IL-6 and GM-CSF. *Arch. Virol.* **124:**21–30.

Kutubuddin, M., Kolaskar, A. S., Galande, S., Gore, M. M., Ghosh, S. N., and Banerjee, K. (1991). Recognition of helper T cell epitopes in envelope (E) glycoprotein of Japanese encephalitis, West Nile and dengue viruses. *Mol. Immunol.* **28:**149–154.

Labrada, L., Liang, X. H., Zheng, W., Johnston, C., and Levine, B. (2002). Age-dependent resistance to lethal alphavirus encephalitis in mice: Analysis of gene expression in the central nervous system and identification of a novel interferon-inducible protective gene, mouse ISG12. *J. Virol.* **76:**688–703.

Labuda, M., Jones, L. D., Williams, T., and Nuttall, P. A. (1993). Enhancement of tick-borne encephalitis virus transmission by tick salivary gland extracts. *Med. Vet. Entomol.* **7:**193–196.

Lancaster, M. U., Hodgetts, S. I., Makenzie, J. S., and Urosevic, N. (1998). Characterization of defective viral RNA produced during persistent infection of Vero cells with Murray Valley encephalitis virus. *J. Virol.* **72:**2474–2482.

Lee, E., and Lobigs, M. (2000). Substitutions at the putative receptor-binding site of an encephalitic flavivirus alter virulence and host cell tropism and reveal a role for glycosaminoglycans in entry. *J. Virol.* **74:**8867–8875.

Lee, E., and Lobigs, M. (2002). Mechanism of virulence attenuation of glycosaminoglycan-binding variants of Japanese encephalitis virus and Murray Valley encephalitis virus. *J. Virol.* **76:**4901–4911.

Lee, H. W., and Scherer, W. F. (1961). The anamnestic antibody response to Japanese encephalitis virus in monkeys and its implications concerning naturally acquired immunity. *J. Immunol.* **86:**151–164.

Leis, A. A., Stokic, D. S., Polk, J. L., Dostrow, V., and Winkelmann, M. (2002). A poliomyelitis-like syndrome from West Nile virus infection. *N. Engl. J. Med.* **347:**1279–1280.

Leport, C., Janowski, M., Brun-Vezinet, F., Rouzioux, C., Rodhain, F., and Vilde, J. L. (1984). West Nile virus meningoencephalitis: value of inteferon assays in primary encephalitis. *Ann. Med. Interne (Paris)* **135:**460–463.

Levenbook, I. S., Pellen, L. J., and Ellisberg, B. L. (1987). The monkey safety test for neurovirulence of YF vaccines: The utility of quantitative clinical evaluation and histological examination. *J. Biol. Stand.* **15:**305–313.

Levine, B. (2002). Apoptosis in viral infections of neurons: A protective or pathologic host response. *Curr. Top. Micro. Immunol.* **265:**95–118.

Levine, B., and Griffin, D. E. (1992). Persistence of viral RNA in mouse brains after recovery from acute alphavirus encephalitis. *J. Virol.* **66:**6429–6435.

Levine, B., and Griffin, D. E. (1993). Molecular analysis of neurovirulent strains of Sindbis virus that evolve during persistent infection of scid mice. *J. Virol.* **67:**6872–6875.

Levine, B., Hardwick, J. M., Trapp, B. D., Crawford, T. O., Bollinger, R. C., and Griffin, D. E. (1991). Antibody-mediated clearance of alphavirus infection from neurons. *Science* **254:**856–860.

Levine, B., Huang, Q., Isaacs, J. T., Reed, J. C., Griffin, D. E., and Hardwick, J. M. (1993). Conversion of lytic to persistent alphavirus infection by the bcl-2 cellular oncogene. *Nature* **361**:739–742.

Lewis, J., Wesselingh, S. L., Griffin, D. E., and Hardwick, J. M. (1996). Alphavirus-induced apoptosis in mouse brains correlates with neurovirulence. *J. Virol.* **70**:1828–1835.

Leyssen, P., Drosten, C., Paning, M., Charlier, N., Paeshuyse, J., De Clercq, E., and Neyts, J. (2003b). Interferons, interferon inducers, and interferon-ribavirin in treatment of flavivirus-induced encephalitis in mice. *Antimicrob. Agents Chemother.* **47**:777–782.

Leyssen, P., Paeshuyse, J., Charlier, N., Van Lommel, A., Drosten, C., De Clercq, E., and Neyts, J. (2003a). Impact of direct virus-induced neuronal dysfunction and immunological damage on the progression of flavivirus (Modoc) encephalitis in a murine model. *J. Neurovirol.* **9**:69–78.

Li, W., Li, Y., Kedersha, N., Anderson, P., Emara, M., Swiderek, K. M., Moreno, G. T., and Brinton, M. A. (2002). Cell proteins TIA-1 and TIAR interact with the 3′ stem-loop of the West Nile virus complementary minus strand RNA and facilitate virus replication. *J. Virol.* **76**:11989–12000.

Liang, X. H., Kleeman, L. K., Jiang, H. H., Gordon, G., Goldman, J. E., Berry, G., Herman, B., and Levine, B. (1998). Protection against fatal Sindbis virus encephalitis by beclin, a novel Bcl-2-interacting protein. *J. Virol.* **72**:8586–8596.

Liao, C.-L., Lin, Y.-L., Shen, S.-C., Shen, Y.-Y., Su, H.-L., Huang, Y.-L., Ma, S.-H., Sun, Y.-C., Chen, K.-P., and Chen, L.-K. (1998). Antiapoptotic but not antiviral function of human bcl-2 assists establishment of Japanese encephalitis virus persistence in cultured cells. *J. Virol.* **72**:9844–9854.

Liao, C. L., Lin, Y. L., Wang, J. J., Huang, Y. L., Yeh, C. T., Ma, S. H., and Chen, L. K. (1997). Effect of enforced expression of human bcl-2 on Japanese encephalitis virus-induced apoptosis in cultured cells. *J. Virol.* **71**:5963–5971.

Libraty, D. H., Pichyangkul, S., Ajariyakhajorn, C., Endy, T. P., and Ennis, F. A. (2001). Human dendritic cells are activated by dengue virus infection: Enhancement by gamma interferon and implications for disease pathogenesis. *J. Virol.* **75**:3501–3508.

Licon Luna, R. M., Lee, E., Müllbacher, A., Blanden, R. V., Langman, R., and Lobigs, M. (2002). Lack of both Fas ligand and perforin protects from flavivirus-mediated encephalitis in mice. *J. Virol.* **76**:3202–3211.

Lin, M. T., Hinton, D. R., Marten, N. W., Bergmann, C. C., and Stohlman, S. A. (1999). Antibody prevents virus reactivation within the central nervous system. *J. Immunol.* **162**:7358.

Lin, Y. L., Huang, Y. L., Ma, S. H., Yeh, C. T., Chiou, S. Y., Chen, L. K., and Liao, C. L. (1997). Inhibition of Japanese encephalitis virus infection by nitric oxide: Antiviral effect of nitric oxide on RNA virus replication. *J. Virol.* **71**:5227–5235.

Lin, Y. L., Liao, C. L., Chen, L. K., Yeh, C. T., Liu, C. I., Ma, S. H., Huang, Y. Y., Huang, Y. L., Kao, C. L., and King, C. C. (1998). Study of dengue virus infection in SCID mice engrafted with human K562 cells. *J. Virol.* **72**:9729–9737.

Liou, M. L., and Hsu, C. Y. (1998). Japanese encephalitis virus is transported across the cerebral blood vessels by endocytosis in mouse brain. *Cell Tissue Res.* **293**:389–394.

Liu, M. T., Armstrong, D., Hamilton, T. A., and Lane, T. E. (2001). Expression of Mig (monokine induced by interferon gamma) is important in T lymphocyte recruitment and host defense following viral infection of the central nervous system. *J. Immunol.* **166**:1790–1795.

Liu, M. T., Chen, B. P., Oertel, P., Buchmeier, M. J., Armstrong, D., Hamilton, T. A., and Lane, T. E. (2000). The T cell chemoattractant IFN-inducible protein 10 is essential in host defense against viral-induced neurological disease. *J. Immunol.* **165:**2327–2330.

Liu, M. T., and Lane, T. E. (2001). Chemokine expression and viral infection of the central nervous system: Regulation of host defense and neuropathology. *Immunol. Res.* **24:**111–119.

Liu, T., and Chambers, T. J. (2001). Yellow fever virus encephalitis: Properties of the brain-associated T-cell response during virus clearance in normal and gamma interferon-deficient mice and requirement for $CD4^+$ lymphocytes. *J. Virol.* **75:**2107–2118.

Liu, Y., Blanden, R. V., and Müllbacher, A. (1989a). Identification of cytolytic lymphocytes in West Nile virus-infected murine central nervous system. *J. Gen. Virol.* **70:**565–573.

Liu, Y., King, N., Kesson, A., Blanden, R. V., and Müllbacher, A. (1988). West Nile virus infection modulates the expression of class I and class II MHC antigens on astrocytes in vitro. *Ann. N. Y. Acad. Sci.* **540:**483–485.

Liu, Y., King, N., Kesson, A., Blanden, R. V., and Müllbacher, A. (1989b). Flavivirus infection up-regulates the expression of class I and class II major histocompatibility antigens on and enhances T cell recognition of astrocytes in vitro. *J. Neuroimmunol.* **21:**157–168.

Lobigs, M., Müllbacher, A., Blanden, R. V., Hammerling, G. J., and Momburg, F. (1999). Antigen presentation in Syrian hamster cells: Substrate selectivity of TAP controlled by polymorphic residues in TAP1 and differential requirements for loading of H2 class I molecules. *Immunogenetics* **49:**931–941.

Lobigs, M., Müllbacher, A., Wang, Y., Pavy, M., and Lee, E. (2003a). Role of type I and type II interferon responses in recovery from infection with an encephalitic flavivirus. *J. Gen. Virol.* **84:**567–572.

Lobigs, M., Pavy, M., and Hall, R. (2003b). Cross-protective and infection-enhancing immunity in mice vaccinated against flaviviruses belonging to the Japanese encephalitis virus serocomplex. *Vaccine* **21:**1572–1579.

Loginova, N. V., Deryabin, P. G., Mikhailova, G. R., Tsareva, A. A., and Buinitskaya, O. B. (1980). Chronic infection of HeLa cells with Japanese encephalitis virus: General characteristics of the system. *Acta. Virol.* **24:**399–405.

Loke, H., Bethell, D. B., Phuong, C. X., Dung, M., Schneider, J., White, N. J., Day, N. P., Farrar, J., and Hill, A. V. (2001). Strong HLA class I-restricted T cell responses in dengue hemorrhagic fever: A double-edged sword? *J. Infect. Dis.* **184:**1369–1373.

Lubiniecki, A. S., Cypess, R. H., and Lucas, J. P. (1974). Synergistic interaction of two agents in mice: Japanese B encephalitis virus and *Trichinella spiralis*. *Am. J. Trop. Med. Hyg.* **23:**235–241.

Luby, J. P., Miller, G., Gardner, P., Pigford, C. A., Henderson, B. E., and Eddins, D. (1967). The epidemiology of St. Louis encephalitis in Houston, Texas, 1964. *Am. J. Epidemiol.* **86:**584–597.

Luby, J. P., Stewart, W. E., 2nd, Sulkin, S. E., and Sanford, J. P. (1969). Interferon in human infections with St. Louis encephalitis virus. *Ann. Intern. Med.* **71:**703–709.

Lucas, M., Mashimo, T., Frenkiel, M. P., Simon-Chazottes, D., Montagutelli, X., Ceccaldi, P. E., Guenet, J. L., and Despres, P. (2003). Infection of mouse neurones by West Nile virus is modulated by the interferon-inducible 2′-5′ oligoadenylate synthetase 1b protein. *Immunol. Cell Biol.* **81:**230–236.

Ludewig, B., Ehl, S., Karrer, U., Odermatt, B., Hengartner, H., and Zinkernagel, R. M. (1998). Dendritic cells efficiently induce protective immunity. *J. Virol.* **72:**3812–3818.

Lustig, S., Danenberg, H. D., Kafri, Y., Kobiler, D., and Ben-Nathan, D. (1992). Viral neuroinvasion and encephalitis induced by lipopolysaccharide and its mediators. *J. Exp. Med.* **176:**707–712.

Malkova, D., and Frankova, V. (1959). The lymphatic system in the development of experimental tick-bone encephalitis in mice. *Acta Virol.* **3:**210–214.

Mandl, C. W., Kroschewski, H., Allison, S. L., Kofler, R., Holzmann, H., Meixner, T., and Heinz, F. X. (2001). Adaptation of tick-borne encephalitis virus to BHK-21 cells results in the formation of multiple heparan sulfate binding sites in the envelope protein and attenuation in vitro. *J. Virol.* **75:**5627–5637.

Manuelidis, E. E. (1956). Neuropathology of experimental West Nile virus infection in monkeys. *J. Neuropathol Exp. Neurol.* **15:**448–460.

Marianneau, P., Steffan, A. M., Royer, C., Drouet, M. T., Jaeck, D., Kirn, A., and Deubel, V. (1999). Infection of primary cultures of human Kupffer cells by dengue virus: No viral progeny synthesis, but cytokine production is evident. *J. Virol.* **73:**5201–5206.

Markoff, L. (2000). Points to consider in the development of a surrogate for efficacy of novel Japanese encephalitis virus vaccines. *Vaccine* **18**(Suppl. 2)**:**26–32.

Martin, D. A., Biggerstaff, B. J., Allen, B., Johnson, A. J., Lanciotti, R. S., and Roehrig, J. T. (2002). Use of immunoglobulin M cross-reactions in differential diagnosis of human flaviviral encephalitis infections in the United States. *Clin. Diagn. Lab. Immunol.* **9:**544–549.

Martinez-Barragan, J. J., and del Angel, R. M. (2001). Identification of a putative coreceptor on Vero cells that participates in dengue 4 virus infection. *J. Virol.* **75:**7818–7827.

Mashimo, T., Lucas, M., Simon-Chazottes, D., Frenkiel, M.-P., Montagutelli, X., Ceccaldi, P.-E., Deubel, V., Guenet, J.-L., and Despres, P. (2002). A nonsense mutation in the gene encoding 2′-5′-oligoadenylate synthetase/L1 isoform is associated with West Nile virus susceptibility in laboratory mice. *Proc. Natl. Acad. Sci. USA* **99:**11311–11316.

Massa, P. T., Whitney, L. W., Wu, C., Ropka, S. L., and Jarosinski, K. W. (1999). A mechanism for selective induction of 2′-5′ oligoadenylate synthetase, anti-viral state, but not MHC class I genes by interferon-beta in neurons. *J. Neurovirol.* **5:**161–171.

Mathews, J. H., Allan, J. E., Roehrig, J. T., Brubaker, J. R., Uren, M. F., and Hunt, A. R. (1991). T-helper cell and associated antibody response to synthetic peptides of the E glycoprotein of Murray Valley encephalitis virus. *J. Virol.* **65:**5141–5148.

Mathews, J. H., Roehrig, J. T., Brubaker, J. R., Hunt, A. R., and Allan, J. E. (1992). A synthetic peptide to the E glycoprotein of Murray Valley encephalitis virus defines multiple virus-reactive T- and B-cell epitopes. *J. Virol.* **66:**6555–6562.

Mathews, V., Robertson, T., Kendrick, T., Abdo, M., Papadimitrou, J., and McMinn, P. (2000). Morphological features of Murray Valley encephalitis virus infection in the central nervous system of Swiss mice. *Int. J. Exp. Pathol.* **81:**31–40.

Mathur, A., Arora, K. L., Rawat, S., and Chaturvedi, U. C. (1986). Persistence, reactivation and latency of Japanese encephalitis virus infection in mice. *J. Gen. Virol.* **67:**381–385.

Mathur, A., and Chaturvedi, U. C. (1992). Breakdown of the blood-brain barrier by virus-induced cytokine during Japanese encephalitis virus infection. *Int. J. Exp. Pathol.* **73:**603–611.

Mathur, A., Kulshreshtha, R., and Chaturvedi, U. C. (1989). Evidence of latency of Japanese encephalitis virus in T lymphocytes. *J. Gen. Virol.* **70:**461–465.

McMinn, P. C. (1997). The molecular basis of virulence of the encephalitogenic flaviviruses. *J. Gen. Virol.* **78:**2711–2722.

McMinn, P. C., Dalgarno, L., and Weir, R. C. (1996). A comparison of the spread of Murray Valley encephalitis viruses of high or low neuroinvasiveness in the tissues of Swiss mice after peripheral inoculation. *Virology* **220:**414–423.

Medana, I., Li, Z., Flugel, A., Tschopp, J., Wekerle, H., and Neumann, H. (2001). Fas ligand (CD95L) protects neurons against perforin-mediated T lymphocyte cytotoxicity. *J. Immunol.* **167:**674–681.

Medana, I. M., Gallimore, A., Oxenius, A., Martinic, M. M. A., Wekerle, H., and Neumann, H. (2000). MHC class I-restricted killing of neurons by virus-specific $CD8^+$ T lymphocytes is effected through the Fas/FasL, but not the perforin pathway. *Eur. J. Immunol.* **30:**3623–3633.

Misra, U. K., and Kalita, J. (1997a). Movement disorders in Japanese encephalitis. *J. Neurol.* **244:**299–303.

Misra, U. K., and Kalita, J. (1997b). Anterior horn cells are also involved in Japanese encephalitis. *Acta Neurol. Scand.* **96:**114–117.

Miyake, M. (1964). The pathology of Japanese encephalitis. *Bull World Health Org.* **30:**153–160.

Momburg, F., Müllbacher, A., and Lobigs, M. (2001). Modulation of transporter associated with antigen processing (TAP)-mediated peptide import into the endoplasmic reticulum by flavivirus infection. *J. Virol.* **75:**5663–5671.

Monath, T. P. (1971). Neutralizing antibody responses in the major immunoglobulin classes to yellow fever 17D vaccination of humans. *Am. J. Epidemiol.* **93:**122–129.

Monath, T. P. (1986). Pathobiology of the flaviviruses. *In* "The Togaviridae and the Flaviviridae" (S. Schlesinger and M. J. Schlesinger, eds.), pp. 375–440. Plenum Press, New York.

Monath, T. P., and Borden, E. C. (1971). Effects of thorotrast on humoral antibody, viral multiplication and interferon during infection with St. Louis encephalitis virus in mice. *J. Infect. Dis.* **123:**297–300.

Monath, T. P., Cropp, C. B., and Harrison, A. K. (1983). Mode of entry of a neurotropic arbovirus into the central nervous system: Reinvestigation of an old controversy. *Lab. Investigation* **48:**399–410.

Monath, T. P., and Heinz, F. X. (1996). Flaviviruses. *In* "Fields Virology" (B. N. Fields, D. M. Knipe, P. M. Howley, *et al.*, eds.), pp. 961–1034. Lippincott-Raven, Philadelphia.

Müllbacher, A., and King, N. J. C. (1989). Natural killer cell lysis of target cells is influenced by β_2-microglobulin expression. *Scand. J. Immunology* **30:**21–29.

Müllbacher, A., and Lobigs, M. (1995). Up-regulation of MHC class I by flavivirus-induced peptide translocation into the endoplasmic reticulum. *Immunity* **3:**207–214.

Munoz, M. L., Cisneros, A., Cruz, J., Das, P., Tovar, R., and Ortega, A. (1998). Putative dengue virus receptors from mosquito cells. *FEMS Microbiol. Lett.* **168:**251–258.

Murali-Krishna, K., Ravi, V., and Manjunath, R. (1996). Protection of adult but not newborn mice against lethal intracerebral challenge with Japanese encephalitis virus by adoptively transferred virus-specific cytotoxic T lymphocytes: Requirement for $L3T4^+$ T cells. *J. Gen. Virol.* **77:**705–714.

Murgod, U. A., Muthane, U. B., Ravi, V., Radhesh, S., and Desai, A. (2001). Persistent movement disorders following Japanese encephalitis. *Neurology* **57:**2313–2315.

Murphy, F. A., Harrison, A. K., Gary, G. W., Jr., Whitfield, S. G., and Forrester, F. T. (1968). St. Louis encephalitis virus infection in mice: Electron microscopic studies of central nervous system. *Lab. Invest.* **19:**652–662.

Myint, K. S., Raengsakulrach, B., Young, G. D., Gettyayacamin, M., Ferguson, L. M., Innis, B. L., Hoke, C. H., Jr., and Vaughn, D. W. (1999). Production of lethal infection that resembles fatal human disease by intranasal inoculation of macaques with Japanese encephalitis virus. *Am. J. Trop. Med. Hyg.* **60:**338–342.

Nargi-Aizenman, J. L., and Griffin, D. E. (2001). Sindbis virus-induced neuronal death is both necrotic and apoptotic and is ameliorated by N-methyl-D-aspartate receptor antagonists. *J. Virol.* **75:**7114–7121.

Nathanson, N., and Cole, G. A. (1970). Fatal Japanese encephalitis virus infection in immunosuppressed spider monkeys. *Clin. Exp. Immunol.* **6:**161–166.

Nathanson, N., Davis, M., Thind, I. S., *et al.* (1966). Histologic studies of the neurovirulence of group B arboviruses. II. Selection of indicator centers. *Am. J. Epidemiol.* **84:**524–535.

Navarro-Sanchez, E., Altmeyer, R., Amara, A., Schwartz, O., Fieschi, F., Virelizier, J. L., Arenzana-Seisdedos, F., and Despres, P. (2003). Dendritic-cell-specific ICAM3-grabbing non-integrin is essential for the productive infection of human dendritic cells by mosquito-cell-derived dengue viruses. *EMBO Rep.* **4**(Suppl)**:**1–6.

Navin, T. R., Krug, E. C., and Pearson, R. D. (1989). Effect of immunoglobulin M from normal human serum on *Leishmania donovani* promastigote agglutination, complement-mediated killing, and phagocytosis by human monocytes. *Infect. Immun.* **57:**1343–1346.

Neogi, D. K., Bhattacharya, N., Chakrabarti, T., and Mukherjee, K. K. (1998). Detection of HIV seropositivity during an outbreak of Japanese encephalitis in Manipur. *J. Commun. Dis.* **30:**113–116.

Neumann, H., Cavalie, A., Jenne, D. E., and Wekerle, H. (1995). Induction of MHC class I genes in neurons. *Science* **269:**549–552.

Ni, H., and Barrett, A. D. T. (1998). Attenuation of Japanese encephalitis virus by selection of its mouse brain membrane receptor preparation escape mutants. *Virology* **241:**30–36.

Ni, H., Ryman, K. D., Wang, H., Saeed, M. F., Hull, R., Wood, D., Minor, P. D., Watowich, S. J., and Barrett, A. D. T. (2000). Interaction of yellow fever virus French neurotropic vaccine strain with monkey brain: Characterization of monkey brain membrane receptor escape variants. *J. Virol.* **74:**2903–2906.

Nir, Y., Beemer, A., and Goldwasser, R. A. (1965). West Nile virus infection in mice following exposure to a viral aerosol. *Br. J. Exp. Pathol.* **46:**443–449.

Njenga, M. K., Pease, L. R., Wettstein, P., Mak, T., and Rodriguez, M. (1997). Interferon alpha/beta mediates early virus-induced expression of H-2D and H-2K in the central nervous system. *Lab. Invest.* **77:**71–84.

Oberhaus, S. M., Smith, R. L., Clayton, G. H., Dermody, T. S., and Tyler, K. L. (1997). Reovirus infection and tissue injury in the mouse central nervous system are associated with apoptosis. *J. Virol.* **71:**2100–2106.

Ochsenbein, A. F., Fehr, T., Lutz, C., Suter, M., Brombacher, F., Hengartner, H., and Zinkernagel, R. M. (1999a). Control of early viral and bacterial distribution and disease by natural antibodies. *Science* **286:**2156–2159.

Ochsenbein, A. F., Pinschewer, D. D., Odermatt, B., Carroll, M. C., Hengartner, H., and Zinkernagel, R. M. (1999b). Protective T cell-independent antiviral antibody responses are dependent on complement. *J. Exp. Med.* **190:**1165–1174.

Ochsenbein, A. F., and Zinkernagel, R. M. (2000). Natural antibodies and complement link innate and acquired immunity. *Immunol. Today* **21**(12)**:**624–630.

Odeola, H. A., and Oduye, O. O. (1977). West Nile virus infection of adult mice by oral route. *Arch. Virol.* **54:**251–253.

Ogata, A., Nagashima, K., Hall, W. W., Ichikawa, M., Kimura-Kuroda, J., and Yasui, K. (1991). Japanese encephalitis virus neurotropism is dependent on the degree of neuronal maturity. *J. Virol.* **65:**880–886.

Ogata, A., Tashiro, K., and Pradhan, S. (2000). Parkinsonism due to predominant involvement of substantia nigra in Japanese encephalitis. *Neurology* **55:**602.

Ogawa, M., Okubo, H., Tsuji, Y., *et al.* (1973). Chronic progressive encephalitis occurring 13 years after Russian spring-summer encephalitis. *J. Neurol. Sci.* **19:**363–373.

Okhuysen, P. C., Crane, J. K., and Pappas, J. (1993). St. Louis encephalitis in patients with human immuodeficiency virus infection. *Clin. Infect. Dis.* **17:**140–141.

O'Leary, J. L., Smith, M. G., and Reames, H. R. (1942). Influence of age on susceptibility of mice to St. Louis encephalitis virus and on the distribution of lesions. *J. Exp. Med.* **75:**233–247.

Oliver, K. R., and Fazakerley, J. K. (1998). Transneuronal spread of Semliki Forest virus in the developing mouse olfactory system is determined by neuronal maturity. *Neuroscience* **82:**867–877.

Oliver, K. R., Scallan, M. F., Dyson, H., and Fazakerley, J. K. (1997). Susceptibility to a neurotropic virus and its distribution in the developing brain is a function of CNS maturity. *J. Neurovirol.* **3:**38–48.

Orange, J. S., Fassett, M. S., Koopman, L. A., Boyson, J. E., and Strominger, J. L. (2002). Viral evasion of natural killer cells. *Nature Immunol.* **3:**1006–1012.

Pan, C.-H., Chen, H.-W., Huang, H.-W., and Tao, M.-H. (2001). Protective mechanisms induced by a Japanese encephalitis virus DNA vaccine: Requirement for antibody but not $CD8^+$ cytotoxic T-cell responses. *J. Virol.* **75:**11457–11463.

Panasiuk, B., Prokopowicz, D., and Panasiuk, A. (2003). Immunological response in HIV-positive patients vaccinated against tick-borne encephalitis. *Infection* **31:**45–46.

Parquet, M. C., Kumatori, A., Hasebe, F., Morita, K., and Igarashi, A. (2001). West Nile virus-induced bax-dependent apoptosis. *FEBS Lett.* **500:**17–24.

Pavri, K. M., Ghalsasi, G. R., Dastur, D. K., Goverdhan, M. K., and Lalitha, V. S. (1975). Dual infections of mice: Visceral larva migrans and sublethal infection with Japanese encephalitis virus. *Trans. R. Soc. Trop. Med. Hyg.* **69:**99–110.

Peiris, J. S. M., Gordon, S., Unkeless, J. C., and Porterfield, J. S. (1981). Monoclonal anti-Fc receptor IgG blocks antibody enhancement of viral replication in macrophages. *Nature* **289:**189–191.

Perelygin, A. A., Scherbik, S. V., Zhulin, I. B., Stockman, B. M., Li, Y., and Brinton, M. A. (2002). Positional cloning of the murine flavivirus resistance gene. *Proc. Natl. Acad. Sci. USA* **99:**9322–9327.

Phillpotts, R. J., Jones, L. D., Lukaszewski, R. A., Lawrie, C., and Brooks, T. J. (2003). Antibody and interleukin-12 treatment in murine models of encephalitogenic flavivirus (St. Louis encephalitis, tick-borne encephalitis) and alphavirus (Venezuelan equine encephalitis) infection. *J. Interferon Cytokine Res.* **23:**47–50.

Pincus, S., Mason, P. W., Konishi, E., Fonseca, B. A., Shope, R. E., Rice, C. M., and Paoletti, E. (1992). Recombinant vaccinia virus producing the prM and E proteins of yellow fever virus protects mice from lethal yellow fever encephalitis. *Virology* **187:**290–297.

Pogodina, V. V., Frolova, M. P., Malenko, G. V., Fokina, G. I., Koreshkova, G. V., Kiseleva, L. L., Bochkova, N. G., and Ralph, N. M. (1983). Study on West Nile virus persistence in monkeys. *Arch. Virol.* **75:**71–86.

Pogodina, V. V., Levina, L. S., Fokina, G. I., *et al.* (1981). Persistence of tick-borne encephalitis virus in monkeys. III. Phenotypes of the persisting virus. *Acta Virol.* **25:**352–360.

Poidinger, M., Coelen, R. J., and Mackenzie, J. S. (1991). Persistent infection of Vero cells by the flavivirus Murray Valley encephalitis virus. *J. Gen. Virol.* **72:**573–578.

Popko, B., Corbin, J. G., Baerwald, K. D., Dupree, J., and Garcia, A. M. (1997). The effects of interferon-gamma on the central nervous system. *Mol. Neurobiol.* **14:**19–35.

Potula, R., Badrinath, S., and Srinivasan, S. (2003). Japanese encephalitis in and around Pondicherry, South India: A clinical appraisal and prognostic indicators for the outcome. *J. Trop. Pediatr.* **49:**48–53.

Powell, K. E., and Kappus, K. D. (1978). Epidemiology of St. Louis encephalitis and other acute encephalitides. *Adv. Neurol.* **19:**197–213.

Pradhan, S., Pandey, N., Shashank, S., Gupta, R. K., and Mathur, A. (1999). Parkinsonism due to predominant involvement of substantia nigra in Japanese encephalitis. *Neurology* **53:**1781–1786.

Prikhod'ko, G. G., Prikhod'ko, E. A., Cohen, J. I., and Pletnev, A. G. (2001). Infection with Langat flavivirus or expression of the envelope protein induces apoptotic cell death. *Virology* **286:**328–335.

Prikhod'ko, G. G., Prikhod'ko, E. A., Pletnev, A. G., and Cohen, J. I. (2002). Langat flavivirus protease NS3 binds caspase-8 and induces apoptosis. *J. Virol.* **76:**5701–5710.

Putnak, J. R., and Schlesinger, J. J. (1990) Protection of mice against yellow fever virus encephalitis by immunization with a vaccinia virus recombinant encoding the yellow fever virus non-structural proteins, NS1, NS2a and NS2b. *J. Gen. Virol.* **71:** 1697–1702.

Raengsakulrach, B., Nisalak, A., Gettayacamin, M., Thirawuth, V., Young, G. D., Myint, K. S., Ferguson, L. M., Hoke, C. H. Jr., Innis, B. L., and Vaughn, D. W. (1999). An intranasal challenge model for testing Japanese encephalitis vaccines in rhesus monkeys. *Am. J. Trop. Med. Hyg.* **60:**329–337.

Raghupathy, R., Chaturvedi, U. C., A1-Sayer, H., Elbishbishi, E. A., Agarwal, R., Nagar, R., Kapoor, S., Misra, A., Mathur, A., Nusrat, H., Azizieh, F., Khan, M. A., and Mustafa, A. S. (1998). Elevated levels of IL-8 in dengue hemorrhagic fever. *J. Med. Virol.* **56:**280–285.

Ramakrishna, C., Ravi, V., Desai, A., Subbakrishna, D. K., Shankar, S. K., and Chandramuki, A. (2003). T helper responses to Japanese encephalitis virus infection are dependent on the route of inoculation and the strain of mouse used. *J. Gen. Virol.* **84:**1559–1567.

Ramos-Castaneda, J., Imbert, J. L., Barron, B. L., and Ramos, C. (1997). A 65-kDa trypsin-sensible membrane cell protein as a possible receptor for dengue virus in cultured neuroblastoma cells. *J. Neurovirol.* **3:**435–440.

Randolph, V. A., and Hardy, J. L. (1988a). Establishment and characterization of St. Louis encephalitis virus persistent infections in *Aedes* and *Culex* mosquito cell lines. *J. Gen. Virol.* **69:**2189–2198.

Randolph, V. A., and Hardy, J. L. (1988b). Phenotypes of St. Louis encephalitis virus mutants produced in persistently infected mosquito cell cultures. *J. Gen Virol.* **69:**2199–2207.

Raung, S. L., Kuo, M. D., Wang, Y. M., and Chen, C. J. (2001). Role of reactive oxygen intermediates in Japanese encephalitis virus infection in murine neuroblastoma cells. *Neurosci. Lett.* **315:**9–12.

Ravi, V., Desai, A. S., Shenoy, P. K., Satishchandra, P., Chandramuki, A., and Gourie-Devi, M. (1993). Persistence of Japanese encephalitis virus in the human nervous system. *J. Med. Virol.* **40:**326–329.

Ravi, V., Parida, S., Desai, A., Chandramuki, A., Gourie-Devi, M., and Grau, G. E. (1997). Correlation of tumor necrosis factor levels in the serum and cerebrospinal fluid with clinical outcome in Japanese encephalitis patients. *J. Med. Virol.* **51:**132–136.

Receveur, M. C., Thiebaut, R., Vedy, S., Malvy, D., Mercie, P., and Bras, M. L. (2000). Yellow fever vaccination of human immunodeficiency virus-infected patients: Report of 2 cases. *Clin. Infect. Dis.* **31:**E7–E8.

Reiss, C. S., Chesler, D. A., Hodges, J., Ireland, D. D. C., and Chen, N. (2002). Innate immune responses in viral encephalitis. *Curr. Top. Microbiol. Immunol.* **265:**63–84.

Reyes, M. G., Gardner, J. J., Poland, J. D., and Monath, T. P. (1981). St. Louis encephalitis: Quantitative histologic and immunofluorescent studies. *Arch. Neurol.* **38:**329–334.

Richter, R. W. (1961). Neurologic sequellae of Japanese B encephalitis. *Neurology* **11:**553–559.

Roehrig, J. T., Hunt, A. R., Johnson, A. J., and Hawkes, R. A. (1989). Synthetic peptides derived from the deduced amino acid sequence of the E-glycoprotein of Murray Valley encephalitis virus elicit antiviral antibody. *Virology* **171:**49–60.

Roehrig, J. T., Johnson, A. J., Hunt, A. R., Beaty, B. J., and Mathews, J. H. (1992). Enhancement of antibody response to flavivirus B-cell epitopes by using homologous or heterologous T-cell epitopes. *J. Virol.* **66:**3385–3390.

Roehrig, J. T., Nash, D., Maldin, B., Labowitz, A., Martin, D. A., Lanciotti, R. S., and Campbell, G. L. (2003). Persistence of virus-reactive serum immunoglobulin M antibody in confirmed West Nile virus encephalitis cases. *Emerg. Infect. Dis.* **9:**376–379.

Roehrig, J. T., Staudinger, L. A., Hunt, A. R., Mathews, J. H., and Blair, C. D. (2001). Antibody prophylaxis and therapy for flaviviral encephalitis infections. *Ann. N.Y. Acad. Sci.* **951:**286–297.

Rojanasuphot, S., Shaffer, N., Chotpitayasunondh, T., Phumiamorn, S., Mock, P., Chearskul, S., Waranawat, N., Yuentrakul, P., Mastro, T. D., and Tsai, T. F. (1998). Response to JE vaccine among HIV-infected children, Bangkok, Thailand. *Southeast Asian J. Trop. Med. Public Health* **29:**443–450.

Rothman, A. L., and Ennis, F. A. (1999). Immunopathogenesis of dengue hemorrhagic fever. *Virology* **257:**1–6.

Sammin, D. J., Butler, D., Atkins, G. J., and Sheahan, B. J. (1999). Cell death mechanisms in the olfactory bulb of rats infected intranasally with Semliki Forest virus. *Neuropathol. Appl. Neurobiol.* **25:**236–243.

Sampson, B. A., Ambrosi, C., Charlot, A., Reiber, K., Veress, J. F., and Armbrustmacher, V. (2000). The pathology of human West Nile Virus infection. *Hum. Path.* **31:**527–531.

Samuel, C. E. (2002). Host genetic variability and West Nile virus susceptibility. *Proc. Natl. Acad. Sci. USA* **99:**11555–11557.

Sangster, M. K., Urosevic, N., Mansfield, J. P., Mackenzie, J. S., and Shellam, G. R. (1994). Mapping the *Flv* locus controlling resistance to flaviviruses on mouse chromosome 5. *J. Virol.* **68:**448–452.

Saxena, S. K., Singh, A., and Mathur, A. (2000). Antiviral effect of nitric oxide during Japanese encephalitis virus infection. *Int. J. Exp. Pathol.* **81:**165–172.

Schlesinger, J. J., and Brandriss, M. W. (1983). 17D yellow fever virus infection of P388D1 cells mediated by monoclonal antibodies: Properties of the macrophage Fc receptor. *J. Gen. Virol.* **64:**1255–1262.

Schlesinger, J. J., Brandriss, M. W., Cropp, C. B., and Monath, T. P. (1986). Protection against yellow fever in monkeys by immunization with yellow fever virus nonstructural protein NS1. *J. Virol.* **60:**1153–1155.

Schlesinger, J. J., Brandriss, M. W., Putnak, J. R., and Walsh, E. E. (1990). Cell surface expression of yellow fever virus non-structural glycoprotein NS1: Consequences of interaction with antibody. *J. Gen. Virol.* **71:**593–599.

Schlesinger, J. J., Brandriss, M. W., and Walsh, E. E. (1985). Protection against 17D yellow fever encephalitis in mice by passive transfer of monoclonal antibodies to the nonstructural glycoprotein gp48 and by active immunization with gp48. *J. Immunol.* **135:**2805–2809.

Schlesinger, J. J., Brandriss, M. W., and Walsh, E. E. (1987). Protection of mice against dengue 2 virus encephalitis by immunization with the dengue 2 virus non-structural glycoprotein NS1. *J. Gen. Virol.* **68:**853–857.

Schlesinger, J. J., and Chapman, S. (1995). Neutralizing F(ab′)2 fragments of protective monoclonal antibodies to yellow fever virus (YF) envelope protein fail to protect mice against lethal YF encephalitis. *J. Gen. Virol.* **76:**217–220.

Schlesinger, J. J., Chapman, S., Nestorowicz, A., Rice, C. M., and Chambers, T. J. (1996). Replication of yellow fever virus in the mouse central nervous system: Comparison of neuroadapted and nonneuroadapted virus and partial sequence analysis of the neuroadapted strain. *J. Gen. Virol.* **77:**1277–1285.

Schlesinger, J. J., Foltzer, M., and Chapman, S. (1993). The Fc portion of antibody to yellow fever virus NS1 is a determinant of protection against YF encephalitis in mice. *Virology* **192:**132–141.

Schmaljohn, C., and Blair, C. D. (1977). Persistent infection of cultured mammalian cells by Japanese encephalitis virus. *J. Virol.* **24:**580–589.

Schmaljohn, C. S., and Blair, C. D. (1979). Clonal analysis of mammalian cell cultures persistently infected with Japanese encephalitis virus. *J. Virol.* **31:**816–822.

Schneider-Schaulies, J., Liebert, U. G., Dorries, R., and ter Meullen, V. (1997). Establishment and control of viral infections in the central nervous system. *In* "Immunology of the Nervous System" (R. W. Keane and W. F. Hickey, eds.), pp. 576–616. Oxford Univ. Press, New York.

Shah, P. S., and Gadkari, D. A. (1987). Persistent infection of porcine kidney cells with Japanese encephalitis virus. *Ind. J. Med Res.* **85:**481–491.

Shahar, A., Lustig, S., Akov, Y., David, Y., Schneider, P., Friedmann, A., and Levin, R. (1990). West Nile virions aligned along myelin lamellae in organotypic spinal cord cultures. *J. Neurosci. Res.* **26:**495–500.

Sharma, S., Mathur, A., Prakash, V., Kulshreshta, R., Kumar, R., and Chaturvedi, U. C. (1991). Japanese encephalitis virus latency in peripheral blood lymphocytes and recurrence of infection in children. *Clin. Exp. Immunol.* **85:**85–91.

Shoji, H., Murakami, T., Murai, I., Kida, H., Sato, Y., Kojima, K., Abe, T., and Okudera, T. (1990). A follow-up study by CT and MRI in 3 cases of Japanese encephalitis. *Neuroradiology* **32:**215–219.

Sibailly, T. S., Wiktor, S. Z., Tsai, T. F., Cropp, C. B., Ekpini, E. R., Adjorlolo-Johnson, G., Gnaore, E., DeCock, K. M., and Greenberg, A. E. (1997). Poor antibody response to yellow fever vaccination in children infected with human immunodeficiency virus type 1. *Pediat. Infect. Dis.* **16:**1177–1179.

Singh, A., Kulshreshta, R., and Mathur, A. (2000). Secretion of the chemokine interleukin-8 during Japanese encephalitis virus infection. *J. Med. Micro.* **49:**607–612.

Slavin, H. B. (1943). Persistence of the virus of St. Louis encephalitis in the central nervous system of mice for over five months. *J. Bacteriol.* **46:**113–116.

Smith, A. L. (1981). Genetic resistance to lethal flavivirus encephalitis: Effect of host age and immune status and route of inoculation on production of interfering Banzi virus in vivo. *Am. J. Trop. Med. Hyg.* **30:**1319–1323.

Smith, H. R., Idris, A. H., and Yokoyama, W. M. (2001). Murine natural killer cell activation receptors. *Immunol. Rev.* **181:**115–125.
Solomon, T., Dung, N. M., Wills, B., Kneen, R., Gainsborough, M., Diet, T. V., Thuy, T. T., Loan, H. T., Khanh, V. C., Vaughn, D. W., White, N. J., and Farrar, J. J. (2003). Interferon alpha-2a in Japanese encephalitis: A randomised double-blind placebo-controlled trial. *Lancet* **361:**821–826.
Solomon, T., Kneen, R., Dung, N. M., Khanh, V. C., Thuy, T. T., HA, D. Q., Day, N. P., Nisalak, A., Vaughn, D. W., and White, N. J. (1998). Poliomyelitis-like illness due to Japanese encephalitis virus. *Lancet* **351:**1094–1097.
Solomon, T., and Vaughn, D. W. (2002). Pathogenesis and clinical features of Japanese encephalitis and West Nile infections. *In* "Japanese Encephalitis and West Nile Viruses" (J. S. Mackenzie, A. D. T. Barrett, and V. Deubel, eds.), pp. 171–194. Springer-Verlag, Berlin.
Southam, C. M., Nojes, W. F., and Mellors, R. (1958). Virus in human cancer cells in vivo. *Virology* **5:**395–400.
Spain-Santana, T. A., Marglin, S., Ennis, F. A., and Rothman, A. L. (2001). MIP-1 alpha and MIP-1 beta induction by dengue virus. *J. Med. Virol.* **65:**324–330.
Steele, K. E., Linn, M. J., Schoepp, R. J., Komar, N., Geisbert, T. W., Manduca, R. M., Calle, P. P., Raphael, B. L., Clippinger, T. L., Larsen, T., Smith, J., Lanciotti, R. S., Panella, N. A., and McNamara, T. S. (2000). Pathology of fatal West Nile virus infections in native and exotic birds during the 1999 outbreak in New York City, New York. *Vet. Pathol.* **37:**208–224.
Stephens, H. A., Klaythong, R., Sirikong, M., Vaughn, D. W., Green, S., Kalayanarooj, S., Endy, T. P., Libraty, D. H., Nisalak, A., Innis, B. L., Rothman, A. L., Ennis, F. A., and Chandanayingyong, D. (2002). HLA-A and -B allele associations with secondary dengue virus infections correlate with disease severity and the infecting viral serotype in ethnic Thais. *Tissue Antigens* **60:**309–318.
Stohlman, S. A., Bergmann, C. C., Lin, M. T., Cua, D. J., and Hinton, D. R. (1998). CTL effector function within the central nervous system requires $CD4^+$ T cells. *J. Immunol.* **160:**2896–2904.
Su, C. M., Liao, C. L., Lee, Y. L., and Lin, Y. L. (2001). Highly sulfated forms of heparin sulfate are involved in Japanese encephalitis virus infection. *Virology* **286:**206–215.
Su, H. L., Liao, C. L., and Lin, Y. L. (2002). Japanese encephalitis virus infection initiates endoplasmic reticulum stress and an unfolded protein response. *J. Virol.* **76:**4162–4171.
Suri, N. K., and Banerjee, K. (1995). Growth and cytopathic effect of Japanese encephalitis virus in astrocyte-enriched cell cultures from neonatal mouse brains. *Acta Virol.* **39:**143–148.
Suzuki, M., and Phillips, C. A. (1966). St. Louis encephalitis: A histopathologic study of the fatal cases from the Houston epidemic in 1964. *Arch. Pathol.* **81:**47–54.
Suzuki, T., Ogata, A., Tashiro, K., Nagashima, K., Tamura, M., Yasui, K., and Nishihira, J. (2000). Japanese encephalitis virus up-regulates expression of macrophage migration inhibitory factor (MIF) mRNA in the mouse brain. *Biochim. Biophys. Acta* **1517:**100–106.
Szilak, I., and Minamoto, G. Y. (2000). West Nile viral encephalitis in an HIV-positive woman in New York. *N. Engl. J. Med.* **42:**59–60.
Ta, M., and Vrati, S. (2000). Mov34 protein from mouse brain interacts with the 3′ noncoding region of Japanese encephalitis virus. *J. Virol.* **74:**5108–5115.

Takada, K., Masaki, H., Konishi, E., Takahashi, M., and Kurane, I. (2000). Definition of an epitope on Japanese encephalitis virus (JEV) envelope protein recognized by JEV-specific murine CD8$^+$ cytotoxic T lymphocytes. *Arch. Virol.* **145:**523–534.

Tassaneetrithep, B., Burgess, T. H., Granelli-Piperno, A., Trumpfheller, C., Finke, J., Sun, W., Eller, M. A., Pattanapanyasat, K., Sarasombath, S., Birx, D. L., Steinman, R. M., Schlesinger, S., and Marovich, M. A. (2003). DC-SIGN (CD209) mediates dengue virus infection of human dendritic cells. *J. Exp. Med.* **197:**823–829.

Tesh, R. B., Travassos da Rosa, A. P., Guzman, H., Araujo, T. P., and Xiao, S. Y. (2002). Immunization with heterologous flaviviruses protective against fatal West Nile encephalitis. *Emerg Infect. Dis.* **8:**245–251.

Thakare, J. P., Wadia, R. S., Banerjee, K., and Ghosh, S. N. (1988). Increased level of myelin basic protein in cerebrospinal fluid of patients of Japanese encephalitis. *Indian J. Med. Res.* **88:**297–300.

Thakare, J. P., Gore, M. M., Risbud, A. R., Banerjee, K., and Ghosh, S. N. (1991). Detection of virus specific IgG subclasses in Japanese encephalitis patients. *Indian J. Med. Res.* **93:**271–276.

Theerasurakarn, S., and Ubul, S. (1998). Apoptosis induction in brain during the fixed strain of rabies virus infection correlates with onset and severity of illness. *J. Neurovirol.* **4:**407–414.

Tschen, S.-I., Bergmann, C. C., Ramakrishna, C., Morales, S., Atkinson, R., and Stohlman, S. A. (2002). Recruitment kinetics and composition of antibody-secreting cells within the central nervous system following viral encephalomyelitis. *J. Immunol.* **168:**2922–2929.

Turell, M. J., and Spielman, A. (1992). Nonvascular delivery of Rift Valley fever virus by infected mosquitoes. *Am. J. Trop. Med. Hyg.* **47:**190–194.

Turell, M. J., Tammariello, R. F., and Spielman, A. (1995). Nonvascular delivery of St. Louis encephalitis and Venezuelan equine encephalitis viruses by infected mosquitoes (Diptera: Culicidae) feeding on a vertebrate host. *J. Med. Entomol.* **32:**563–568.

Tyor, W. R., Wesselingh, S., Levine, B., and Griffin, D. E. (1992). Long term intraparenchymal Ig secretion after acute viral encephalitis in mice. *J. Immunol.* **149:**4016–4020.

Uren, M. F., Doherty, P. C., and Allan, J. E. (1987). Flavivirus-specific murine L3T4$^+$ T cell clones: Induction, characterization and cross-reactivity. *J. Gen. Virol.* **68:**2655–2663.

Urosevic, N., van Maanen, M., Mansfield, J. P., Mackenzie, J. S., and Shellam, G. R. (1997). Molecular characterization of virus-specific RNA produced in the brains of flavivirus-susceptible and -resistant mice after challenge with Murray Valley encephalitis virus. *J. Gen. Virol.* **78:**23–29.

van der Most, R. G., Harrington, L. E., Giuggio, V., Mahar, P. L., and Ahmed, R. (2002). Yellow fever virus 17D envelope and NS3 proteins are major targets of the antiviral T cell response in mice. *Virology* **296:**117–124.

van der Most, R. G., Murali-Krishna, K., and Ahmed, R. (2003). Prolonged presence of effector-memory CD8 T cells in the central nervous system after dengue virus encephalitis. *Int. Immunol.* **15:**119–125.

van der Most, R. G., Murali-Krishna, K., Ahmed, R., and Strauss, J. H. (2000). Chimeric yellow fever/dengue virus as a candidate dengue vaccine: Quantitation of the dengue virus-specific CD8 T cell response. *J. Virol.* **74:**8094–8101.

Vaneeva, G. G. (1969). Long-term observation over children having sustained tick-borne encephalitis. *Pediatrics* **48:**40–48.

Vargin, V. V., and Semenov, B. F. (1986). Changes of natural killer cell activity in different mouse lines by acute and asymptomatic flavivirus infections. *Acta Virol.* **30:**303–308.

Varnavski, A. N., and Khromykh, A. A. (1999). Noncytopathic flavivirus replicon RNA-based system for expression and delivery of heterologous genes. *Virology* **255:**366–375.

Vince, V., and Grevic, N. (1969). Development of morphological changes in experimental tick-borne meningoencephalitis induced in mice by different virus doses. *J. Neurol. Sci.* **9:**109–130.

Vlaycheva, L., and Chambers, T. J. (2002). Neuroblastoma cell-adapted yellow fever 17D: Characterization of a viral variant associated with persistent infection and decreased virus spread. *J. Virol.* **75:**10912–10922.

Volanakis, J. E. (2002). The role of complement in innate and adaptive immunity. *Curr. Top. Microbiol. Immunol.* **266:**41–56.

Waldvogel, K., Bossart, W., Huisman, T., Boltshauser, E., and Nadal, D. (1996). Severe tick-borne encephalitis following passive immunization. *Eur. J. Pediatr.* **155:**775–779.

Wang, J. J., Liao, C. L., Chiou, Y. W., Chiou, C. T., Huang, Y. L., and Chen, L. K. (1997). Ultrastructure and localization of E proteins in cultured neuron cells infected with Japanese encephalitis virus. *Virology* **238:**30–39.

Wasay, M., Diaz-Arrastia, R., Suss, R. A., Kojan, S., Haq, A., Burns, D., and Van Ness, P. (2000). St Louis encephalitis: A review of 11 cases in a 1995 Dallas, Tex, epidemic. *Arch. Neurol.* **57:**114–118.

Weiner, L. P., Cole, G. A., and Nathanson, N. (1970). Experimental encephalitis following peripheral inoculation of West Nile virus in mice of different ages. *J. Hyg.* **68:**435–446.

Wesselingh, S. L., Levine, B., Fox, R. J., Choi, S., and Griffin, D. E. (1994). Intracerebral cytokine mRNA expression during fatal and nonfatal alphavirus encephalitis suggests a predominant type 2 T cell response. *J. Immunol.* **152:**1289–1297.

Wu, S. J., Grouard-Vogel, G., Sun, W., Mascola, J. R., Brachtel, E., Putvatana, R., Louder, M. K., Filgueira, L., Marovich, M. A., Wong, H. K., Blauvelt, A., Murphy, G. S., Robb, M. L., Innes, B. L., Birx, D. L., Hayes, C. G., and Frankel, S. S. (2000). Human skin Langerhans cells are targets of dengue virus infection. *Nature Med.* **6:**816–820.

Xiao, S. Y., Guzman, H., Zhang, H., Travassos da Rosa, A. P., and Tesh, R. B. (2001). West Nile virus infection in the golden hamster (*Mesocricetus auratus*): A model for West Nile encephalitis. *Emerg. Infect. Dis.* **7:**714–721.

Yang, J. S., Ramanathan, M. P., Muthumani, K., Choo, A. Y., Jin, S. H., Yu, Q. C., Hwang, D. S., Choo, D. K., Lee, M. D., Dang, K., Tang, W., and Kim, J. J. (2002). Induction of inflammation by West Nile virus capsid through the caspase-9 apoptotic pathway. *Emerg. Infect. Dis.* **8:**1379–1384.

Yazi Mendoza, M., Salas-Benito, J. S., Lanz-Mendoza, H., Hernandez-Martinez, S., and del Angel, R. M. (2002). A putative receptor for dengue virus in mosquito tissues: Localization of a 45-kDa glycoprotein. *Am. J. Trop. Med. Hyg.* **67:**76–84.

Yocupicio-Monroy, R. M., Medina, F., Reyes del Valle, J., and del angel, R. M. (2003). Cellular protein from human monocytes bind to dengue 4 virus minus strand 3′ untranslated RNA. *J. Virol.* **77:**3067–3076.

Zeidner, N. S., Higgs, S., Happ, C. M., Beaty, B. J., and Miller, B. R. (1999). Mosquito feeding modulates Th1 and Th2 cytokines in flavivirus susceptible mice: An effect mimicked by injection of sialokinins, but not demonstrated in flavivirus resistant mice. *Parasite Immunol.* **21:**35–44.

Zhang, M. J., Wang, M. J., Jiang, S. Z., and Ma, W. Y. (1989). Passive protection of mice, goats, and monkeys against Japanese encephalitis with monoclonal antibodies. *J. Med. Virol.* **29:**133–138.

Zhang, P. F., Klutch, M., Muller, J., and Marcus-Sekura, C. J. (1993). St. Louis encephalitis virus establishes a productive, cytopathic and persistent infection of Sf9 cells. *J. Gen. Virol.* **74:**1703–1708.

Zimmerman, H. M. (1946). The pathology of Japanese B encephalitis. *Am. J. Pathol.* **22:**965–991.

Zisman, B., Wheelock, E. F., and Allison, A. C. (1971). Role of macrophages and antibody in resistance of mice against yellow fever virus. *J. Immunol.* **107:**236–243.

Zlotnik, I., Carter, G. B., and Grant, D. P. (1971). The persistence of louping ill virus in immunosuppressed guinea-pigs. *Br. J. Exp. Pathol.* **52:**395–407.

Zlotnik, I., Grant, D. P., and Carter, G. B. (1976). Experimental infection of monkeys with viruses of the tick-borne encephalitis complex: Degenerative cerebellar lesions following inapparent forms of the disease or recovery from clinical encephalitis. *Br. J. Exp. Pathol.* **57:**200–210.

Zocher, M., Czub, S., Schulte-Monting, J., de La Torre, J. C., and Sauder, C. (2000). Alterations in neurotrophin and neurotrophin receptor gene expression patterns in the rat central nervous system following perinatal Borna disease virus infection. *J. Neurovirol.* **6:**462–467.

ADVANCES IN VIRUS RESEARCH, VOL 60

PATHOGENESIS AND PATHOPHYSIOLOGY OF YELLOW FEVER

Thomas P. Monath* and Alan D. T. Barrett†

*Acambis Inc. and Department of Microbiology and Immunology
Harvard School of Public Health
Cambridge, Massachusetts 02139
†Department of Pathology
University of Texas Medical Branch
Galveston, Texas 77555

I. Introduction

Yellow fever (YF) virus causes a systemic illness characterized by high viremia, hepatic, renal, and myocardial injury, hemorrhage, and high lethality. YF is only found in tropical South America

0065-3527/03 $35.00

and sub-Saharan Africa, where the enzootic transmission cycle involves tree-hole breeding mosquitoes and nonhuman primates. Two distinct YF virus genotypes in South America and five genotypes in Africa have been distinguished by nucleotide sequencing. However, the genotypes represent a single serotype defined by neutralization. Between 1985 and 1999, 25,846 cases and 7118 deaths were reported, of which 22,952 cases (89%) occurred in Africa. Most cases escape official notification. It is estimated that the true incidence may be 200,000 cases and 40,000 deaths annually (Robertson *et al.*, 1996).

YF has many features in common with other viral hemorrhagic fevers but stands apart by causing the most severe injury to liver and a clinical syndrome in which fulminating hepatitis predominates over other features. The hepatic injury induced by YF virus is pathologically distinct, and the pathogenesis is different from other viral hepatitides.

The YF virus was first isolated in 1927. The complete genome sequence of the YF virus (the 17D-204 vaccine substrain) was reported in 1985 (Rice *et al.*, 1985), and some information has accumulated about the virus-specified virulence factors by comparing the sequences of virulent and attenuated strains. Overall, however, knowledge about pathogenesis from the perspectives of virus and host remains descriptive and fragmentary, and the complex pathophysiological disturbances leading to death of the host remain largely obscure.

This review approaches the subject in the following way. The disease syndrome is presented briefly. The virus-specified determinants identified at the molecular level and associated with virulence are then discussed. We then describe, in sequence, the interactions of virus and host at the level of the cell, organized tissues (organs), and intact host, including innate and adaptive immunological mechanisms responsible for recovery from infection. The response to the YF 17D vaccine and the occurrence and pathogenesis of severe adverse events associated with the YF 17D vaccine are then discussed.

II. Disease Syndrome

YF infection may be subclinical or abortive with a nonspecific grippe-like illness. On the other end of the spectrum, it causes potentially lethal pansystemic disease with fever, jaundice, renal failure, and hemorrhage. This variability in response is associated with differences in the pathogenicity of virus strains and genetic and acquired host resistance factors.

After an incubation period of 3–6 days, the onset of illness is abrupt, with fever, chills, and headache. Three stages of disease have been described in humans (Beeuwkes, 1936; Berry and Kitchen, 1931; Monath, 1987). The initial "period of infection" lasts several days, during which virus is present in blood. Few data are available on the levels of viremia in humans, and no data exist on viremia during the incubation period. In one report (MacNamara, 1955), viremia peaked on day 2–3 after the onset of illness, with titers of up to 5.6 $\log_{10}$ mouse intracerebral (IC) LD_{50}/ml (approx 7 $\log_{10}$ PFU/ml).

Symptoms during the "period of infection" include generalized malaise, headache, photophobia, lumbosacral pain, generalized myalgia, nausea, vomiting, restlessness, irritability, and dizziness. On examination the patient appears acutely ill, with hyperemia of the skin, conjunctival injection, and tenderness and enlargement of the liver. The heart rate is slow relative to fever (Faget's sign). The average fever is 102–103 °F and lasts 3.3 days, but the temperature may rise as high as 105 °F—a poor prognostic sign. Clinical laboratory abnormalities include leukopenia ($1.5–2.5 \times 10^9$ cells/liter) with a relative neutropenia and minimal elevations of serum aspartate aminotransferase (AST) and alanine aminotransferase (ALT).

The "period of infection" may be followed by a "period of remission" with disappearance of fever and symptoms lasting up to 48 h. In cases of abortive infection, the patient may recover at this stage.

The third stage, the "period of intoxication," begins on the third to sixth day after onset. Approximately one in seven persons infected with YF virus progress to this stage, developing moderate to severe hemorrhagic fever and dysfunction of multiple organs (MacNamara, 1957; Monath, 1987). Symptoms and signs include fever, severe asthenia, nausea, vomiting, epigastric pain, jaundice, oliguria, cardiovascular instability, and hemorrhage. Virus disappears from blood, and antibodies appear. The patient is jaundiced, and the liver is typically enlarged and tender. In contrast to other viral hepatitides, AST levels exceed ALT, presumably due to viral injury to the myocardium and skeletal muscle, but also possibly due to increased mitochondrial permeability associated with apoptotic cell death. The rise in serum transaminase and bilirubin is proportional to disease severity, and very high levels carry a poor prognosis (Elton *et al.*, 1955; Oudart and Rey, 1970). The serum ammonia level may be elevated.

Renal dysfunction is characterized by albuminuria, oliguria, azotemia, and rising blood urea nitrogen and creatinine. The concentration of albumin in the urine may reach high levels, reflecting an increase in glomerular permeability. Urinalysis shows elevated albumin, microscopic hematuria, and proteinaceous casts. Specific gravity is high, in part due to the presence of albumin. Urine pH is low, reflecting reabsorption of bicarbonate as a response to metabolic acidosis. Edema, ascites, and pleural effusion are not described despite the severity of renal failure, but most patients have not received intensive care that may result in fluid overload. In some patients who survive the acute phase, renal failure requiring dialysis predominates as a late manifestation (Boulos *et al.*, 1988).

Hemorrhagic manifestations include coffee-grounds hematemesis ("black vomit"), melena, hematuria, metrorrhagia, petechiae, ecchymoses, bleeding from the gums and nose, and prolonged bleeding at needle puncture sites. Laboratory findings include thrombocytopenia, prolonged clotting and prothrombin times, and reductions in clotting factors synthesized by the liver (factors II, V, VII, IX, and X). Tests indicating disseminated intravascular coagulation include diminished fibrinogen and factor VIII and elevated fibrin split products (Borges *et al.*, 1973; Santos *et al.*, 1973).

The heart is also affected in YF, presumably the result of virus replication in the myocardial cells. Electrocardiograpic abnormalities include sinus bradycardia without conduction defects and ST-T changes, particularly elevated T waves (Chagas and De Freitas, 1929). Bradycardia may contribute to the physiological decompensation associated with hypotension, reduced perfusion, and metabolic acidosis in severe cases. Acute cardiac enlargement indicative of heart failure may occur during the course of YF infection (Berry and Kitchen, 1931).

Central nervous system (CNS) signs include delirium, agitation, convulsions, stupor, and coma. In severe cases, the cerebrospinal fluid is under increased pressure and may contain elevated protein without inflammatory cells. In patients dying of YF, CNS signs appear to result from cerebral edema or metabolic factors, based on the virtual absence of inflammatory changes in brain tissue. Pathological changes include petechiae, perivascular hemorrhages, and edema (Stevenson, 1939). Neuroinvasion and YF encephalitis are extremely rare, with few reports of paralysis, optic neuritis, and cranial nerve palsy suggesting neurological infection (Stefanopoulo and Mollaret, 1934). If neuroinvasion of YF virus occurred during the acute phase, late encephalitis (after recovery from hepatitis) would be the likely outcome, but this is not observed. Significantly, nonhuman primates succumb to viscerotropic disease even when virus is inoculated directly into the brain.

The critical phase of the illness occurs between the fifth and the tenth day, at which point the patient either dies or recovers rapidly. The terminal illness is characterized by hypotension, shock, hemorrhage, and metabolic acidosis. Case–fatality rates in various reports vary widely between 5% and 70%, principally due to differences in reporting. In recent outbreaks, the case–fatality rate in patients with jaundice approximated 20% (Monath *et al.*, 1980; Nasidi *et al.*, 1989). Severity of illness and case fatality rates are highest in infants and the elderly (Beeuwkes, 1936; Hanson, 1929).

III. Virus-Specific Virulence Factors

YF virus is genetically distant from other flaviviruses and is the most distantly related agent among mosquito-borne flaviviruses based on its antigenic relationships and nucleotide sequence studies (Kuno *et al.*, 1998; Zanotto *et al.*, 1996). In the most recent taxonomic classification based on a synthesis of epidemiological, antigenic, and genetic relationships, YF virus is the type species in a complex that includes Wesselsbron, Sepik, Edge Hill, Bouboui, Uganda S, Banzi, Jugra, Saboya, and Potiskum viruses. Genetically, the closest relative to YF virus is Sepik virus, which is found only in Asia. There is very limited nucleotide sequence information for YF virus. To date, only four wild-type strains have been sequenced: three from west Africa and one from Trinidad. As seven genotypes of YF virus have been described, there is still much to be learned about the genetics of YF virus. It is noteworthy that Wesselsbron virus induces a disease in sheep that resembles many features of YF, but this is the sole example of shared pathogenesis across members of the virus complex. Members of other mosquito-borne flavivirus complexes very rarely cause illness resembling YF. These include West Nile hepatitis (Georges *et al.*, 1987) and dengue hemorrhagic fever (DHF). It is likely that liver cells are involved in viral replication of many heterologous flaviviruses, but that dysfunction of the liver is typically subclinical.

From the perspective of pathogenesis, YF virus elicits two distinct and separate patterns of infection and injury: viscerotropism and neurotropism. "Viscerotropism" refers to the ability of YF virus to infect and cause damage to extraneural organs, including liver, spleen, heart, and kidneys, whereas "neurotropism" refers to the ability to infect the brain parenchyma and cause encephalitis. Wild-type YF virus strains are predominantly viscerotropic in nonhuman primates and humans, but are neurotropic following intracerebral challenge of other species, including rodents and rabbits. Neurotropism in rodent hosts is a feature of virtually all flaviviruses, although many cause lethal encephalitis only in infant animals. Neuroinvasiveness in rodents following peripheral inoculation is a charactistic of the encephalitic flaviviruses, and wild-type strains of YF virus are rarely neuroinvasive in rodents aged 10 days or older. As discussed later, neurotropism and viscerotropism are distinct features of YF virus at the molecular level.

Although no systematic studies have been performed, it has long been noted that wild-type YF virus strains differ with respect to viscerotropism and lethality for nonhuman primates. The rhesus monkey was widely investigated as a model of the human disease virus. The lethality of South American virus strains for monkeys ranged between <10% and 100% (Laemmert, 1944; Sawyer *et al.*, 1930). Similar variability was observed across African isolates (T. P. Monath, unpublished data). A systematic study in a large number of South American marmosets (*Callithrix jacchus*) showed that South American YF virus strains were more lethal than African viruses (Laemmert, 1944). These observations indicate that YF virus virulence is highly dependent on the pairing of virus strain and host species. A confounding factor in all early studies is the laboratory passage history of virus strains being compared, which was often not controlled carefully. In particular, the passage of wild-type strains in cell culture results in the attenuation of viscerotropism in nonhuman primates (A. D. T. Barrett, unpublished data).

The genome of YF virus was the first flavivirus to be fully sequenced (Rice *et al.*, 1985). The 5′ and 3′ noncoding regions (NCR) of the genome have secondary structure and complementary sequences and serve as promoters for negative and positive RNA strands during replication. The 3′ NCR contains conserved consensus sequences that pair with core gene sequences during cyclization and has a secondary structure characterized by a pseudoknot (Olsthoorn and Bol, 2001). Mutations or deletions in the NCRs may thus modulate virus replication, and such changes have been shown to attenuate dengue-4 virus. Genomic sequencing of wild-type YF virus strains has demonstrated variability of the 3′ NCR structure due to the presence of one to three repeat sequence elements (Wang *et al.*, 1996), but there is no evidence that this variability is associated with differences in virulence.

As for other flaviviruses, the envelope (E) protein of YF virus contains the most important determinants for cell tropism, virulence, and immunity (Lindenbach and Rice, 2001). The crystallographic structure of the YF E glycoprotein has not been studied directly, but is predicted to be similar to that revealed for tick-borne encephalitis virus. Molecular coordinates of YF virus virulence factors are discussed with reference to this generalized flavivirus structure.

The Asibi and French viscerotropic strains represent the first two YF viruses isolated in 1927. The entire genomes of these strains have been sequenced, as well as the genomes of two live-attenuated vaccines derived from them: the 17D and the French neurotropic vaccines (Hahn *et al.*, 1987; Jennings *et al.*, 1993; Rice *et al.*, 1985; Wang *et al.*, 1995). Two substrains of YF 17D (17D-204 and 17DD) used for vaccine production and variants of 17D-204 virus (ATCC, France, and ARILVAX vaccine produced in the United Kingdom, and the 17D-213 WHO vaccine seed) have also been sequenced partially or fully (Duarte dos Santos *et al.*, 1995; Dupuy *et al.*, 1989; Jennings *et al.*, 1993; Post *et al.*, 1992; Pugachev *et al.*, 2002; Ryman *et al.*, 1997b).

The YF 17D virus has reduced neurotropism after intracerebral (IC) inoculation of mice compared to parental Asibi and reduced neurotropism and viscerotropism after inoculation of rhesus monkeys. Because the vaccine strain has lost the ability to induce tissue injury and clinical disease, genomic comparisons of the parental and vaccine strains provide clues to the molecular basis of virulence. However, many mutations occurred during the >230 passages that separate parental and vaccine strains, and the specific determinants encoding virulence properties remain uncertain. Some virulence determinants on the E glycoprotein have been identified, as discussed later. It is clear that virulence is multigenic, involving both structural and nonstructural genes of the virus. It is also important to note that nearly all studies on the molecular determinants of virulence have employed mouse models, which reveal only one of the three major biological properties of the virus (neurotropism versus viscerotropism and vector competence). Hamsters have been shown to be susceptible to a lethal disease with hepatic dysfunction and necrosis resembling wild-type YF after infection with virus strains adapted by serial passage in hamster liver (Tesh *et al.*, 2002; Xiao *et al.*, 2001). This model will permit dissection of the molecular determinants associated with viscerotropism (at least for the hamster).

Comparisons of the wild-type Asibi and attenuated 17D vaccine genomes have revealed 20 amino acid differences (0.59%) and four nucleotide differences in the 3′ NCR (Table I). The coding differences are distributed disproportionately within the E and NS2A proteins.

Because the E protein is implicated in tropism for and virus entry into cells, the eight amino acid differences between Asibi and vaccine strains are suspected to play a role in attenuation. The role of the flavivirus E glycoprotein in neurovirulence was demonstrated by studies in which the E gene of a nonneurovirulent dengue virus was replaced by the corresponding gene of a neurovirulent tick-borne encephalitis virus, resulting in a conversion to neurovirulent phenotype (Pletnev *et al.*, 1993), by heterologous YF chimeras having donor genes from viruses with different neurovirulence profiles (Chambers and Nickells, 2001), and by site-directed mutagenesis (Arroyo *et al.*, 2001). Although the possibilities have been reduced to a limited number of mutations (Table I), it has not been possible to pinpoint precisely which underlie the difference in neurovirulence between parental and vaccine strains.

A number of studies with monoclonal antibodies have identified E protein epitopes that are vaccine specific (17D-204 and 17DD substrains and FNV), wild-type specific, and 17D-204 substrain specific (Gould *et al.*, 1985, 1989; Schlesinger *et al.*, 1983; Sil *et al.*, 2000). Ryman *et al.* (1997b, 1998) mapped one of the wild-type epitopes to E173 and a 17D-204 substrain-specific epitope to E305 and E325. Such epitopes have been predicted to be involved in the virulence phenotype of YF virus, but to date only the 17D-204 substrain-specific epitope has been found to alter the phenotype of the virus (Ryman *et al.*, 1998).

Studies of YF and heterologous flaviviruses have identified three areas of the crystallographic structure of the E glycoprotein in which mutational changes alter virulence properties. These include the tip of the fusion domain (domain II), the hinge region between domains I and II, and the upper lateral surface of domain III containing the receptor-binding site (Rey *et al.*, 1995; Mandl *et al.*, 2000). Locations of the eight amino acids that distinguish Asibi and 17D viruses are shown in Table I. Four are nonconservative changes (E52 Gly → Arg; E200 Lys → Thr; E305 Ser → Phe; and E380 Thr → Arg). At least three wild-type yellow fever strains with different passage histories or geographic origins [Asibi (Hahn *et al.*, 1987), the French viscerotropic virus (Wang *et al.*, 1995), and Peruvian strain 1899/81 (Ballinger-Crabtree and Miller, 1990)] are identical at these amino acid residues, suggesting that one or more of the mutations in these four determinants in the 17D vaccine strain are responsible for attenuation (Duarte dos Santos *et al.*, 1995).

Amino acid residues E52, E173, and E200 are in the hinge region between domains I and II, and mutations could alter the acid-dependent conformational change in the endosome during virus internalization. Neuroadaptation of YF17D virus by brain–brain passage resulted in an increase in neurovirulence of the virus and reversion (Ile → Thr)

TABLE I
AMINO ACID DIFFERENCES BETWEEN PARENTAL ASIBI VIRUS AND ATTENUATED 17D VACCINES

Gene	Amino acid	Asibi	17D (17D-204 and 17DD substrain) vaccines
M	36	Leu	Phe
E	52	Gly	Arg
	170	Ala	Val
	173	Thr	Ile
	200	Lys	Thr
	299	Met	Ile
	305	Ser	Phe
	380	Thr	Arg
	407	Ala	Val
NS1	307	Ile	Val
NS2A	61	Met	Val
	110	Thr	Ala
	115	Thr	Ala
	126	Ser	Phe
NS2B	109	Ile	Leu
NS3	485	Asp	Asn
NS4A	146	Val	Ala
NS4B	95	Ile	Met
NS5	836	Glu	Lys
	900	Pro	Leu
(3′ NCR)		U	C
		U	C
		G	A
		A	C

at E173 (Chambers and Nickells, 2001; Schlesinger *et al.*, 1996). The Thr → Ile mutation at E173 corresponds to a site in the tick-borne encephalitis virus at which a neutralization escape mutant reduced neuroinvasiveness in mice (Holzmann *et al.*, 1997). Mutations in the hinge region at nearby locations E176 and E177 of a YF/Japanese encephalitis chimera were also implicated in attenuation (Arroyo *et al.*, 2001). Ryman *et al.* (1997b) obtained further evidence for the importance of residue E173, showing that it encodes an epitope recognized by a wild-type-specific monoclonal antibody (Mab 117; Gould *et al.*, 1989) and that reversion at this site may have contributed to the neurovirulence phenotype of a variant [17D(wt+)] recovered from a 17D-204 vaccine.

Studies of heterologous flaviviruses also suggest that the hinge region may also play a role in the viscerotropic properties of YF virus. This was suggested by a study of the attenuated YF/Japanese encephalitis chimeric virus in which a reversion to the wild-type JE residue at E279 (in the domain I–II interface) caused an increase in neurovirulence for mice and a significant decrease in viscerotropism for monkeys (Monath *et al.*, 2002a). Results indicated that molecular determinants for mouse neurovirulence have little relevance to viscerotropic virulence of YF virus strains for humans. This is not surprising, as the French neurotropic vaccine (used widely for human immunization between 1933 and 1982) was developed by over 128 sequential passages in mouse brains, resulting in an adapted virus that was highly neurotropic for mice but had lost viscerotropism for monkeys and humans. Dengue type 2 virus adapted by mouse brain passage was also attenuated for humans (Bray *et al.*, 1998; Wisseman *et al.*, 1963). A mutation at E126 in the E protein hinge region from a negatively charged to a positively charged amino acid (Glu → Lys) has been implicated in the attenuation of dengue "viscerotropism" (Bray *et al.*, 1998; Gualano *et al.*, 1998). In a separate study, dengue-1 virus adapted by mouse brain passage had increased neurovirulence and caused apoptosis of neural cells but reduced apoptosis in human hepatocytes, again suggesting that mutational changes caused opposing effects on neurotropic and viscerotropic properties of the virus (Duarte dos Santos *et al.*, 2000). The adapted virus contained mutations in regions of the genome affecting virus replication and assembly: the hinge region at E196 (Met → Val), the interface between domains I and III at E365 (Val → Ile), and two mutations in the proximal stem–anchor of E (E405, Thr → Ile) and in the NS3 helicase region.

Mutations at E305 and E380 of YF Asibi virus are located in domain III and E299 at the interface of domains I and III. Domain III is proposed to contain determinants for tropism and cell–receptor interactions. E305 is in the upper lateral surface of domain III, and residue E380 is in a highly conserved region implicated in virus–receptor interactions, as well as hemagglutination (Rey *et al.*, 1995). Changes in the cell attachment motif could alter tropism of the virus. Mutations in region E308–311 of tick-borne encephalitis virus resulted in significant attenuation (Mandl *et al.*, 2000). Residue E305 was implicated in the attenuation of 17D vaccine by the sequence analysis of virus recovered from the brain of a child with in postvaccinal encephalitis (Anonymous, 1966; Jennings *et al.*, 1994). The brain virus was found to have increased neurovirulence for mice and monkeys, indicating that a mutation increasing neurovirulence may have occurred during replication in the affected patient (Jennings *et al.*, 1994). The virus differed from 17D vaccine at E303 Glu → Lys, at a position in domain III very near the 17D-204 substrain-specific E305 residue, Two other mutations (at E155 and NS4B76) were also present in the brain isolate and could have contributed to the phenotypic change (Jennings *et al.*, 1994; E. Wang *et al.*, unpublished results).

The change at residue E380 changes the putative cell (integrin) attachment motif from Thr-Gly-Asp in Asibi to Arg-Gly-Asp (RGD) in 17D vaccine (Post *et al.*, 1992). In one study, mutations in the RGD sequence predicted to alter integrin binding did not interfere with YF virus replication (van der Most *et al.*, 1999). Mutations in the RGD sequence of Murray Valley encephalitis attenuated neurovirulence for mice (Hurrelbrink and McMinn, 2001; Lobigs *et al.*, 1990). However, the mutants were not significantly inhibited by heparin, suggesting that receptors other than integrins were responsible for interacting with viral ligands. Heparan sulfate and other carbohydrate-containing molecules on cell surfaces have been identified as cell-binding sites or receptors for flaviviruses (Martinez-Barragan and del Angel, 2001), including YF virus (Germi *et al.*, 2002), but it is clear that receptors for flaviviruses differ across cell lines and virus strains (Bielefeldt-Ohhmann *et al.*, 2001). Attenuation of Murray Valley and Japanese encephalitis virus neuroinvasiveness by passage in adenocarcinoma cells was associated with mutations in the cell receptor-binding domain at E306 and E390, respectively (Lee and Lobigs, 2002). The mutants had enhanced susceptibility to inhibition by heparin and enhanced affinity for glycosaminoglycan (GAG) receptors on extraneural cells. These observations suggested that changes in charged amino acids that enhanced affinity for GAG receptors reduced viremia and neuroinvasion.

The (Val → Ala) mutation at position E407 of Asibi virus occurs in helix I of the N-terminal stem–anchor region of the E protein. Mutations in the stem–anchor region may alter the structural integrity and spatial characteristics of the prM–E heterodimer. Such mutations have been associated with the attenuation of multiple flaviviruses, including dengue (Bray *et al.*, 1998; Duarte dos Santos *et al.*, 2000), tick-borne encephalitis group virus (Holbrook *et al.*, 2001), Japanese encephalitis (Ni and Barrett, 1988), and YF/Japanese encephalitis chimeras (Arroyo *et al.*, 2001). The stem–anchor region is involved in reconfiguration of the E protein from a dimeric to a trimeric structure during acid-induced fusion (Allison *et al.*, 1999; Wang *et al.*, 1999).

The 17D vaccine is not "fixed" with respect to neurovirulence so that sequential mouse brain passage of the vaccine results in increasing mouse virulence (Collier *et al.*, 1959). The neuroadapted 17D virus reverted to the wild-type (Asibi) sequence at amino acid residues E52 and 173 and also had other mutations, including one at the putative virulence determinant at residue E305 (Ser → Val) and amino acid mutations in nonstructural genes (NS1, NS2A, NS4A, NS4B, and NS5) (Chambers and Nickells, 2001; Schlesinger *et al.*, 1996). A chimera with the 17D backbone and the E protein from the neuroadapted YF virus did not have increased neurovirulence compared to 17D. However, introduction of all mutations in E and NS genes into the 17D infectious clone increased neurovirulence. This observation demonstrated that multiple genes were involved in virulence and implicated mutations in the NS proteins or the 3′ noncoding region of the virus. Studies with other flaviviruses have also shown that mutations in the NS coding region may reduce neurovirulence (Duarte dos Santos *et al.*, 2000; McMinn *et al.*, 1995), presumably by restricting the rate of viral RNA and protein production.

There are 11 amino acid changes in the nonstructural proteins of Asibi virus associated with the attenuation of 17D (Table I). One change occurs in the NS1 protein; 4 in NS2A; 1 each in NS2B, NS3, NS4A, and NS4B; and 2 in NS5. The mutations in NS2A may affect assembly or release of YF virus particles (Kümmerer and Rice, 2002). The change in NS3 occurs at residue 485, in a region of the protein with RNA helicase and triphosphatase activities involved in unwinding RNA during replication. The two mutations in the RNA polymerase (NS5) may affect replication efficiency and may contribute to the attenuation of 17D. A plaque variant purified from a 17D vaccine had reduced mouse neurovirulence and differed at an amino acid residue in NS5 137 Pro → Ser (Xie *et al.*, 1998). This mutation is in a region encoding the methyltransferase activity of NS5.

The 3′ NCR plays a critical role in replication and virulence. The 3′ NCR is divided into a proximal region that is variable in length among YF virus strains and contains one to three repeat sequence elements (depending on genotype) (Wang *et al.*, 1996) and a 3′-terminal region that contains a 90–120 nucleotide conserved region. The latter is involved in folding of the stem–loop structure and serves as a promoter for minus strand synthesis during replication. Studies with other flaviviruses have shown that mutations in the stem–loop region may affect replication adversely (Mandl *et al.*, 1998; Zeng *et al.*, 1998). In the case of dengue-4 virus, the proximal region of 3′ NCR does not appear to be critical for replication but mutations or deletions in this region may nevertheless attenuate virulence (Men *et al.*, 1996). Thus, one or more of the mutations in the 3′ NCR of 17D present in both the variable and the conserved proximal region may contribute to virulence determinants of YF virus. The change at nucleotide 10367 might be an exception, as one 17D vaccine contains a heterogeneous mixture, including the wild-type sequence (Pugachev *et al.*, 2002).

Less is known about the determinants of YF virus viscerotropism, principally because of the difficulty of assessing this property in nonhuman primates. As pointed out earlier, the hinge region of E might contain residues implicated in viscerotropism. To address this question, Wang *et al.* (1995) compared the sequence of the French viscerotropic strain with that of the French neurotropic vaccine (FNV) from which it was derived by >128 mouse brain passages. The principal phenotypic change in FNV is loss of viscerotropism for monkeys and humans, whereas FNV remains enhanced in neurovirulence for mice and monkeys. Comparison of parental and vaccine strains revealed 35 (1%) amino acid changes in multiple genes, with the highest frequency of mutations in C, M, E, NS2A, and NS4B. The large number of differences and lack of biological data on the role of these mutations preclude speculations on the genetic basis of viscerotropism. Sequence comparison of FNV with 17DD and 17D-204 vaccines (both of which have markedly attenuated viscerotropism) revealed only two shared differences from the parental and other wild-type YF viruses. These common differences, which evolved during the development of vaccine strains by adaptation to different hosts, were in the M protein (M36 Leu → Phe) and NS4B (95 Ile → Met). Unfortunately, little was revealed with respect to the molecular structures involved in viscerotropism.

The discovery that Syrian hamsters develop an illness and pathological changes resembling human YF (Tesh *et al.*, 2001) represents a potential breakthrough in the identification of molecular determinants responsible for viscerotropism. To induce disease in the hamster, it is necessary to adapt most strains of wild-type YF virus by serial liver–liver passage in hamsters. For example, Asibi virus becomes hamster virulent between the six and seventh passage. The genome of the viscerotropic hamster-adapted P7 virus has 14 nucleotide changes encoding seven amino acid substitutions (McArthur *et al.*, 2003). Five of the seven substitutions are in the E glycoprotein at positions E27 (Gln → His), E28 (Asp → Gly), E155 (Asp → Ala), E323 (Lys → Arg), E331 (Lys → Arg). None of these mutations represent reversions to wild-type Asibi sequence. However, two changes (E323 and E331) are in the upper lateral surface of domain III involved in binding to cell receptors and are suspected to play a role in tropism for hamster liver. In addition to the mutations in E, there were two substitutions leading to amino acid changes in NS2A48 (Thr → Ala) and NS4B98 (Val → Ile). A mutation near the latter site in NS4B (at residue 95) is associated with attenuation of YF 17D and FNV vaccines. In addition, a mutation at E155 was associated with reversion to virulence (neurotropism) of YF 17D (Jennings *et al.*, 1994).

In another study, a mutation at E279 in the hinge 4 region of a YF/JE chimeric virus reduced viscerotropism of the virus in nonhuman primates (Monath *et al.*, 2002a), suggesting that mutations in the hinge region of the YF genome during derivation of 17D virus (Table I) could be responsible for the attenuation of viscerotropism. The Met → Lys mutation at position E279 in the YF/JE chimera is within the β strand of the secondary structure, increases the net positive charge of the protein, and results in a shortening of the β strand secondary structure.

Both Asibi and the French viscerotropic virus (FVV) lost both neurotropism and viscerotropism after a few passages in HeLa cells (Barrett *et al.*, 1990; Dunster *et al.*, 1999). The HeLa-passaged virus accumulated 10 amino acid mutations, including 5 in the E protein, 1 in NS2A, and 3 in NS2B. The NS4B (95 Ile → Met) associated with attenuation of Asibi to 17D by passage in chick embryo tissue and of FVV to FNV by passage in mouse brain also appeared during HeLa cell passage, indicating an important role in the attenuation of YF vaccines. All three attenuation processes share two phenotypic markers: loss of viscerotropism and loss of vector competence for *Aedes aegypti*. To date the potential role of NS4B in viscerotropism or vector competence has not been investigated. The changes in the HeLa-passed virus E protein are at residues E27 (Gln → His), E155 (Asp → Ala), E228 (Met → Lys), E331 (Lys → Arg), and E390 (His → Pro). Interestingly, three of these mutations (E27, E155, and E331) were also associated with adaptation in the hamster (McArthur *et al.*, 2003), suggesting that they could be viscerotropism virulence factors. The reversion at E155 was present in the virus recovered from a fatal case of postvaccinal encephalitis (Jennings *et al.*, 1994), suggesting that this locus (possibly in concert with other mutations) may also play a role in neurovirulence. However, the mutation at E155 may not be important in neurovirulence, as some 17D strains have a wild-type residue at this locus (Post *et al.*, 1992) and as a neutralization escape mutant of 17D at E155 did not show any change in neurovirulence (Ryman *et al.*, 1997a).

IV. Virus–Cell Interactions

YF virus infects cells of diverse origin, including those derived from mosquitoes, ticks, birds, and mammals. AP61 *A. pseudoscutellaris* mosquito cells are exquisitely susceptible to infection. Among a number of tick cell lines, only *Dermacentor variabilis* was susceptible (Pudney, 1987), but cells from *Amblyomma* ticks implicated in YF virus transmission in nature (Germain *et al.*, 1979) have not been evaluated. Primary chick and duck embryo cells replicate YF viruses and develop cytopathic effects. Mammalian cell cultures, such as baby hamster kidney (BHK), monkey kidney (Vero, CV-1, MA-104, LLC-MK2), porcine kidney (PS-1), human adrenocortical adenocarcinoma cells (SW-13), and human histiocytic lymphoma cells (U937), are also efficient host cells for YF virus replication. Unadapted wild-type YF virus strains generally replicate to lower titers than 17D vaccine. This was illustrated by comparative studies in primary human mononuclear cell cultures (Liprandi and Walder, 1983). Among the cell types represented in peripheral blood mononuclear cells, $CD14^{+}$ monocytes were most susceptible to infection with dengue virus (Sydow *et al.*, 2000) and may also be the preferred targets for YF virus.

Virus ligand–host cell receptor interactions, insofar as they are known, were discussed earlier in Section III. Mechanisms by which flaviviruses enter cells include receptor-mediated endocytosis in clathrin-coated vesicles, phagocytosis by monocyte/macrophages, and direct penetration following fusion of the viral envelope with the cell membrane, as seen in mosquito cells.

The response of cells to YF virus infection is understood principally at the descriptive level. Cells propagate new virus particles, accumulate viral antigen, and undergo cytopathic effects. Morphogenesis of YF and other flaviviruses in cell culture, in mouse brain, monkey liver, and in mosquitoes have been studied by electron microscopy (David-West *et al.*, 1972; McGavran and White, 1964; Murphy, 1980). Mature virus particles accumulate in membrane-bound cisternae of the rough endoplasmic reticulum (ER). Controversy still surrounds how virions mature in these spaces, as a nucleocapsid budding process has been difficult to confirm. Virus particles are released from infected cells by exocytosis from plasma membranes or by cell rupture when cytopathic effects are at a very advanced stage. Virus infection proceeds rapidly, with intracellular virus reaching maximal levels in approximately 24 h. The most prominent morphological changes in the infected cell are proliferation and hypertrophy of the ER to accommodate accumulating virions. It is not known how flavivirus infection stimulates the synthesis of ER membranes. As the infection progresses, damage to mitochondria is evident, large vacuoles containing virus particles and ribosomes at the margins appear, and there is an increase in lysosomal bodies. As the cell disintegrates, the cytoplasm becomes rarified and subsequently condensed. Cytopathic effects are relatively retarded in flavivirus infections compared to many other cytopathic virus infections and develop over a period of 3–5 days after infection, with the time course being highly dependent on multiplicity of infection.

Host cell macromolecular synthesis is also affected relatively slowly by YF and other flaviviruses. This is not unexpected, as posttranslational processing of viral proteins is highly dependent on host cell proteases. The mechanism or mechanisms whereby flaviviruses shut down cellular protein translation are not known. The host cell protein elongation factor-1α interacts with *cis*-acting sequences in the 3′ stem-loop structure of flavivirus RNA (Blackwell and Brinton, 1997). Flavivirus C protein has also been shown to interact with heterogeneous nuclear ribonucleoprotein K (Chang *et al.*, 2001). Similar interactions with host cell proteins that are normally involved in cellular mRNA translation or regulation of gene expression have been described for other virus families and may represent a common mechanism for shutting down host cell protein synthesis.

Various lines of evidence from morphological and biochemical studies suggest that flaviviruses typically induce cell death by apoptosis rather than necrosis. Thus, flavivirus-infected cells are characterized by shrinkage and condensation of cytoplasmic and nuclear material, surface membrane blebbing, and fragmentation of cellular DNA detectable as ladders in agarose gels. YF and dengue virus-induced apoptotic cell death of cultured human hepatocyte cells has been reported (Marianneau *et al.*, 1997, 1998). Both viruses activate the transcription factor NF-κB, which in turn induces apoptosis. YF virus caused higher replication and antigen accumulation than dengue virus and was not associated with CPE or apoptosis until late in infection (Marianneau *et al.*, 1998). Apoptosis represents a protective mechanism whereby the host eliminates virus-infected cells, and the propensity of dengue virus to program cell death at an early stage of the infection may explain the minimal damage to liver tissue in human infection compared to that seen in YF. Bcl-2 and bcl-X gene expression appears to control flavivirus-induced apoptosis in a cell-specific manner (Su *et al.*, 2001), but hepatocytes have not yet been investigated in this regard. In the case of another flavivirus (West Nile), the

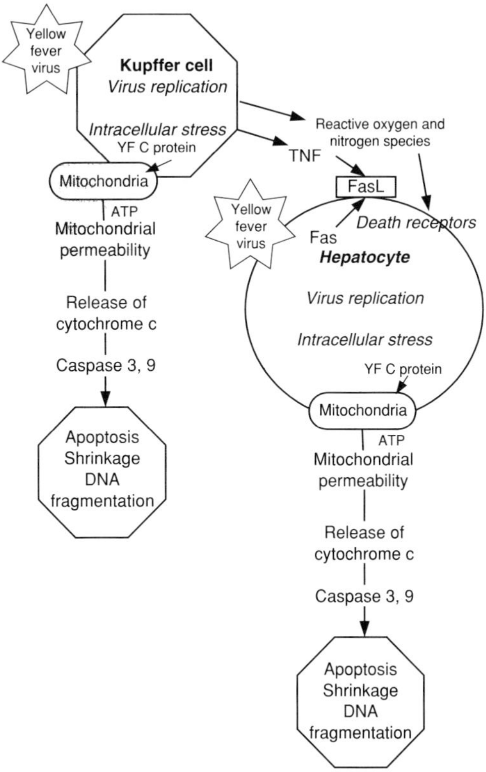

FIG 1. Hypothetical pathway of yellow fever virus-induced apoptotic cell death.

capsid C protein destabilizes mitochondrial membranes and is responsible for inducing apoptosis through caspase-9 induction (Yang *et al.*, 2002). It is uncertain whether the C protein of YF virus plays a similar role (Fig. 1).

An increase in cell membrane permeability accompanies cytopathology associated with infection by many viruses, and flaviviruses appear to be no exception. Membrane permeability changes are dependent on the synthesis of viral proteins, in particular the small hydrophobic proteins NS2A, NS2B, NS4A, and NS4B. Chang *et al.* (1999) expressed flavivirus nonstructural proteins in *Escherichia coli* and found that the protease NS2B-NS3 increased cell membrane permeability. However, it is uncertain how the NS proteins function to modify membrane permeability.

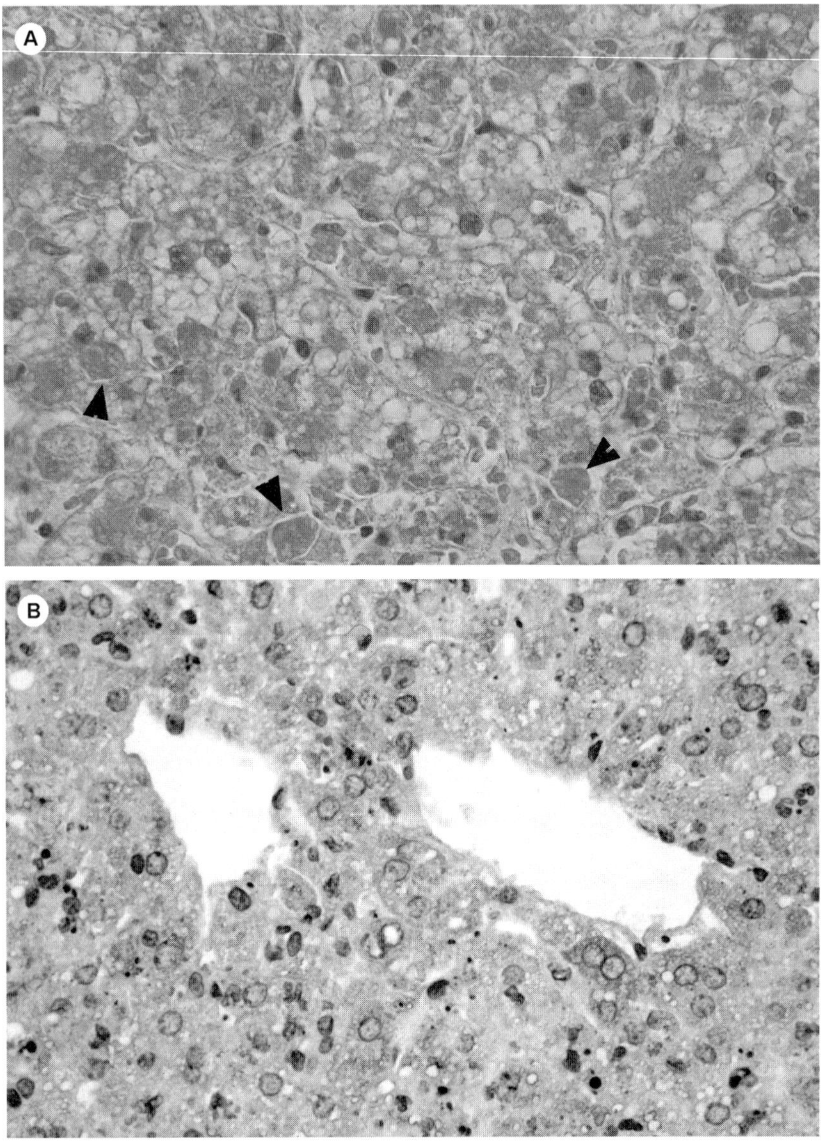

FIG 2. Histopathology of liver from a YF patient. (A) Hepatocytes exhibit cytoplasmic condensation and degeneration, many of which still retain a nucleus. Numerous councilman bodies are present in this field (arrowheads). Residual microvesicular steatosis is evident. Magnification 100×. (B) Immunohistochemical stain demonstrating the YFV antigen in the cytoplasm of many hepatocytes. The midzonal distribution of cell damage and antigen staining is less evident in this patient than in most cases; the figure is meant

V. Infection of Organized Tissues

The liver, spleen, lymph nodes, heart, and kidneys are the organs most affected by YF virus. Pathological changes in other organs are characterized by capillary hemorrhage (petechiae) and bile staining rather than parenchymal damage or by secondary bacterial superinfection (e.g., pneumonia). As noted previously, the brain typically shows swelling without evidence of viral infection.

A. Skin

Resident dendritic cells in the epidermis are permissive to infection with dengue virus and are believed to play an important role in initiating infection after a mosquito bite (Marovich *et al.*, 2001). It is likely that YF infection follows a similar pathway, but no data are currently available.

B. Liver

There are five central features of YF liver pathology: (1) Eosinophilic degeneration of hepatocytes and Kupffer cells; (2) midzonal hepatocellular necrosis; (3) absence of inflammation; (4) microvesicular fatty changes; and (5) retention of the reticulin structure, with return to normal histology without fibroblastic response (scarring) in surviving cases.

Injury to hepatocytes is manifest in the period of intoxication, approximately 10–15 days after infection. In fatal human cases, an average of 80% of hepatocytes are affected (Klotz and Belt, 1930a). Injury is characterized by the eosinophilic degeneration of Kupffer cells and hepatocytes, with condensation of nuclear chromatin (so-called "Councilman bodies"), a manifestation of apoptotic cell death (Fig. 2). A unique feature of YF is the predominance of hepatocellular necrapoptosis in the midzone (zone 2) of the liver lobule, with sparing of cells bordering the central vein and portal tracts (Fig. 2). Hepatocytes undergo cloudy swelling, accumulate fat, and subsequently collapse as Councilman bodies. Ballooning necrosis of cells as seen in other viral hepatitides is not a feature of YF (Vieira *et al.*, 1983). The virtual absence of

to illustrate the similarity of pathological effects to that seen in the hamster model (Fig. 3) Magnification 100×. Contributed by Shu-Yuan Xiao, University of Texas Medical Branch, Galveston, TX. (See Color Insert.)

inflammation is another feature of YF and is consistent with apoptotic cell death, whereas ballooning necrosis caused by viral hepatitis is associated with diffuse mononuclear cell infiltration. Microvesicular fatty changes are a consistent finding in YF and are seen throughout the liver lobule. The changes are characterized by the presence of tiny fat globules within the cytoplasm of hepatocytes, without displacement of the cell nucleus, and are quite distinct from the large intracellular fat bodies seen in alcoholic steatosis or kwashiorkor. The microsteatosis in YF probably reflects decreased apoprotein synthesis by hepatocytes, leading to reduced coupling with lipid and secretion of lipoproteins. Finally, YF hepatitis is rarely associated with significant hemorrhage or extravasation of erythrocytes.

The reason for the peculiar midzonal distribution of hepatic cell injury is unknown. Dengue, a close relative and cause of hemorrhagic fever rarely manifest by jaundice, generally causes minimal liver lesions that are paracentral (around the central vein) rather than midzonal, although exceptions occur, with extensive midzonal necrosis and microsteatosis as seen in YF (Huerre *et al.*, 2001). Midzonal necrosis has been described in low-flow hypoxia due to ATP depletion and oxidative stress of marginally oxygenated cells at the border between anoxic and normoxic cells (Marotto *et al.*, 1988). Oxidative stress induces a variety of cytokines, such as NF-κB, interleukin (IL)-6, and RANTES, as shown for dengue virus-infected human hepatocytes (Lin *et al.*, 2000). Because the last phase of illness in YF is characterized by shock, oxidative stress might contribute to midzonal injury in YF infection. However, YF virus antigen and RNA have been observed principally in hepatocytes in the midzone (Monath *et al.*, 1989), suggesting that the predilection of these cells to virus replication is responsible for the midzonal distribution of pathological changes. Why the midzone should be preferential sites for virus replication is, nevertheless, unclear.

In studies employing the monkey model, it was suggested that endotoxin might contribute to the pathogenesis of YF (Monath *et al.*, 1981a). In other liver diseases, such as carbon tetrachloride intoxication, hepatic damage may be accentuated by enterically derived endotoxin (Nolan and Leibowitz, 1978). Damage to organized intestinal lymphoid tissue (e.g., Peyer's patches) in YF may provide a portal of entry and reduce the clearance of enterically derived endotoxin. This hypothesis has not been explored experimentally, nor is there empirical evidence that such a mechanism operates in human YF.

C. Spleen and Lymph Nodes

These tissues undergo profound changes in YF infection, characterized by the appearance of large mononuclear or histiocytic cells in and distension of the follicles and the disappearance of small lymphocytes (Klotz and Belt, 1930b). Lymphocytic elements in the germinal centers of spleen, lymph nodes, tonsils, and Peyer's patches are depleted, and the large mononuclear or histiocytic cells accumulate in the splenic follicles. Necrosis of B-cell germinal centers is more striking in monkeys than in humans (Monath *et al.*, 1981a).

D. Kidneys

On gross examination the kidneys appear swollen. Light microscopic changes are characterized by severe eosinophilic degeneration and a microvesicular fatty change of renal tubular epithelium without inflammation, analogous to the injury sustained by hepatocytes. Antigen is demonstrable by immunocytochemistry in renal tubular cells of fatal human cases (DeBrito *et al.*, 1992), suggesting direct viral injury. Despite the heavy loss of albumin in the urine indicating glomerular damage, little attention has been paid to these structures. Glomerular lesions [Schiff-positive changes in the basement membrane (Barbareschi, 1957) and cloudy swelling and degeneration of cells lining Bowman's capsule] and the presence of YF antigen in glomerulae 2–3 days after infection of monkeys (T. P. Monath, unpublished observations) indicate that direct viral injury accounts for albuminuria observed in advance of renal failure. Ultrastructural studies of glomeruli would be illuminating but have not been reported.

E. Heart

Myocardial cells undergo similar necrapoptotic changes as seen in liver and kidney, including microsteatosis. As in other organs, damage may be discontinuous and patchy and there is no significant inflammatory response. Lesions have been noted in the sinoauricular node and bundle of His (Lloyd, 1931), suggesting a basis for the paradoxical bradycardia (Faget's sign) observed in YF, and possibly for late deaths attributed to disturbances of cardiac rhythm. Antigen is found by immunocytochemistry, indicating direct viral damage to myocardial fibers. As in the liver, the reticulin structure is preserved, and no fibrosis occurs after recovery.

F. Blood Vessels and Endothelium

Damage to and plasma leakage from capillaries are characteristic of many viral hemorrhagic fevers, and YF is no exception. Gross findings at autopsy include widespread petechial hemorrhages of skin and mucosae, moderate pleural and peritoneal effusions, and pulmonary edema. Edema of the brain and other organs, particularly around blood vessels, is seen on microsopic examination. The pathogenesis of vascular leakage in YF is unknown. In the case of DHF, endothelial cells in culture are permissive to virus replication. Activation of NF-κB, which regulates the expression of proinflammatory cytokines, especially IL-8, is suspected to play a role in plasma leakage *in vivo* (Bosch *et al.*, 2002). Elevated plasma levels of IL-8 (presumably released from endothelial or mononuclear cells) have been observed in patients with DHF. Soluble vascular adhesion molecule −1 (sVCAM-1) levels in serum are elevated in DHF, a marker for endothelial damage (Murgue *et al.*, 2001). While there are no reports on cytokine dysregulation in vascular leak in cases of YF, it most certainly is a factor.

VI. Pathogenesis and Pathophysiology in the Intact Host

A. Experimental Animal Models

In rodents (mice, hamsters, and guinea pigs), the unadapted wild-type YF virus is principally neurotropic. Suckling mice are susceptible to lethal encephalitis after intracerebral (IC) or intraperitoneal (IP) inoculation, whereas adult animals develop overt encephalitis only after Ic, intraocular, or intranasal inoculation. The time to death is typically 7–10 days, depending on the virus strain and the history of sequential brain passage of the virus. Encephalitis can be induced in adult mice following infection by the peripheral route if the blood-brain barrier is disturbed by sham IC inoculation, allowing virus entry from the bloodstream (Sawyer and Lloyd, 1931). Examination of the mouse neuroinvasive phenotype of strains of YF virus has revealed that 17D vaccine is not virulent when inoculated by the IP route in mice older than 5 days, whereas mouse-adapted FNV is virulent in mice under 3 weeks of age. Wild-type strains differ in mouse neuroinvasiveness when 8-day-old mice are used. East and central African strains are attenuated, whereas west African and South American strains are virulent (Deubel *et al.*, 1987; Fitzgeorge and Bradish, 1980).

The inverse relationship between age and susceptibility to neuroinvasion after peripheral infection seen in mice is also observed in humans following vaccination with 17D or FNV vaccines. This observation underlies the contraindication of vaccine for very young infants. Zisman *et al.* (1971) suggested that the development of resistance with age was related to the maturation of macrophages involved in clearance of YF virus. The mechanism for age-related intrinsic resistance to infection of macrophages with YF viruses has not been studied, but may relate to the secretion of antiviral factors, such as interferons-α and -β, as shown for many other viruses. Macrophages from infant mice produce little endogenous interferon compared to cells from older animals.

Nonhuman primates develop viscerotropic infection characterized by hepatitis. Asian monkey species, such as rhesus and cynomolgus macaques, develop fulminant hepatitis resembling the human disease, whereas African and many New World nonhuman primates have silent infections (Bugher, 1951; Monath, 1988). However, the course of infection in susceptible monkey species is highly dependent on the YF virus strain. Experimentally infected European hedgehogs (*Erinaceus europaeus*, an insectivore), but not the African hedgehog, develop viscerotropic infections (Theiler, 1951), but these animals are not accessible for laboratory studies. Tesh and colleagues (2001; Xiao *et al.*, 2001) developed a hamster model of viscerotropic YF. Hamsters (*Mesocricetus auratus*) developed lethal infection characterized by high viremia, elevated serum aminotransferases, and bilirubinemia. Pathological changes resembled human YF (Figs. 2 and 3), including hepatocellular necrosis with apoptosis [eosinophilic degeneration (Councilman bodies)], microvesicular steatosis, and lymphoid hyperplasia in the spleen, with antigen demonstrable by immunocytochemistry. This model will be useful for future research, including comparison of wild-type YF strains, dissection of the molecular basis for virulence, and studies of pathogenesis. Lethal hepatitis in the hamster required adaptation by serial passage. The molecular changes that occurred with conversion from an avirulent to a hamster-lethal virus strain were described earlier in Section III.

The pathogenesis of YF in the rhesus monkey resembles that in humans and is characterized by jaundice, renal failure, coagulopathy, and shock (Dennis *et al.*, 1969; Monath *et al.*, 1981a; Tigertt *et al.*, 1960). However, these animals sustain a more fulminating illness than humans, lasting only 3–4 days between onset and death, and have viral loads 100-fold higher than humans based on viremia levels. Organ dysfunction and failure appear approximately 24 h before death. Pathological changes closely resemble those in humans, except that the liver and splenic and lymph node changes are more severe.

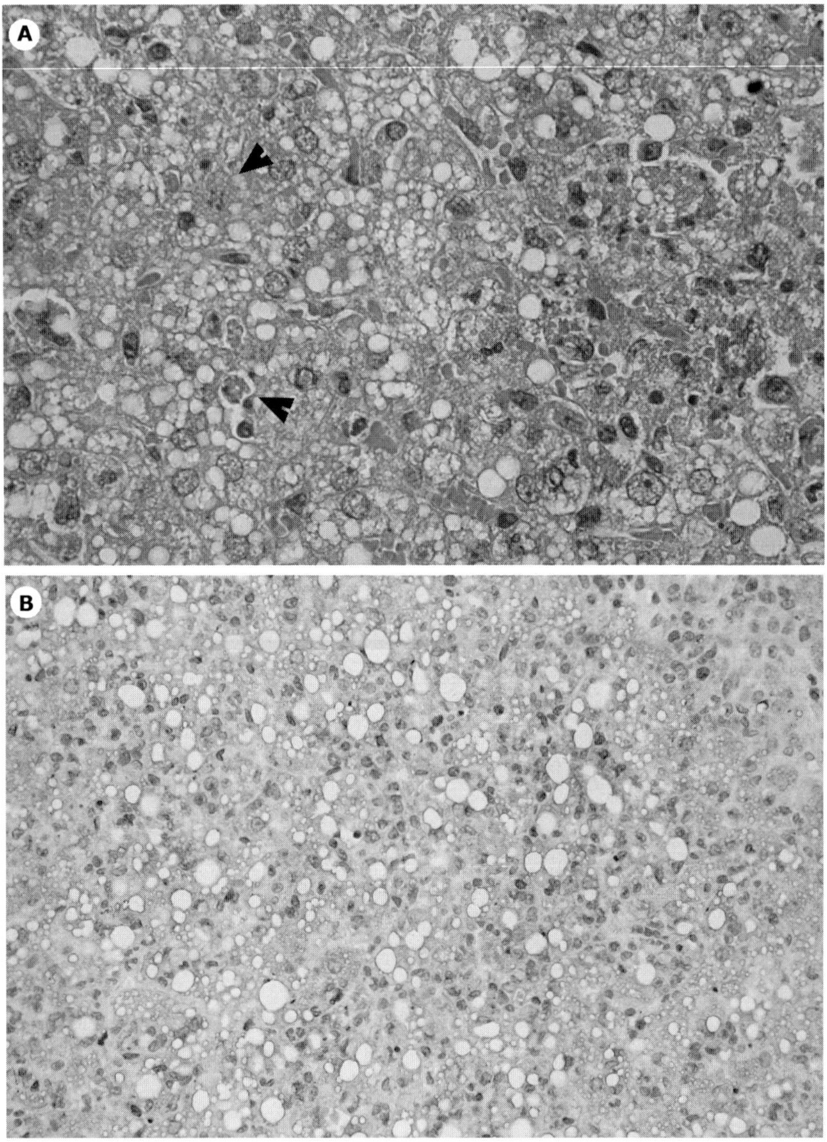

FIG 3. Histopathology of liver from a hamster infected by the Jimenez strain of yellow fever virus (postinfection day 6). (A) Hepatocytes contain small and large lipid droplets in the cytoplasm (vacuoles). Many hepatocytes are undergoing necrapoptosis (arrowhead). Hematoxylin and eosin stain. Magnification 100×. (B) Immunohistochemical staining for YFV antigen using the anti-YFV antibody. Magnification 50×. Contributed by Shu-Yuan Xiao, University of Texas Medical Branch, Galveston, TX. (See Color Insert.)

The rhesus monkey model is discussed further in the section on pathophysiology.

B. *Spread of Virus within the Host*

The course of YF virus infection in the host is believed to follow the generalized scheme described by Mims (1987) (Fig. 4). The virus is introduced into the epidermis or dermis in saliva from a blood-feeding mosquito. The dose of virus inoculated by the mosquito under conditions of natural infection is not known exactly, but is probably in the range of 3–4 $\log_{10}$ PFU. Rhesus monkeys develop lethal infection after inoculation of doses <1 PFU, with higher doses simply shortening the incubation period but not affecting outcome. After inoculation, the virus is disseminated via lymphatic channels to the draining lymph nodes. These structures release extracellular virus into the bloodstream, seeding multiple organized tissues, including the reticuloendothelial system. The latter releases virus, causing a secondary viremia with subsequent involvement of other target tissues such as kidney and heart.

In point of fact, few empirical data are available allowing confirmation of the hypothetical life cycle of YF virus. The cells involved in early infection at the cutaneous portal of entry are unknown, but observations on dengue viruses suggest that dendritic cells may play a primary role (Marovich *et al.*, 2001). Evidence from experimental animals (Monath *et al.*, 1981a; Tigertt *et al.*, 1960) suggests that lymphoid cells are important targets for YF replication (Fig. 4). After IP inoculation of rhesus monkeys, which may cause infection of the liver via the portal circulation, fixed macrophages (Kupffer cells) in the liver were infected first, within 24 h after inoculation. Viremia was detected on day 2, presumably representing the secondary viremia shed from the liver. Virus appeared in hepatocytes and renal tubules on day 2 and in bone marrow, spleen, and lymph nodes on day 3. Early infection and injury to Kupffer cells were also noted after a subcutaneous infection of monkeys. It is hypothesized that Kupffer cells play a critical role in extrinsic resistance, meaning that they protect the hepatocyte from YF infection, which progresses only after these cells undergo necraptosis (Figs. 1 and 4). In the rhesus monkey model, liver biopsies revealed only Kupffer cell degeneration, without pathological changes in hepatocytes or elevated liver enzymes until the last 24 h of life, at which point coagulative necraptosis began and chemical dysfunction occurred (Monath *et al.*, 1981a). Interferon produced by these cells (Wheelock and Edelman, 1969) may inhibit the viral infection of hepatocytes and superoxide anions, or TNF-α

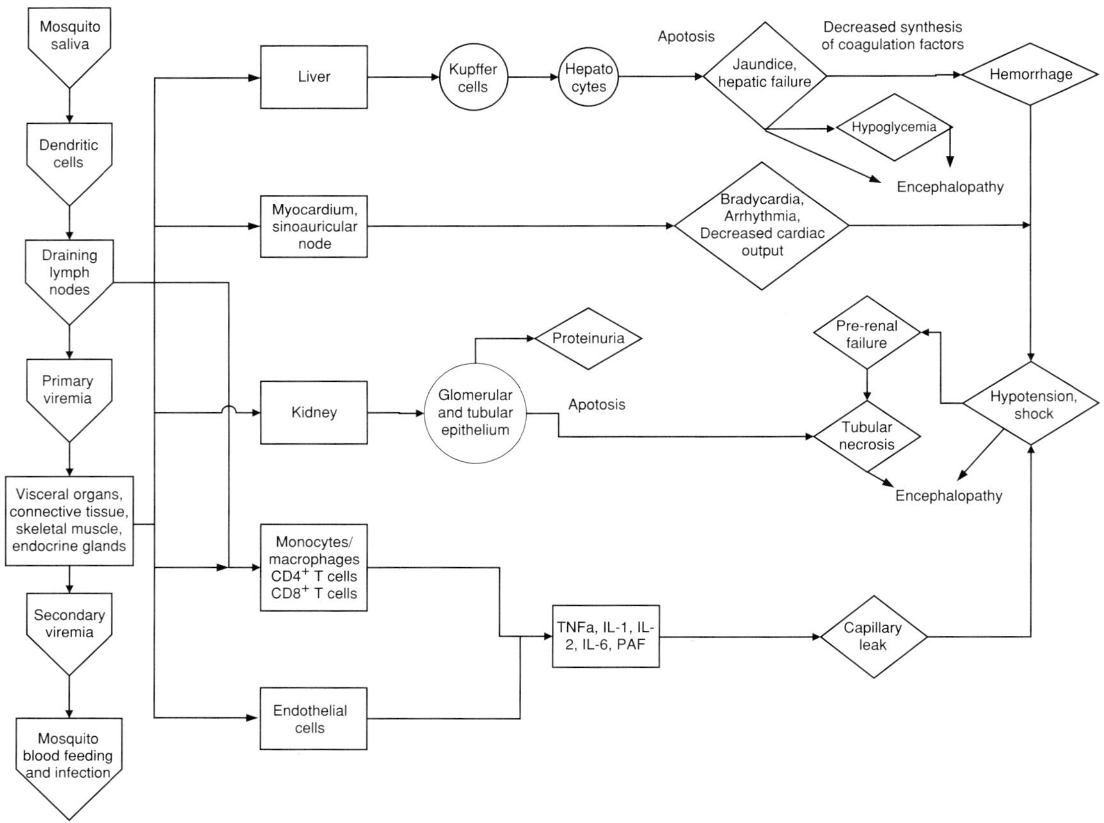

FIG 4. Pathogenesis of yellow fever based principally on hypothetical considerations and parallels with other flavivirus infections and hemorrhagic fevers.

produced by Kupffer cells may contribute to bystander effects on surrounding cells or killing of infected hepatocytes that display TNF-α receptors (CD120a) (Baker and Reddy, 1998). However, no mechanistic studies have been reported in viscerotropic models on the role of macrophages and their mediators.

C. Fever and Nonspecific Signs and Symptoms

Although patients with YF have not been studied, several lines of evidence suggest that immune activation and release of cytokines are responsible for nonspecific symptoms associated with the infection. Approximately 20% of persons inoculated with YF 17D vaccine develop similar systemic symptoms, which, however, are mild and well tolerated (Monath *et al.*, 2002c). These symptoms correspond temporally to the course of viremia, elevated serum markers of T-cell activation (neopterin and β2-microglobulin) (Reinhardt *et al.*, 1998), interferon-α (Wheelock and Sibley, 1965), TNF-α, and cytokines induced by TNF-α (IL-1 receptor antagonist and IL-6) (Hacker *et al.*, 1998). The acute-phase symptoms in YF and those observed in patients treated with interferon overlap (fever, chills, headache, myalgia/arthralgia, anorexia, nausea, and vomiting).

D. Pathogenesis of Hepatic Failure

Whereas other viral hepatitides damage the liver indirectly via the host's antigen-specific cellular immune response, YF causes direct cytopathic infection of hepatocytes. Indeed, the liver affected by YF virus is virtually devoid of inflammatory cells, whereas the liver infected with hepatitis B virus is characterized by a dramatic influx of class I-restricted cytotoxic T lymphocytes (CTLs) and secondary waves of lymphocytes and neutrophils attracted by cytokines released by the CTLs. The extent of direct acute liver injury in YF is severe and sufficient to explain the clinical course of the disease. Markers of fulminant hepatic dysfunction, such as the level of bilirubin and transaminase enzymes in blood, correlate directly with disease severity and prognosis (Elton *et al.*, 1955; Oudart and Rey, 1970).

E. Pathogenesis of Renal Failure

The pathogenesis of renal injury is poorly understood. In one study of experimental YF in rhesus macaques, the disturbance in function was prerenal until the terminal 24 h of life (Monath *et al.*, 1981a). Prerenal failure was characterized by normal histology on needle biopsy, oliguria, and diminished urine Na^+ excretion, indicative of reduced renal perfusion. During the terminal phase of infection, oliguria worsened and was accompanied by proteinuria, cylindruria, azotemia, acidosis, hyperkalemia, and shock. At necropsy, severe tubular necrosis was observed. Although direct injury by YF virus may have contributed, the observations in this model suggested that hemodynamic factors and renal ischemia caused renal failure (Fig. 4). Disordered renal function and salt and water regulation may be secondary effects of hepatic injury in YF, as well as other diseases. As seen in the rhesus model of YF, functional renal failure characterized by normal renal histology, renal retention of Na^+, low K^+ excretion, and reduced renal blood flow (measured by *p*-aminohippurate clearance) has been noted in fulminant hepatic failure when glomerular filtration rates are maintained above 3 ml/min, while below that level, acute tubular necrosis occurred (Wilkinson *et al.*, 1976).

F. Coagulation Defects

Few human patients have been studied with respect to coagulation defects (Bendersky *et al.*, 1980; Borges *et al.*, 1973; Santos *et al.*, 1973). Global reduction in clotting factors and prolonged prothrombin, partial thromboplastin, and clotting times have been demonstrated. Hemorrhage in YF occurs in the setting of marked liver impairment, and reduced synthesis of clotting factors by the liver is probably the preeminent explanation for abnormal bleeding.

However, in patients with YF, there is also some evidence for disseminated intravascular coagulation, including elevated fibrin split products and thrombocytopenia in YF (Borges *et al.*, 1973; Santos *et al.*, 1973). Thus, both coagulation and fibrinolysis are activated by the viral infection, and the balance of these countervailing effects determines hemostasis. These effects have been studied much more thoroughly in patients with DHF (Huang *et al.*, 2001), where liver function and synthesis of clotting factors are relatively well maintained compared to YF. Activation markers of coagulation (prolonged activated partial thromboplastin time, F1+2 and TATc) and of fibrinolysis (t-PA, PAPc, and D-dimer) have been found in blood from DHF patients and are probably activated by cytokines, particularly TNF-α, IL-6, and IL-1β (Huang *et al.*, 2001; Suharti *et al.*, 2002). In patients with DHF, the intrinsic pathway of coagulation is activated, involving factors XII, XI, IX, and VIII. However, contact activation triggered by the exposure of collagen, elastin, and basement membrane (as in acute injury to as blood vessel) is not responsible for activation of the intrinsic pathway in flavivirus infection, as kallikrein-C1 inhibitor levels were not detectable (van Gorp *et al.*, 2001). Inhibition of fibrinolysis is impaired in dengue [due to decreased thrombin-activatable fibrinolysis inhibitor (TAFI)] (van Gorp *et al.*, 2002) and may contribute to the severity of bleeding. Several reports suggest that dengue virus E glycoprotein sequences with homology to plasminogen may bind and activate this molecule, leading to fibrinolysis (Markoff *et al.*, 1993; Monroy and Ruiz, 2000). Some authors have suggested that virus-induced fibrinolysis is the primary event in DHF, with secondary activation of procoagulant pathways (as opposed to primary disseminated intravascular coagulation). It is unclear at this time whether the mechanisms that operate in DHF also underlie the pathogenesis of YF.

The coagulation defects in experimental YF of monkeys have been investigated. Animals developing severe and lethal infection also had global coagulation abnormalities (Dennis *et al.*, 1969; Monath *et al.*, 1981a). All clotting factors declined to very low levels, particularly fibrinogen and factors V, VII, VIII, and X, indicating that, unlike DHF, YF infection did not specifically activate the intrinsic pathway. Evidence for consumption coagulopathy was not striking, as fibrin-split products were elevated modestly and platelet counts were depressed only mildly. Heparin therapy did not change the outcome of experimental YF (Dennis *et al.*, 1969). Thus, it may be concluded that the preeminent contributor to bleeding in YF is decreased synthesis of clotting factors by the diseased liver (Fig. 4).

In two studies in rhesus monkeys, abnormalities of platelet function (reduced activation by ADP and collagen and reduced aggregation by arachidonic acid and thrombin) were demonstrated (S. Fisher-Hoch and T. P. Monath, unpublished results). Thus, in addition to reduced clotting factor synthesis, impairment of hemostasis by platelet dysfunction contributes to bleeding.

G. Blood Cells and Bone Marrow

Leukopenia, particularly neutropenia, is a feature of the early phase of YF infection in humans. Thrombocytopenia has been reported in humans, but is generally modest. In fatal experimental YF of nonhuman primates, platelet counts may also be depressed, but generally remain within normal limits (Dennis *et al.*, 1969; Monath *et al.*, 1981a). Bone marrow was examined in one study of experimental YF in rhesus monkeys (Cosgriff, unpublished data). The most prominent findings were hyperplasia and degeneration of megakaryocytes, suggesting that platelet production may have been impaired. Interferons, which are elevated following YF vaccination and may be induced more markedly by wild-type virus, may contribute to leukopenia and thrombocytopenia by bone marrow suppression and induction of a wide array of cytokines as described in dengue (La Russa and Innis, 1995; Rothwell *et al.*, 1996). Neutropenia and thrombocytopenia may also be caused by cytokines (e.g., IL-1, IL-6, TNF-α) released during the acute-phase response, upregulation of adhesion molecules on vascular endothelium, and margination of white cells and platelets as described in dengue. In dengue, infection of mononuclear cells was associated with a release of MIP-1, which was postulated to play a role in bone marrow suppression (Spain-Santana *et al.*, 2001).

H. Metabolic Changes

Limited experimental studies in monkeys have revealed an array of acid–base and electrolyte disturbances, principally occurring in the last 24–48 h of life (Liu and Griffin, 1982; Monath *et al.*, 1981a). Significant findings include increased plasma and total circulatory K^+, increased extracellular water, Na^+ and K^+, reflecting modification of cell transport mechanisms, and increased cell membrane permeability. As noted previously, these effects are likely the result of both direct YF viral injury [possibly mediated by nonstructural protein synthesis (Chang *et al.*, 1999)] and pathophysiological factors, low perfusion and hypoxia. During the preterminal period, a decrease in arterial pCO_2 was noted (reflecting respiratory compensation for metabolic acidosis). In the last few hours of life, severe metabolic acidosis and respiratory and acidosis are accompanied by hypotension and shock.

YF 17D vaccine administered to children 4–11 years old recovering from malnutrition induced a significant catabolic effect lasting up to 12 days, in the absence of fever or other signs of increased metabolism (Gandra and Scrimshaw, 1961). It is likely that wild-type YF virus would have more dramatic effects that could result in a net loss of body nitrogen and aggravate the clinical effects of protein malnutrition. Further studies in fasted adults, however, showed that metabolic changes and mobilization of protein associated with YF 17D were affected by diet and were not observed in subjects consuming a high protein diet (Bistrian *et al.*, 1977, 1981).

I. Hypotension and Shock

Endothelial cells are hypothesized to play a central role in the pathogenesis of many hemorrhagic fevers (Peters and Zaki, 2002). As noted previously, primary endothelial cells are susceptible to infection by dengue virus and release proinflammatory cytokines, IL-6 and IL-8, which are also found in sera from cases with DHF (Huang *et al.*, 2001). Infection of endothelial cells and monocyte/macrophages, leading to a cascade of events characterized by the release of cytokines and oxygen-free radicals, may trigger a series of events, including coagulation, fibrinolysis, plasma leakage, and hypotension. Cytokine dysregulation has not been studied in wild-type YF infections, nor are there studies of direct endothelial infection and injury by YF virus, but it may be conjectured that these are important factors in pathogenesis (Fig. 4).

J. Encephalopathy

In the terminal stage of infection, humans and monkeys develop depression, stupor, and coma, which is probably metabolic in origin. Contributing factors include hypoglycemia (due to impaired glycogenolysis by the liver), hypotension, and acidosis. Changes in tissue concentrations of water and electrolytes in various compartments of the central nervous system have been described, particularly intracellular dehydration of the medulla, cerebellum, and spinal cord (Liu and Griffin, 1982). Liver failure probably contributes to encephalopathy. Elevated plasma ammonia levels have been noted in one human case (Colebunders *et al.*, 2002). As noted previously, neuroinvasion by YF virus and true encephalitis are extremely rare events in viscerotropic YF infection.

K. Interactions with the Immune System

Approximately 3–4 $\log_{10}$ of YF virus are inoculated by the hematophagous female mosquito (Turell, 1988). The virus is deposited mainly in the extravascular tissues of the host during probing, as saliva injected intravascularly is apparently reingested by the mosquito during blood feeding (Turell and Tammariello, 1993). If YF and dengue share similar infection patterns, dendritic cells in the epidermis may be the first cells to undergo productive infection (Marovich *et al.*, 2001). In the immunized host, the small virus inoculum would encounter antibodies in extracellular transudate and lymph. This suggests that a low level of immunity would be sufficient to protect the host against disease, but it is not known whether immunity is sufficient to sterilize the mosquito inoculum. Individuals immunized previously with YF 17D vaccine (Monath *et al.*, 2002b; Sweet *et al.*, 1962) or persons to whom a vaccine virus is inoculated together with immune serum (Sawyer *et al.*, 1932) often show a booster or primary immune response, respectively, indicating that preformed antibody is insufficient for sterilizing immunity.

In the immunologically naive human host infected with YF virus, nonspecific innate immune responses provide the first line of defense against infection. There are few specific studies on the innate immune response to YF virus, but there is no reason to doubt that the mechanisms differ from those described for other flaviviruses. Natural killer (NK) cells and type I interferons protect the host during the early phase of virus replication, before the appearance of specific cytotoxic T cells and antibodies. Peripheral blood mononuclear cells from individuals inoculated with YF 17D had cytotoxic activity against uninfected K562 cell targets and were presumed to represent NK cell antiviral activity (Fagaeus *et al.*, 1982). Sabin (1952) demonstrated that inoculation of dengue virus simultaneously or shortly after inoculation of YF 17D vaccine delayed the onset and ameliorated dengue illness. This is not the result of cross-protective adaptive immunity, as interference between YF and a completely unrelated orbivirus (UGMP-359) was demonstrated *in vivo* (David-West, 1975). Moreover, nonhuman primates inoculated with YF 17D vaccine and challenged with virulent virus 1–3 days later (prior to the appearance of antibodies) were partially protected (Smithburn and Mahaffy, 1945). These interference phenomena were mediated by interferons, other cytokines, NK cells, or other nonspecific host resistance factors.

Because one in approximately seven persons infected with YF virus develops clinical illness (Monath *et al.*, 1980), it is clear that the innate immune system may fail to protect the host. In flavivirus-infected cells, TAP-dependent peptide transport into the ER is enhanced (Momburg *et al.*, 2001; Müllbacher and Lobigs, 1995). Unlike other viruses that downregulate MHC I protein processing, flaviviruses may increase the expression of MHC I proteins on the cell surface through the increased supply of peptides and through the action of interferon. Because MHC I proteins protect cells against attack by NK cells, the flavivirus-infected cell may avoid clearance by this important innate immune mechanism.

In vitro replication of YF is inhibited *in vitro* by type I interferons. Rhesus monkeys treated with a potent inducer of interferon-α [poly(I)–poly(C)] before or shortly after infection developed modestly elevated serum interferon levels and were protected against lethal YF infection (Stephen *et al.*, 1977). Vaccination of humans with 17D virus results in a serum interferon response (Wheelock and Sibley, 1965). During the early phase of infection with 17D virus, elevated levels of the interferon-dependent enzyme 2′,5′-oligoadenylate synthetase were found in T and B cells (Bonnevie-Nielsen *et al.*, 1995). Because interferon appears shortly after viremia and interferon may be effective if given shortly after infection (Stephen *et al.*, 1977), it could play a role in recovery from natural infection. Studies indicate that the 2′,5′-oligoadenylate synthetase system is under genetic control, at least in the mouse. This host resistance mechanism could be very important in determining susceptibility to YF virus.

Interferon-γ activates nonspecific antiviral host defense mechanisms, including NK cells and $CD14^+$ monocyte/macrophages, and upregulates MHC class I and II, Th1-dependent immunoglobulin synthesis, and CTL activity (Kurane *et al.*, 1989a, 1989b). Monkeys treated with interferon-γ had lower YF viremia and liver injury (Arroyo *et al.*, 1988), and interferon-γ knockout mice demonstrated deficient YF viral clearance and had reduced numbers of inflammatory cells after intracerebral inoculation compared to parental mice (Liu and Chambers, 2001). It is uncertain whether the *in vivo* effects were due to the immunomodulatory activity of interferon-γ, cytokines induced, or a direct antiviral effect (Kundig *et al.*, 1993).

In addition to interferon-γ, a variety of proinflammatory cytokines and markers of T-cell activation have been found during the early viremic period following vaccination in humans receiving YF vaccine, including TNF-α, IL-1Ra, neopterin, β_2-microglobulin, and circulating $CD8^+$ T cells (Hacker *et al.*, 1998, 2001; Reinhardt *et al.*, 1998). It is likely that similar (if not more pronounced) responses occur in persons infected with the wild-type YF virus.

As mentioned previously, at least seven persons are infected for every one that develops illness, in contrast to the highly susceptible rhesus macaque that develops uniformly fatal infection with some YF virus strains. Following IP inoculation, the LD_{50} of the Asibi strain in the monkey is 0.01 mouse intracerebral LD_{50}, or approximately 0.2 PFU. Higher virus doses shorten the incubation period but not the lethal outcome of infection, implying that NK cells, interferons, and other innate immune responses are insufficient to clear even a minimal infection in highly susceptible monkey species.

1. Adaptive Immune Response

Infection with YF virus or vaccine is followed by rapid development of a specific immune response. Neutralizing antibodies, cytolytic antibodies against viral proteins on the surface of infected cells, antibody-dependent cell-mediated cytotoxicity (ADCC), and cytotoxic T cells are presumed to mediate the clearance of primary infection. However, there are few data on responses other than humoral immunity in YF. Neutralizing antibodies appear 7–8 days after YF virus infection (4–5 days after disease onset) (Berry and Kitchen, 1931). The humoral response to YF virus is characterized by the appearance of IgM-neutralizing antibodies during the first week of illness (Lhuiller and Sarthou, 1983; Monath *et al.*, 1981b). Virus-specific cellular immune responses have not been measured in cases of natural infection with YF virus, but probably occur at the same time antibodies appear or even earlier. Thus, the appearance of adaptive immunity coincides with the clinical crisis and visceral injury during the period of intoxication. Circulating infectious virus or viral genomes detected by polymerase chain reaction, (PCR), as well as hemagglutinating, complement-fixing, or immunoprecipitating antigen (Hughes, 1933; Tigertt *et al.*, 1960), may be found in blood together with antibody during the period of intoxication. This suggests that immune clearance (virus–antibody complexes, complement activation, CTL-mediated clearance of infected cells, and release of cytokines) may contribute to pathogenesis, capillary leak, and shock. Although there is no direct evidence to support the notion of an immunopathological mechanism in acute YF infection, this is certainly an area for future investigation. It is worth mentioning here reports in the literature of antibody-mediated enhancement of YF (Barrett and Gould, 1986; Gould *et al.*, 1987) and other flavivirus encephalitis in rodent models. Acceleration of death occurs when the IgG antibody directed against the E protein is administered to animals that already have large amounts of viral antigen in the brain parenchyma. Immunopathological events may also contribute to naturally acquired flavivirus encephalitis, as neuroinvasion occurs late after infection. While there is no evidence that indirect immunopathological mechanisms underlie visceral tissue damage in YF, there have been no studies of immune suppression of hamsters or monkeys undergoing viscerotropic infection to dissect the role of immunity in liver injury.

Individuals with preexisting immunity to heterologous flaviviruses develop broadly cross-reactive antibody responses following infection with YF virus. In contrast to dengue, where the heterologous (enhancing) antibody may aggravate disease, heterologous flavivirus immunity may cross-protect against YF. Rhesus monkeys infected previously with dengue virus were protected against YF virus challenge, whereas passive transfer of the dengue antibody had no effect, suggesting that cellular immunity was responsible (Snijders *et al.*, 1934; Theiler and Anderson, 1975). Monkeys immunized actively with Zika and Wesselsbron viruses were also protected against YF virus challenge (Henderson *et al.*, 1970). Other studies indicated that humans with prior exposure to unspecified African flaviviruses experienced a lower incidence of YF disease than individuals with primary YF infections (Monath *et al.*, 1980). These observations indicate that prior flavivirus immunity provide protection rather than immunological enhancement of YF infection.

Antibodies are generated against the YF virus NS1 protein, which is present in the cytoplasm and on the surface of infected cells and is also secreted from infected cells. Mice and monkeys immunized against NS1 developed complement-mediated cytolytic antibodies and were protected against YF challenge (Gould *et al.*, 1986; Putnak and Schlesinger, 1990; Schlesinger *et al.*, 1986). Monoclonal antibodies against NS1 having high complement-fixation (CF) activity protected animals after passive administration (Gould *et al.*, 1986; Schlesinger *et al.*, 1985). It was suggested that antibodies against NS1 expressed on cell membranes lead to complement-mediated cytolysis and clearance of infected cells (Schlesinger *et al.*, 1995). In humans with acute YF virus infections, a soluble complement-fixing antigen (presumably NS1) is present in serum (Tigertt *et al.*, 1960), and strong CF antibody responses (presumably directed against NS1) are found by early convalescence. It is presently uncertain whether anti-NS1 contributes to protective immunity against reinfection, plays a role in virus clearance and recovery from infection, or both.

Cytotoxic T lymphocytes mediate killing of flavivirus-infected cells and contribute to viral clearance during primary infection. In studies of T- and B-cell knockout mice with YF virus encephalitis induced by IC inoculation of virus, both $CD4^+$ cells and antibodies were shown to be critical mediators of protection (Liu and Chambers, 2001). Very little is known about the human cellular immune responses to YF. There are no data on cellular responses to wild-type YF virus, and only four human subjects vaccinated with YF 17D have been studied (Co *et al.*, 2002). All subjects responded to vaccination by lymphoproliferation and cytotoxic T-cell and interferon-γ production in ELISPOT assays. Antigen-specific responses by interferon-γ ELISPOT assay were observed at the earliest time point (14 days) and the latest time point (19 months) examined after vaccination. $CD8^+$ T-cell lines were established from the subjects, most being HLA B35 restricted. $CD8^+$ T-cell epitopes were found in E, NS1, NS2B, and NS3, with T-cell frequencies specific for these epitopes in the range of 10^{-4}. van der Most *et al.* (2002) used the mouse model to show that NS3 contained a $CD8^+$ T-cell epitope, whereas the E protein contained $CD8^+$ epitopes and a $CD4^+$ epitope.

2. *Antibody-Dependent Enhancement*

Antibody-dependent enhancement of YF replication in monocyte–macrophages has been demonstrated *in vitro*. YF virus replicated to higher titers in U937 cells in the presence of serum IgG from human subjects immunized previously with 17D vaccine (Schlesinger and Brandriss, 1981). Monoclonal antibodies with YF virus-specific or flavivirus group reactivity enhanced YF virus growth in murine macrophage (P388D1) cells. Enhancement is mediated by virus complexed with a nonneutralizing antibody, which attaches to Fcγ type I or II receptors on cells. However, enhancement in macrophage-like cells was demonstrated in the presence of neutralizing antibody. This suggested that Fcγ receptor-bearing cells may facilitate YF virus infection in an immune host, whereas cells with other virus receptors are protected (Schlesinger and Brandriss, 1981). This observation may explain why individuals with YF-neutralizing antibodies who are revaccinated with YF 17D vaccine develop increases in serum antibodies (Monath *et al.*, 2002b; Wisseman and Sweet, 1962).

Evidence for *in vivo* enhancement mediated by YF antibodies is mixed. In one study, human volunteers immunized previously with YF 17D had increased immune responses to live dengue-2 vaccine, possibly due to an antibody-mediated enhancement of dengue virus replication (Bancroft *et al.*, 1981; Scott *et al.*, 1983) or to a rapid expansion of group-reactive memory T- and B-cell clones. Subjects with prior YF immunity who were inoculated with a live chimeric YF/Japanese encephalitis vaccine had slightly higher viremias than naïve subjects, also suggesting the possibility of immune enhancement (Monath *et al.*, 2002b). As noted earlier, there is no evidence for enhancement of YF infection by heterologous flavivirus immunity. However, it is possible that natural or artificial YF immunity may modulate disease expression by dengue viruses. The "natural experiment" is underway in South America, where dengue viruses have recently invaded the region endemic for YF.

3. *Other Host Factors Affecting Susceptibility*

YF virus appears to be more lethal at the extremes of age, in infants and in the elderly (Beeuwkes, 1936; Hanson, 1929; Jones and Wilson, 1972; MacNamara, 1955). YF occurs at higher incidence in males than in females. The gender difference is determined by occupational exposure to infected mosquitoes in South America, but an epidemiological explanation for the excess of cases in males is less clear in Africa. In a large clinical trial of YF 17D vaccines, the neutralizing antibody response was statistically significantly higher in males than in females (Monath *et al.*, 2002c). Moreover, the incidence of postvaccinal encephalitis caused by FNV vaccine (Rey *et al.*, 1966) and 17D vaccine (see later) was higher in males than in females. These data suggest that males undergo a more active infection with YF viruses than females.

Genetic determinants are known to determine the susceptibility of mice to flavivirus infections and to influence immune responses. Some evidence suggests that genetic factors determine the human responses to YF infection. In older reports of YF epidemics, the lethality of YF was noted to be lower in blacks than in whites, but the role of acquired immunity in resistance could not be separated from genetic factors. However, whites had a higher incidence of DHF than blacks during the epidemic in Cuba in 1981, while there was no racial difference in the background of immunity (Bravo *et al.*, 1987). Reports of YF vaccine-associated viscerotropic adverse events (see later) also indicate that genetic factors control susceptibility.

Protein-calorie malnutrition has been associated with a failure to respond immunologically to 17D vaccine. In a study of eight children (mean age 2 years) with kwashiorkor, only one seroconverted after 17D vaccination, compared to five of six controls (Brown and Katz, 1966). Although there are no reports, it is likely that malnourished individuals would have more severe disease caused by wild-type YF virus.

The neutralizing antibody responses to 17D vaccine were impaired significantly in pregnant women compared to nonpregnant females (Nasidi *et al.*, 1993). The difference was attributed to the immunosuppression associated with pregnancy. Surprisingly, pregnancy has not been reported as a risk factor for the severity of wild-type YF disease.

The rising prevalence of HIV infection in YF endemic areas represents an important acquired risk factor affecting viremia and interhuman transmission of YF by mosquitoes, the severity of disease expression, and susceptibility to YF 17D vaccine-associated adverse events. Reduced seroconversion rates following YF 17D vaccination of HIV-infected subjects have been reported (Sibailly *et al.*, 1997). Because both HIV and 17D target for human lymphoid cells, it is possible that HIV infection interferes with replication of 17D vaccine. However, it is significant that there has been no recorded increase in the incidence of reversion to virulence of 17D vaccine in HIV-infected subjects who are potentially immunosuppressed.

Smoking enhanced the immune response to YF 17D vaccine (Monath *et al.*, 2002c). There are few studies on the effect of smoking on the immune system, but most report a suppressive effect, for example, after hepatitis B vaccine (Roome *et al.*, 1993). In the case of 17D vaccine, it was postulated that the adaptive immune response might be enhanced by the suppression of innate immune responses (e.g., NK cell function) attributable to smoking.

VII. Responses to Live Yellow Fever (YF) 17D Vaccine in Humans

The YF 17D vaccine was developed in 1936 by empirical passage of the prototype Asibi virus in substrates that were restrictive for growth (minced mouse embryo followed by minced whole chick embryo). Loss of viscerotropism for monkeys occurred between the 89th and the 114th passages, and a marked reduction in mouse neurovirulence occurred between the 114th and the 176th passages (Theiler, 1951). Monkeys developed solid immunity against challenge with wild-type YF virus. The vaccine came into wide-scale use in the late 1930s. YF 17D vaccines are not plaque purified and contain heterogeneous subpopulations of plaque-size variants with differing biological properties, including mouse neurovirulence (Gould *et al.*, 1989; Liprandi, 1981; Xie *et al.*, 1998). Nucleotide sequence heterogeneities within structural and nonstructural genes and the 3′ noncoding region indicate the quasispecies nature of the YF 17D vaccine (Pugachev *et al.*, 2002). Despite these observations suggesting the possibility of genetic instability, the vaccine has a long history of safe and effective use. No individual plaque variant found in the commercial 17D vaccine has shown a mouse neurovirulence phenotype that exceeds the vaccine itself. Xie *et al.* (1998) sequenced 17D virus strains isolated from the sera of six subjects given the YF 17D vaccine and found the genomes to be identical to the vaccine virus administered to the recipients except for no more than two nucleotide changes that were found in NS5. Similarly, virus strains recovered from sera of monkeys 30 days after the intracerebral inoculation of 17D-204 vaccine contained no mutations or a single silent mutation in NS5 (Xie *et al.*, 1998). In addition, the truly clonal chimeric YF 17D virus used as a vector for foreign genes has been safe and immunogenic (Monath *et al.*, 2002b), indicating that the quasispecies nature of the 17D vaccine is not critical to vaccine performance.

This being said, rare mutational events in YF 17D vaccine have been shown to alter virulence and pathogenicity in humans. The first of two recorded fatal cases of encephalitis occurred in 1965 in a 3-year-old child who received the 17D vaccine (Anonymous, 1966). Jennings *et al.* (1994) defined the characteristics of the virus from brain tissue and commercial 17D vaccines at a similar laboratory passage level. The virus was more neurovirulent for mice and monkeys than commercial YF 17D, reacted with a wild-type specific monoclonal antibody, and contained mutations at E155 and E303 and a mutation in NS4B. The potential role of these mutations in neurovirulence, particularly the change at position E-303, was discussed in Section III.

Viremia is a measure of virus replication in extraneural tissues and is reflective of viscerotropism of the virus. Whereas the wild-type YF virus causes high viremias in monkeys and humans, the 17D vaccine induces minimal virus titers in circulating blood. Viremia has been measured in a number of studies, but may be illustrated in three reports using modern methods for infectivity determinations. In one study of 15 young adults (Wheelock and Sibley, 1965), 14 (93.3%) had low-titer viremias 3–6 days after vaccination. In a second study, 30–60% of the subjects had detectable viremias between days 4 and 6 after vaccination. (Monath *et al.*, 2002b). The mean duration of viremia was approximately 2 days; the mean peak virus titer in serum was <20 PFU/ml, and in no case did viremia exceed 2 $\log_{10}$ PFU/ml. Similarly, Reinhardt *et al.* (1998) studied viremia in 12 naïve adults 18–50 years of age, and by means of plaque assay and reverse transcription-polymerase chain reaction (RT-PCR). The highest incidence of viremia by the plaque assay was on days 5 (42%) and 6 (33%), the duration of viremia was 1–3 days, and the peak titer was 97 PFU/ml with most values <20 PFU/ml. All 12 subjects, including 5 who had no viremia by plaque assay, were positive by RT-PCR, and sera contained detectable viral RNA at least 1 day longer than by the infectivity assay. There are no data on viremia levels in infants or in the elderly or in immunosuppressed subjects who are at higher risk of neuroinvasion with YF 17D virus.

Cessation of viremia occurs at the time of appearance of neutralizing antibodies. Neutralizing antibodies following inoculation of YF 17D are found in 10% of subjects by day 7 and in 90% 10–14 days after immunization (Bonnevie-Nielsen *et al.*, 1995; Lang *et al.*, 1999; Monath, 1971; Smithburn and Mahaffy, 1945). Immunity is remarkably long lasting. This was confirmed by a study of World War II veterans conducted 30–35 years after vaccination with a single dose of 17D vaccine (Poland *et al.*, 1981). The reason for durable immunity following YF 17D is uncertain. Persistent YF infections in animals have been reported, including virus recovery from brain up to 159 days (Penna and Bittencorp, 1943), from sera 4 weeks after IC inoculation of monkeys (Xie, 1997), and from brain tissue of mice 4 weeks after inoculation (Gould *et al.*, 1989). These observations, as well as the prolonged synthesis of IgM antibodies in persons immunized with 17D virus (Monath, 1971), suggest that chronic persistent infection or storage of antigen *in vivo*, possibly in follicular dendritic cells, may explain the durability of the human immune response.

A. *YF Vaccine-Associated Serious Adverse Events*

1. *Vaccine-Associated Neurotropic Adverse Events (Postvaccinal Encephalitis)*

Cases of encephalitis caused by YF 17D vaccine have occurred principally, but not exclusively, in very young infants. A total of 26 cases are recorded in the published literature, of which 15 (13 in infants $\leq$ 4 months of old) occurred during the 1950s, before restriction of use of the vaccine in infants under 6–9 months (the age restriction varies according to different published guidelines) (Centers for Disease Control, 2002; Monath, 1999).

The incidence of postvaccinal encephalitis in very young infants has been estimated at 0.5–4/1000 based on two reports that provide denominator data (Louis *et al.*, 1981; Stuart, 1956). In contrast, the risk of developing encephalitis in persons over 9 months of age (the current minimum age recommended for routine immunization in the United States) is estimated as $\sim$1/400,000 based on the number of doses distributed in the United States in 2001–2002 and reports to the Vaccine Adverse Event Reporting System (Centers for Disease Control, 2002). As might be expected from the attenuated phenotype of YF 17D after IC inoculation of rhesus monkeys, YF 17D encephalitis in humans is generally self-limited. The clinical course is typically brief, and recovery is usually complete. One 3-year-old patient died (Anonymous, 1966), and a 29 year old (Merlo *et al.*, 1993) had residual mild ataxia 11 months after onset. A report of fatal encephalitis in an adult male with immune suppression due to HIV was presumably caused by a diminished immune response, allowing neuroinvasion and unrestrained replication of the vaccine virus in the CNS (Kengsakul *et al.*, 2002).

The increased risk of encephalitis in young infants parallels the increased susceptibility of suckling mice to neuroinvasion and neurovirulence of YF 17D virus. The possible reasons for the age-related susceptibility include (1) immaturity of the blood–brain barrier; (2) prolonged or higher viremia; and (3) immaturity of the immune system and delayed clearance of the infection. There are no data on 17D viremia levels or on the kinetics of the immune response in infants. The incidence of encephalitis following 17D vaccine has been higher in males (18/25 cases of known gender, 72%) than in females.

2. Vaccine-Associated Viscerotropic Adverse Events

This represents a newly recognized rare complication of 17D vaccines. Cases have been associated with vaccines manufactured in Brazil (17DD substrain), France, England, and the United States (17D-204 substrain). At least 13 cases (7 fatal, 54%) have been described of a syndrome closely resembling wild-type yellow fever; 9 case histories have been reported (Centers for Disease Control, 2002; Chan *et al.*, 2001; Martin *et al.*, 2001; Vasconcelos *et al.*, 2001). Eleven cases occurred in adults immunized for travel and 2 were in children and young adults living in an endemic region. Six cases were in persons >60 years, who had a complex clinical presentation labeled "multiorgan failure," reflecting some uncertainty as to the role of YF 17D in direct viral injury, and who did not have sufficient postmortem evaluation to clarify pathogenesis. In persons surviving long enough to evaluate the immune response, antibody titers to yellow fever were much higher than expected (≥1:10,240) (Martin *et al.*, 2001), consistent with an overwhelming infection (although a secondary response in the setting of prior heterologous flavivirus exposure was not ruled out). In contrast, virological evidence in cases occurring in Brazil and Australia supported the conclusion that an overwhelming infection with 17D virus was responsible (Chan *et al.*, 2001; Vasconcelos *et al.*, 2001). The syndrome was essentially identical to wild-type YF, with rapid onset of fever and malaise 2–5 days of vaccination, jaundice, oliguria, cardiovascular instability, hemorrhage, and midzonal necrosis of the liver at autopsy. Large amounts of YF viral antigen were found in liver, heart, and other affected organs. The incidence is still uncertain, but may be as high as 1/300,000–400,000.

YF vaccine-associated viscerotropic adverse events are apparently not caused by mutations arising in the virus, but appear to be related to genetically determined host susceptibility. Analyses of the vaccine lots, seed viruses, and isolates recovered from patients revealed no mutations in the vaccine that could explain the adverse events (Galler *et al.*, 2001). Virus recovered from tissues of a fatal case was inoculated into the livers of rhesus monkeys. The animals had typical YF 17D viremias and showed no signs of hepatic dysfunction or anatomical evidence of liver injury.

VIII. Summary

It will be apparent to the reader that there is much to learn about the pathogenesis of YF. The role of specific genes and molecular determinants of neurotropism and viscerotropism has been defined only partially. The availability of infectious clones and a small animal (hamster) model should allow dissection of virulence factors, which can then be tested in the more difficult monkey model. The marked differences between wild-type YF strains should be evaluated by evaluating the relationships between virulence and genome sequence. The role of cytokine dysregulation and endothelial injury in YF will be elucidated as access to patients and of patients to more sophisticated medical care improves. The number of cases of YF in unvaccinated travelers hospitalized after return from the tropics has unfortunately increased, but such cases afford unique opportunities to study the pathogenesis of renal failure, coagulopathy, vascular instability, and shock, as well as new treatment modalities. At the cellular level, there are also important opportunities for research on YF virus–cell receptor interactions, the control of apoptotic cell death, and the predilection for cells of the midzone of the liver lobule. The role of dendritic cells in the early stage of YF infection is deserving of study. Finally, the role of the immune response to infection, particularly cellular immunity, is poorly characterized, and the suggestion that immune clearance may aggravate the condition of the host during the period of intoxication should be evaluated in appropriate animal models.

References

Allison, S. L., Stiasny, K., Stadler, K., *et al.* (1999). Mapping of functional elements in the stem-anchor region of tick-borne encephalitis virus envelope protein. *E. J. Virol.* **73:**5605–5612.

Anonymous (1966). Fatal viral encephalitis following 17D yellow fever vaccine inoculation. *J. Am. Med. Soc.* **198:**671.

Arroyo, J., Guirakhoo, F., Fenner, S., *et al.* (2001). Molecular basis for attenuation of neurovirulence of a yellow fever virus/Japanese encephalitis virus chimera vaccine (ChimeriVax-JE). *J. Virol.* **75:**934–942.

Arroyo, J. I., Apperson, S. A., Cropp, C. B., *et al.* (1988). Effect of human gamma interferon on yellow fever virus infection. *Am. J. Trop. Med. Hyg.* **38:**647–650.

Baker, S. J., and Reddy, E. P. (1998). Modulation of life and death by the TNF receptor superfamily. *Oncogene* **17:**3261–3270.

Ballinger-Crabtree, M., and Miller, B. R. (1990). Partial nucleotide sequence of South American yellow fever strain 1899/91: Structural proteins and NS1. *J. Gen. Virol.* **71:**2115–2121.

Bancroft, W. H., Jr., Top, F. H., Jr., Eckels, K. H., *et al.* (1981). Dengue-2 vaccine: Virological, immunological, and clinical responses of six yellow fever immune recipients. *Infect. Immun.* **31**:698–703.

Barbareschi, G. (1957). Glomerulosi tossica in fiebre gialla. *Rev. Biol. Trop.* **5**:201.

Barrett, A. D. T., and Gould, E. A. (1986). Comparison of neurovirulence of different strains of yellow fever virus in mice. *J. Gen. Virol.* **67**:631–637.

Barrett, A. D. T., Monath, T. P., Cropp, C. B., Adkins, J. A., Ledger, T. N., Gould, E. A., Schlesinger, J. J., Kinney, R. M., and Trent, D. W. (1990). Attenuation of wild-type yellow fever virus by passage in HeLa cells. *J. Gen. Virol.* **71**:2301–2306.

Beeuwkes, H. (1936). Clinical manifestations of yellow fever in the West African native as observed during four extensive epidemics of the disease in the Gold Coast and Nigeria. *Trans. R. Soc. Trop. Med. Hyg.* **1**:61.

Bendersky, N., Carlet, J., Ricomme, J. L., Souied, G., Belaiche, J., Lange, F., Cattan, D., Lafaix, C., and Rapin, M. (1980). Deux cas mortels de fièvre jaune observés en France et contractés au Senegal. *Bull. Soc. Pathol. Exot.* **73**:54–61.

Berry, G. P., and Kitchen, S. F. (1931). Yellow fever accidentally contracted in the laboratory: A study of seven cases. *Am. J. Trop. Med. Hyg.* **11**:365.

Bielefeldt-Ohhmann, H., Meyer, M., Fitzpatrick, D. R., and Mackenzie, J. S. (2001). Dengue virus binding to human leukocyte cell lines: Receptor usage differs between cell types and virus strains. *Virus Res.* **73**:81–89.

Bistrian, B. R., George, D. T., Blackburn, G. L., *et al.* (1981). The metabolic response to yellow fever immunization: Protein-sparing modified fast. *Am. J. Clin. Nutr.* **34**:229–237.

Bistrian, B. R., Winterer, J. C., Blackburn, G. L., *et al.* (1977). Failure of yellow fever immunization to produce a catabolic response in individuals fully adapted to a protein-sparing modified fast. *Am. J. Clin. Nut.* **30**:1518–1522.

Blackwell, J. L., and Brinton, M. A. (1997). Translation elongation factor-1 alpha interacts with the 3′ stem-loop region of West Nile virus genomic RNA. *J. Virol.* **71**:6433–6444.

Bonnevie-Nielsen, V., Heron, I., *et al.* (1995). Lymphocytic 2′,5′-oligoadenylate synthetase activity increases prior to the appearance of neutralizing antibodies and immunoglobulin M and immunoglobulin G antibodies after primary and secondary immunization with yellow fever vaccine. *Clin. Diag. Lab. Immunol.* **2**:302–306.

Borges, A. P. A., Oliveira, G. S. C., and Almeida Netto, J. C. (1973). Estudo da coagulaçao sanguinea na febre amarela. *Rev. Patologia Trop.* **2**:143.

Bosch, I., Xhaja, K., Estevez, L., Raines, G., Melichar, H., Warke, R. V., Fournier, M. V., Ennis, F. A., and Rothman, A. L. (2002). Increased production of interleukin-8 in primary human monocytes and in human epithelial and endothelial cell lines after dengue virus challenge. *J. Virol.* **76**:5588–5597.

Boulos, M., Segurado, A. A., and Shirome, M. (1988). Severe yellow fever with 23-day survival. *Trop. Geog. Med.* **40**:356–358.

Bravo, J. R., Guzman, M. G., and Kouri, G. P. (1987). Why dengue haemorrhagic fever in Cuba? I. Individual risk factors for dengue haemorrhagic fever/dengue shock syndrome. *Trans. R. Soc. Trop. Med. Hyg.* **81**:816–820.

Bray, M., Men, R., Tokimatsu, I., *et al.* (1998). Genetic determinants responsible for acquisition of dengue 2 virus mouse neurovirulence. *J. Virol.* **72**:1647–1651.

Brown, R. E., and Katz, M. (1966). Failure of antibody production to yellow fever vaccine in children with Kwashiorkor. *Trop. Geogr. Med.* **18**:125–128.

Bugher, J. C. (1951). The mammalian host in yellow fever. *In* "Yellow Fever" (G. K. Strode, ed.),pp. 299–384. McGraw-Hill, New York.

Centers for Disease Control (2002). Adverse events associated with 17D-derived yellow fever vaccination—United States, 2001–2002. *Morb. Mortal Wkly. Rep.* **51:**989–993.
Chagas, E., and De Freitas, L. (1929). Electrocardiogramma na febre amarela. *Mem. Inst. Oswaldo Cruz (Rio de Janeiro)* **7:**72 Suppl. No.
Chambers, T. J., and Nickells, M. (2001). Neuroadapted yellow fever virus 17D: Genetic and biological characterization of a highly mouse-neurovirulent virus and its infectious molecular clone. *J. Virol.* **75:**10912–10922.
Chan, R. C., Penney, D. J., Litele, D., *et al.* (2001). Hepatitis and death following vaccination with 17D-204 yellow fever vaccine. *Lancet* **358:**121–122.
Chang, C. J., Luh, H. W., Wang, S. H., Lin, H. J., Lee, S. C., and Hu, S. T. (2001). The heterogeneous nuclear ribonucleoprotein K (hnRNP K) interacts with dengue virus core protein. *DNA Cell Biol.* **20:**569–577.
Chang, Y. S., Liao, C. L., Tsao, C. H., Chen, M. C., Liu, C. I., Chen, L. K., and Lin, Y. L. (1999). Membrane permeabilization by small hydrophobic nonstructural proteins of Japanese encephalitis virus. *J. Virol.* **73:**6257–6264.
Co, M. D., Terajima, M., Cruz, J., *et al.* (2002). Human cytotoxic T lymphocyte responses to live attenuated 17D yellow fever vaccine: Identification of HLA-B35-restricted CTL epitopes on nonstructural proteins NS1, NS2b, NS3, and the structural protein E. *Virology* **293:**151–163.
Colebunders, R., Mariage, J.-L., Coche, J.-Ch., Pirenne, B., Kempinaire, S., Hantson, P., Van Gompel, A., Niedrig, M., Van Esbroeck, M., Bailey, R., and Schmitz, H. (2002). A Belgian traveller who acquired yellow fever in the Gambia. *Clin. Infect. Dis.* **35:**113–116.
Collier, W. A., De Roever-Bonnet, H., and Hoekstra, J. (1959). A neurotropic variety of the vaccine strain 17D. *Trop. Geogr. Med.* **11:**80.
David-West, T. S., Labzoffsky, N. A., and Hamvas, J. J. (1972). Morphogenesis of yellow fever virus in mouse brain. *Arch. Gesamte Virusforsch.* **36:**372–379.
David-West, T. S. (1975). Concurrent and consecutive infection and immunization with yellow fever and UGMP-359 virus. *Arch. Virol.* **48:**21–28.
De Brito, T., Siqueira, S. A. C., Santos, R. T. M., *et al.* (1992). Human fatal yellow fever: Immunohistochemical detection of viral antigens in the liver, kidney and heart. *Path. Res. Pract.* **188:**177–181.
Dennis, L. H., Reisberg, B. E., Crosbie, J., *et al.* (1969). The original haemorrhagic fever: Yellow fever. *Br. J. Haematol* **17:**455–462.
Deubel, V., Schlesinger, J. J., Digoutte, J. P., and Girard, M. (1987). Comparative immunochemical and biological analysis of African and South American yellow fever viruses. *Arch. Virol.* **94:**331–338.
Duarte dos Santos, C. N., Post, P. R., Carvalho, R., *et al.* (1995). Complete nucleotide sequence of yellow fever virus vaccine strains 17DD and 17D-213. *Virus Res.* **95:**35–41.
Duarte dos Santos, C. N., Frenkiel, M.-P., Courageot, M.-P., *et al.* (2000). Determinants in the envelope E protein and viral RNA helicase NS3 that influence the induction of apoptosis in response to infection with dengue type 1 virus. *Virology* **274:**292–308.
Dunster, L. M., Wang, H., Ryman, K. D., *et al.* (1999). Molecular and biological changes associated with HeLa cell attenuation of wild-type yellow fever virus. *Virology* **261:**309–318.
Dupuy, A., Despres, P., Cahour, A., *et al.* (1989). Nucleotide sequence comparison of the genome of two 7D-204 yellow fever vaccines. *Nucleic Acids Res.* **17:**3989.
Elton, N. W., Romero, A., and Trejos, A. (1955). Clinical pathology of yellow fever. *Am. J. Clin. Pathol.* **25:**135.

Fagaeus, A., Ehrnst, A., Klein, E., *et al.* (1982). Characterization of blood mononuclear cells reacting with K 562 cells after yellow fever vaccination. *Cell. Immunol.* **67:**37–48.

Fitzgeorge, R., and Bradish, C. J. (1980). The in vivo differentiation of strains of yellow fever virus in mice. *J. Gen. Virol.* **46:**1–13.

Galler, R., Pugachev, K. V., Santos, C. L. S., *et al.* (2001). Phenotypic and molecular analyses of yellow fever 17DD vaccine virus associated with serious adverse events in Brazil. *Virology* **290:**309–319.

Gandra, Y. R., and Scrimshaw, N. S. (1961). Infection and nutritional status. II. Effect of mild virus infection induced by 17-D yellow fever vaccine on nitrogen metabolism in children. *Am. J. Clin. Nutr.* **9:**159–163.

Georges, A. J., Lesbordes, J. L., Georges-Courbot, M. C., *et al.* (1987). Fatal hepatitis from West Nile virus. *Ann Inst. Pasteur Virol.* **138:**237–244.

Germain, M., Saluzzo, J.-F., Cornet, J. P., *et al.* (1979). Isôlement du virus de la fièvre jaune à partir de la ponte et larves d'une tique. *Amblyomma variegatum. C. R. Acad. Sci. Paris D* **289:**635–637.

Germi, R., Crance, J. M., Garin, D., Guimet, J., Lortat-Jacob, H., Ruigrok, R. W., Zarski, J. P., and Drouet, E. (2002). Heparan sulfate-mediated binding of infectious dengue virus type 2 and yellow fever virus. *Virology* **292:**162–168.

Gould, E. A., Buckley, A., Cammack, N., Barrett, A. D. T., Clegg, J. C. S., Ishak, R., and Varma, M. G. R. (1985). Examination of the immunological relationships between flaviviruses using yellow fever virus monoclonal antibodies. *J. Gen. Virol.* **66:**1369–1382.

Gould, E. A., Buckley, A., Barrett, A. D. T., and Cammack, N. (1986). Neutralizing (54k) and nonneutralizing (54k and 48k) monoclonal antibodies against structural and non-structural yellow fever virus proteins confer immunity in mice. *J. Gen. Virol.* **67:**591–595.

Gould, E. A., Buckley, A., Cane, P. A., *et al.* (1989). Use of a monoclonal antibody specific for wild-type yellow fever virus to identify a wild-type antigenic variant in 17D vaccine pools. *J. Gen. Virol.* **70:**1889–1894.

Gould, E. A., Buckley, A., Groeger, B. K., *et al.* (1987). Immune enhancement of yellow fever virus neurovirulence for mice: Studies of mechanisms involved. *J. Gen. Virol.* **68:**3105–3112.

Gualano, R. C., Pryor, M. J., Cauch, M. R., *et al.* (1998). Identification of a major determinant of mouse neurovirulence of dengue virus type 2 using stably cloned genomic-length cDNA. *J. Gen. Virol.* **79:**437–446.

Hacker, U. T., Erhardt, S., Tschop, K., *et al.* (2001). Influence of the IL-1Ra gene polymorphism on in vivo synthesis of IL-1Ra and IL-1beta after live yellow fever vaccination. *Clin. Exp. Immunol.* **125:**465–469.

Hacker, U. T., Jelinek, T., Erhardt, S., *et al.* (1998). In vivo synthesis of tumor necrosis factor-alpha in healthy humans after live yellow fever vaccination. *J. Infect. Dis.* **177:**774–778.

Hahn, C. H., Dalrymple, J. M., Strauss, J. H., *et al.* (1987). Comparison of the virulent Asibi strain of yellow fever virus with the 17D vaccine strain derived from it. *Proc. Natl. Acad. Sci. USA* **84:**2019–2023.

Hanson, H. (1929). Observations on the age and sex incidence of deaths and recoveries in the yellow fever epidemic in the department of Lambayeque, Peru, in 1921. *Am. J. Trop. Med. Hyg.* **9:**233.

Henderson, B. E., Cheshire, P. P., Kirya, G. B., *et al.* (1970). Immunologic studies with yellow fever and selected African group B arboviruses in rhesus and vervet monkeys. *Am. J. Trop. Med. Hyg.* **19:**110–118.

Holbrook, M. R., Ni, H., Shope, R. E., *et al.* (2001). Amino acid substitution(s) in the stem-anchor region of Langat virus envelope protein attenuates mouse neurovirulence. *Virology* **286:**54–61.

Holzmann, H., Stiasny, K., Ecker, M., *et al.* (1997). Characterization of monoclonal antibody-escape mutants of tick-borne encephalitis virus with reduced neuroinvasiveness in mice. *J. Gen. Virol.* **78:**31–37.

Huang, Y. H., Liu, C. C., Wang, S. T., Lei, H. Y., Liu, H. L., Lin, Y. S., Wu, H. L., and Yeh, T. M. (2001). Activation of coagulation and fibrinolysis during dengue virus infection. *J. Med. Virol.* **63:**247–251.

Huerre, M. R., Lan, N. T., Marianneau, P., Hue, N. B., Khun, H., Hung, N. T., Khen, N. T., Drouet, M. T., Huong, V. T., Ba Ha, D. Q., Buisson, Y., and Deubel, V. (2001). Liver histopathology and biological correlates in five cases of fatal dengue fever in Vietnamese children. *Virch. Arch.* **438:**107–115.

Hughes, T. P. (1933). Precipitin reaction in yellow fever. *J. Immunol.* **25:**275.

Hurrelbrink, R. J., and McMinn, P. C. (2001). Attenuation of Murray Valley encephalitis virus by site-directed mutagenesis of the hinge and putative receptor-binding regions of the envelope protein. *J. Virol.* **75:**7692–7702.

Jennings, A. D., Gibson, C. A., Miller, B. R., Mathews, J. H., Mitchell, C. J., Roehrig, J. T., Wood, D. J., Taffs, F., Sil, B. K., Whitby, S. N., *et al.* (1994). Analysis of a yellow fever virus isolated from a fatal case of vaccine-associated human encephalitis. *J. Infect. Dis* **169:**512–518.

Jennings, A. D., Whitby, J. E., Minor, P. D., *et al.* (1993). Comparison of the nucleotide and deduced amino acid sequences of the structural protein genes of the yellow fever 17DD vaccine strain from Senegal with those of other yellow fever vaccine viruses. *Vaccine* **11:**679–681.

Jones, M. M., and Wilson, D. C. (1972). Clinical features of yellow fever cases at Vom Christian Hospital during the 1969 epidemic on the Jos Plateau, Nigeria. *Bull. World Health Organ.* **46:**653–657.

Kengsakul, K., Sathirapongsasuti, K., and Punyagupta, S. (2002). Fatal myeloencephalitis following yellow fever vaccination in a case with HIV infection. *J. Med. Assoc. Thai* **85:**131–134.

Klotz, O., and Belt, T. H. (1930a). Pathology in spleen in yellow fever. *Am. J. Pathol.* **6:**655.

Klotz, O., and Belt, T. H. (1930b). Pathology of the liver in yellow fever. *Am. J. Pathol.* **6:**663.

Kümmerer, B. M., and Rice, C. M. (2002). Mutations in the yellow fever virus nonstructural protein 2A selectively block production of infectious virus. *J. Virol.* **76:**4773–4784.

Kundig, T. M., Hengartner, H., and Zinkernagel, R. M. (1993). T-cell dependent interferon-γ exerts an antiviral effect in the central nervous system but not in peripheral solid organs. *J. Immunol.* **150:**2316–2321.

Kuno, G., Chang, G.-J., Tsuchiya, K. R., *et al.* (1998). Phylogeny of the genus. *Flavivirus. J. Virol.* **72:**73–83.

Kurane, I., Innis, B. L., Nisalak, A., *et al.* (1989a). Human T cell responses to dengue antigens: Proliferative responses and interferon-gamma production. *J. Clin. Invest.* **83:**506–513.

Kurane, I., Meager, A., and Ennis, F. A. (1989b). Dengue virus-specific human T-cell clones: Serotype cross-reactive proliferation, interferon-gamma production, cytotoxic activity. *J. Exp. Med.* **170:**763–775.

La Russa, V. F., and Innis, B. L. (1995). Mechanisms of dengue virus-induced bone marrow suppression. *Baillieres Clin. Haematol.* **8:**249–270.

Laemmert, H. W., Jr. (1944). Susceptibility of marmosets to different strains of yellow fever virus. *Am. J. Trop. Med.* **24:**71.

Lang, J., Zuckerman, J., Clarke, P., *et al.* (1999). Comparison of the immunogenicity and safety of two 17D yellow fever vaccines. *Am. J. Trop. Med. Hyg.* **60:**1045–1050.

Lee, E., and Lobigs, M. (2002). Mechanism of virulence attenuation of glycosaminoglycan-binding variants of Japanese encephalitis virus and Murray Valley encephalitis virus. *J. Virol.* **76:**4901–4911.

Lhullier, M., and Sarthou, J. L. (1983). Intérêt des IgM antiamariles dans le diagnostic et la surveillance épidémiologique de la fièvre jaune. *Ann. Virol. (Inst. Pasteur)* **134E:**349.

Lin, Y. L., Liu, C. C., Chuang, J. I., Lei, H. Y., Yeh, T. M., Lin, Y. S., Huang, Y. H., and Liu, H. S. (2000). Involvement of oxidative stress, NF-IL-6, and RANTES expression in dengue-2-virus-infected human liver cells. *Virology* **276:**114–126.

Lindenbach, B. D., and Rice, C. M. (2001). *Flaviviridae;* the viruses and their replication. *In* "Fields' Virology" (D. M. Knipe and P. M. Howley, eds.), Vol. I., p. 991. Lippincott Williams & Wilkins, Philadelphia.

Liprandi, F. (1981). Isolation of plaque variants differing in virulence from the 17D strain of yellow fever virus. *J. Gen. Virol.* **56:**363–370.

Liprandi, F., and Walder, R. (1983). Replication of virulent and attenuated strains of yellow fever virus in human monocytes and macrophage-like cells (U937). *Arch. Virol.* **76:**51–61.

Liu, C. T., and Griffin, M. J. (1982). Changes in body fluid compartments, tissue water and electrolyte distribution, and lipid concentration in rhesus macaques with yellow fever. *Am. J. Vet. Res.* **43:**2013–2018.

Liu, T., and Chambers, T. J. (2001). Yellow fever virus encephalitis: Properties of the brain-associated T-cell response during virus clearance in normal and gamma interferon-deficient mice and requirement for $CD4^+$ lymphocytes. *J. Virol.* **75:**2107–2118.

Lloyd, W. (1931). The myocardium in yellow fever. II. The myocardial lesions in experimental yellow fever. *Am. Heart J.* **6:**504.

Lobigs, M., Usha, R., Nestorowicz, A., *et al.* (1990). Host cell selection of Murray Valley encephalitis virus variants altered at the RGD sequence in the envelope protein and in mouse virulence. *Virology* **176:**587–595.

Louis, J. J., Chopard, P., and Larbre, F. (1981). Un cas d'encephalite après vaccination anti-amarile par la souche 17 D. *Pédiatrie* **36:**547–550.

MacNamara, F. N. (1955). "Man as the Host of the Yellow Fever Virus," pp. 1–140. MD Thesis. Cambridge Univ. Press, Cambridge, UK.

MacNamara, F. N. (1957). A clinico-pathological study of yellow fever in Nigeria. *W. Afr. Med. J.* **6:**137.

Mandl, C. W., Allison, S. L., Holzmann, H., *et al.* (2000). Attenuation of tick-borne encephalitis virus by structure-based site-specific mutagenesis of a putative flavivirus receptor binding site. *J. Virol.* **74:**9601–9609.

Mandl, C. W., Holzmann, H., Meixner, T., *et al.* (1998). Spontaneous and engineered deletions in the 3′ noncoding region of tick-borne encephalitis virus: Construction of highly attenuated mutants of a flavivirus. *J. Virol.* **72:**2132–2140.

Marianneau, P., Cardona, A., Edelman, L., *et al.* (1997). Dengue virus replication in human hepatoma cells activates NF-kappa B which in turn induces apoptotic cell death. *J. Virol.* **71:**3244–3249.

Marianneau, P., Steffan, A.-M., Royer, C., *et al.* (1998). Differing infection patterns of dengue and yellow fever viruses in a human hepatoma cell line. *J. Infect. Dis.* **178:**1270–1278.

Markoff, L. S., Innis, B. L., Houghton, R., and Henchal, L. S. (1993). Development of cross-reactive antibodies to plasminogen during the immune response to dengue. *J. Infect. Dis.* **64:**294–301.

Marotto, M. E., Thurman, R. G., and Lemasters, J. J. (1988). Early midzonal cell death during low-flow hypoxia in the isolated, perfused rat liver: Protection by allopurinol. *Hepatology* **8:**585–590.

Marovich, M., Grouard-Vogel, G., Louder, M., Eller, M., Sun, W., Wu, S. J., Putvatana, R., Murphy, G., Tassaneetrithep, B., Burgess, T., Birx, D., Hayes, C., Bat Schlesinger-Frankel, S., and Mascola, J. (2001). Human dendritic cells as targets of dengue virus infection. *J. Invest. Dermatol. Symp. Proc.* **6:**219–224.

Martin, M., Tsai, T. F., Cropp, C. B., *et al.* (2001). Fever and multisystem organ failure associated with 17D-204 vaccination: a report on four cases. *Lancet* **358:**98–104.

Martinez-Barragan, J. J., and del Angel, R. M. (2001). Identification of a putative coreceptor on Vero cells that participate in dengue-4 virus infection. *J. Virol.* **75:**7818–7827.

McArthur, M. A., Suderman, M. T., Mutebi, J.-P., Xiao, S.-Y., and Barrett, A. D. T. (2003). Molecular characterization of a hamster viscerotropic strain of yellow fever virus. *J. Virol* **77:**1462–1468.

McGavran, M. H., and White, J. D. (1964). Electron microscopic and immunofluorescent observations on monkey liver and tissue culture cells infected with the Asibi strain of yellow fever virus. *Am. J. Pathol.* **45:**501–517.

McMinn, P. C., Marshall, I. D., and Dalgarno, L. (1995). Neurovirulence and neuroinvasiveness of Murray Valley encephalitis virus mutants selected by passage in a monkey kidney cell line. *J. Gen. Virol.* **76:**865–872.

Men, R., Bray, M., Clark, D., *et al.* (1996). Dengue type 4 virus mutants containing deletions in the 3′ noncoding region of the RNA genome: Analysis of growth restriction in cell culture and altered viremia pattern and immunogenicity in rhesus monkeys. *J. Virol.* **70:**3930–3937.

Merlo, C., Steffen, R., Landis, T., *et al.* (1993). Possible association of encephalitis and 17D yellow fever vaccination in a 29-year-old traveller. *Vaccine* **11:**691.

Mims, C. A. (1987). "The Pathogenesis of Infectious Disease." Academic Press, London.

Momburg, F., Müllbacher, A., and Lobigs, M. (2001). Modulation of transporter associated with antigen processing (TAP)-mediated peptide transport into the endoplasmic reticulum by flavivirus infection. *J. Virol.* **75:**5663–5671.

Monath, T. P. (1971). Neutralizing antibody responses in the major immunoglobulin classes to yellow fever 17D vaccination of humans. *Am. J. Epidemiol.* **93:**122–129.

Monath, T. P. (1987). Yellow fever: A medically neglected disease. *Rev. Infect Dis.* **9:**165–175.

Monath, T. P. (ed.) (1988). Yellow fever. *In* "The Arboviruses: Ecology and Epidemiology," Vol. V, pp. 139–231. CRC Press, Boca Raton, Fla.

Monath, T. P. (1999). Yellow fever vaccine. *In* "Vaccines" (S. A. Plotkin and W. A. Orenstien, eds.),3rd Ed., pp. 815–879. Saunders, Philadelphia.

Monath, T. P., Arroyo, J., Levenbook, I., *et al.* (2002a). Single mutation in the flavivirus envelope protein hinge region increases neurovirulence for mice and monkeys but decreases viscerotropism for monkeys: Relevance to development and safety testing of live, attenuated vaccines. *J. Virol.* **76:**1932–1943.

Monath, T. P., Ballinger, M. E., Miller, B. R., *et al.* (1989). Detection of yellow fever viral RNA by nucleic acid hybridization and viral antigen by immunocytochemistry in fixed human liver. *Am. J. Trop. Med. Hyg.* **40:**663–668.

Monath, T. P., Brinker, K. R., Chandler, F. W., *et al.* (1981a). Pathophysiologic correlations in a rhesus monkey model of yellow fever. *Am. J. Trop. Med. Hyg.* **30:**431–433.

Monath, T. P., Craven, R. B., Adjukiewicz, A., *et al.* (1980). Yellow fever in the Gambia, 1978–1979: Epidemiologic aspects with observations on the occurrence of Orungo virus infections. *Am. J. Trop. Med. Hyg.* **29:**929–940.

Monath, T. P., Cropp, C. P., and Muth, D. J. (1981b). Indirect fluorescent antibody test for the diagnosis of yellow fever. *Trans. R. Soc. Trop. Med. Hyg.* **75:**282–286.

Monath, T. P., McCarthy, K., Bedford, P., *et al.* (2002b). Clinical proof of principle for ChimeriVax: Recombinant live, attenuated vaccines against flavivirus infections. *Vaccine* **20:**1004–1018.

Monath, T. P., Nichols, R., Archambault, W. T., *et al.* (2002c). Comparative safety and immunogenicity of two yellow fever 17D vaccines (ARILVAX and YF-VAX) in a phase III multicenter, double-blind clinical trial. *Am. J. Trop. Med. Hyg.* **66:**533–541.

Monroy, V., and Ruiz, B. H. (2000). Participation of the dengue virus in the fibrinolytic process. *Virus Genes* **21:**197–208.

Müllbacher, A., and Lobigs, M. (1995). Up-regulation of MHC class I by flavivirus-induced peptide translocation into the endoplasmic reticulum. *Immunity* **3:**207–214.

Murgue, B., Cassar, O., and Deparis, X. (2001). Plasma concentrations of sVCAM-1 and severity of dengue infections. *J. Med. Virol.* **65:**97–104.

Murphy, F. A. (1980). Morphology and morphogenesis. *In* "St. Louis Encephalitis" (T. P. Monath, ed.),pp. 65–103. Amer. Pub. Hlth. Assoc., Washington, DC.

Nasidi, A., Monath, T. P., and DeCock, K. (1989). Urban yellow fever epidemic in western Nigeria, 1987. *Trans. R. Soc. Trop. Med. Hyg.* **83:**401–406.

Nasidi, A., Monath, T. P., Vandenberg, J., *et al.* (1993). Yellow fever vaccination and pregnancy: A four-year prospective study. *Trans. R. Soc. Trop. Med. Hyg.* **87:**337–339.

Ni, H., and Barrett, A. D. T. (1998). Attenuation of Japanese encephalitis virus by selection of its mouse brain receptor preparation escape mutants. *Virology* **241:**30–36.

Nolan, J. P., and Leibowitz, M. D. (1978). Endotoxin and the liver. III. Modification of acute carbon tetrachloride injury by polymyxin B: An antiendotoxin. *Gastroenterology* **75:**445–449.

Olsthoorn, R. C., and Bol, J. F. (2001). Sequence comparison and secondary structure analysis of the 3′ noncoding region of flavivirus genomes reveals multiple pseudoknots. *RNA* **7:**1370–1377.

Oudart, J.-L., and Rey, M. (1970). Protéinurie, protéinémie et transaminasémies dans 23 cas de fièvre jaune confirmée. *Bull. World Health Org.* **42:**95.

Penna, H. A., and Bittencourt, A. (1943). Persistence of yellow fever virus in brains of monkeys immunized by cerebral inoculation. *Science* **97:**448.

Peters, C. J., and Zaki, S. R. (2002). Role of the endothelium in viral hemorrhagic fevers. *Crit. Care Med.* **30:**S268–S273.

Pletnev, A. G., Bray, M., and Lai, C.-J. (1993). Chimeric tick-borne encephalitis and dengue type 4 viruses: Effects of mutations on neurovirulence in mice. *J. Virol.* **67:**4956–4963.

Poland, J. D., Calisher, C. H., Monath, T. P., *et al.* (1981). Persistence of neutralizing antibody 30–35 years after immunization with 17D yellow fever vaccine. *Bull. World Health Org.* **59:**895–900.

Post, P. R., Santos, C. N. D., Carvalho, R., *et al.* (1992). Heterogeneity in envelope protein sequence and N-linked glycosylation among yellow fever virus vaccine strains. *Virology* **188:**160–167.

Pudney, M. (1987). Tick cell lines for the isolation and assay of arboviruses. *In* "Arboviruses in Arthropod Cells in Vitro" (C. E. Yunker, ed.), Vol. I, pp. 87–101. CRC Press, Boca Raton, Fla.

Pugachev, K. V., Ocran, S. W., Guirakhoo, F., *et al.* (2002). Heterogeneous nature of the genome of the ARILVAX yellow fever 17D vaccine revealed by consensus sequencing. *Vaccine* **20:**996–999.

Putnak, J. R., and Schlesinger, J. J. (1990). Protection of mice against yellow fever virus encephalitis by immunization with a vaccinia virus recombinant encoding the yellow fever virus non-structural proteins, NS1, NS2a and NS2b. *J. Gen. Virol.* **71:**1697–1702.

Reinhardt, B., Jaspert, R., Niedrig, M., *et al.* (1998). Development of viremia and humoral and cellular parameters of immune activation after vaccination with yellow fever virus strain 17D: A model of human flavivirus infection. *J. Med. Virol.* **56:**159–167.

Rey, F. A., Heinz, F. X., Mandl, C., *et al.* (1995). The envelope glycoprotein from the tick-borne encephalitis virus at 2Å resolution. *Nature* **375:**291–298.

Rey, M., Satge, P., Collomb, H., *et al.* (1966). Aspects épidémiologiques et cliniques des encéphalites consécutives à la vaccination antiamarile (d'après 248 cas observés dans quatre services hospitaliers de Dakar à la suite de la campagne 1965). *Bull. Soc. Méd. Afr. Noire Langue Fr.* **11:**560.

Rice, C. M., Lenches, E. M., Eddy, S. R., *et al.* (1985). Nucleotide sequence of yellow fever virus: Implications for flavivirus gene expression and evolution. *Science* **229:**726–733.

Robertson, S. E., Hull, B. P., Tomori, O., *et al.* (1996). Yellow fever: A decade of reemergence. *J. Am. Med. Soc.* **276:**1157–1162.

Roome, A. J., Walsh, S. J., Cartter, M. L., *et al.* (1993). Hepatitis B vaccine responsiveness in Connecticut public safety personnel. *J. Am. Med. Soc.* **270:**2931–2934.

Rothwell, S. W., Putnak, R., and La Russa, V. F. (1996). Dengue-2 virus infection of human bone marrow: Characterization of dengue-2 antigen-positive stromal cells. *Am. J. Trop. Med. Hyg.* **54:**503–510.

Ryman, K. D., Ledger, T. N., Campbell, G. A., Watowich, S. J., and Barrett, A. D. T. (1998). Mutation in a 17D-204 vaccine substrain-specific envelope protein epitope alters the pathogenis of yellow fever virus in mice. *Virology* **244:**59–65.

Ryman, K. D., Ledger, T. N., Weir, R. C., Jr., *et al.* (1997a). Yellow fever virus envelope protein has two discrete type-specific neutralizing epitopes. *J. Gen. Virol.* **78:**1353–1356.

Ryman, K. D., Xie, H., Ledger, T. N., *et al.* (1997b). Antigenic variants of yellow fever virus with an altered neurovirulence phenotype in mice. *Virology* **230:**376–380.

Sabin, A. (1952). Research on dengue during World War II. *Am. J. Trop. Med. Hyg.* **1:**30.

Santos, F., Pereira Lima, C., Paiva, P., *et al.* (1973). Coagulaçao intravascular disseminada aguda na febre amarela: dosagem dos factores da coagulaçao. *Brasilia Med.* **9:**9.

Sawyer, W. A., Kitchen, S. F., Frobisher, M., Jr., and Lloyd, W. (1930). Relationship of yellow fever of Western Hemisphere to that of Africa and to leptospiral jaundice. *J. Exp. Med.* **51:**493–517.

Sawyer, W. A., Kitchen, S. F., and Lloyd, W. (1932). Vaccination against yellow fever with immune serum and virus fixed for mice. *J. Exp. Med* **55:**945.

Sawyer, W. A., and Lloyd, W. (1931). The use of mice in tests of immunity against yellow fever. *J. Exp. Med.* **54:**533–555.

Schlesinger, J. J., and Brandriss, M. W. (1981). Growth of 17D yellow fever virus in a macrophage-like cell line, U937: Role of Fc and viral receptors in antibody-mediated infection. *J. Immunol* **127:**659–665.

Schlesinger, J. J., Brandriss, M. W., Cropp, C. B., *et al.* (1986). Protection against yellow fever in monkeys by immunization with yellow fever virus nonstructural protein NS1. *J. Virol.* **60:**1153–1155.

Schlesinger, J. J., Brandriss, M. W., and Monath, T. P. (1983). Monoclonal antibodies distinguish between wild and vaccine strains of yellow fever virus by neutralization, hemagglutination inhibition, and immune precipitation of the virus envelope. *Virology* **125:**8–17.

Schlesinger, J. J., Brandriss, M. W., and Walsh, E. E. (1985). Protection against 17D yellow fever encephalitis in mice by passive transfer of monoclonal antibodies to the nonstructural glycoprotein gp48 and by active immunization with gp48. *J. Immunol.* **135:**2805–2809.

Schlesinger, J. J., and Chapman, S. (1995). Neutralizing F(ab')2 fragments of protective monoclonal antibodies to yellow fever virus (YF) envelope protein fail to protect mice against lethal YF encephalitis. *J. Gen. Virol.* **76:**217–220.

Schlesinger, J. J., Chapman, S., Nestorowicz, A., *et al.* (1996). Replication of yellow fever virus in the mouse central nervous system: Comparison of neuroadapted and nonneuroadapted virus and partial sequence analysis of the neuroadapted strain. *J. Gen. Virol.* **77:**1277–1285.

Scott, R., McN, Eckels, K. H., and Bancroft, W. H. (1983). Dengue 2 vaccine: Dose response in volunteers in relation to yellow fever immune status. *J. Infect. Dis* **148:**1055–1060.

Sibailly, T. S., Wiktor, S. Z., Tsai, T. F., *et al.* (1997). Poor antibody response to yellow fever vaccination in children infected with human immunodeficiency virus type 1. *Pediat. Infect. Dis. J.* **16:**1177–1179.

Sil, B. K., Dunster, L. M., Ledger, T. N., Wills, M. R., Minor, P. D., and Barrett, A. D. T. (2000). Identification of envelope protein epitopes that are important in the attenuation process of wild-type yellow fever virus. *J. Virol.* **66:**4265–4270.

Smithburn, K. C., and Mahaffy, A. F. (1945). Immunization against yellow fever. *Am. J. Trop. Med.* **45:**217.

Snijders, E. P., Postmus, S., and Schüffner, W. (1934). On the protective power of yellow fever sera and dengue sera against yellow fever virus. *Am. J. Trop. Med.* **14:**519.

Spain-Santana, T. A., Marglin, S., Ennis, F. A., and Rothman, A. L. (2001). MIP-1 alpha and MIP-1 beta induction by dengue virus. *J. Med. Virol.* **65:**324–330.

Stefanopoulo, G. J., and Mollaret, P. (1934). Hémiplégie d'origine cérébrale et névrite optique au cours d'un cas de fièvre jaune. *Bull. Mém. Soc. Med. Hôp. Paris* **50:**1463.

Stephen, E. L., Sammons, M. L., Pannier, W. L., *et al.* (1977). Effect of a nuclease-resistant derivative of polyriboinosinic-polyribocytidylic acid complex on yellow fever in rhesus monkeys (*Macaca mulatta*). *J. Infect. Dis.* **136:**122–126.

Stevenson, L. D. (1939). Pathological changes in the central nervous system in yellow fever. *Arch. Pathol.* **27:**249.

Stuart, G. (1956). Reactions following vaccination against yellow fever. *In* "Yellow Fever Vaccination" (K. C. Smithburns, *et al.*, ed.),p. 143. World Health Organization, Geneva.

Su, H. L., Lin, Y. L., Yu, H. P., Tsao, C. H., Chen, L. K., Liu, Y. T., and Liao, C. L. (2001). The effect of human bcl-2 and bcl-X genes on dengue virus-induced apoptosis in cultured cells. *Virology* **282:**141–153.

Suharti, C., van Gorp, E. C., Setiati, T. E., Dolmans, W. M., Djokomoeljanto, R. J., Hack, C. E., Ten, C. H., and van der Meer, J. W. (2002). The role of cytokines in activation of coagulation and fibrinolysis in dengue shock syndrome. *Thromb Haemost* **87:**42–46.

Sweet, B. H., Wisseman, C. J., Jr., and Kitaoka, M. (1962). Immunological studies with group B arthropod-borne viruses. II. Effect of prior infection with Japanese encephalitis virus on the viremia in human subject following administration of 17D yellow fever vaccine. *Am. J. Trop. Med. Hyg.* **11:**562.

Sydow, F. F., Santiago, M. A., Neves-Souza, P. C., Cerqueira, D. I., Gouvea, A. S., Lavatori, M. F., Bertho, A. L., and Kubelka, C. F. (2000). Comparison of dengue infection in human mononuclear leukocytes with mosquito C6/36 and mammalian Vero cells using flow cytometry to detect virus antigen. *Mem. Inst. Oswaldo Cruz.* **95:**483–489.

Tesh, R. B., Guzman, H., da Rosa, A. P., *et al.* (2001). Experimental yellow fever virus infection in the golden hamster (*Mesocricetus auratus*). I. Virologic, biochemical, and immunologic studies. *J. Infect. Dis.* **183:**1431–1436.

Theiler, M. (1951). The virus. *In* "Yellow Fever" (G. K. Strode, ed.),pp. 46–136. McGraw Hill, New York.

Theiler, M., and Anderson, C. R. (1975). The relative resistance of dengue-immune monkeys to yellow fever virus. *Am. J. Trop. Med. Hyg.* **24:**115–117.

Tigertt, W. D., Berge, T. O., Gochenour, W. S., *et al.* (1960). Experimental yellow fever. *Trans. N. Y. Acad. Sci.* **22:**323–333.

Turell, M. J. (1988). Horizontal and vertical transmission of viruses by insect and tick vectors. *In* "The Arboviruses; Epidemiology and Ecology" (T. P. Monath, ed.), Vol. 1, pp. 127–152. CRC Press, Boca Raton, Fla.

Turell, M. J., Tammariello, R. F., and Spielman, A. (1995). Nonvascular delivery of St. Louis encephalitis and Venezuelan equine encephalitis by infected mosquitoes (Diptera Culicidae) during feeding on a vertebrate host. *J. Med. Entomol.* **32:**563–568.

van der Most, R. G., Corver, J., and Strauss, J. H. (1999). Mutagenesis of the RGD motif in the yellow fever virus 17D envelope protein. *Virology* **265:**83–95.

van der Most, R. G., Murati-Krishna, K., Ahmed, R., *et al.* (2000). Chimeric yellow fever/dengue virus as a candidate dengue vaccine: Quantitation of the dengue-specific $CD8^+$ T-cell response. *J. Virol.* **74:**8094–8101.

van Gorp, E. C., Minnema, M. C., Suharti, C., Mairuhu, A. T., Brandjes, D. P., ten Cate, H., Hack, C. E., and Meijers, J. C. (2001). Activation of coagulation factor XI, without detectable contact activation in dengue haemorrhagic fever. *Br. J. Haematol.* **113:**94–99.

van Gorp, E. C., Setiati, T. E., Mairuhu, C., Cate, Ht. H., Dolmans, W. M., Van Der Meer, J. W., Hack, C. E., and Brandjes, D. P. (2002). Impaired fibrinolysis in the pathogenesis of dengue hemorrhagic fever. *J. Med. Virol.* **67:**549–554.

Vasconcelos, P. F. C., Luna, E. J., Galler, R., *et al.* (2001). Serious adverse events associated with yellow fever 17DD vaccine in Brazil: Report of two cases. *Lancet* **358:**91–97.

Vieira, W. T., Gayotto, L. C., De Lima, C. P., *et al.* (1983). Histopathology of the human liver in yellow fever with special emphasis on the diagnostic role of the Councilman body. *Histopathology* **7:**195–208.

Wang, E., Ryman, K. D., Jennings, A. D., *et al.* (1995). Comparison of the genomes of the wild-type French viscerotropic strain of yellow fever virus with its vaccine derivative French neurotropic vaccine. *J. Gen. Virol.* **76:**2749–2755.

Wang, E., Weaver, S. C., Shope, R. E., *et al.* (1996). Genetic variation in yellow fever virus: Duplication in the 3′ noncoding region of strains from Africa. *Virology* **225:**274.

Wang, S., He, R., and Anderson, R. (1999). prM-and cell-binding activities domains of the dengue virus E protein. *J. Virol.* **73:**2547–2551.

Wheelock, E. F., and Edelman, R. (1969). Specific role of each human leukocyte type in viral infections. III. 17D yellow fever virus replication and interferon production in homogeneous leukocyte cultures treated with phytohemagglutinin. *J. Immunol* **103:**429–436.

Wheelock, E. F., and Sibley, W. A. (1965). Circulating virus, interferon and antibody after vaccination with the 17-D strain of yellow-fever virus. *N. Engl. J. Med* **273:**194–198.

Wilkinson, S. P., Arroyo, V. A., Moodie, H., Blendis, L. M., and Williams, R. (1976). Abnormalities of sodium excretion and other disorders of renal function in fulminant hepatic failure. *Gut* **17:**501–505.

Wisseman, C. L., Jr., and Sweet, B. (1962). Immunological studies with group B anthropod-borne viruses. III. Response of human subjects to revaccination with 17D strain yellow fever vaccine. *Am. J. Trop. Med. Hyg.* **11:**570.

Wisseman, C. L., Jr., Sweet, B. H., Rosenzweig, E. C., *et al.* (1963). Attenuated living type 1 dengue vaccines. *Am. J. Trop. Med. Hyg.* **12:**620–623.

Xiao, S. Y., Zhang, H., Guzman, H., and Tesh, R. B. (2001). Experimental yellow fever virus infection in the golden hamster (*Mesocricetus auratus*). II. Pathology. *J. Infect. Dis.* **183:**1437–1444.

Xie, H. (1997). Mutations in the Genome of Yellow Fever 17D-204 Vaccine Virus Accumulate in the Nonstructural Protein Genes. Ph.D. Dissertation, University of Texas Medical Branch, Galveston, Tex.

Xie, H., Ryman, H. D., Campbell, G. A., *et al.* (1998). Mutation in NS5 protein attenuates mouse neurovirulence of yellow fever 17D vaccine virus. *J. Gen. Virol.* **79:**1895–1899.

Yang, J. -S., Ramanathan, M. P., Muthumani, K., *et al.* (2002). Induction of inflammation by West Nile virus capsid through the caspase-9 apoptotic pathway. *Emerg. Infect. Dis.* **8:**1379–1384.

Zanotto, P. M., Gould, E. A., Gao, G. F., *et al.* (1996). Population dynamics of flaviviruses revealed by molecular phylogenies. *Proc. Natl. Acad. Sci. USA* **93:**548–553.

Zeng, L., Falgout, B., and Markoff, L. (1998). Identification of specific nucleotide sequences within the conserved 3′-SL in the dengue type 2 virus genome required for replication. *J. Virol.* **72:**7510–7522.

Zisman, B., Wheelock, E. F., and Allison, A. C. (1971). Role of macrophages and antibody in resistance of mice against yellow fever virus. *J. Immunol.* **107:**236–243.

ADVANCES IN VIRUS RESEARCH, VOL 60

IMMUNOLOGY AND IMMUNOPATHOGENESIS OF DENGUE DISEASE

Alan L. Rothman

Center for Infectious Disease and Vaccine Research
University of Massachusetts Medical School
Worcester, Massachusetts 01655

The pathophysiological basis of severe dengue disease (i.e., dengue hemorrhagic fever [DHF]), appears to be multifactorial, involving complex interactions among viral factors, host genetics, and the immunologic background of the host, principally prior exposure to dengue virus. Analysis of these processes has been limited to observational studies of naturally infected humans because there have not been useful animal models of dengue disease. Substantial evidence points to dengue virus-reactive T cells as a critical effector in the development of DHF. We are beginning to define the critical elements of T-cell epitope specificity and functional responses that contribute to DHF. Additional studies in well-characterized patient cohorts from different geographic regions will be needed to advance this research and guide new approaches to prevention and treatment of DHF.

0065-3527/03 $35.00

I. Introduction

The focus of this review is on the cellular immune responses to dengue viruses, specifically the responses of "classical" T lymphocytes, and the evidence for their participation in the pathogenesis of dengue disease. Other chapters in this volume have described in detail the basic biology of dengue and other flaviviruses as well as the typical clinical features of dengue disease. However, several of these characteristics are critical to the understanding of the immune response to dengue viruses and to the model of T lymphocyte-mediated immunopathogenesis of dengue disease, and therefore a brief review is in order.

II. Characteristics of Dengue Viruses

As noted elsewhere in this volume, the designation dengue virus (DEN) refers to not a single member of the flavivirus genus but rather to a complex of four closely related viruses. Initial studies found these viruses to cause identical clinical syndromes but to be distinguishable based on the ability of convalescent serum to prevent infection (Sabin, 1952), and these viruses were therefore designated as DEN serotypes 1, 2, 3, and 4.

Molecular analysis of DEN has supported the serologic definition of four distinct viruses. The overall amino acid sequence homology among the DEN serotypes is ~70% (Irie *et al.*, 1989). Sequence differences are not distributed evenly across the viral genome. Sequence differences are more frequent in the region encoding the major envelope (E) protein, as compared to other gene regions, such as the NS3 region, which are more highly conserved.

III. Sequential Infections with Dengue Viruses

Early challenge studies in human volunteers demonstrated that infection with one DEN serotype induced solid immunity to other strains of the same (homologous) DEN serotype (Sabin, 1952). However, protection against other (heterologous) DEN serotypes was short-lived, lasting no more than ~2 months, after which individuals showed no reduction in clinical disease compared to nonimmune (DEN-naïve) individuals.

Based on these challenge studies and observations of individuals who acquire natural infections in endemic regions, it can be said that each individual can develop multiple (theoretically, up to four) sequential symptomatic DEN infections during his or her lifetime. The first exposure to any of the four DEN serotypes is termed a primary DEN infection. The next infection, which must involve a DEN serotype different from the primary infection, is termed a secondary DEN infection.

Tertiary or quaternary DEN infections have been documented only in a limited number of cases, and it has been suggested that such infections are less often symptomatic because of partial immunity from the multiple prior DEN infections. However, difficulty in obtaining virologic confirmation of three separate DEN infections in endemic regions and inability to distinguish secondary and later DEN infections on the basis of serologic test results confound the interpretation of epidemiologic data with regard to this point.

IV. Clinical/Epidemiological Observations

A. *Significance and Characteristics of the Plasma Leakage Syndrome*

Efforts to achieve a fuller understanding of the pathogenesis of dengue disease rely on consistency in the description of dengue disease. Although its name suggests that clinically evident bleeding is pathognomonic for dengue hemorrhagic fever (DHF), the consensus opinion based on clinical experience is that morbidity and mortality related to dengue virus infection are more related to the associated plasma leakage syndrome and its potential to cause circulatory shock (Anonymous, 1997). In keeping with this opinion, adequate volume repletion has proved to be a highly successful strategy for treatment of DHF, reducing the case–fatality rate to less than 1% in experienced hands.

Although plasma leakage has been thought to clearly distinguish DHF from dengue fever, (DF), recent studies have called this assumption into question. Detailed prospective observation of children with acute dengue disease has shown evidence of plasma leakage in nearly half of cases, often consisting of small pleural effusions with <20% hemoconcentration (Kalayanarooj *et al.*, 1997). Furthermore, calf plethysmography of children with dengue virus infection demonstrated abnormal microvascular permeability in many children without clinical shock (Bethell *et al.*, 2001). These studies suggest that recognized DHF may represent cases at one end of a wide spectrum of disease severity, with the same underlying pathogenesis, rather than a disease with a distinct pathogenesis.

Plasma leakage in DHF follows a stereotypical pattern that itself sheds light on the mechanism of disease. Clinically overt plasma leakage develops and resolves rapidly over a short period of time in the absence of complications of systemic hypotension. Further, plasma leakage nearly always occurs late during the acute illness, at or near defervescence. This has been shown to be approximately coincident with clearance of plasma viremia and falls several days after peak plasma viral titers (Vaughn *et al.*, 1997, 2000). These features have suggested that plasma leakage is induced by immunologic responses rather than direct viral effects and is likely to relate to vascular dysfunction rather than direct tissue damage, characteristics that have been observed in cytokine-mediated plasma leakage syndromes.

B. Association of Dengue Hemorrhagic Fever (DHF) with Secondary Infections

One of the most striking epidemiologic observations is the association of DHF with secondary DEN infections (Halstead, 1980; Vaughn *et al.*, 2000). DHF clearly can occur during a primary DEN infection. This has been observed in infants during the first year of life who have residual anti-DEN antibody from transfer across the placenta (Kliks *et al.*, 1988), but also in DEN-naïve older children and adults. However, observational studies, as well as several well-designed prospective cohort studies, have demonstrated convincingly that the risk of DHF is 15–80 times higher during secondary DEN infections (Burke *et al.*, 1988; Guzman *et al.*, 1987; Sangkawibha *et al.*, 1984; Thein *et al.*, 1997). The effect of prior immunity on DHF risk may also depend on characteristics of the infecting virus. DEN-2 viruses belonging to the so-called "American" genotype have not been associated with DHF, even during secondary infections (Watts *et al.*, 1999). However, the vast majority of DHF cases in highly endemic areas, such as southeast Asia, are associated with secondary DEN infections. This epidemiologic association supports the hypothesis that plasma leakage is mediated immunologically.

V. Cellular Immunology of Dengue Viruses

A. Populations Studied

T lymphocyte responses to DEN have been studied both in humans, the natural vertebrate host for DEN, and in experimentally infected nonhuman primates and mice. However, there is only limited published information on T lymphocyte responses to DEN in these animal models. The similarity to humans in the course of viremia after experimental inoculation is a distinct advantage of the nonhuman primate model of DEN infection (Halstead *et al.*, 1973a, 1973b). However, the limited availability and expense of this animal model and the limited characterization of genes influencing T lymphocyte responses, such as major histocompatibility antigens, represent significant obstacles to the use of this model for studies of cellular immune responses to DEN. These same factors represent advantages for use of inbred mice, and more information has been published about T lymphocyte responses to DEN in that species (Roehrig *et al.*, 1992; Rothman *et al.*, 1989, 1993, 1996; Spaulding *et al.*, 1999). However, there is no measurable DEN replication after systemic infection in immunocompromised mice, and therefore attempts to develop a mouse model to define the relationship between immunology and disease pathogenesis have been unsuccessful to date.

There is a substantial body of published data on the T lymphocyte responses to DEN in humans. Important limitations on these studies are the genetic heterogeneity of the population and safety restraints on experimentation. Although techniques exist for the characterization of important immune response genes and genes controlling susceptibility to other diseases, it is not possible to prospectively control for such potential influences on the risk for severe dengue disease. The safety considerations largely preclude experimental challenge studies designed to have a highest risk of severe disease, such as sequential inoculation with virus strains known to be associated with severe disease. Subjects with severe dengue disease have been accessible only in field studies in endemic regions, in which cases the exact timing and sequence of DEN infections have been difficult to define, virus strains have often been unavailable (especially from the primary DEN infections), and most cases have occurred in children, limiting the size of specimens available for analysis.

Clinical trials of candidate live-attenuated DEN vaccines have proved to be a useful setting in which to study T lymphocyte responses to DEN. Many such studies have been conducted in nonendemic regions and in flavivirus-naïve adults. This has greatly simplified the characterization of flavivirus exposure both before and after immunization, although it has largely limited these studies to the characterization of immune responses to primary DEN infections. A more significant limitation on these studies has been that the virus strains used have been modified from naturally circulating viral strains, usually by serial *in vitro* propagation. While some candidate vaccines have proved to be insufficiently attenuated, producing clinical illness indistinguishable from naturally acquired dengue fever (Innis *et al.*, 1988), it is still possible that viral characteristics acquired during *in vitro* propagation could have modified the immune responses measured. In addition, the subjects who have participated in these vaccine trials have not been followed through secondary DEN infections, and no cases of DHF have been reported in vaccine recipients. Therefore, it has not been possible to establish clear associations between the immune responses detected and the pathogenesis of severe dengue disease during secondary DEN infections.

B. Immunologic Techniques Used

A wide variety of techniques have been used to study T lymphocyte responses to DEN and their potential roles in disease pathogenesis. These have included *in vitro* functional assays of T lymphocyte reactivity to DEN antigens, as well as *ex vivo* assays of T-cell activation occurring during acute DEN infections.

Functional responses of T lymphocytes to DEN antigens have been characterized using bulk culture techniques and studies of antigen-specific T-cell clones. Lymphocyte proliferation, *in vitro* cytokine production, and cytotoxicity assays using peripheral blood mononuclear cells (PBMC) from DEN-immune individuals have provided a basic characterization of T-cell responses to homologous and heterologous DEN, although these assays provide an imperfect quantitation of antigen-specific T cells. Studies of DEN-reactive T-cell clones have described in detail the specificity of T-cell recognition of DEN epitopes, including cross-reactivity against heterologous DEN and other flaviviruses. However, the significance of these studies is somewhat limited by their focus on a small subpopulation of the full DEN-responsive T-cell repertoire. Several novel techniques to detect and enumerate antigen-specific T cells in peripheral blood, such as cytokine ELISPOT assays, intracellular cytokine staining, and staining with HLA-peptide tetramers (McMichael and O'Callaghan, 1998), are being applied to study DEN-reactive T cells, and it is hoped that these studies will significantly enhance our understanding of DEN immunology and pathogenesis.

C. T-Cell Recognition of DEN Proteins

Immunogenic epitopes of DEN have been defined using both T-cell clones and PBMC (Gagnon *et al.*, 1996; Kurane *et al.*, 1993, 1995, 1998; Livingston *et al.*, 1995; Mathew *et al.*, 1998; Okamoto *et al.*,

1998; Zeng *et al.*, 1996; Zivna *et al.*, 2002; Zivny *et al.*, 1999). T-cell epitopes are widely distributed on the DEN polyprotein. T-cell epitopes have been found on nearly all of the DEN proteins (Table I). There appears to be a concentration of T-cell epitopes on the DEN NS3 protein (Kurane *et al.*, 1991a; Lobigs *et al.*, 1994), although this may reflect biases related to technique. Characteristics of NS3 that might explain its immunogenicity have not been defined; however, the high homology among flaviviruses suggests that there may be functional constraints on the amino acid sequence of this protein that could counterbalance immunologic pressure.

The specific epitopes recognized by DEN-reactive T lymphocytes have varied considerably between individuals in keeping with the heterogeneity of HLA class I and class II alleles. Zivna *et al.* (2002) reported, however, that T cells specific for an epitope in the NS3 protein (amino acids 222–230) recognized in the context of HLA-B*07 were detectable in 9 of 10 HLA-B*07-positive Thai subjects, suggesting that this epitope is immunodominant in individuals with this HLA class I allele. Identification of additional immunodominant T-cell epitopes will provide important tools to further define the immune response to DEN and its role in disease pathogenesis.

TABLE I

T-Cell Epitopes on Dengue Viral Proteins Defined Using T-Cell Clones

Epitope	HLA allele(s)	Serotypes recognized[a]	Reference
C (47–55)	DPw4	4	Gagnon *et al.* (1996)
C (83–92)	DR1, DPw4	2,4	Gagnon *et al.* (1996)
NS3 (71–79)	B62	2,3	Zivny *et al.* (1999)
NS3 (146–154)	DR15	1,2,3,4	Kurane *et al.* (1995)
NS3 (202–211)	DR15	1,2,3	Zeng *et al.* (1996)
NS3 (222–231)	B7	1,2,3,4	Mathew *et al.* (1998)
NS3 (224–234)	DR15	2,3,4	Kurane *et al.* (1998)
NS3 (235–243)	B62	2,3	Zivny *et al.* (1999)
NS3 (241–249)	DR15	3	Zeng *et al.* (1996)
NS3 (255–264)	DPw2	1,2,3,4	Okamoto *et al.* (1998)
NS3 (351–361)	DR15	1,3	Zeng *et al.* (1996)
NS3 (500–508)	B35	1,2,3,4	Livingston *et al.* (1995)
NS3 (527–535)	B7	1,2,3,4	Zivna *et al.* (2002)

[a] The dengue serotype that each subject was infected or immunized with is underlined.

D. Effector Functions of DEN-Reactive T Cells

Memory DEN-reactive T lymphocytes detected after primary DEN infection display a range of responses to *in vitro* stimulation with DEN virus or viral antigens. Proliferation responses of PBMC from DEN-immune donors or DEN-specific T-cell clones have been measured based on [^{3}H]-thymidine incorporation (Kurane *et al.*, 1989a). Proliferation of individual T cells in PBMC using flow cytometry, based on the dilution of the fluoresent label CFSE, has also been detected. DEN-specific proliferation has been observed in both CD4^{+} and CD8^{+} T-cell populations, although under most conditions the [^{3}H]-thymidine incorporation assay predominantly measures CD4^{+} T-cell proliferation.

Cytokine production in response to stimulation with DEN has been detected in culture supernatants of stimulated PBMC or established T-cell clones (Gagnon *et al.*, 1999; Kurane *et al.*, 1989a, 1989b). In addition, individual CD4^{+} and CD8^{+} cytokine-producing cells have been enumerated by cytokine ELISPOT assays as well as intracellular staining with anticytokine antibodies, both in slide preparations and by flow cytometry (Mori *et al.*, 1997). Gagnon *et al.* (1999) found a predominant type 1 cytokine response in bulk PBMC as well as T-cell clones involving the production of interferon (IFN)-γ, as well as tumor necrosis factor (TNF-α), interleukin (IL)-2, and TNF-β. Interestingly, minimal production of type 2 cytokines such as IL-4, except in a few subjects, has been detected even under *in vitro* conditions favoring type 2 cytokine responses.

Cytolytic responses to DEN have also been detected in both CD4^{+} and CD8^{+} T-cell populations, although CD8^{+} cytotoxic T cells contribute the majority of activity in short-term stimulated T-cell lines (Mathew *et al.*, 1996). Gagnon *et al.* (1999) studied the mechanisms of target cell lysis by a panel of DEN-reactive CD4^{+} cytotoxic T-cell clones. In addition to lysis of cognate target cells (autologous cells expressing DEN antigens), most of the CD4^{+} T-cell clones also lysed uninfected bystander target cells once they were activated by contact with cognate target cells. Lysis of cognate target cells was predominantly perforin mediated, whereas lysis of bystander target cells involved Fas–FasL interactions. The ability of DEN-reactive CD4^{+} T cells to lyse cells not expressing viral antigens could result in damage to uninfected cells such as hepatocytes *in vivo*.

E. T-Cell Cross-Reactivity

The relatively high sequence homology among the four DEN serotypes, as well as other flaviviruses, contributes to substantial cross-reactivity among DEN-reactive memory T cells. In assays of bulk PBMC responses, proliferation, cytokine production, and cytolytic responses to one or more of the heterologous DEN serotypes have been detected after primary DEN infection in nearly all subjects studied, although the responses to the homologous DEN serotype nearly always have been significantly higher than the responses to the heterologous serotypes (Dharakul *et al.*, 1994; Kurane *et al.*, 1989a; Mathew *et al.*, 1996). Studies of T-cell clones have demonstrated that the pattern of cross-reactivity in bulk PBMC reflects the presence of individual T-cell clones specific for the homologous serotype, as well as clones that recognize one of more of the heterologous serotypes. This has been demonstrated directly in PBMC using flow cytometry to measure IFN-γ production at the single cell level.

Individual T-cell clones have been identified that recognize other flaviviruses, including Kunjin, Japanese encephalitis, West Nile, and yellow fever viruses (Kurane *et al.*, 1995; Spaulding *et al.*, 1999). Responses to these flaviviruses in bulk PBMC have generally been low or undetectable in the absence of exposure to these viruses. However, T-cell cross-reactivity between DEN and other flaviviruses may explain the observed enhanced antibody responses to candidate live-attenuated DEN vaccines in yellow fever virus-immunized individuals (Scott *et al.*, 1983).

Patterns of cross-reactivity with heterologous DEN serotypes and other flaviviruses have been somewhat predictable in bulk PBMC in that greater cross-reactivity has been noted between serotypes 2 and 4 or between serotypes 1 and 3 than for other pairs (Green *et al.*, 1993). However, cross-reactivity becomes less predictable at the level of individual T-cell clones because the effect of specific amino acid changes on T-cell recognition depends on their position within the epitope. Individual T-cell clones also display different sensitivity to amino acid substitutions, even within the same epitope. Thus, T-cell clones have been identified that recognize other flaviviruses but do not recognize all DEN serotypes.

Cross-reactivity to heterologous DEN serotypes also appears to be diverse with regard to specific effector functions. Zivny *et al.* (1999) noted that $CD8^+$ T cells from a DEN-3-immunized individual specific for an epitope on the NS3 protein (amino acids 71–79) could recognize cells expressing the corresponding DEN-2 sequence in cytotoxicity assays, but did not demonstrate IFN-γ production or proliferation responses to the DEN-2 sequence. The DEN-3 (SVKKDLISY) and DEN-2 (DVKKDLISY) sequences differ by a single amino acid. It has been suggested that the heterologous DEN-2 sequence does not fully activate the $CD8^+$ T cells and therefore induces only a part of the available effector response profile of these cells.

F. Responses to Primary versus Secondary DEN Infections

As noted earlier, primary DEN infections induce a combination of T-cell responses specific to the homologous serotype, as well as T-cell responses cross-reactive to one or more heterologous serotypes. During secondary DEN infections, these memory DEN-reactive T cells would be expected to be present at a higher frequency than naïve T cells reactive only with the novel DEN serotype. In addition, because memory T cells have a lower threshold for activation than naïve T cells, the preexisting memory DEN-reactive T cells would be expected to respond more rapidly than naïve DEN-reactive T cells. The competition from memory T cells would likely alter the profile of the T-cell response to secondary DEN infection.

Although limited, available data on T-cell responses to DEN following secondary DEN infection support these predictions. Mathew *et al.* (1998) found that the predominant cytolytic responses in bulk PBMC obtained after secondary DEN infection and in T-cell clones isolated from these PBMC were broadly serotype cross-reactive and were also directed predominantly at epitopes on nonstructural DEN proteins.

VI. Immunopathogenesis of DHF

A. Mechanisms of Plasma Leakage

Based on the aforementioned *in vitro* observations regarding memory T lymphocyte responses after primary DEN infection, Rothman and Ennis (1999) proposed a model of immunopathogenesis of plasma leakage in DHF mediated through interactions between DEN-infected cells and memory $CD4^+$ and $CD8^+$ DEN-reactive T lymphocytes (Fig. 1). It was hypothesized that viral entry into target cells (principally monocytes and/or macrophages) through the putative viral receptor, as well as through the uptake of virus–antibody complexes via Fcγ receptors, represents the initial, afferent phase of disease. Subsequently, these target cells present DEN peptides in the context of both HLA class I and class II molecules. This stimulates the efferent, response phase of disease, involving the activation of DEN-reactive memory T lymphocytes to proliferate and produce proinflammatory type 1 cytokines, such

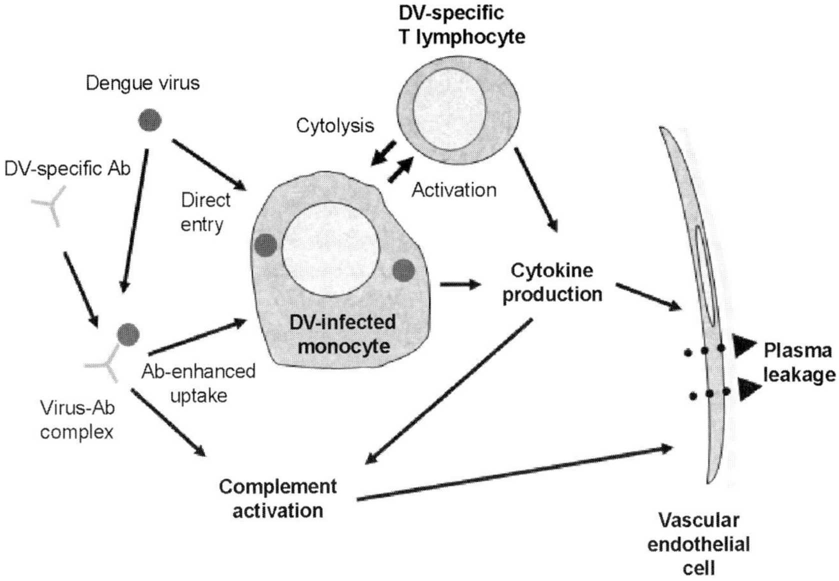

FIG 1. Model of T-cell-mediated immunopathogenesis of plasma leakage in dengue hemorrhagic fever. This model proposes that plasma leakage is caused directly by effects of cytokines such as IFN-γ and TNF-α and products of complement activation on vascular endothelial cells. In this model, memory dengue virus-specific T cells induced by an earlier primary dengue virus infection are capable of recognizing cells infected with the new dengue virus serotype; memory T cells are activated more rapidly than naïve T cells, resulting in greater cytokine production in the secondary dengue virus infection. Antibody-enhanced viral uptake in monocytes (and other Fcγ receptor-bearing cells) results in a greater burden of virus-infected cells, which express viral antigens, further augmenting activation of virus-specific T cells. Source: www.umassmed.edu/cidvr/faculty/rothman.cfm.

as IFN-γ and TNF-α, which directly affect vascular endothelial cells to produce plasma leakage.

The increased incidence of DHF in secondary DEN infections is explained in this model by the greater magnitude and more rapid kinetics of activation of memory T lymphocytes already present at the time of secondary DEN infections. Therefore, the major predictions of this model are (1) that there is a higher level of T-cell activation *in vivo* in individuals with DHF than in individuals with mild disease; (2) that the levels of proinflammatory cytokines produced *in vivo* in individuals with DHF are higher than in individuals with mild disease; and (3) that memory T lymphocytes that recognize serotype cross-reactive epitopes (found more commonly in the more highly conserved non-structural proteins) will be expanded preferentially during secondary DEN infection, particularly in patients with DHF.

1. T-Cell Activation

These predictions have been tested in patients with acute dengue virus infection. Circulating levels of soluble CD4 (sCD4), sCD8, soluble IL-2 receptor α chain (sIL-2R), and soluble TNF receptors (sTNFR) have been measured (Bethell *et al.*, 1998; Green *et al.*, 1999b; Hober *et al.*, 1996; Kurane *et al.*, 1991b). These products are released from activated T lymphocytes and serve as markers of T-cell activation, although some, such as sTNFR, are not completely specific for T lymphocyte activation. Elevated serum or plasma levels of these molecules have been reported. Elevated levels of sCD4 were found in one study (Kurane *et al.*, 1991b) but not in a separate cohort of subjects (Green *et al.*, 1999b). Elevations in sTNFR, particularly the 75-kDa form (sTNFR-II), was the most striking finding (Green *et al.*, 1999b; Hober *et al.*, 1996). Interestingly, elevations in plasma levels of sTNFR-II correlated with the degree of plasma leakage as early as 2 days prior to the onset of plasma leakage (Green *et al.*, 1999b).

Green *et al.* (1999a) also directly demonstrated high levels of T-cell activation in DHF using flow cytometry to examine the expression of activation markers on circulating T cells. In particular, a high percentage of $CD8^+$ T cells expressing the early activation marker CD69 was noted. The percentage of $CD69^+$ cells among $CD8^+$ T cells was significantly higher in individuals with DHF than in those with dengue fever. The percentage of $CD69^+$ cells was highest at the earliest time point studied, during the febrile period, indicating that T-cell activation preceded plasma leakage.

2. Cytokine Production

The aforementioned evidence of T-cell activation is consistent with the model of immunopathogenesis, but does not identify the specific mediators of plasma leakage. Therefore, levels of various cytokines in serum or plasma of patients with acute DEN infections were measured (Bethell *et al.*, 1998; Green *et al.*, 1999b, 1999c; Hober *et al.*, 1993; Kurane *et al.*, 1991b). Interpretation of the results of these studies is subject to several limitations, however. If cytokine production occurs predominantly in extravascular sites, cytokine levels in the circulation may sample a small fraction of the total cytokine production. Furthermore, the half-life of cytokines in the circulation is short; therefore, analysis of single samples or even of samples collected once daily may not accurately reflect *in vivo* cytokine production during all phases of disease.

Elevated levels of several proinflammatory type 1 cytokines, such as TNF-α, IFN-γ, and IL-18, have been detected in acute-phase sera or plasma of patients with DHF (Table II) (Green *et al.*, 1999b; Hober *et al.*, 1993; Mustafa *et al.*, 2001). Green *et al.* (1999b) found that plasma levels of IFN-γ were also higher in patients with DF than in normal controls; however, IFN-γ elevations were more striking in patients with DHF than in those with DF. Some groups have also reported elevations in serum levels of IL-1β and IL-6; however, these have not been confirmed in other studies (Green *et al.*, 1999b; Hober *et al.*, 1993).

Interestingly, elevated levels of type 2 cytokines, which have predominantly anti-inflammatory activity in animal models, have also been detected (Chaturvedi *et al.*, 2000; Green *et al.*, 1999c). In particular, plasma levels of IL-10 were significantly higher in patients with DHF than in those with DF, and the levels of IL-10 correlated with the degree of plasma leakage (Green *et al.*, 1999c). This apparent contradiction may reflect normal homeostatic mechanisms providing negative feedback for proinflammatory cytokine production.

Gagnon *et al.* (2002) have demonstrated cytokine production *in vivo* by PBMC of patients with acute DEN infections. Expression of mRNA for IFN-γ, TNF-α, and TNF-β was detected *ex vivo* by RT–PCR. Furthermore, intracellular IFN-γ, TNF-α, and TNF-β, as well as IL-2 and IL-4, proteins were detected by antibody staining of cytospin preparations of PBMC without *in vitro* restimulation. Although these data

TABLE II
ALTERED CYTOKINE LEVELS IN PATIENTS WITH ACUTE DENGUE VIRUS INFECTIONS

Cytokine	Findings[a]	Reference
TNF-α	DHF > normal	Hober *et al.* (1993)
	DHF > DF = OFI = normal	Green *et al.* (1999b)
IL-2	DHF = DF > normal	Kurane *et al.* (1991)
IL-6	DHF > normal	Hober *et al.* (1993)
IL-10	DHF > DF	Green *et al.* (1999c)
IL-12	DHF < DF	Green *et al.* (1999c)
IL-18	DHF > DF > normal	Mustafa *et al.* (2001)
IFN-γ	DHF > DF > OFI = normal	Green *et al.* (1999b)

[a] DHF, dengue hemorrhagic fever; DF, dengue fever; OFI, other (nondengue) febrile illnesses.

indicate that PBMC contribute to cytokine production *in vivo*, the frequency of cytokine-containing cells in PBMC was extremely low. This finding may reflect localization of cytokine-secreting cells in tissues rather than in the circulation, as suggested earlier.

3. Expansion of DEN-Reactive T Cells

Both molecular and functional approaches have been used to define the T-cell population that is responding during acute DEN infection. The molecular approach takes advantage of the unique pattern of DNA rearrangements that occur in the genes for the T-cell receptor (TCR) α and β chains during T-cell maturation in the thymus. Individual T-cell clones can therefore be distinguished from one another in the circulation. Gagnon *et al.* (2001) analyzed the pattern of TCR β chain gene V region usage in PBMC obtained at several time points during and after acute DEN infection using a semiquantitative RT–PCR method. Marked oligoclonal T-cell expansions, evidenced by shifts in TCR Vβ gene usage, were detected during the acute DEN infection. Because nonspecific T-cell activation and expansion would be expected to affect all T-cell populations equally, these data provide evidence of an antigen-driven expansion of T cells by the acute DEN infection. Specificity of the expanded T-cell population for DEN must be inferred, however, because the molecular approach does not provide functional information about antigen specificity.

To demonstrate formally that *in vivo* cytokine production and T-cell activation reflect activation of memory DEN-reactive T cells, it is therefore essential to study these antigen-specific T cells directly. Direct detection of antigen-specific cells *ex vivo* is difficult; however, such work is now progressing using techniques developed to study other viral infections. The techniques of cytokine ELISPOT assays, intracellular cytokine staining, and HLA-peptide tetramer staining have been adapted to study the frequency of DEN-reactive T cells at various time points in patients with DHF and DF.

Loke *et al.* (2001) used an IFN-γ ELISPOT assay to examine the T-cell response to a series of peptides predicted to bind well to common HLA alleles in Vietnam. They detected peptide-specific T cells in PBMC of several subjects with DHF. Interestingly, some of the individuals in whom peptide-specific T cells were detected did not have the HLA allele predicted to bind that peptide. These results demonstrated the presence of DEN-reactive T cells in PBMC at the time of plasma leakage.

Zivna *et al.* (2002) analyzed the response to a highly conserved HLA-B*07-restricted epitope (NS3 222–230) in 10 HLA-B*07+ subjects, of whom five had DHF and five had DF. Peptide-specific T cells were detected by an IFN-γ ELISPOT assay in PBMC obtained after the acute DEN infection from nine of the 10 subjects; the nine responders had experienced secondary DEN infections, whereas the one nonresponder had experienced a primary DEN infection. The frequency of peptide-specific T cells was significantly higher in the subjects with DHF than in those with DF. It was possible to study PBMC obtained during the acute phase of illness from five subjects; high frequencies of peptide-specific T cells were detected in PBMC from two of three subjects with DHF but not in PBMC from two subjects with DF. In one subject with DF but no detectable response to this epitope, DEN-reactive T cells were directed at a different viral epitope. These data suggest that the epitope specificity of the T-cell response to DEN may influence the risk for DHF.

B. *Factors Influencing DHF Risk*

1. *HLA*

The possibility that specific T-cell responses may be more likely to induce DHF has raised interest in associations between DHF and specific immune response genes, particularly HLA class I and class II alleles. Earlier studies from Thailand (Chiewsilp *et al.*, 1981) and Cuba (Paradoa Perez *et al.*, 1987) identified several candidate HLA class I associations with disease, as did a study from Vietnam (Loke *et al.*, 2001). All three studies followed a similar approach, comparing HLA allele frequencies in subjects with DHF to those in the general population of the same country. However, the specific HLA alleles found to be associated positively or negatively with DHF differed in each study (Table III).

The differences in results among these studies could be attributed to differences in the genetic background of the population studied or to differences in the viruses that predominated in the study population. Experiments in inbred laboratory mice demonstrated that the infecting DEN serotype can influence the T-cell response to a highly conserved epitope even in genetically identical subjects (Spaulding *et al.*, 1999). To analyze the associations of DHF with both HLA class I alleles and DEN serotype, Stephens *et al.* (2002) studied a prospectively enrolled cohort of Thai children, among whom the infecting DEN serotype was known in >95% of cases. This study cohort also included children with DF to provide additional comparisons to confirm HLA associations with disease severity. Results of this analysis showed that specific HLA alleles, such as A*0207, were associated with DHF only during secondary DEN infections. Interestingly, other HLA

TABLE III
HLA CLASS I ASSOCIATIONS WITH DENGUE HEMORRHAGIC FEVER

Reference	Alleles showing positive association	Alleles showing negative association
Chiewsilp *et al.* (1981)	A2, B blank	B13
Paradoa-Perez *et al.* (1987)	A1, B blank, Cw1	A29
Loke *et al.* (2001)	A24	A33
Stephens *et al.* (2002)		
Any serotype	A*0207, B*51	A*0203, B*52, B4, B62, B76, B77
Dengue-1	A*0207, B*51	A*0203
Dengue-2	A*0207	B*52
Dengue-3	B*51	A*0203

alleles, such as A*0203, showed an association with DF rather than DHF. Furthermore, the strength of association of specific HLA class I alleles with DHF depended on the DEN serotype. These findings of complex associations among HLA, viral serotype, and disease will require confirmation in other patient cohorts.

2. Viral Burden

Several studies have demonstrated that *in vivo* viral replication, measured as viremia, is greater in subjects with DHF than in those with DF (Murgue *et al.*, 2000; Vaughn *et al.*, 2000). As shown in the model of DHF immunopathogenesis described earlier, presentation of DEN antigens by infected cells is the driving force for activation of DEN-reactive memory T cells. If viral replication is higher in DHF than in DF, this could itself induce greater activation of T cells *in vivo*. To address this possibility, Libraty *et al.* (2002) analyzed levels of both viral RNA and key cytokines and immune activation markers in serial plasma samples obtained from a group of Thai children with acute secondary DEN-3 infections. These data demonstrated a correlation between viremia and immune activation. By multivariate logistic regression analysis, the viral RNA level was the only factor that independently correlated with plasma leakage, as measured by the size of the pleural effusion. Alternatively, viral RNA, IL-10, and sTNFR-II levels were each independently (negatively) correlated with the lowest platelet count during illness, and sIL-2R was the only factor that showed an independent correlation with the highest levels of hepatic transaminases.

This analysis also showed a characteristic timing of peak viremia, innate immune responses, and type 1 and type 2 adaptive immune responses during illness. Plasma viremia was typically highest on the day of presentation. IFN-α levels were also highest at presentation. IFN-γ and IL-10 reached their highest levels a median of 2 days after the peak of viremia, whereas sIL-2R and sTNFR-II levels reached their highest levels 3 days after the peak of viremia. The peak of plasma IFN-γ levels occurred significantly earlier in subjects with DHF than in those with DF; in all patients with DHF but in only 50% of those with DF, the IFN-γ peak occurred within 2 days of the peak of viremia. No significant difference was noted in the timing of peak IL-10 levels relative to peak viremia. These data suggest that the timing of the type 1 (proinflammatory) and the counterbalancing type 2 (anti-inflammatory) responses relative to viremia may be critical in the development of DHF.

3. Chemokines

The association between viremia and immune activation may involve other mechanisms than antigen presentation alone. Interactions between the innate and adaptive immune responses might also contribute to the immunopathogenesis of DHF. Studies have demonstrated that DEN infection induces the production of various chemoattractant cytokines (chemokines), including IL-8, RANTES, MIP-1α, and MIP-1β (Avirutnan *et al.*, 1998; Bosch *et al.*, 2002; Chen and Wang, 2002; Huang *et al.*, 2000; King *et al.*, 2002; Lin *et al.*, 2000; Murgue *et al.*, 1998; Spain-Santana *et al.*, 2001). This effect has been demonstrated in many different cell types, including primary human monocytes and monocyte-derived macrophages, myelomonocytic cell lines, mastocyte cell lines, endothelial cells, and epithelial cell lines. Expression of chemokines *in vivo* during acute DEN infection has also been demonstrated by ELISA and ribonuclease protection assays (Avirutnan *et al.*, 1998; Juffrie *et al.*, 2000; Raghupathy *et al.*, 1998; Spain-Santana *et al.*, 2001). Chemokines attract T lymphocytes and other inflammatory cell types (Luster, 1998), often with a greater effect on memory T cells than on naïve T cells. This would be expected to result in a greater recruitment of DEN-reactive T cells during secondary DEN infections than during primary DEN infections, enhancing the opportunity for activation of DEN-reactive T cells.

4. Characteristics of Preinfection T-Cell Responses

As noted earlier, considerable differences exist between individuals in the T-cell response to primary DEN infections, which reflect differences in viral serotypes, HLA alleles, and perhaps other uncharacterized factors. To what extent then do these differences influence the risk for DHF? The answer to this question has enormous implications for DEN vaccine development. This question is being addressed through prospective cohort studies. As an initial approach, Mangada *et al.* (2002) analyzed *in vitro* proliferation and cytokine responses to DEN antigens in PBMC collected prior to secondary DEN-3 infections in 22 Thai children. While no differences were found in mean proliferation or cytokine (IFN-γ and TNF-α) responses to DEN antigens between subjects who were hospitalized during the acute infection and those who did not require hospitalization, several patterns of T-cell responses appeared to be associated with increased or decreased disease severity. Four subjects demonstrated *in vitro* TNF-α production in response to one or more DEN serotypes; all four were hospitalized during the acute DEN infection. In contrast, four subjects did not have detectable proliferation or cytokine responses to any DEN serotype, despite serologic evidence of prior DEN infection; none of these four subjects were hospitalized during the acute DEN infection. Subjects with broadly serotype cross-reactive IFN-γ responses were also less likely to be hospitalized during acute DEN infection.

VII. Needs for Future Research

The emerging picture based on recent research findings is of complex interactions between functional characteristics of the memory T-cell response after primary DEN infection and the viral burden early in secondary DEN infection contributing to the pathogenetic cascade that presents clinically as DHF. However, this model needs to be viewed cautiously in light of regional differences in the genetic makeup of the human population and the circulating viral strains (and perhaps the mosquito vector?). We are only beginning to scratch the surface of these interactions, and current understanding is not yet sufficient to guide vaccination or therapeutic strategies or to define DHF risk on an individual level. It is to be hoped that further research in the context of ongoing prospective studies, using newer immunologic techniques, will achieve this potential.

References

Anonymous (1997). "Dengue Haemorrhagic Fever: Diagnosis, Treatment and Control." World Health Organization, Geneva.

Avirutnan, P., Malasit, P., Seliger, B., Bhakdi, S., and Husmann, M. (1998). Dengue virus infection of human endothelial cells leads to chemokine production, complement activation, and apoptosis. *J. Immunol.* **161**(11)**:**6338–6346.

Bethell, D. B., Flobbe, K., Cao, X. T., Day, N. P., Pham, T. P., Buurman, W. A., Cardosa, M. J., White, N. J., and Kwiatkowski, D. (1998). Pathophysiologic and prognostic role of cytokines in dengue hemorrhagic fever. *J. Infect. Dis.* **177**(3)**:**778–782.

Bethell, D. B., Gamble, J., Pham, P. L., Nguyen, M. D., Tran, T. H., Ha, T. H., Tran, T. N., Dong, T. H., Gartside, I. B., White, N. J., and Day, N. P. (2001). Noninvasive measurement of microvascular leakage in patients with dengue hemorrhagic fever. *Clin. Infect. Dis.* **32**(2)**:**243–253.

Bosch, I., Xhaja, K., Estevez, L., Raines, G., Melichar, H., Warke, R. V., Fournier, M. V., Ennis, F. A., and Rothman, A. L. (2002). Increased production of IL-8 in primary human monocytes, and in human epithelial and endothelial cell lines after dengue virus challenge. *J. Virol.* **76:**5588–5597.

Burke, D. S., Nisalak, A., Johnson, D. E., and Scott, R. M. (1988). A prospective study of dengue infections in Bangkok. *Am. J. Trop. Med. Hyg.* **38**(1)**:**172–180.

Chaturvedi, U. C., Agarwal, R., Elbishbishi, E. A., and Mustafa, A. S. (2000). Cytokine cascade in dengue hemorrhagic fever: Implications for pathogenesis. *FEMS Immunol. Med. Microbiol.* **28**(3)**:**183–188.

Chen, Y. C., and Wang, S. Y. (2002). Activation of terminally differentiated human monocytes/macrophages by dengue virus: Productive infection, hierarchical production of innate cytokines and chemokines, and the synergistic effect of lipopolysaccharide. *J. Virol.* **76**(19)**:**9877–9887.

Chiewsilp, P., Scott, R. M., and Bhamarapravati, N. (1981). Histocompatibility antigens and dengue hemorrhagic fever. *Am. J. Trop. Med. Hyg.* **30**(5)**:**1100–1105.

Dharakul, T., Kurane, I., Bhamarapravati, N., Yoksan, S., Vaughn, D. W., Hoke, C. H., and Ennis, F. A. (1994). Dengue virus-specific memory T cell responses in human volunteers receiving a live attenuated dengue virus type 2 candidate vaccine. *J. Infect. Dis.* **170:**27–33.

Gagnon, S. J., Ennis, F. A., and Rothman, A. L. (1999). Bystander target cell lysis and cytokine production by dengue virus-specific human $CD4^+$ cytotoxic T lymphocyte clones. *J. Virol.* **73**(5)**:**3623–3629.

Gagnon, S. J., Leporati, A., Green, S., Kalayanarooj, S., Vaughn, D. W., Stephens, H. A., Suntayakorn, S., Kurane, I., Ennis, F. A., and Rothman, A. L. (2001). T cell receptor V beta gene usage in Thai children with dengue virus infection. *Am. J. Trop. Med. Hyg.* **64**(1–2)**:**41–48.

Gagnon, S. J., Mori, M., Kurane, I., Green, S., Vaughn, D. W., Kalayanarooj, S., Suntayakorn, S., Ennis, F. A., and Rothman, A. L. (2002). Cytokine gene expression and protein production in peripheral blood mononuclear cells of children with acute dengue virus infections. *J. Med. Virol.* **67:**41–46.

Gagnon, S. J., Zeng, W., Kurane, I., and Ennis, F. A. (1996). Identification of two epitopes on the dengue 4 virus capsid protein recognized by a serotype-specific and a panel of serotype-cross-reactive human $CD4^+$ cytotoxic T-lymphocyte clones. *J. Virol.* **70**(1)**:**141–147.

Green, S., Kurane, I., Edelman, R., Tacket, C. O., Eckels, K. H., Vaughn, D. W., Hoke, C. H., Jr., and Ennis, F. A. (1993). Dengue virus-specific human $CD4^+$ T-lymphocyte responses in a recipient of an experimental live-attenuated dengue virus type 1 vaccine: Bulk culture proliferation, clonal analysis, and precursor frequency determination. *J. Virol.* **67**(10)**:**5962–5967.

Green, S., Pichyangkul, S., Vaughn, D. W., Kalayanarooj, S., Nimmannitya, S., Nisalak, A., Kurane, I., Rothman, A. L., and Ennis, F. A. (1999a). Early CD69 expression on peripheral blood lymphocytes from children with dengue hemorrhagic fever. *J. Infect. Dis.* **180:**1429–1435.

Green, S., Vaughn, D. W., Kalayanarooj, S., Nimmannitya, S., Suntayakorn, S., Nisalak, A., Lew, R., Innis, B. L., Kurane, I., Rothman, A. L., and Ennis, F. A. (1999b). Early immune activation in acute dengue is related to development of plasma leakage and disease severity. *J. Infect. Dis.* **179**(4)**:**755–762.

Green, S., Vaughn, D. W., Kalayanarooj, S., Nimmannitya, S., Suntayakorn, S., Nisalak, A., Rothman, A. L., and Ennis, F. A. (1999c). Elevated plasma interleukin-10 levels in acute dengue correlate with disease severity. *J. Med. Virol.* **59:**329–334.

Guzman, M. G., Kouri, G., Martinez, E., Bravo, J., Riveron, R., Soler, M., Vazquez, S., and Morier, L. (1987). Clinical and serologic study of Cuban children with dengue hemorrhagic fever/dengue shock syndrome (DHF/DSS). *PAHO Bull.* **21**(3)**:**270–279.

Halstead, S. B. (1980). Immunological parameters of togavirus disease syndromes. *In* "The Togaviruses: Biology, Structure, Replication" (R. W. Schlesinger, ed.), pp. 107–173. Academic Press, New York.

Halstead, S. B., Shotwell, H., and Casals, J. (1973a). Studies on the pathogenesis of dengue infection in monkeys. I. Clinical laboratory responses to primary infection. *J. Infect. Dis.* **128**(1)**:**7–14.

Halstead, S. B., Shotwell, H., and Casals, J. (1973b). Studies on the pathogenesis of dengue infection in monkeys. II. Clinical laboratory responses to heterologous infection. *J. Infect. Dis.* **128**(1)**:**15–22.

Hober, D., Delannoy, A. S., Benyoucef, S., De Groote, D., and Wattre, P. (1996). High levels of sTNFR p75 and TNF alpha in dengue-infected patients. *Microbiol. Immunol.* **40**(8)**:**569–573.

Hober, D., Poli, L., Roblin, B., Gestas, P., Chungue, E., Granic, G., Imbert, P., Pecarere, J. L., Vergez-Pascal, R., Wattre, P., and Maniez-Montreuil, M. (1993). Serum levels of tumor necrosis factor-α (TNF-α), interleukin-6 (IL-6), and interleukin-1β (IL-1β) in dengue-infected patients. *Am. J. Trop. Med. Hyg.* **48**(3)**:**324–331.

Huang, Y. H., Lei, H. Y., Liu, H. S., Lin, Y. S., Liu, C. C., and Yeh, T. M. (2000). Dengue virus infects human endothelial cells and induces IL-6 and IL-8 production. *Am. J. Trop. Med. Hyg.* **63**(1–2)**:**71–75.

Innis, B. L., Eckels, K. H., Kraiselburd, E., Dubois, D. R., Meadors, G. F., Gubler, D. J., Burke, D. S., and Bancroft, W. H. (1988). Virulence of a live dengue virus vaccine candidate: A possible new marker of dengue virus attenuation. *J. Infect. Dis.* **158**(4)**:**876–880.

Irie, K., Mohan, P. M., Sasaguri, Y., Putnak, R., and Padmanabhan, R. (1989). Sequence analysis of cloned dengue virus type 2 genome (New Guinea-C strain). *Gene* **75**(2)**:**197–211.

Juffrie, M., van Der Meer, G. M., Hack, C. E., Haasnoot, K., Sutaryo, Veerman, A. J., and Thijs, L. G. (2000). Inflammatory mediators in dengue virus infection in children: Interleukin-8 and its relationship to neutrophil degranulation. *Infect. Immun.* **68**(2)**:**702–707.

Kalayanarooj, S., Vaughn, D. W., Nimmannitya, S., Green, S., Suntayakorn, S., Kunentrasai, N., Viramitrachai, W., Ratanachu-eke, S., Kiatpolpoj, S., Innis, B. L., Rothman, A. L., Nisalak, A., and Ennis, F. A. (1997). Early clinical and laboratory indicators of acute dengue illness. *J. Infect. Dis.* **176**(2)**:**313–321.

King, C. A., Anderson, R., and Marshall, J. S. (2002). Dengue virus selectively induces human mast cell chemokine production. *J. Virol.* **76**(16)**:**8408–8419.

Kliks, S. C., Nimmanitya, S., Nisalak, A., and Burke, D. S. (1988). Evidence that maternal dengue antibodies are important in the development of dengue hemorrhagic fever in infants. *Am. J. Trop. Med. Hyg.* **38**(2)**:**411–419.

Kurane, I., Brinton, M. A., Samson, A. L., and Ennis, F. A. (1991a). Dengue virus-specific, human $CD4^+$ CD8- cytotoxic T-cell clones: Multiple patterns of virus cross-reactivity recognized by NS3-specific T-cell clones. *J. Virol.* **65**(4)**:**1823–1828.

Kurane, I., Dai, L. C., Livingston, P. G., Reed, E., and Ennis, F. A. (1993). Definition of an HLA-DPw2-restricted epitope on NS3, recognized by a dengue virus serotype-cross-reactive human $CD4^+$ CD8- cytotoxic T-cell clone. *J. Virol.* **67**(10)**:**6285–6288.

Kurane, I., Innis, B. L., Nimmannitya, S., Nisalak, A., Meager, A., Janus, J., and Ennis, F. A. (1991b). Activation of T lymphocytes in dengue virus infections: High levels of soluble interleukin 2 receptor, soluble CD4, soluble CD8, interleukin 2, and interferon-gamma in sera of children with dengue. *J. Clin. Invest.* **88:**1473–1480.

Kurane, I., Innis, B. L., Nisalak, A., Hoke, C., Nimmannitya, S., Meager, A., and Ennis, F. A. (1989a). Human T cell responses to dengue virus antigens: Proliferative responses and interferon gamma production. *J. Clin. Invest.* **83:**506–513.

Kurane, I., Meager, A., and Ennis, F. A. (1989b). Dengue virus-specific human T cell clones: Serotype crossreactive proliferation, interferon gamma production, and cytotoxic activity. *J. Exp. Med.* **170**(3)**:**763–775.

Kurane, I., Okamoto, Y., Dai, L. C., Zeng, L. L., Brinton, M. A., and Ennis, F. A. (1995). Flavivirus-cross-reactive, HLA-DR15-restricted epitope on NS3 recognized by human $CD4^+$ CD8- cytotoxic T lymphocyte clones. *J. Gen. Virol.* **76:**2243–2249.

Kurane, I., Zeng, L., Brinton, M. A., and Ennis, F. A. (1998). Definition of an epitope on NS3 recognized by human $CD4^+$ cytotoxic T lymphocyte clones cross-reactive for dengue virus types 2, 3, and 4. *Virology* **240**(2)**:**169–174.

Libraty, D. H., Endy, T. P., Houng, H. H., Green, S., Kalayanarooj, S., Suntayakorn, S., Vaughn, D. W., Nisalak, A., Rothman, A. L., and Ennis, F. A. (2002). Differing influences of viral burden and immune activation on disease severity in secondary dengue 3 virus infections. *J. Infect. Dis.* **185:**1213–1221.

Lin, Y. L., Liu, C. C., Chuang, J. I., Lei, H. Y., Yeh, T. M., Lin, Y. S., Huang, Y. H., and Liu, H. S. (2000). Involvement of oxidative stress, NF-IL-6, and RANTES expression in dengue-2-virus-infected human liver cells. *Virology* **276**(1)**:**114–126.

Livingston, P. G., Kurane, I., Dai, L. C., Okamoto, Y., Lai, C. J., Men, R., Karaki, S., Takiguchi, M., and Ennis, F. A. (1995). Dengue virus-specific, HLA-B35-restricted, human $CD8^+$ cytotoxic T lymphocyte (CTL) clones: Recognition of NS3 amino acids 500 to 508 by CTL clones of two different serotype specificities. *J. Immunol.* **154**(3)**:**1287–1295.

Lobigs, M., Arthur, C. E., Müllbacher, A., and Blanden, R. V. (1994). The flavivirus nonstructural protein NS3 is a dominant source of cytotoxic T cell peptide determinants. *Virology* **202:**195–201.

Loke, H., Bethell, D. B., Phuong, C. X., Dung, M., Schneider, J., White, N. J., Day, N. P., Farrar, J., and Hill, A. V. (2001). Strong HLA class I-restricted T cell responses in dengue hemorrhagic fever: A double-edged sword? *J. Infect. Dis.* **184**(11)**:**1369–1373.

Luster, A. D. (1998). Chemokines: Chemotactic cytokines that mediate inflammation. *N. Engl. J. Med.* **338**(7)**:**436–445.

Mangada, M. M., Endy, T. P., Nisalak, A., Chunsuttiwat, S., Vaughn, D. W., Libraty, D. H., Green, S., Ennis, F. A., and Rothman, A. L. (2002). Dengue-specific T cell responses in peripheral blood mononuclear cells obtained prior to secondary dengue virus infections in Thai schoolchildren. *J. Infect. Dis.* **185**(12):1697–1703.

Mathew, A., Kurane, I., Green, S., Stephens, H. A. F., Vaughn, D. W., Kalayanarooj, S., Suntayakorn, S., Chandanayingyong, D., Ennis, F. A., and Rothman, A. L. (1998). Predominance of HLA-restricted CTL responses to serotype crossreactive epitopes on nonstructural proteins after natural dengue virus infections. *J. Virol.* **72**(5):3999–4004.

Mathew, A., Kurane, I., Rothman, A. L., Zeng, L. L., Brinton, M. A., and Ennis, F. A. (1996). Dominant recognition by human $CD8^+$ cytotoxic T lymphocytes of dengue virus nonstructural proteins NS3 and NS1.2a. *J. Clin. Invest.* **98**(7):1684–1694.

McMichael, A. J., and O'Callaghan, C. A. (1998). A new look at T cells. *J. Exp. Med.* **187**(9):1367–1371.

Mori, M., Kurane, I., Janus, J., and Ennis, F. A. (1997). Cytokine production by dengue virus antigen-responsive human T lymphocytes in vitro examined using a double immunocytochemical technique. *J. Leukocyte Biol.* **61**:338–345.

Murgue, B., Cassar, O., Deparis, X., Guigon, M., and Chungue, E. (1998). Implication of macrophage inflammatory protein-1 alpha in the inhibition of human haematopoietic progenitor growth by dengue virus. *J. Gen. Virol.* **79**(Pt 8):1889–1893.

Murgue, B., Roche, C., Chungue, E., and Deparis, X. (2000). Prospective study of the duration and magnitude of viraemia in children hospitalised during the 1996–1997 dengue-2 outbreak in French Polynesia. *J. Med. Virol.* **60**(4):432–438.

Mustafa, A. S., Elbishbishi, E. A., Agarwal, R., and Chaturvedi, U. C. (2001). Elevated levels of interleukin-13 and IL-18 in patients with dengue hemorrhagic fever. *FEMS Immunol. Med. Microbiol.* **30**(3):229–233.

Okamoto, Y., Kurane, I., Leporati, A. M., and Ennis, F. A. (1998). Definition of the region on NS3 which contains multiple epitopes recognized by dengue virus serotype-cross-reactive and flavivirus-cross-reactive, HLA-DPw2-restricted $CD4^+$ T cell clones. *J. Gen. Virol.* **79**:697–704.

Paradoa Perez, M. L., Trujillo, Y., and Basanta, P. (1987). Association of dengue hemorrhagic fever with the HLA system. *Haematologia* **20**(2):83–87.

Raghupathy, R., Chaturvedi, U. C., Al-Sayer, H., Elbishbishi, E. A., Agarwal, R., Nagar, R., Kapoor, S., Misra, A., Mathur, A., Nusrat, H., Azizieh, F., Khan, M. A., and Mustafa, A. S. (1998). Elevated levels of IL-8 in dengue hemorrhagic fever. *J. Med. Virol.* **56**(3):280–285.

Roehrig, J. T., Johnson, A. J., Hunt, A. R., Beaty, B. J., and Mathews, J. H. (1992). Enhancement of the antibody response to flavivirus B-cell epitopes by using homologous or heterologous T-cell epitopes. *J. Virol.* **66**(6):3385–3390.

Rothman, A. L., and Ennis, F. A. (1999). Immunopathogenesis of dengue hemorrhagic fever. *Virology* **257**(1):1–6.

Rothman, A. L., Kurane, I., and Ennis, F. A. (1996). Multiple specificities in the murine $CD4^+$ and $CD8^+$ T-cell response to dengue virus. *J. Virol.* **70**(10):6540–6546.

Rothman, A. L., Kurane, I., Lai, C. J., Bray, M., Falgout, B., Men, R., and Ennis, F. A. (1993). Dengue virus protein recognition by virus-specific murine $CD8^+$ cytotoxic T lymphocytes. *J. Virol.* **67**(2):801–806.

Rothman, A. L., Kurane, I., Zhang, Y. M., Lai, C. J., and Ennis, F. A. (1989). Dengue virus-specific murine T-lymphocyte proliferation: Serotype specificity and response to recombinant viral proteins. *J. Virol.* **63**(6):2486–2491.

Sabin, A. B. (1952). Research on dengue during World War II. *Am. J. Trop. Med. Hyg.* **1**:30–50.

Sangkawibha, N., Rojanasuphot, S., Ahandrik, S., Viriyapongse, S., Jatanasen, S., Salitul, V., Phanthumachinda, B., and Halstead, S. B. (1984). Risk factors for dengue shock syndrome: A prospective epidemiologic study in Rayong, Thailand. I. The 1980 outbreak. *Am. J. Epidemiol.* **120**(5)**:**653–669.

Scott, R. M., Eckels, K. H., Bancroft, W. H., Summers, P. L., McCown, J. M., Anderson, J. H., and Russell, P. K. (1983). Dengue 2 vaccine: Dose response in volunteers in relation to yellow fever immune status. *J. Infect. Dis.* **148**(6)**:**1055–1060.

Spain-Santana, T. A., Marglin, S., Ennis, F. A., and Rothman, A. L. (2001). MIP-1alpha and MIP-1beta induction by dengue virus. *J. Med. Virol.* **65**(2)**:**324–330.

Spaulding, A. C., Kurane, I., Ennis, F. A., and Rothman, A. L. (1999). Analysis of murine $CD8^+$ T-cell clones specific for the dengue virus NS3 protein: Flavivirus cross-reactivity and influence of infecting serotype. *J. Virol.* **73**(1)**:**398–403.

Stephens, H. A., Klaythong, R., Sirikong, M., Vaughn, D. W., Green, S., Kalayanarooj, S., Endy, T. P., Libraty, D. H., Nisalak, A., Innis, B. L., Rothman, A. L., Ennis, F. A., and Chandanayingyong, D. (2002). HLA-A and -B allele associations with secondary dengue virus infections correlate with disease severity and the infecting viral serotype in ethnic Thais. *Tissue Antigens* **60**(4)**:**309–318.

Thein, S., Aung, M. M., Shwe, T. N., Aye, M., Zaw, A., Aye, K., Aye, K. M., and Aaskov, J. (1997). Risk factors in dengue shock syndrome. *Am. J. Trop. Med. Hyg.* **56**(5)**:**566–572.

Vaughn, D. W., Green, S., Kalayanarooj, S., Innis, B. L., Nimmannitya, S., Suntayakorn, S., Endy, T. P., Raengsakulrach, B., Rothman, A. L., Ennis, F. A., and Nisalak, A. (2000). Dengue viremia titer, antibody response pattern and virus serotype correlate with disease severity. *J. Infect. Dis.* **181**(1)**:**2–9.

Vaughn, D. W., Green, S., Kalayanarooj, S., Innis, B. L., Nimmannitya, S., Suntayakorn, S., Rothman, A. L., Ennis, F. A., and Nisalak, A. (1997). Dengue in the early febrile phase: viremia and antibody responses. *J. Infect. Dis.* **176**(2)**:**322–330.

Watts, D. M., Porter, K. R., Putvatana, P., Vasquez, B., Calampa, C., Hayes, C. G., and Halstead, S. B. (1999). Failure of secondary infection with American genotype dengue 2 to cause dengue haemorrhagic fever. *Lancet* **354**(9188)**:** 1431–1434.

Zeng, L., Kurane, I., Okamoto, Y., Ennis, F. A., and Brinton, M. A. (1996). Identification of amino acids involved in recognition by dengue virus NS3-specific, HLA-DR15-restricted cytotoxic $CD4^+$ T-cell clones. *J. Virol.* **70**(5)**:**3108–3117.

Zivna, I., Green, S., Vaughn, D. W., Kalayanarooj, S., Stephens, H. A. F., Chandanayingyong, D., Nisalak, A., Ennis, F. A., and Rothman, A. L. (2002). T cell responses to an HLA B^*07-restricted epitope on the dengue NS3 protein correlate with disease severity. *J. Immunol.* **168:**5959–5965.

Zivny, J., DeFronzo, M., Jarry, W., Jameson, J., Cruz, J., Ennis, F. A., and Rothman, A. L. (1999). Partial agonist effect influences the CTL response to a heterologous dengue virus serotype. *J. Immunol.* **163**(5)**:**2754–2760.

ADVANCES IN VIRUS RESEARCH, VOL 60

NEUTRALIZATION AND ANTIBODY-DEPENDENT ENHANCEMENT OF DENGUE VIRUSES

Scott B. Halstead

Department of Preventive Medicine and Biometrics
Uniformed Services University of the Health Sciences
Bethesda, Maryland 20814

0065-3527/03 $35.00

I. Introduction

Dengue (DEN) viruses cause a wide spectrum of diseases from dengue hemorrhagic fever/dengue shock syndrome (DHF/DSS) to dengue fever and undifferentiated febrile illnesses. The four dengue viruses are unique among human pathogens in that the natural dengue immunity status can modify disease in two directions: lessened or increased severity (Vaughn *et al.*, 2000). While many genetic as well as acquired factors affect the severity of human dengue virus infections, the most important of these is the dual role of IgG1 dengue antibodies. The regulatory roles of dengue antibodies appear to derive from two attributes of the dengue virus group: (1) four types have evolved from a common ancestor (Wang *et al.*, 2000), resulting in viruses with many common antigens but a sufficient critical structural difference(s) to permit sequential infections, and (2) dengue viruses have a tropism for and ability to replicate in cells of the mononuclear phagocyte system. These cells support antibody-dependent enhanced (ADE) infection.

Demonstrations of dengue virus–antibody interactions are to be seen in many natural experiments in human beings, as well as in laboratory animal studies. In this review, studies on the neutralization and protection against dengue infection by antibodies and many relevant clinical and epidemiological as well as experimental studies that support the phenomenon of ADE are discussed. Epidemiological studies provide the background evidence that shows ADE to be a biologically plausible hypothesis.

For further details on host factors that control dengue infection, on the growth of dengue viruses in mononuclear phagocytes, dengue ADE in humans, and ADE *in vivo* and *in vitro* involving viruses from a wide spectrum of taxons and many species of animals, the reader is referred to published reviews (Halstead, 1980, 1982, 1988, 1994; Mascola, 1993; Porterfield, 1986; Sullivan, 2001; Wu *et al.*, 2000).

II. Neutralization (NT) and Protection by Antibodies

Current understanding concerning *in vivo* protection against dengue infection by antibodies is sound epidemiologically, but organized research has been limited. There have been no systematic studies on the natural history of dengue antibody responses in humans following first or heterotypic infections. A critical unknown for designed research is the precise identity of the cells that support human dengue infection and their relative contribution to disease. The absence of such data means that little is known about the biologically relevant receptor systems for attachment, entry, and replication of dengue viruses in human cells. The literature on dengue viral neutralization by antibodies is composed largely of *in vitro* observations on common laboratory cell systems. The most important observations have been made on other viruses.

A. Mechanisms of Neutralization in Vitro

Antibodies are a major defense against arboviral infections. Several mechanisms have been proposed for the neutralization of viruses *in vivo*: (1) aggregation of viruses resulting in elimination by phagocytic cells; (2) blocked attachment to one or another cell receptor by (a) stearic interference, (b) capsid stabilization, or (c) structural changes; or (3) neutralization of uncoating due to (a) capsid stabilization or (b) interference with fusion (Smith, 2001).

B. Mechanisms of Attachment and Internalization of Dengue Viruses

The consensus of researchers is that antibodies, enhanced significantly by *in vivo* amplifying accessory systems, such as complement, block the attachment and entry of flaviviruses into cells. The steps involved in cell entry appear to be initial attachment to the plasma membrane followed by uptake via receptor-mediated endocytosis with pH-dependent fusion to the endocytic vesicle membrane (Gollins and Porterfield, 1986). There may be a requirement for attachment to another primary receptor and/or coreceptors prior to endocytosis. Exposure of dengue viruses to acidic pH in endosome vesicles results in major structural modifications at the virion surface, including trimerization of the E protein and a rearrangement of domain II exposing the putative flavivirus-conserved fusion domain (Allison *et al.*, 1995; Stiasny *et al.*, 1996). The initial attachment of dengue and other flaviviruses *in vitro* to plasma membranes of cell lines of mammalian and mosquito origin has been reported to be dependent on the presence of glycosoaminoglycans, particularly heparan sulfate (Chen *et al.*, 1997; Hung *et al.*, 1999). Binding of viruses is reported to be inhibited by preincubation or coincubation with heparin or treatment with heparinase. The consensus of these and other authors is that this is a generic binding mechanism followed by viral attachment to another cell type-specific, possibly multicomponent receptor or receptor complex (Bielefeldt-Ohmann, 1998).

Specific binding of dengue virus to proteins from cells of mammalian and arthropod origin has been identified to several proteins with molecular masses ranging from approximately 30 to 80 kDa. None have been fully characterized (Bielefeldt-Ohmann, 1998; Bielefeldt-Ohmann *et al.*, 2001; Ramos-Castaneda *et al.*, 1997). Cell receptor sites on dengue viruses are also not identified, although an important determinant of pathogenicity for the related Murray Valley encephalitis virus (MVE) E protein involves residues 388–390. This is close to the Mab 3H5-binding site identified by Trirawatanapong *et al.*, (1992). Neutralization sites on the dengue 2 E protein are reasonably well identified, with small domains at amino acids 35–55, 352–368, and 386–397, whereas larger domains have been identified using Mabs and truncated recombinant fusion proteins, 60–205 and 298–397 (Megret *et al.*, 1992; Roehrig *et al.*, 1990). The lipid envelope of dengue viruses is derived from host cell membranes and constitutes 15–20% of the total weight of the virus particle. Carbohydrates represent 9–10% of the weight of virus particles and are found as glycolipids and glycoproteins. Therefore, the cells used to prepare dengue viral seeds contribute critically to the measurement of neutralization *in vitro*. In retrovirus research, the antibodies to host cells used to prepare viral seeds resulted in false-positive vaccine efficacy because antibodies directed against host cell antigens protected monkeys subsequently challenged with homologous virus (Hoxie *et al.*, 1987).

C. Observations on Neutralization and Protection in Vivo

There is abundant epidemiological evidence of lifelong homotypic protective immunity. For example, older Cubans were protected against DEN-2 infections in 1981 because they had been infected with the same virus in the early 1940s (Guzman *et al.*, 1990). Challenges with homologous strains performed by Sabin (1952) resulted in no viremia. Heterologous infection is more complicated. As described by Sabin, adult human volunteers infected with DEN-1 or DEN-2 viruses were protected from clinical illness when challenged with heterologous virus within "2 months" of primary infections. However, viremic infections occurred in monoimmune volunteers challenged with heterologous virus between 2 and 3 months after primary infection. These infections "gave rise to malaise and slight fever." Mild illnesses accompanies cross-challenges up to 9 months following primary infection and classical dengue fever occurs thereafter (Sabin, 1952). Other than the studies of Sabin, cross-protection among various dengue viruses and strains has not been studied systematically. Third and fourth dengue infections are documented in prospective seroepidemiological studies, but seldom result in clinical illness (Graham *et al.*, 1999; Sangkawibha *et al.*, 1984).

The ADE phenomenon implies that undiluted sera from individuals convalescent from a single dengue infection must contain more than 40–50 dengue antibody molecules capable of attaching to virions (see later). Despite this high concentration of cross-reactive antibodies, second dengue infections occur regularly. Most of these are silent. In a Thai school children cohort, 85% who experienced documented secondary DEN-2 infections were not sick. Undiluted preinfection sera from these children prevented DEN-2 infection of human monocyte cultures. However, the smaller group of hospitalized children circulated dengue antibodies that did not neutralize DEN-2 in this culture system (Kliks *et al.*, 1989). In a more recent study, Thai children acquired clinically overt secondary DEN-3 infections even when their preillness sera contained DEN-3 neutralizing antibodies measured in LLC-MK_2 cells, but when an autologous DEN-3 strain was used, neutralizing antibody titers were significantly lower than with a heterologous strain. The severity of illnesses associated with secondary DEN-3 infections was inversely related to titers of autologous DEN-3 neutralizing antibodies (T. Endy and D. Libraty, personal communication). It seems likely that solid nonviremic protection resides in avid antibodies that are directed against a viral attachment receptor unique to each serotype.

III. Studies in Monkeys: Definition of the "Protected" Antibody Response

Hemagglutination-inhibition (HI) and neutralizing responses to DEN-1–4 viruses were studied in a large group of rhesus monkeys that were subsequently challenged with heterologous or homologous viruses at varying intervals (Halstead and Palumbo, 1973; Halstead *et al.*, 1973a). Viremia was detected reliably following infection of susceptible monkeys but never following challenge with homologous virus. Animals given a homologous challenge developed transient four- to eightfold HI antibody responses over a period of 10 days, notably to the homologous hemagglutination (HA) antigen. However, neutralizing antibody titers did not change. This was defined as a "protected" response and suggested that the inoculum itself stimulated an antibody response, possibly to noninfectious immune complexes and that no viral replication occurred in the monkey. When anamnestic HI responses were observed following heterologous challenge, it was possible to document infection by the recovery of viruses from tissues (Halstead *et al.*, 1973a).

A. Affinity Maturation

During an immune response, the affinity of antibodies that react with the antigen that triggered the response increases with time, a phenomenon known as affinity maturation. Generally, this is assumed to be a phenomenon that occurs during the early stages of the immune response. As an example, IgG rhesus monkey antibodies obtained on day 7 following a primary dengue 2 infection only enhanced DEN-2 infection in monocytes and did not neutralize homologous virus *in vitro* (Halstead and O'Rourke, 1977b). Despite current doctrine, the affinity maturation of an antibody response may progress for years following flavivirus infections. This was demonstrated by rising log neutralization indices over a 4-year period in sera from American servicemen infected subclinically with Japanese encephalitis virus while stationed in Korea. They were bled at 1 and 5 years after infection (Halstead and Grosz, 1962). The phenomenon has not been studied in dengue but could explain why neutralizing antibody responses following inapparent primary dengue infections become relatively monotypic over the period of 1 year (S. B. Halstead, unpublished observation). Long-term affinity maturation may explain the observation that DSS case fatality rates were significantly greater when viruses of the same DEN-2 genotype were involved in secondary infections at 20 compared with 4 years after DEN-1 infection (i.e., at the longer interval after primary infection there was diminished heterotypic neutralization as antibody was directed more avidly at the type-specific neutralization site[s] permitting ADE antibodies to operate) (Guzman *et al.*, 2002a, 2002b).

B. Number of Antibody Molecules Required to Neutralize Dengue Viruses

Two major hypotheses of the neutralization of animal viruses have been proposed: single-hit and multiple-hit neutralization (Della-Porta and Westaway, 1977; Dulbecco *et al.*, 1956). The latter authors, who favor the multi-hit theory, have summarized experimental evidence that reconciles single-hit and multi-hit kinetics. Their hypothesis might be termed the "straw that breaks the camel's back." Multiple molecules of neutralizing antibody attach to critical sites on animal virions and when the final critical site is covered, neutralization occurs (Della-Porta and Westaway, 1977). The initial attachment of antibodies to virions occurs extremely rapidly, but covering of the final neutralizing site occurs more slowly and obeys single-hit kinetics.

Stoichiometric studies of antibody binding have established a regular relationship between virion size and the number of antibody molecules required to neutralize. For picornaviruses, with a diameter of 30 nm, 5–6 antibody molecules are required. It should be recognized that antibodies are large relative to virion, 150 Å, or equal to the radius of a picornavirus particle (Burton *et al.*, 2001; Smith, 2001). Different antibodies may attach with unique stearic orientations. Effective IgG antibodies bind primarily as the result of electrostatic interactions. Bivalent binding of IgG is both more energetic and more avid than monovalent binding. Papilloma viruses, spherical particles 40–55 nm in diameter, require 38 antibody molecules to neutralize, whereas influenza A particles, irregular in shape with a diameter of 80–120 nm, require 70 antibody molecules (Burton *et al.*, 2001). Antibody saturation of virions is considerably greater than the minimal binding for neutralization. Picorna, papilloma, and influenza viruses, respectively, accommodate 30, 150, and 400 antibody molecules at saturation. It can be predicted that complete neutralization of dengue viruses will require 40–50 molecules of avid antibody.

C. Data from Dengue Monoclonal Antibody Studies: Ratio between NT and Antibody-Dependent Enhancement (ADE)

High concentrations of type-specific and group-specific dengue 2-derived monoclonal and polyclonal antibodies neutralize many different strains of dengue 2 viruses both in non-Fc receptor and in Fc receptor-bearing cells (Halstead *et al.*, 1984; Morens *et al.*, 1987). Generally speaking, peak enhancement titers were 100- to 1000-fold higher than the highest dilution, demonstrating significant neutralization in non-Fc receptor-bearing cells (Halstead *et al.*, 1984; Morens *et al.*, 1987, 1991).

D. Measuring Neutralizing Antibodies in Vitro

The first dengue neutralization tests were introduced by Sabin, who adopted DEN-1 and -2 viruses to kill weanling mice and measured antibody usually by the serum-constant virus dilution method (Sabin, 1952; Sabin and Schlesinger, 1945). Sabin (1952) also developed an ingenious human *in vivo* neutralization test based on the regular appearance of an indurated erythematous lesion at the site of intracutaneous inoculation of DEN-1 and -2 viruses within 2-3 days. The seed viruses used for these studies were viremic human sera. To perform the test viremic and immune test, sera were mixed and held at 37 °C for 2 h and then inoculated intracutaneously, often in a row on the backs of volunteers. The absence of the lesion was read as neutralization.

The classical neutralization reaction as defined by Dulbecco *et al.* (1956) consists of a reaction mixture (usually high concentrations of virus and antibody) and a measuring system (usually plaque assay). After varying periods of incubation of reactants, virus and antibody are diluted in chilled medium and residual virus assayed. When virus survival (defined as the logarithm of the ratio of residual virus to virus in control tube) is plotted as a function of time, the kinetic curve descends as a straight line, the slope decreasing and finally ending in a horizontal line. The descending part of the kinetic curve is first order; the slope is the same irrespective of the concentration of virus (percentage law). The horizontal portion of the curve is a function of the concentration of antibody compared with virus. At very high concentrations of antibody, final survival (persistent fraction) is a constant for the assay system. The same concentration of reactants inoculated into different assay systems may give difference percentage persistent fractions. When virus in the persistent fraction is incubated with antiantibody, complete neutralization usually results (Ashe and Notkins, 1966).

After the introduction of a dengue plaque assay, plaque reduction has been the preferred method of measuring dengue antibodies (Sukhavachana *et al.*, 1966). The classic description of the 50% plaque reduction neutralization test (PRNT) is in LLC-MK2 cells (Russell *et al.*, 1967). This test was modified to a single overlay, and cells were incubated in the dark to prevent photoinactivation by neutral red (Halstead *et al.*, 1970b). Tests were adapted to micro methods without loss of accuracy and simplified by the introduction of the suspension assay using BHK21 clone 15 cells (Morens *et al.*, 1985a, 1985b). Given the rapid changes in dengue virus genomes affected by even a single passage in cells and the fact that cell antigens can be integrated into virions during such passages, the huge number of cell systems, virus strains, and methods of preparing virus seeds used to measure neutralizing antibodies must be of concern and argues urgently for greater standardization.

Statistical considerations of Poisson-distributed particles support the use of a 50% end point for the most accurate quantification of antibody titers rather than the 90% end points adopted by many in the field (Russell *et al.*, 1967). Use of a computer algorithm or plotting data on log probit graph paper makes possible the determination of 50% end points using only three dilutions (e.g., 1:10, 1:100, and 1:1000) Measurement of PRNT antibodies depends on the accurate quantitation of plaques in both control and serum dilutions. It is important to avoid plaque overlap (usually below 15–20 plaques in each well of a 12-well plate). It is also important to have sufficient replicates, usually three per serum dilution. For best power, plaque counts can be summed rather than averaged (Detre and White, 1970).

IV. Antibody-Dependent Enhancement: Clinical and Epidemiological Evidence of Immunologically Enhanced Disease

A. *Dengue Hemorrhagic Fever (DHF)/Dengue Shock Syndrome (DDS) Is Associated with Heterotypic Dengue Infections*

That a previous dengue infection can modify the severity of a subsequent dengue infection was inferred from a highly significant association between severe illness and secondary-type flavivirus antibody responses (Halstead *et al.*, 1967). Relevant data from 1962 to 1964 hospital and epidemiological studies are summarized in Table I. Despite many independent documentations of the high frequency of secondary antibody responses and severe dengue disease, some critics have suggested that countries endemic for several dengue serotypes

TABLE I

Correlation between Primary and Secondary Dengue Infections and Syndromes of Varying Severity in Children Ages <1–14 Years, Bangkok and Thonburi, 1962, Dengue-Infected Outpatients, and Dengue-Infected Inpatients, Bangkok Children's Hospital, 1962–1964

Severity group	Primary	Secondary	% Secondary
All children	165,794	126,728	29[a]
FUO/OPD	33	61	65[b]
FUO/Hospital	13	21	64[c]
DHF/no shock	65	262	80[c]
DSS	2	160	99[c]

[a] After Halstead (1980).
[b] After Halstead *et al.* (1969a).
[c] After Nimmannitya *et al.* (1969).

must have more secondary than primary dengue infections (Rosen, 1977, 1986, 1989, 1999). From the standpoint of the individual child, it is obvious that a second can only follow a first dengue infection and, therefore, second cannot be more common than first infections.

However, neutralizing antibody profiles on sera from residents of dengue-hyperendemic areas do provide evidence that infections with third and fourth dengue serotypes (also referred to as "secondary" infections) occur. If so, then both the prevalence of multitypic antibodies and the incidence of second, third, and fourth infections are likely to be greater than primary infections across older childhood. However, "secondary" infections of high parity (third or fourth infections) are largely inapparent, there being little evidence that such infections are associated with clinical disease (see later).

Mathematical models of age-specific sequential dengue infection rates fit observed age-specific DHF hospitalization rates in dengue endemic countries best with second but not third or fourth sequential infections (Fischer and Halstead, 1970). Of interest, this model predicted there would be 58.5 DHF/DSS cases per 1000 secondary dengue infections, nearly the same ratio found in prospective studies (see later).

B. DHF Occurs with Passively Acquired Dengue Antibodies

The strongest evidence that antibodies are sufficient to enhance dengue infections is the regular occurrence of DHF/DSS in infants less than 1 year during their first dengue infection (Halstead *et al.*, 1970a). Indeed, it was the bimodal age distribution of children hospitalized with DHF that first suggested there might be two different causal immunological mechanisms (Fig. 1). The unique distribution of ages at which infants were hospitalized for DHF pointed to an etiological role for maternal antibody (Fig. 2). As illustrated in Fig. 3, maternal antibodies can be viewed as acting in two ways: initially protecting infants from clinical disease but later enhancing dengue infections. Evidence for this hypothesis was found in a study of 13 infants with DHF accompanying primary DEN-2 infections (Kliks *et al.*, 1988). Infants were hospitalized between 4 and 12 months of age, all born to mothers whose dengue serum neutralizing antibody profiles suggested multiple earlier dengue infections. In laboratory studies, maternal dengue antibodies were used as surrogates for cord blood. All infants received antibodies that neutralized DEN-2 (Table II). When these antibodies were catabolically degraded below protective levels, infants were at risk of developing DHF during a DEN-2 infection. As shown in Table II, the

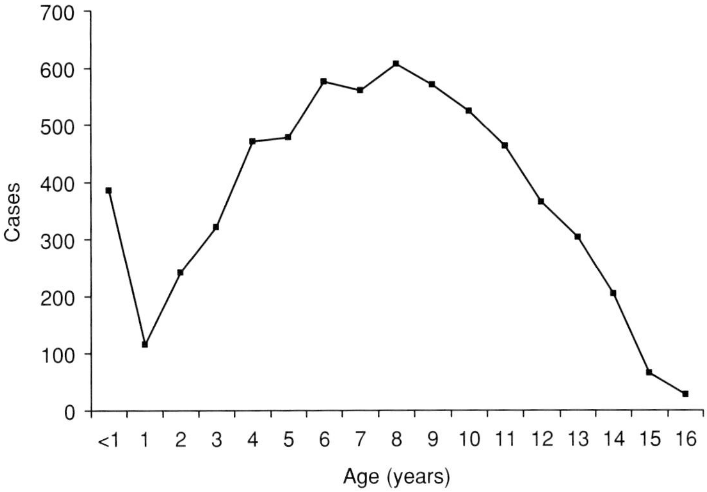

FIG 1. Age distribution of serologically confirmed dengue hemorrhagic fever/dengue shock syndrome cases hospitalized at the Queen Sirikit National Institute for Child Health (Bangkok Children's Hospital), Thailand 1990–1999 (courtesy Dr. Ananda Nisalak, AFRIMS, Bangkok).

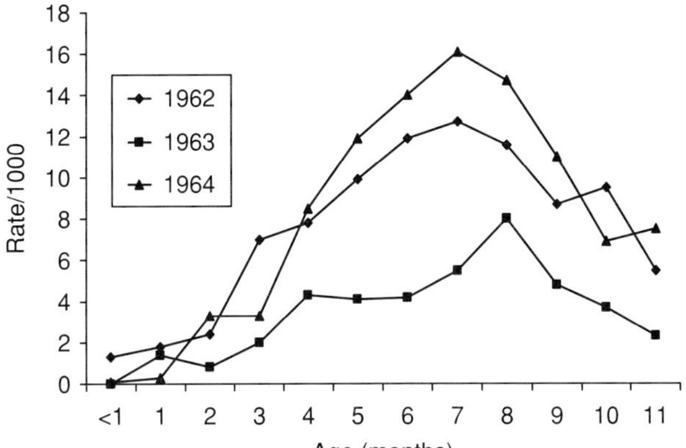

FIG 2. Hospitalization rate by month of age of infants with DHF/DSS during the first year of life at Bangkok and Thonburi hospitals, Thailand, 1962–1964 (after Halstead *et al.*, 1969b).

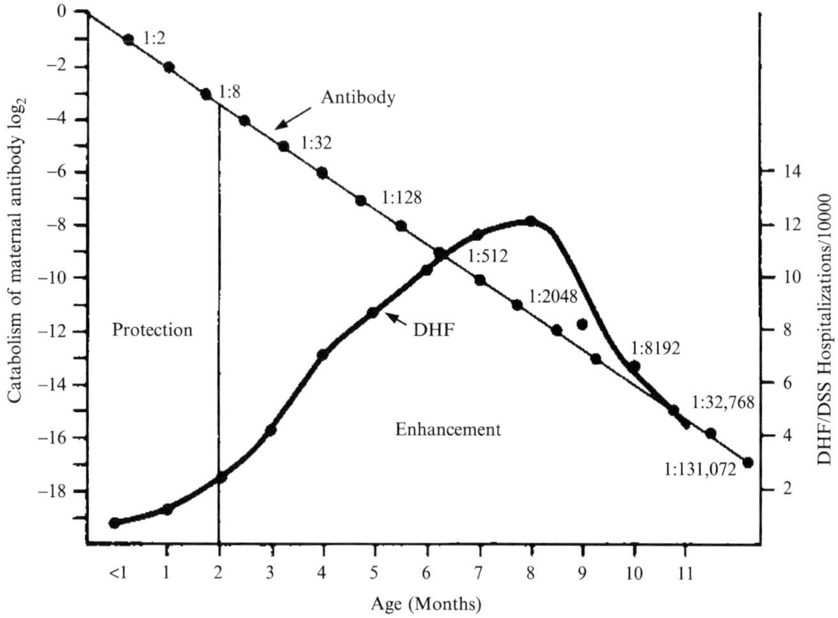

FIG 3. Explanation of shape of age-specific hospitalizations of infants with DHF/DSS: infants acquire protective levels of maternal antibody at birth, which degrade to below protective levels, permitting antibody-dependent enhancement. At the end of 1 year, ADE antibodies decline. After Halstead (1981), with permission.

higher the PRNT of maternal anti-DEN-2 at birth, the older infants were when hospitalized for DEN-2 infections.

Despite a lack of research attention, infants in Bangkok are at an especially high risk of developing DHF during primary dengue infections. Hospitalization rates were estimated for infant DHF/DSS from a random sample of infant DHF/DSS cases studied virologically at Bangkok Children's Hospital in 1962 and extrapolated to the entire urban population (Halstead *et al.*, 1969b). It was estimated that 12.3 infants were hospitalized per 1000 primary infections (Table III) or about one half the secondary infection DHF rate (see Table VII). If, as suggested by Kliks *et al.* (1988), infants are at risk to DHF for only a 1-month period, the rate becomes 110.6/1000 or four times the DHF hospitalization rate for Bangkok children during a second dengue infection (Halstead, 1980). This rate is higher than the 42.7/1000 mean rate of childhood DHF hospitalizations/secondary infections (see later), suggesting that the passive antibody is more effective than actively acquired postinfection immunity in enhancing dengue disease.

TABLE II
RELATIONSHIP BETWEEN DEN-2 NEUTRALIZING ANTIBODY TITERS IN MOTHER'S SERA AND AGE OF ONSET OF DHF/DSS IN THEIR INFANTS[a]

Case number	DEN-2 $PRNT_{50}$ mother's serum[b]	Estimated DEN-2 $PRNT_{50}$ at onset of DHF[c]	Age at onset of DHF (months)
1	30	1.9	4
2	50	6.2	3
3	80	2.5	6
4	90	0.7	8
5	200	3.1	7
6	290	4.5	7
7	350	2.7	8
8	360	45.0	4
9	420	3.3	8
10	500	15.2	6
11	720	5.6	8
12	2000	3.9	11
13	8200	4.0	12

[a] After Kliks *et al.* (1988).
[b] Reciprocal plaque reduction neutralization titer (PRNT) as determined by the method of Russell and Nisalak (1967).
[c] Predicted assuming half-life of maternal IgG dengue antibody of 35 days.

C. DHF in Prospective Seroepidemiological Studies

The role for second dengue infections has been documented in seven prospective population-based seroepidemiological studies (Table IV), and the role of second, not third or fourth, heterotypic infections is documented in prospective studies in which DHF cases were hospitalized among a prebled cohort (Table V). In three of the cohorts, sera taken from children prior to hospitalization for DHF had neutralizing antibody profiles consistent with a single antecedent dengue infection. In all cohort studies combined, 21 children were hospitalized with DHF/DSS among 379 who experienced a second dengue infection, whereas no DHF was observed among 312 children who experienced a primary infection ($p < 0.0001$). Clear evidence that a second dengue infection produces DHF was provided by two epidemics in Cuba where DHF/DSS occurred following 1981 and 1997 introductions of DEN-2 in a population partially immune to DEN-1 (Guzman *et al.*, 1990, 2000b). These data are discussed in greater detail later.

TABLE III
HOSPITALIZATION RATES FOR DHF IN INFANTS LESS THAN 1 YEAR OLD PER 1000 PRIMARY DENGUE INFECTIONS ESTIMATED USING AT RISK PERIODS OF 9 OR 1 MONTH

Place, date	At-risk period	At risk	DHF cases	DHF/1000	Reference
Bangkok, 1962	9 months	26,519[a]	326	12.3	Halstead (1980)
	1 month	2,947[b]	326	110.6	

[a] Estimated dengue infections in infants, ages 3–11 months (see Fig. 1) based on 1962 infection rate of 41%.
[b] Estimated dengue infections assuming infant is at risk to DHF/DSS for period of 4 weeks. See text.

The role of antibody in changing the clinical expression of dengue may be dramatic. It now appears that primary DEN-2 and -4 infections, certainly in children, are almost entirely silent, clinically overt disease accompanying only second dengue infections (Guzman *et al.*, 2000b; Vaughn *et al.*, 2000). Table VI summarizes evidence for adults and children from the 1997 epidemic in Santiago de Cuba in which primary infections with Asian genotype DEN-2 were nearly all inapparent. During this same outbreak, a very high proportion of DEN-1-immune adults infected with DEN-2 developed either dengue fever or DHF/DSS.

From cohort studies and prospective seroepidemiological studies of defined populations, it has been possible to calculate hospitalization rates for individuals experiencing secondary dengue infections, with most being in the range of 2–3% (Table VII). In seven studies, serologically characterized DHF cases were observed in a community in which primary and secondary dengue infection rates were measured by testing antibodies in representative pre- and postrainy season serum pairs (Table IV). Two of these were in Cuba. In the first, DEN-2 was introduced in 1981 following a 1977–1979 DEN-1 outbreak. Prior to 1977, there had been no dengue in Cuba since before World War II (Guzman *et al.*, 1990). After 1981, years of strict mosquito control prevented the introduction of dengue viruses. However, in 1997, DEN-2 was again introduced, this time producing an outbreak restricted to Santiago de Cuba. Only individuals older than 18 years were immune to DEN-1; correspondingly, in 1997, all DHF/DSS cases were restricted to the 18 years plus age group.

Hospitalization rates during secondary dengue infection have been measured in the dengue-endemic countries of Thailand and Myanmar (Burma) (Table VII). Among seven population-based studies on dengue

TABLE IV

PRIMARY AND SECONDARY DENGUE INFECTIONS MEASURED IN SELECTED OPEN POPULATIONS AND AMONG DHF/DSS CASES IN CHILDREN AGE 1 YEAR OR OLDER FROM EIGHT PROSPECTIVE SEROEPIDEMIOLOGICAL STUDIES

Place	Year	Serol. test	Age (years)	Total population	Infections		DHF/DSS		Reference
					Primary	Secondary	Primary	Secondary	
Bangkok	1962	HI	1–15	2097	412	313	n.d	n.d.	Halstead *et al.* (1969b)
				842,451	165,794	125,728	297[a] (DHF)	2528[a] (DHF)	Nimmannitya *et al.* (1969) Halstead (1980)
							12[a] (DSS)	1428[a] (DSS)	
Koh Samui	1966	HI	1–15	13,975	1,900	2,700	1	33[b]	Russell *et al.* (1968)
Koh Samui	1967	HI	1–15	13,975	n.d.	n.d.	0	15[c]	Winter *et al.* (1969)
Rayong	1980	PRNT	<1–10	8,885	1717	920	0	18[d]	Sangkawibha *et al.* (1984)
Yangon, Myanmar	1984–1988	PRNT	2–7	3,579 (25,320)	472 (8,512)	448 (4,181)	7[e]	138[f]	Thein *et al.* (1997)
Havana	1981	PRNT	3–14	359,879	20,393	59,875	18	1213[g]	Guzman *et al.* (1990)
Santiago de Cuba	1997	PRNT	20+	325,310	13,116	4810	3	202	Guzman *et al.* (2000b)

[a] Children's Hospital study only: DEN-1 (10), DEN-2 (10), and DEN-3 (4).
[b] DEN-1 (1), DEN-2 (16), and DEN-3 (7).
[c] Only from DF cases: DEN-4 (7).
[d] DEN-2 (7).
[e] DEN-1 (1), DEN-2 (1), and DEN-3 (1).
[f] DEN-1 (3), DEN-2 (23), DEN-3 (4), and DEN-4 (1).
[g] DEN-2 isolated only.

TABLE V

DHF HOSPITALIZATIONS IN CHILDREN WITH AND WITHOUT PREILLNESS DENGUE ANTIBODY IN FOUR PROSPECTIVE COHORT STUDIES

Place	Year	Serol. test[a]	Age (years)	Cohort total	Infections		DHF cases		Reference
					Primary	Secondary	Primary	Secondary	
Koh Samui	1966	HI	2–12	336	26	83	0	3	Russell *et al.* (1968)
Rayong	1980	PRNT	4–10	881	93	112	0	4	Sangkawibha *et al.* (1984)
Bangkok	1980	PRNT[1]	4–15	1757	31	59	0	7	Burke *et al.* (1988)
Jogjakarta	1995–1996	PRNT	4–9	1837	162	120	0	7	Graham *et al.* (1999)

[a] HI, hemagglutination-inhibition; PRNT, plaque reduction neutralization test.

[1] Paired sera screened by HI and then tested by PRNT.

TABLE VI
DENGUE 2 EPIDEMIC IN SANTIAGO DE CUBA, 1997[a]

Cases	Total	Primary	%	Secondary	%
Deaths	12	1	0.008	11	0.2
DHF/DSS	193	2	0.015	191	1.9
Dengue fever	5003	395	3.0	4608	95.8
Total	5208	398	3.0	4810	100
Inapparent	12,718	12,718	97.0	0	0
Total dengue infections	17,926	13,116		4810	

[a] High rate of inapparent infections in susceptible persons. Overt disease in persons with antibodies from previous dengue 1 infection. Primary infections included persons of all ages. Secondary infections occurred only in persons age 18 and above. After Guzman *et al.* (2000b).

in children, DHF rates ranged from 12.2 to 118.6/1000 secondary dengue infections, with a mean of 42.7/1000. Four independent studies in Thailand, Myanmar, and Cuba produced tightly clustered values of 20.1–36.1/1000 secondary dengue infections. An eighth measurement was made on Cuban adults infected with DEN-1 and then DEN-2 at an interval of 18–20 years. Four studies estimated attack rates per 1000 secondary dengue infections separately for dengue shock syndrome (Table VII). The highest DSS rate, 208.0 per 1000 secondary dengue infections, was calculated for infections occurring in the specific infection sequence of DEN-1 and then DEN-2. The same infection sequence was thought to cause most DHF cases in the 1980 Bangkok school study (Burke *et al.*, 1988) and also yielded high case to infection ratios.

All large studies include a few DHF/DSS cases in children older than 1 year associated with infections classified as primary type (Table IV). In Bangkok, 1962, Yangon 1984–1988, Havana, 1981, and Santiago, 1997, primary infections comprised 7.3, 5.0, 1.5, and 1.5% of all hospitalized DHF cases, respectively. DHF hospitalizations per 1000 primary infections in children in Bangkok, Yangon, Havana, and Santiago were 1.9, 0.9, 0.9, and 0.23/1000, respectively (data derived from Table IV). When means of primary (1.0) and secondary infection population-based hospitalization rates are compared, DHF/DSS occurs about 40 times more frequently during secondary than primary infections. These comparisons, of course, exclude infants under the age of 1 year.

TABLE VII

DHF/DSS or DSS Hospitalization Rates per 1000 Secondary Dengue Infections Calculated from Data Derived from Prospective Cohort Studies or from Epidemiologically Defined Populations (for data, See Table IV)[a]

Place	Year	Age	Second infection	DHF/DSS	DHF/10^4 second infection	DSS/10^4 second infection	Reference
Bangkok	1962	1–15	125,728	2528/1428	20.1[a]	11.4[a]	Halstead (1980)
Koh Samui	1966	2–12	83	3	36.1		Russell *et al.* (1968)
Koh Samui	1966	1–15	2,700	33	12.2		Russell *et al.* (1968)
Rayong	1980	4–10	112	4	—	35.7[a]	Sangkawibha *et al.* (1984)
Rayong	1980	<1–10	920	18	—	19.6[a]	Sangkawibha *et al.* (1984)
Rayong	1980	<1–10	48	10	—	208.0[a,b]	Sangkawibha *et al.* (1984)
Bangkok	1980	4–15	59	7	118.6[a,c]		Burke *et al.* (1988)
Havana	1981	3–14	59,875	1213	20.3		Guzman *et al.* (1990)
Yangon	1984–1998	2–6	4,181	138	33.0[a]		Thein *et al.* (1997)
Jogyakarta	1995–1996	4–9	120	7	58.3[a]		Graham *et al.* (1999)
Santiago de Cuba	1997	18+	4,810	202	42.0[a]		Guzman *et al.* (2000b)

[a] Cases graded by WHO criteria.
[b] DSS cases among secondary DEN-2 infections only.
[c] Denominator likely to be too small as secondary infections determined by insensitive HI test.

D. *Factors Controlling Illness Severity during Secondary Infections*

Many factors have been identified that influence the clinical expression of secondary dengue infections. Factors intrinsic to the host, such as HLA types, have been described (Chiewsilp *et al.*, 1981; Loke *et al.*, 2001; Paradoa Perez *et al.*, 1987). The role of sex, age, genetic, and nutritional status are reviewed or described elsewhere (Halstead, 1997). Antibody and viral factors are discussed here.

1. *Heterotypic Neutralizing Antibodies*

Varying amounts and kinds of heterotypic neutralizing antibodies are raised following primary dengue infections. As shown in Table VIII, when preinfection sera were studied at low dilutions in human monocyte cultures, those from children whose secondary DEN-2 infections were inapparent consistently neutralized DEN-2, whereas sera from children who developed an illness requiring hospitalization lacked heterotypic DEN-2 neutralizing antibodies (Kliks *et al.*, 1989). This observation shows that low levels of heterotypic neutralizing antibodies (most were anti-DEN-1) do not prevent but downregulate DEN-2 infections. Around one fifth of monotypically dengue-immune school children lacked heterotypic DEN-2 antibodies and developed DHF when infected by DEN-2. This is virtually the same ratio for DSS during sequential DEN-1 and then DEN-2 in Rayong, Thailand (see Table VII).

The degree to which heterotypic antibodies are raised to antigens common to two or more dengue viruses or as a result of a specific antibody response repertoire controlled by the host is unknown. However, evidence shows that DEN-2 viruses may evolve to escape from heterotypic neutralization (see later).

2. *Sequence of Viral Infection*

The sequence of infection may be highly determinative of disease severity. Although infections in all possible sequences were documented, only secondary DEN-2 infections were pathogenic in the 1980 cohort study at Rayong, Thailand (Table IX) (Sangkawibha *et al.*, 1984). Infection sequences associated with DSS cases were known from virus isolations from acute-phase sera or neutralizing antibodies in preillness sera interpreted by the original antigenic sin phenomenon (Halstead *et al.*, 1983). Burmese workers came to a similar conclusion in their 1984–1988 longitudinal seroepidemiological study in Yangon, Myanmar (Thein *et al.*, 1997). DHF/DSS is not universally associated with secondary DEN-2 infection. In an Indonesian study (Table IX),

TABLE VIII

MODERATING EFFECT OF HETEROTYPIC NEUTRALIZING ANTIBODIES DURING SECONDARY-DENGUE INFECTIONS[a]

Dengue 2 NT antibodies in undiluted sera, June 1980	Admitted to hospital for DHF	No school absences
Yes	1	29
No	6	4
	7	33

[a] Undiluted preillness sera from 40 children experiencing secondary dengue 2 infections were incubated with dengue 2 virus and added to human monocyte cutures. No virus was scored as neutralization, growth of virus as enhancement. Sera from children with inapparent infections neutralized, whereas sera from clinically ill children enhanced dengue 2 infection. After Kliks *et al.* (1989) and Burke *et al.* (1988).

TABLE IX

INCIDENCE OF DHF/DSS CASES AMONG SECONDARY DENGUE INFECTIONS BY SPECIFIC SEQUENCE IN CHILDHOOD COHORTS STUDIED IN THAILAND AND INDONESIA[a]

		Second dengue infection			
Place, year, age	First dengue infection	Dengue 1	Dengue 2	Dengue 3	Dengue 4
Rayong, Thailand, 1980, ages <1–10	Dengue 1	—	10/48	0/9	0/20
	Dengue 2	0/257	—	0/34	0/82
	Dengue 3	0/125	6/92	—	0/40
	Dengue 4	0/114	2/84	0/15	—
Jogyakarta, Indonesia, 1995–1996, ages 4–9	Dengue 1	—	0/31	1/6	0/6
	Dengue 2	3/41	—	0/7	1/8
	Dengue 3	0/9	0/2	—	0/4
	Dengue 4	0/4	0/5	0/1	—

[a] After Sangkawibha *et al.* (1984) and Graham *et al.* (1999).

DSS was associated with sequences ending in DEN-1, -3, and -4, but not DEN-2, even though secondary DEN-2 infections were common (Graham *et al.*, 1999). Dengue 3 was associated with an outbreak of DHF/DSS on Tahiti in a population that had prior infection experience with DEN-1 and DEN-2 (Chungue *et al.*, 1990). In 2001 in Tahiti, DEN-1 produced a large DHF outbreak in children exposed to DEN-2 and -3 viruses (Hubert, personal communication). The pathogenicity of secondary infections with various dengue viruses appears to change over time as illustrated by the changing dominance of different viruses recovered from DHF cases over a 30-year observation at the Bangkok Children's Hospital (Fig. 4).

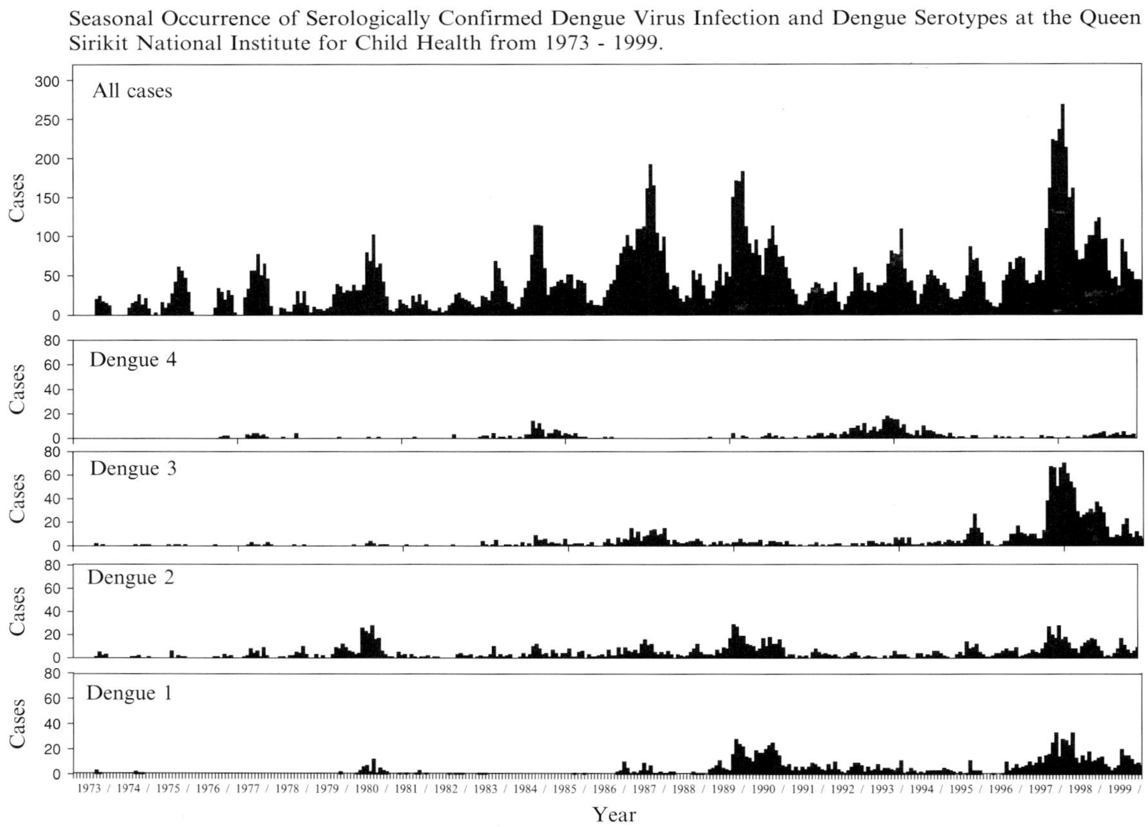

FIG 4. Virus isolations from DHF/DSS patients hospitalized at Bangkok Children's Hospital, 1973–1999 (courtesy Dr. T. Endy, AFRIMS, Bangkok, Thailand).

3. Interval between First and Second Dengue Infections

DHF/DSS was observed when DEN-1 and then DEN-2 infections occurred at intervals of 4 and 20 years in Cubans. In both the nationwide and Santiago epidemics, severe disease was caused by Asian genotype dengue 2 viruses (Guzman *et al.*, 1995). Dengue 1 was transmitted in 1977–1979 and is the only known introduction of this virus into Cuba. After each epidemic, comprehensive vector control was instituted and dengue transmission was terminated. The absence of new dengue infections permitted an accurate study of epidemic infection rates by measuring neutralizing antibody prevalence in randomly collected sera (Guzman *et al.*, 1990, 2000b). Age-specific secondary DEN-2 infection rates were used to estimate infections. DHF cases and deaths reported by age from Havana in 1981 and Santiago in 1997 were then used to calculate DHF and death rates per 10,000 secondary DEN-2 infections (Guzman *et al.*, 2002a). When DHF and death rates were compared in identical age groups of adults (essentially 20–65+ years), the 1997 epidemic was 8 and 25 times more severe than that of 1981, respectively (Table X). Case fatality rates among adults who experienced secondary DEN-2 infections at intervals of 4 and 20 years were 0.6 and 2.6%, respectively, differing by a factor of 4.7-fold.

E. Factors Controlling Illness Severity during Primary Infections in the Presence of Passively Acquired Antibody

It was noted (Table II) that sera from mothers whose infants acquired primary infection DHF had variable amounts of DEN-2 neutralizing antibodies. Figure 5 demonstrates the positive correlation between the DEN-2 $PRNT_{50}$ titer of maternal antibodies (presumably the titer in infant sera at birth) and the age of the infant at hospitalization

TABLE X

INCREASED SEVERITY OF CLINICAL DISEASE IN ADULTS EXPERIENCING DENGUE 2 INFECTIONS 4 AND 20 YEARS FOLLOWING DENGUE 1 INFECTION[a]

Place	Date	Secondary DEN-2 infections	DHF cases	Deaths	$DHF/10^4$ secondary DEN-2	$Deaths/10^4$ secondary DEN-2	Case fatality rate (%)
Havana	1981	242,070	1329	21	54.9	0.9	1.58
Santiago	1997	4,810	202	11	419.9	22.8	5.4

[a] After Guzman *et al.* (2002b).

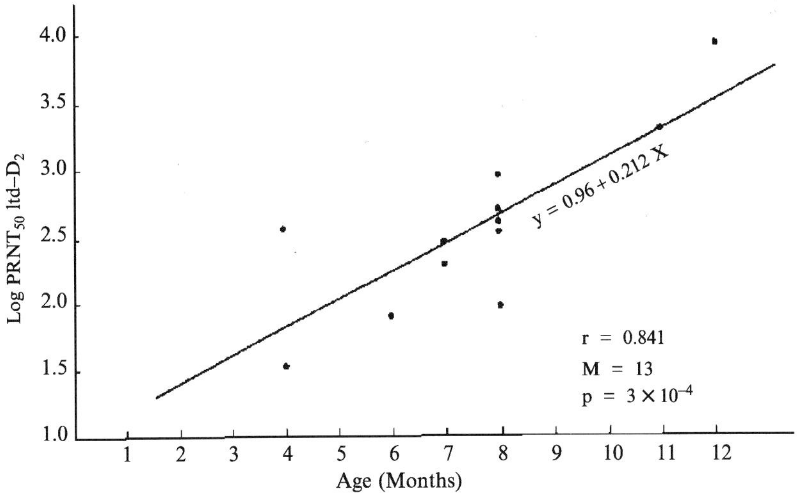

FIG 5. Correlation between maternal dengue 2 neutralizing antibody titers and age their infants were hospitalized for DHF/DSS caused by primary dengue 2 infections. After Kliks *et al.* (1988), with permission.

for DHF. Most infants were admitted within 3 weeks of the time that maternal antibodies were expected to fall to a titer of 1:10 (Kliks *et al.*, 1988). Infants are at risk when neutralizing antibodies have fallen below a protective threshold but titers of ADE antibodies are maximal (Fig. 3).

V. VIRAL FACTORS

A. Genotypes

The concept that DHF/DSS is caused by intrinsic differences in dengue viruses has been a favored explanation for the "emergence" of this syndrome in Southeast Asia. When DEN-3 and -4 were associated with DHF/DSS in the Philippines in 1956, Hammon *et al.* (1960) speculated that these viruses might be responsible for the "new" disease. Later, putative DEN-5 and -6 viruses were thought to be DHF associated (Hammon and Sather, 1964). When DEN-1 and -2 viruses were shown to cause dengue fever as well as DHF (Halstead *et al.*, 1969c), researchers looked for virulent and nonvirulent dengue viruses circulating in different biological niches, for example, *Aedes aegypti*- versus *Aedes albopictus*-transmitted viruses (Rudnick, 1965, 1967).

It seemed logical to expect that intrinsic virulence would have a genetic basis. Many dengue viruses recovered from mild and severe dengue infections have been partially sequenced and compared (Table XI), but no reproducible viral genetic differences have been found as yet within DHF-endemic areas.

However, a disease severity-related genetic difference is well established. The first dengue virus recovered in the American hemisphere was DEN-2 TR 1751 (Anderson and Downs, 1956). Reports of outbreaks of dengue prior to World War II are attributed to DEN-2 by serological studies in Panama and Cuba (Guzman *et al.*, 1990; Rosen, 1958). In 1963, DEN-3 was introduced into the hemisphere and was first recognized in Puerto Rico (Neff *et al.*, 1967). It is not known how widely these two viruses were transmitted, but conditions for sequential infection existed. This epidemiological situation was not accompanied by DHF/DSS. In 1977, DEN-1 was introduced into the Caribbean and spread quickly throughout the region (Anonymous, 1978). Again, sequential infections were possible, but there were no reports of DHF/DSS. However, DHF/DSS immediately accompanied the introduction of an Asian genotype DEN-2 in Cuba in 1981 following an earlier 1977–1979 virgin soil epidemic of DEN-1 (Guzman *et al.*, 1995; Kouri *et al.*, 1989). Similar sequential introductions into Peru of DEN-1 in 1990 and an American genotype DEN-2 in 1995 *did not* result in DHF/DSS (Watts *et al.*, 1999). This occurred in the

TABLE XI
SUMMARY OF STUDIES LOOKING FOR GENETIC SEQUENCES THAT CORRELATE WITH DENGUE DISEASE SEVERITY

Serotype	Gene sequenced	DF	DHF, I&II	DSS	Observation	Reference
2	E	4	2	2	No differences	Blok *et al.* (1989)
3	E (amino acids 25–89)	9	11	1	No differences	Chungue *et al.* (1993)
3	PrM–E	1	2	2	No differences	Lee *et al.* (1993)
2	E	1	1	1	No differences	Lee *et al.* (1993)
2	E/NS-1 (240 nucleotides)	1	2	1	No differences	Lee *et al.* (1993)
2	E/NS-1	0	2	1	No differences	Thant *et al.* (1995)
2	C–PrM–E	1	2	1	1–2 amino acids	Thant *et al.* (1996)
2	Entire	4	3	1	No differences	Mangada and Igarashi (1998)
2	E/NS-1 (240 nucleotides)	30	462	—	No differences	Rico-Hesse *et al.* (1998)

Amazonian city of Iquitos, population 344,686, where ongoing serological monitoring made it possible to measure secondary DEN-2 infections. Using rates of DHF during secondary infections cited earlier (Table VII), the estimated 49,000 secondary DEN-2 infections should have resulted in thousands of DHF/DSS hospitalizations and hundreds of deaths. Careful examination of hospital records revealed no DHF/DSS-like disease. In fact, secondary infections were accompanied only by mild disease with low attack rates of overt disease despite high infection rates. When Asian and American DEN-2 genomes were sequenced directly from viremic sera, a total of six encoded amino acid differences were found in prM, E, NS4b, and NS 5 genes along with mutations conferring predicted structural changes in the 5′- and 3′-nontranslated regions (Leitmeyer *et al.*, 1999).

B. Antigenic Structure

A significant correlate of genetic differences can be in the structure of virion antigens. When Peruvian anti-DEN-1 sera were tested, they significantly neutralized the American genotype DEN-2, whereas Asian DEN-2 was much less neutralized (Table XII) (Kochel *et al.*, 2002). This observation points to the existence on the envelope of American genotype DEN-2 viruses of DEN-1-like structure(s) and the absence of this structure on Asian genotype DEN-2 viruses.

C. Escape Mutants

During the 1981 and the 1997 DHF/DSS epidemics in Cuba, month-to-month increases were observed in the fraction of severe compared with mild cases and also in case fatality rates (Table XIII) (Guzman *et al.*, 2000a). The possibility of the rapid selection of dengue virus

TABLE XII

GEOMETRIC MEAN NEUTRALIZING ANTIBODY TITERS IN 44 SERA FROM DEN-1-IMMUNE RESIDENTS OF IQUITOS, PERU, WHEN TESTED BY HOMOLOGOUS DENGUE VIRUS (DEN-1) OR DEN-2 STRAINS REPRESENTING EITHER AMERICAN OR ASIAN GENOTYPES[a]

Genotype	Asian	American	Asian	Asian
Virus	DEN-1	DEN-2	DEN-2	DEN-2
Strain	16007	IQT 2913	16881	OBS 8041
Isolated, place, date	Thailand, 1964	Peru, 1996	Thailand, 1964	Venezuela, 2000
GMT, $N = 44$	578	549	56	29

[a] After Kochel *et al.* (2002).

TABLE XIII
MONTHLY DHF/DSS ATTACK RATES, CASE FATALITY RATES, AND DEATH RATES DURING DEN-2 EPIDEMICS IN CUBA, 1981, AND SANTIAGO DE CUBA, 1997[a]

	Cuba—1981				Santiago—1997			
	June	July	August	P^b	May	June	July	P^b
DHF/DSS	1,881	6,223	2,120		37	132	29	
Deaths	38	77	40		1	6	5	
Total dengue	96,664	183,443	43,315		705	1785	244	
DHF/DSS/dengue (%)	1.9	3.4	4.9	$<10^{-7}$	5.2	7.4	11.9	<0.01
Deaths/DHF/DSS (%)	2.0	1.2	1.9	>0.05	2.7	4.5	17.2	<0.05
Deaths/dengue (%)	0.04	0.04	009	<0.01	0.14	0.34	2.05	<0.01

[a] After Guzman *et al.* (2000a).
[b] First month versus third month.

neutralization escape mutants seemed an obvious explanation. It was hypothesized that the serial transfer of DEN-2 viruses in DEN-1-immune hosts might result in the selection of mutants that lacked their DEN-1-like antigenic structure(s). These viruses should have a selection advantage being enhanced by antibodies directed against noncritical structural sites.

VI. ADE *IN VIVO*

A. *Animal Studies*

Crucial to the biological relevance of ADE is the fact that the enhancement phenomenon was discovered in an *in vivo* model (Halstead *et al.*, 1973c; Marchette *et al.*, 1973). Three key papers were published in 1973. The first, describing the enhanced growth of dengue virus *in vitro*, reported laboratory work performed earlier that year (Halstead *et al.*, 1973b). However, observations on rhesus monkey viremia reported as "clinical laboratory studies" on primary and heterologous dengue infections had been in progress from 1965 (Halstead *et al.*, 1973d). It was also observed that dengue viruses were abundant in tissues of monkeys experiencing secondary infections (Marchette *et al.*, 1973).

B. *Viremia in Primary versus Secondary Dengue Infections in Monkeys*

Primary infections and all 12 combinations of sequential infections with four dengue viruses (i.e., 1-2, 1-3, 1-4, 2-1, 2-3, 2-4, 3-1, 3-2, 3-4,

4-1, 4-2, 4-3) were studied in groups of rhesus monkeys (Halstead *et al.*, 1973d). After primary and secondary infections, monkeys were bled daily for 10 days to measure viremia, blood platelets, prothrombin time, serum proteins, hematocrit, and, in some cases, total serum complement. A single animal, infected in the sequence DEN-4 and then DEN-2, had a physiological response consistent with DHF. Animals given dengue 2 as their second infection had peak viremias that were more than 10-fold higher than those in animals infected for the first time using the same strain, dose, and route of infection (Table XIV). The mean duration of secondary DEN-2 viremias (3.4 days) was slightly shorter than that of primary viremias (4.0 days). Smaller numbers of animals with primary and secondary DEN-1 and DEN-4 viremias were studied. Viremias did not differ quantitatively, although the onset of viremia was a little later during secondary compared with primary infections.

C. *Viremia in Monkeys Receiving Passive Dengue Antibodies*

In the only study on this phenomenon, two groups of five rhesus monkeys were given dengue-immune or nonimmune human cord blood serum diluted to a final dilution in blood of 1:300 (Halstead, 1979). Animals were infected with 1000–9000 pfu of DEN-2 and were then viremia quantitated daily for 10 days. The dengue-immune cord blood pool was from infants born to mothers of southeast Asian origin and contained neutralizing antibodies to all four dengue viruses. Control monkeys received dengue antibody negative cord blood. The DEN-2 $PRNT_{50}$ of pooled cord blood was 1:140, whereas the enhancing titer, measured in nonimmune human monocytes, was greater than $1{:}10^{-6}$. Following inoculation of antibodies, monkeys had no detectable dengue neutralizing antibodies, but enhancing antibodies were detected to dilutions of $1{:}10^{-4}$. Viremia levels were consistently higher in animals receiving dengue antibodies (Fig. 6).

TABLE XIV
DENGUE 2 VIREMIA AND INFECTION PARITY IN MONKEYS[a]

Infection parity	n	Viremia onset day (mean)	P	Duration (days) (mean)	P	Mean peak viremia (pfu $\log_{10}$/1.0 ml)	P
Primary	24	3.0		4.0		2.7	
Secondary	44	3.6	0.05	3.4	0.5	3.7	<0.001

[a] After Halstead *et al.* (1973d).

D. ADE during Human Dengue Infections

Studies on DHF in Indonesian children provided the first evidence that mean viremias were higher during secondary than primary dengue 3 infections (Table XV) (Gubler *et al.*, 1979). Because these were single point determinations rather than serial bleedings, only the range provides relevant data. Better data are now available. DEN-1 and -2 peak viremia titers measured early during secondary dengue infections predict imminent disease severity (Table XVI). These data were obtained from children admitted to study soon after

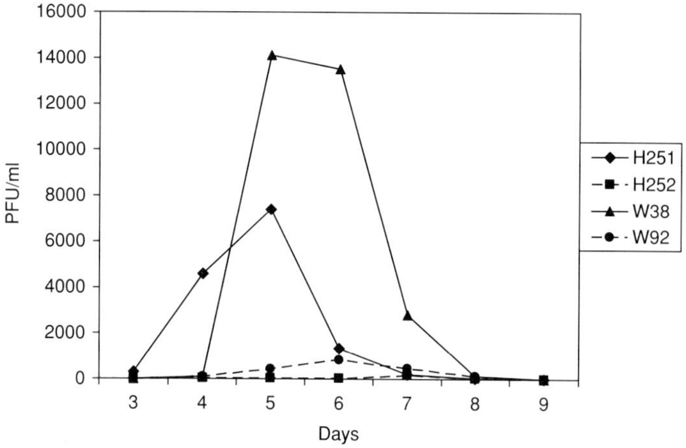

FIG 6. Daily reciprocal viremia titers in H-251(◆) and W-38 (▲) susceptible rhesus monkeys given human dengue immune cord blood serum and H-252 (■) and W-92 (•) given normal cord blood serum intravenously and then infected with dengue 2 (16681 strain). After Halstead (1979).

TABLE XV

VIRUS LOAD IN ACUTE PHASE SERA, DISEASE SYNDROME, AND INFECTION PARITY[a]

		Primary infection			Secondary infection		
	Syndrome	n	Mean ($\log_{10}$)	Range	n	Mean ($\log_{10}$)	Range
Dengue 3	Dengue fever	5	5.5	4.3–7.0	5	6.6	4.2–8.0
	DHF I&II	1	7.3	7.3	3	6.0	2.8–8.2
	DSS	2	5.4	4.6–6.4	9	6.1	3.8–8.2

[a] After Gubler *et al.* (1979).

TABLE XVI
"PEAK" VIRUS LOAD IN ACUTE PHASE SERA FROM SECONDARY DENGUE INFECTIONS IN RELATION TO INFECTION SEVERITY[a]

	Dengue 1		Dengue 2	
Syndrome	n	Peak mean ($\log_{10}$)	n	Peak mean ($\log_{10}$)
Dengue fever	13	5.5[b]	16	6.5[c]
DHF I&II	10	7.3[b]	26	7.6[c]
DSS	3	8.6[b]	5	8.5[c]

[a] After Halstead *et al.* (1973d).
[b] $R = 0.5; p = 0.01$.
[c] $R = 0.5; p < 0.01$.

the onset of fever and were bled for 5 consecutive days, making it possible to identify peak viremia (Vaughn *et al.*, 2000). Thus, viremia correlates strongly with disease severity. Disease severity, in turn, is correlated with secondary dengue infections. Because most primary dengue infections in children are mild or inapparent, it may never be possible to assay viremias in this immunological group. There can be little doubt that antibody-dependent enhancement of human dengue viral infections has been documented and correlates with, even predicts, disease severity.

VII. ADE *IN VITRO*

In vitro antibody-enhanced infections, or underneutralization of viruses probably due to antibody-dependent enhancement, have been described in the literature since the 1930s (Halstead, 1982). The first explicit descriptions of antibody-related plaque enhancement were by Hawkes (1964) and Hawkes and Lafferty (1967). These authors conjectured that enhanced plaque formation of MVE by avian anti-MVE on chick embryo "fibroblasts" (CEF) was due to the stabilization of viral infectivity by antibody. It remained for Kliks and Halstead (1980, 1983) to demonstrate that CEF monolayers were "contaminated" with 2% functional mononuclear phagocytes. At a relatively low multiplicity of infection (MOI), MVE virions did not infect mononuclear phagocytes. However, some of the other 98% chick fibroblastic cells were permissive to MVE infection. The addition of MVE antibody at low dilutions directed viruses to infect mononuclear phagocytes. Efficient antibody-mediated infections in this small population accounted for the 20–30% increase in infected cells observed.

The explanation of the Hawkes–Lafferty phenomenon had been made possible by studies on the immune enhancement of dengue virus infection in peripheral blood mononuclear phagocytes (Halstead *et al.*, 1973b; Halstead and O'Rourke, 1977a, 1977b). These studies had defined the phenomena of immune-enhanced viral infection of cells as possessing several possible attributes: (1) cells are infected in the presence of antibody or cells are infected that are derived from an immune host. *In vivo*, this may translate to an enhanced susceptibility to infection, which is decoupled from the number of infectious particles in an infectious inoculum. (2) Cells may be infected more quickly. This can be illustrated most clearly at low incubation temperatures. *In vivo*, the incubation period from infection to onset of disease may be shortened in the presence of enhancing antibodies. (3) There are increased numbers of productively infected cells. This phenomenon may be derivative of attributes 1 and/or 2. *In vivo*, this could translate into increased viremia and/or increased severity of disease. (4) Altered cell tropism is a hypothetical but unexplored attribute of enhanced infection. The *in vivo* correlate might be an altered disease course (e.g., atypical measles or neurological AIDS).

A. *Factors Controlling ADE* in Vitro

In vitro and *in vivo* studies demonstrate that indirect virus–cell entry is controlled by many factors, some interconnected. Each is discussed briefly.

1. Ligands

Infection enhancement may result when bivalent ligands bind to receptors on viruses and on cells. Ligand binding increases the forces that bring viruses in close approximation to other usually normal cell–virus receptors. For the most part, in well-studied systems, virus entry into cells follows the normal route by mechanisms described by Marsh and Helenius (1989). Bivalent ligands described to date include immunoglobulin G (IgG), immunoglobulin M (IgM) plus complement, IgG plus complement, complement alone, and an unidentified factor in unheated plasma.

2. Antibody

The best studied and possibly the most powerful and ubiquitous ligand that enhances the infection of animal viruses is antibody. In both natural and experimental settings, there is a corresponding requirement for an appropriate receptor on the cell surface—for example,

Fc receptors for IgG immunoglobulins and complement receptors when IgM serves as an intermediate ligand.

a. Immunoglobulin Class Most published studies on ADE have used antiviral IgG as a source of enhancing antibody. It is known that a number of cell types also carry FcR for IgA, IgE, and IgM. Whether any of these latter immunoglobulin types contribute significantly to ADE is unknown. In early studies on rhesus monkeys, whole sera obtained 14 days after dengue 2 infection contained neutralizing antibodies (PRNT) when assayed in continuous rhesus monkey kidney cells, but when assayed on rhesus peripheral blood mononuclear cells (PBMC), weak ADE was noted at dilutions between 1:160 and 1:2560 (Halstead, 1982). When IgM and IgG were assayed separately, the IgG fraction exhibited *only* ADE to a serum-equivalent dilution of 1:10,240, while decomplemented IgM only neutralized virus. However, when fresh complement was added, the IgM fraction neutralized dengue 2 to a serum equivalent dilution of 1:40, but enhanced infection in rhesus PBMC to a serum-equivalent dilution of 1:800. ADE, due to IgG, was unaffected by complement. When whole serum was assayed in PBMC in the presence of complement, virus was neutralized at low dilution (1:10 and 1:40) and weak ADE was noted at higher dilutions (1:160–1:2560). These studies show that early IgM is very efficient at neutralizing dengue virus, but at high dilutions in the presence of complement, IgM can mediate ADE. Cardosa *et al.* (1983) studied this phenomenon and showed that the IgM–virus–complement complex attaches to C3 receptors and initiates virus entry into cells via this ligand–receptor interaction.

While low concentration of IgM antibodies can mediate ADE, in an autologous system, it is well to remember that *in vivo*, at least in the dengue system, infection enhancement by IgM could be expected to occur only *very* early in the immune response and then only transiently. In whole serum, antidengue IgM effectively ablated the pro-ADE effect of IgG (Halstead, 1982; Halstead and O'Rourke, 1977b). The reason why early antidengue IgG enhanced but failed to neutralize dengue 2 has not been investigated further. It is likely that this reflects the rather poor avidity and conformational "fit" of early IgG idiotypes to critical virion epitopes.

b. Ig Subclass In humans, principally IgG1 and IgG3 are responsible for ADE, whereas in mice it is IgG2a. This reflects receptors on the mononuclear phagocyte cell system used: human PBMC have IgG1 and IgG3 receptors. There is a large family of Fc receptors that have parallel functions but are only loosely related on a structural basis. These are discussed later. Mouse macrophages have at least three types of FcR: FcR I, which binds to IgG2a; FcR II, which binds to IgG2b; and FcR III, which binds to IgG3 (Porterfield, 1986).

c. Phylogenetic Class of Fc Terminus Hawkes (1964) and Hawkes and Lafferty (1967) observed that avian antisera enhanced viral plaque formation on CEF and chorioallantoic membranes (CAM). However, enhancement was not observed when the same reagents were assayed on pig kidney cell cultures or in suckling mice. Mouse and rabbit antisera failed to enhance viral plaque or pock formation on CEF and CAM. This latter phenomenon is explained by the structural similarities between Fc termini and FcR receptors at the level of phylogenetic class. Fc–FcR interactions do not occur regularly when the antibody source and cells providing the FcR differ between avians and mammals. Curiously, guinea pig anti-MVE enhanced MVE plaque formation on CEF (Kliks and Halstead, 1983).

d. Idiotype Specificity Controversy continues to swirl around the idiotype specificity that mediates ADE and the corresponding epitopes on viral surfaces. A fundamental fact established repeatedly, both with polyclonal and monoclonal antibodies, is that antibodies directed at epitopes that neutralize viruses efficiently can, in general, produce ADE when tested at concentrations below the neutralization threshold (Halstead, 1982; Porterfield, 1986). The importance of antibody concentration is discussed later.

It is well established that ADE occurs when antibodies, even at relatively high concentrations, are directed against virion epitopes that are not thought to mediate attachment to and entry through the plasma membrane. The classical example is infections with dengue 1, 3, or 4 viruses, which raise antibodies against dengue subgroup or flavivirus group epitopes, which can enhance dengue 2 infection *in vitro* and are thought to enhance infection and disease *in vivo* (Halstead, 1982). In fact, the most powerful risk factor ever identified for the occurrence of DHF/DSS in humans was the circulation of antibodies raised by heterotypic dengue viral infections, which exhibited dengue 2 ADE when undiluted serum was tested in human PBMC *in vitro* (Kliks *et al.*, 1989). In contrast, even small amounts of cross-reactive dengue 2 neutralizing antibody raised during primary infections due to dengue 1, 3, or 4 viruses downregulated secondary dengue 2 infections to produce inapparent disease. *In vitro*, a wide range of heterotypic polyclonal and monoclonal antibodies have been shown to produce ADE in the flavivirus family (Halstead, 1980; Halstead *et al.*, 1980; Porterfield, 1986).

e. Concentration A large number of studies have shown that homotypic neutralizing antibodies, polyclonal or monoclonal, produce ADE at dilutions above the neutralization end point. Dengue neutralizing antibodies transferred passively to human infants by mothers with multiple previous dengue infections can produce severe disease in such infants when primary infections occur in the first few months of life (Kliks *et al.*, 1988).

f. Enhancing Titer As illustrated in Fig. 7, measurements of viral infections in PBMC cultures have three parameters. The highest dilution of antibody that produces significant and reproducibly enhanced infections is called the "enhancing titer." With respect to *in vivo* phenomena, it is important to point out that many sera, when tested at low dilutions, neutralize; infection enhancement is only demonstrable at higher dilutions. Many authors confuse the highest dilution at which ADE can be demonstrated as a measure of biological potency.

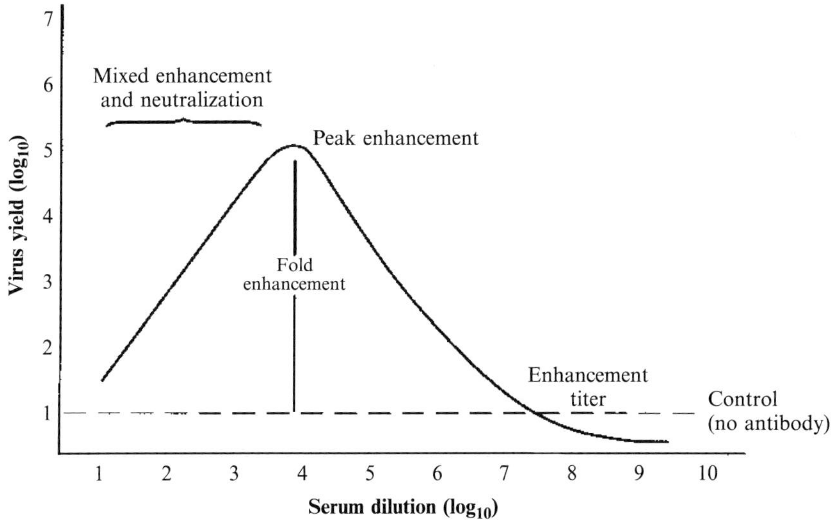

FIG 7. Diagram defining the various parameters of antibody-dependent enhancement. The test is performed by adding virus plus antibody dilutions to mononuclear phagocyte cultures. At low dilution of serum, little or no virus growth is observed; at a higher dilution, enhanced viral growth occurs (peak enhancement). At a still higher dilution, the enhancement effect is gradually lost (enhancement titer). From Halstead (1994).

It is not. *In vivo*, antibodies that neutralize in undiluted serum will neutralize regardless of the enhancing titer of this same serum when diluted. It is a general rule that sera that competently neutralize to relatively high titer will demonstrate extremely high enhancing titers.

g. Peak Enhancement Titer The dilution of sera at which fold enhancement is greatest is labeled the peak enhancement titer (Fig. 7). Those infants who developed severe disease during primary dengue 2 infections acquired their infections at a time when it was estimated that peak infection enhancement would occur.

h. Enhancing Power The fold increase in the number of cells infected or the amount of virus produced at peak enhancement in antibody-supplemented assays as compared with controls is expressed as "power." As illustrated in Fig. 7, over many dilutions at the low range, enhancing and neutralizing antibodies are mixed. Under these conditions, infection enhancement is suboptimal. At some dilution of antibody, optimal viral infections occur; at still higher dilutions, the concentration of infectious antibody–virus complexes is not great enough to saturate the system. Power must also reflect the abundance of virus receptors on FcR-bearing cells, the efficiency of entry of virus, the viability of virus after reaching the cytosol, and competence to complete all steps to achieve assembly and final release of virions.

3. Cells

a. FcR Bearing With the few exceptions noted, the *sine qua non* of antibody-dependent and noncomplement-dependent enhancement of viral infection is the existence of *permissive* FcR-bearing cells. FcR can be distributed naturally or can be induced following infection by one of several members of the herpes virus family (Bell *et al.*, 1990; McKeating *et al.*, 1990). A large family of distinct classes of Fc receptors exists. They have been defined partially by monoclonal antibodies. Each receptor is represented differentially on cells in various tissues. Among the reasonably characterized immunoglobulin Fc receptors, there are three for IgG (FcR I, FcR II, FcR III), at least one for IgM, one for IgE, and one for IgA (Unkeless *et al.*, 1988). In humans, Fc receptors have been detected on T, B, and natural killer lymphocytes, polymorphonuclear leukocytes, mononuclear phagocytes, and dendritic cells, including Langerhans cells. Monocytes bear two Fc receptors. FcR I is found on monocytes/macrophages and binds human IgG_1 and IgG_3 with high avidity. Fc receptors are also found at high density on epithelial cells of the intestinal tract (Kobayashi *et al.*, 1989). FcR I numbers are augmented by interferon-γ, and such augmentation can be shown to increase dengue viral ADE (Kontny *et al.*, 1988).

b. Cytophilia While noncomplexed monomeric IgG1 and IgG3 occupy FcR furtively, this weak and brief interaction merely reflects the concentration of immunoglobulins in the extracellular medium. Attempts to "stick" a monomeric-enhancing antibody on human or monkey PBMC prior to the addition of virus have failed (Halstead and O'Rourke, 1977b). When antibody was incubated with PBMC from nonimmune donors for 1 h at 37 °C and cells were washed in a different IgG-containing medium, no ADE occurred unless antibody was added back to the reaction mixture. In contrast, despite extensive washing to remove original serum or loosely bound IgG, PBMC from dengue-immune rhesus monkeys or human beings are permissive to dengue infection *in vitro* (Halstead *et al.*, 1973a, 1973b; Halstead and O'Rourke, 1977b). Using silica, these cells have been identified as monocytes (Halstead, 1982; Halstead and O'Rourke, 1977b). Enhanced infection was reduced or ablated by treatment with trypsin or anti-IgG prior to the addition of dengue virus.

The incubation of human anti–yellow fever serum with U937 continuous human monocytoid cells resulted in an irreversible cytophilic attachment of antibody that could not be eluted off at 37 °C (Schlesinger and Brandriss, 1981). This may be related to the finding of Anderson and Abraham (1980) that monomeric IgG myeloma immunoglobulins showed a higher affinity for U937 Fc receptors than aggregated immunoglobulin of the same subclass. Thus, in U937, but not PBMC, monomeric immunoglobulins may become cell associated *in vitro*.

c. Age of Cells In the dengue system, ADE is optimal when human or rhesus PBMC are held in culture for 24 h at 37 °C before adding virus and antibody (Halstead and O'Rourke, 1977b). In cells infected at 48–96 h, a sharp decline in viral replication was observed. Similar results were obtained in PBMC, which were plated onto glass or plastic surfaces. Under both conditions, but especially the latter, monocytes differentiate into macrophages. Splenic, thymic, and lymph mode macrophages from rhesus monkeys showed a similar reduction in permissiveness when held in culture for 2–4 days prior to infection (Halstead *et al.*, 1977). Bone marrow mononuclear phagocytes showed no such maturational phenomenon, being equally permissive on days 0 and 4 after removal from their rhesus monkey host. This may reflect the fact that the progeny of stem cells *in vitro* are relatively immature cells.

d. Activation The effects of macrophage activation on viral replication have been examined in a number of studies, the majority of which report that activated macrophages are more restrictive (Mogensen, 1979; Morahan *et al.*, 1985). There are some contrary reports. van der Groen *et al.* (1976) found that Semliki forest virus replicated better in protease-peptone-elicited macrophages than in resident macrophages, and Hotta and colleagues (Hotta and Hotta, 1982; Wiharta *et al.*, 1985) reported increased yields of dengue viruses in macrophages activated by extracts of bacterial and parasitic cell walls and peptidoglycans. Cardosa *et al.* (1986) showed that IgG–West Nile virus complexes grew better in BCG-infected mouse peritoneal macrophages than in resident or thioglycolate-elicited macrophages. In contrast, in the IgM–complement–West Nile virus system, virus grew better in thioglycolate-elicited macrophages than in others (Cardosa *et al.*, 1986).

The effect of macrophage activation has even been demonstrated *in vivo* when two susceptible rhesus monkeys were inoculated with *Cornybacterium parvum* and an additional two animals were inoculated with pertussis vaccine intravenously. Three days later, the four animals were infected with dengue 2 (S. B. Halstead, unpublished results). Viremias were measured daily and compared with sham-inoculated controls. Viremia levels were significantly higher in animals given substances known to stimulate and activate the mononuclear phagocyte system.

Clearly, age or differentiation of cells *in vitro* plus activation (a separate process) of monocytes/macrophages has a profound effect on ADE, although not in predictable or yet understandable ways.

e. Tissue Source For many of the animal diseases for which ADE may be operative *in vivo*, the sites of replication of virus in the intact animal are known to be one or more varieties of tissue macrophages (Halstead, 1982; Porterfield, 1986). For reviews of nonarboviruses, see Burke (1992), McGuire *et al.* (1990), and Stott *et al.* (1984). *In vitro* systems rarely employ these biologically relevant cells. Usually, monocytes are obtained from peripheral blood. Several workers have used macrophages obtained by lavage of the peritoneal cavity. Most such techniques activate these cells. Whether such activation affects viral growth and the ability to demonstrate ADE has been carefully studied in some but not all systems (see earlier discussion).

f. Primary Cells versus Cell Line Reference was made earlier to differences in FcR on primary versus monocytoid cell lines. Finally, primary cells and cell lines may differ in the distribution and density of viral receptors. The number of human and other mammalian mononuclear phagocyte cell lines is too large to be reviewed here. Suffice it to say, great care must be taken in extrapolating results from surrogate models to predict or interpret *in vivo* phenomena.

4. *Virus Receptor*

Elegant studies in the West Nile model (Gollins and Porterfield, 1986) have demonstrated that in ADE the entry of the virus into the cytosol is critically dependent on the normal virus receptor and the process of endocytosis. Bispecific Fab dengue antibodies linked chemically to antibodies to the three Fc receptors and to β_2-microglobulin all enhanced dengue infection equally, suggesting that infection did not occur via a phagocytic event (Mady *et al.*, 1991). In other words, antibody attracts virus to the cell surface, after which virus attaches and enters via normal cell receptors.

Convincing experiments demonstrating the mode of entry of virus into cells have not been published for most viruses that participate in ADE. Most authors assume infection is via Fc-mediated endocytosis, yet several different kinds of outcomes of the attachment to cells of virus–IgG complexes to FcR are evident. For a review, see Halstead (1994).

5. *Viruses*

a. Infective Agent (Sensitizing versus Eliciting Viruses) In systems requiring two infections or an immunization process to sensitize to ADE, the virus whose cell entry is enhanced is referred to as the eliciting virus or infection, whereas the initial infection or immunization is referred to as the sensitizing virus or infection or the sensitizing immunization.

b. Eliciting Virus MOI has been studied carefully in the dengue system (Halstead and O'Rourke, 1977b). Serial 10-fold dilutions of virus were tested in human and monkey PBMC. When MOI ranged between 0.001 and 0.1, the infection of cells occurred regularly in the presence of antibody with 3- to 10-fold lower infection in the absence of antibody. When the MOI was raised to between 1 and 100, the differential effect of adding antibody was often lost. The day of peak infection is related directly to MOI. In the dengue system, MOI in the range 0.01–0.05 resulted in peak extracellular virus yields between days 2 and 4 after infection.

Hawkes and Lafferty (1967) found that infection enhancement could be optimized when reactants were held at 4 °C for 30–90 min. At 37 °C, enhancement was optimal if reactants were incubated on PBMC for less than 6 min. Longer periods of incubation pushed the reaction in the direction of neutralization.

i. Nonenhanceable Viruses. Burke *et al.* (1988) published data suggesting that prior dengue infection in human beings does not enhance the severity of a secondary dengue 1 infection (Burke *et al.*, 1988). It is important to note that there is no *in vitro* correlate of this observation. Dengue 1 infections can be enhanced *in vitro* by dengue antibodies when assayed in a macrophage-like cell line (Morens and Halstead, 1990).

ii. Intratypic Variants. Genetic but not antigenic studies have shown that DHF/DSS epidemics have resulted when the eliciting infection is due to a southeast Asian genotype (Rico-Hesse, 1990). Preliminary data from Kliks (personal communication, 1990) suggest that in the presence of a polyclonal enhancing antibody, southeast Asian dengue 2 topotypes grow in primary human PBMC better than Caribbean dengue 2 strains under the same antibody and cell culture conditions. Morens *et al.* (1991) reported preliminary evidence that the ability of dengue 2 strains to grow in PBMC varied directly with the severity of infection in the hosts from whom virus strains were recovered. Similar results were obtained when DEN-2 viruses recovered from varying syndromes were enhanced by dengue 4 monoclonal antibodies (Morens and Halstead, 1987).

c. Sensitizing Infections or Immunization (Passive/Active)

i. Intragenus Viruses. So far as is known, DHF/DSS is not regularly caused by sequential flavivirus infection outside the dengue subgroup. Yellow fever, Japanese encephalitis, and West Nile viruses circulate in dengue-endemic areas, but are not associated epidemiologically with DHF/DSS. There has been one report of a DHF-like illness in a dengue-infected individual given Japanese encephalitis vaccine previously (Okuno *et al.*, 1989). Adult males who were yellow fever immune when given the S-1 live-attenuated dengue 2 virus had enhanced vaccine infectivity (higher seroconversion rates) but were not more symptomatic than in nonimmune dengue vaccine recipients (Bancroft *et al.*, 1984).

ii. Intragroup Viruses. In the dengue system, sequential infections with DEN 1–2, DEN 3–2, and DEN 4–2 produced DHF/DSS, but the attack rate decreased in descending order (Sangkawibha *et al.*, 1984).

The large Cuban and Venezuelan DHF/DSS epidemics were due to dengue 1–2 infection sequences. Secondary dengue 3 and secondary dengue 4 infections result in DHF/DSS, but which sensitizing dengue types or strains play in the outcome of secondary dengue 3 and 4 infections is not well studied.

VIII. Discussion

A. *Mechanisms Regulating Severity of Dengue Illnesses*

1. *Infected Cell Mass*

Although this review focuses on viral factors that contribute to the host's preillness immunological response, the mechanisms controlling disease severity should not be ignored. It is self-evident that the most important regulator of disease severity is the total infected cell mass and possibly infection kinetics. Dengue infection enhancement has been demonstrated in sequentially infected rhesus monkeys (Halstead *et al.*, 1973d). Why do monkeys not get sick with dengue? Secondary DEN-2 infections in monkeys produced enhanced peak viremia, but monkey viremias were five logs lower than in humans (Tables XIV and XV). Studies in Indonesia provided the first evidence that mean viremias were higher during secondary than primary dengue 3 infections (Table XVI) (Gubler *et al.*, 1979). Because these were single point determinations rather than serial bleedings, only the data range is of interest. Evidence that high DEN-1 and -2 viremias actually predict disease severity is shown in Table XV. Studied children were bled early after onset of fever and then for 5 consecutive days. This permitted the identification of peak viremias (Vaughn *et al.*, 2000). The highest concentration of extracellular virions is the best known surrogate measure of infected cell mass. This value is now known for clinically apparent secondary dengue infections. There may be problems in ever knowing the range of primary dengue infections. Primary DEN-2 and -4 infections are usually mild or inapparent, making it unlikely that sequential bleedings will ever be available to study infection kinetics. Can there be any doubt that peak viremia accompanying mild primary infections will be far lower than those accompanying secondary dengue infections?

B. *Emergence of Neutralization Escape Mutants*

A model for a mechanism resulting in ADE in the dengue group of viruses and, in turn, controlling the emergence of DHF in dengue hyperendemic regions can be made. The model derives from longitudinal studies on dengue infections in humans that were able to detect minor disease (Watts *et al.*, 1999). A DEN-2 strain was identified that does not produce DHF/DSS following a DEN-1 infection. The reason appears to be that this virus is partially neutralized by DEN-1 antibodies (Kochel *et al.*, 2002). It is likely that the American genotype DEN-2 circulated relatively silently in the American tropics for a long period of time. During most of this time, the strain was not placed under selection pressure from antibodies raised to heterotypic dengue serotypes. In contrast, those DEN-2 viruses that have cocirculated with other serotypes have lost a structure(s) highly reactive with DEN-1 antibodies. Can it be that the American genotype of DEN-2 represents a structure close to that of ancestral DEN-2 viruses?

Might this same phenomenon be occurring in southeast Asia? Dengue types circulating silently (with cross-reactive epitopes) may be selected as escape mutants to circulate without cross-reactive epitopes, suddenly allowing a new virus to emerge to cause enhanced infections and DHF. These American genotype dengue strains, susceptible to cross-neutralization, might be found in persons with very mild illnesses or in wild-caught mosquitoes. Observations in Cuba suggest that strains of "increased virulence" emerge rapidly and reproducibly within one dengue serotype. These may be due not to a major mutation, but to another escape phenomenon occurring when a clonal virus is passaged in a population immune to a different serotype. Research on dengue should shift forcefully to population-based studies directed at understanding the mechanisms driving the emergence of severe disease in dengue-hyperendemic regions.

References

Allison, S. L., Schalich, J., Stiasny, K., Mandl, C. W., Kunz, C., and Heinz, F. X. (1995). Oligomeric rearrangement of tick-borne encephalitis virus envelope proteins induced by an acidic pH. *J. Virol.* **69:**695–700.

Anderson, C. L., and Abraham, G. N. (1980). Characterization of the Fc receptor for IgG on a human macrophage cell line, U937. *J. Immunol.* **125:**2735–2741.

Anderson, C. R., and Downs, W. G. (1956). Isolation of dengue virus from a human being in Trinidad. *Science* **124:**224–225.

Anonymous (1978). Dengue in the Caribbean, 1977. Workshop, pp. 1–186.

Ashe, W. K., and Notkins, A. L. (1966). Neutralization of an infectious herpes simplex virus-antibody complex by anti-gammaglobulin. *Proc. Natl. Acad. Sci. USA* **56:**447–451.

Bancroft, W. H., Scott, R. M., Eckels, K. H., Hoke, C. H., Jr., Simms, T. E., Jesrani, K. D., Summers, P. L., Dubois, D. R., Tsoulos, D., and Russell, P. K. (1984). Dengue virus type 2 vaccine: Reactogenicity and immunogenicity in soldiers. *J. Infect. Dis.* **149:**1005–1010.

Bell, S., Cranage, M., Borysiewicz, L., and Minson, T. (1990). Induction of immunoglobulin G Fc receptors by recombinant vaccine viruses expressing glycoproteins E and I of herpes simplex virus type 1. *J. Virol.* **64:**2181–2186.

Bielefeldt-Ohmann, H. (1998). Analysis of antibody-independent binding of dengue viruses and dengue virus envelope protein to human myelomonocytic cells and B lymphocytes. *Virus Res.* **57:**63–79.

Bielefeldt-Ohmann, H., Meyer, M., Fitzpatrick, D. R., and MacKenzie, A. E. (2001). Dengue virus binding to human leukocyte cell lines: Receptor usage differs between cell types and virus strains. *Virus Res.* **73:**81–89.

Blok, J., Samuel, S., Gibbs, A. J., and Vitarana, U. T. (1989). Variation of the nucleotide and encoded amino acid sequences of the envelope gene from eight dengue-2 viruses. *Arch. Virol.* **105:**39–53.

Burke, D. S. (1992). Human HIV vaccine trials: Does antibody-dependent enhancement pose a genuine risk? *Perspect. Biol. Med.* **35:**511–530.

Burke, D. S., Nisalak, A., Johnson, D. E., and Scott, R. M. (1988). A prospective study of dengue infections in Bangkok. *Am. J. Trop. Med. Hyg.* **38:**172–180.

Burton, D. R., Saphire, E. O., and Parren, P. W. H. I. (2001). A model for neutralization of viruses based on antibody coating of the virion surface. *In* "Antibodies in Viral Infection" (D. R. Burton, ed.), pp. 109–143. Springer-Verlag, New York.

Cardosa, M. J., Gordon, S., Hirsch, S., Springer, T. A., and Porterfield, J. S. (1986). Interaction of West Nile virus with primary murine macrophages: Role of cell activation and receptors for antibody and complement. *J. Virol.* **57:**952–959.

Cardosa, M. J., Porterfield, J. S., and Gordon, S. (1983). Complement receptor mediates enhanced flavivirus replication in macrophages. *J. Exp. Med.* **158:**258–263.

Chen, Y., Maguire, T., Hileman, R. E., Fromm, J. R., Esko, J. D., Linhardt, R. J., and Marks, R. M. (1997). Dengue virus infectivity depends on envelope protein binding to target cell heparan sulfate. *Nature Med.* **3:**866–871.

Chiewsilp, P., Scott, R. M., and Bhamarapravati, N. (1981). Histocompatibility antigens and dengue hemorrhagic fever. *Am. J. Trop. Med. Hyg.* **30:**1100–1105.

Chungue, E., Deubel, V., Cassar, O., Laille, M., and Martin, P. M. V. (1993). Molecular epidemiology of dengue-3 viruses and genetic relatedness among dengue-3 strains isolated from patients with mild or severe form of dengue fever in French-Polynesia. *J. Gen. Virol.* **74:**2765–2770.

Chungue, E., Spiegel, A., Roux, J., Laudon, F., and Cardines, R. (1990). Dengue-3 in French Polynesia: Preliminary data. *Med. J. Aust.* **152:**557–558.

Della-Porta, A. J., and Westaway, E. G. (1977). A multi-hit model for the neutralization of animal viruses. *J. Gen. Virol.* **38:**1–19.

Detre, K., and White, C. (1970). The comparison of two Poisson-distributed observations. *Biometrics* **26:**851–854.

Dulbecco, R. M., Vogt, R. M., and Strickland, B. G. (1956). A study of the basic aspects of neutralization of two animal viruses, Western equine encephalitis virus and poliomyelitis virus. *Virology* **2:**162–205.

Fischer, D. B., and Halstead, S. B. (1970). Observations related to pathogenesis of dengue hemorrhagic fever. V. Examination of age specific sequential infection rates using a mathematical model. *Yale J. Biol. Med.* **42:**329–349.

Gollins, S. W., and Porterfield, J. S. (1986). A new mechanism for the neutralization of enveloped viruses by antiviral antibody. *Nature* **321:**244–246.

Graham, R. R., Juffrie, M., Tan, R., Hayes, C. G., Laksono, I., Ma'roef, C., Erlin, Sutaryo, Porter, K. R., and Halstead, S. B. (1999). A prospective seroepidemiologic study on dengue in children four to nine years of age in Yogyakarta, Indonesia. I. Studies in 1995–1996. *Am. J. Trop. Med. Hyg.* **61:**412–419.

Gubler, D. J., Suharyono, W., Lubis, I., Eram, S., and Suliantisaroso, J. (1979). Epidemic dengue hemorrhagic fever in rural Indonesia. I. Virological and epidemiological studies. *Am. J. Trop. Med. Hyg.* **28:**701–710.

Guzman, M. G., Deubel, V., Pelegrino, J. L., Rosario, D., Marrero, M., Sariol, C., and Kouri, G. (1995). Partial nucleotide and amino acid sequences of the envelope and the envelope/nonstructural protein-1 gene junction of four dengue-2 virus strains isolated during the 1981 Cuban epidemic. *Am. J. Trop. Med. Hyg.* **52:**241–246.

Guzman, M. G., Kouri, G., Bravo, J., Valdes, L., Vazquez, S., and Halstead, S. B. (2002a). Effect of age on outcome of secondary dengue 2 infections. *Int. J. Infect. Dis.* **6:**118–124.

Guzman, M. G., Kouri, G., and Halstead, S. B. (2000a). Do escape mutants explain rapid increases in dengue case-fatality rates within epidemics? *Lancet* **355:**1902–1903.

Guzman, M. G., Kouri, G., Valdes, L., Bravo, J., Alvarez, M., Vazquez, S., Delgado, I., and Halstead, S. B. (2000b). Epidemiologic studies on dengue in Santiago de Cuba, 1997. *Am. J. Epidemiol.* **152:**793–799.

Guzman, M. G., Kouri, G., Valdes, L., Bravo, J., Vazquez, S., and Halstead, S. B. (2002b). Enhanced severity of secondary dengue 2 infections occurring at an interval of 20 compared with 4 years after dengue 1 infection. *Panam. J. Epidemiol.* **81:**223–227.

Guzman, M. G., Kouri, G. P., Bravo, J., Soler, M., Vazquez, S., and Morier, L. (1990). Dengue hemorrhagic fever in Cuba, 1981: A retrospective seroepidemiologic study. *Am. J. Trop. Med. Hyg.* **42:**179–184.

Halstead, S. B. (1979). In vivo enhancement of dengue virus infection in rhesus monkeys by passively transferred antibody. *J. Infect. Dis.* **140:**527–533.

Halstead, S. B. (1980). Immunological parameters of Togavirus disease syndromes. *In* "The Togaviruses: Biology, Structure, Replication" (R. W. Schlesinger, ed.), pp. 107–173. Academic Press, New York.

Halstead, S. B. (1981). The pathogenesis of dengue: Molecular epidemiology in infectious disease. *Am. J. Epidemiol.* **114:**632–648.

Halstead, S. B. (1982). Immune enhancement of viral infection. *Prog. Allergy* **31:**301–364.

Halstead, S. B. (1988). Pathogenesis of dengue: Challenges to molecular biology. *Science* **239:**476–481.

Halstead, S. B. (1994). Antibody-dependent enhancement of infection: A mechanism for indirect virus entry into cells. *In* "Cellular Receptors for Animal Viruses" pp. 493–515. Cold Spring Harbor Laboratory Press, Cold Spring Harbor, NY.

Halstead, S. B. (1997). Epidemiology of dengue and dengue hemorrhagic fever. *In* "Dengue and Dengue Hemorrhagic Fever" (D. J. Gubler and G. Kuno, eds.), pp. 23–44. CAB, Wallingford, UK.

Halstead, S. B., Casals, J., Shotwell, H., and Palumbo, N. (1973a). Studies on the immunization of monkeys against dengue. I. Protection derived from single and sequential virus infections. *Am. J. Trop. Med. Hyg.* **22:**365–374.

Halstead, S. B., Chow, J., and Marchette, N. J. (1973b). Immunologic enhancement of dengue virus replication. *Nature New Biol.* **243:**24–26.

Halstead, S. B., and Grosz, C. R. (1962). Subclinical Japanese encephalitis. I. Infection of Americans with limited residence in Korea. *Am. J. Hyg.* **75:**190–201.

Halstead, S. B., Nimmannitya, S., and Cohen, S. N. (1970a). Observations related to pathogenesis of dengue hemorrhagic fever. IV. Relation of disease severity to antibody response and virus recovered. *Yale J. Biol. Med.* **42:**311–328.

Halstead, S. B., Nimmannitya, S., and Margiotta, M. R. (1969a). Dengue and chikungunya virus infection in man in Thailand, 1962–1964. II. Observations on disease in outpatients. *Am. J. Trop. Med. Hyg.* **18:**972–983.

Halstead, S. B., Nimmannitya, S., Yamarat, C., and Russell, P. K. (1967). Hemorrhagic fever in Thailand: recent knowledge regarding etiology. *Jpn. J. Med. Sci. Biol.* **20:**96–103.

Halstead, S. B., and O'Rourke, E. J. (1977a). Antibody-enhanced dengue virus infection in primate leukocytes. *Nature* **265:**739–741.

Halstead, S. B., and O'Rourke, E. J. (1977b). Dengue viruses and mononuclear phagocytes. I. Infection enhancement by non-neutralizing antibody. *J. Exp. Med.* **146:**201–217.

Halstead, S. B., O'Rourke, E. J., and Allison, A. C. (1977). Dengue viruses and mononuclear phagocytes. II. Identity of blood and tissue leukocytes supporting in vitro infection. *J. Exp. Med.* **146:**218–228.

Halstead, S. B., and Palumbo, N. E. (1973). Studies on the immunization of monkeys against dengue. II. Protection following inoculation of combinations of viruses. *Am. J. Trop. Med. Hyg.* **22:**375–381.

Halstead, S. B., Porterfield, J. S., and O'Rourke, E. J. (1980). Enhancement of dengue virus infection in monocytes by flavivirus antisera. *Am. J. Trop. Med. Hyg.* **29:**638–642.

Halstead, S. B., Rojanasuphot, S., and Sangkawibha, N. (1983). Original antigenic sin in dengue. *Am. J. Trop. Med. Hyg.* **32:**154–156.

Halstead, S. B., Scanlon, J. E., Umpaivit, P., and Udomsakdi, S. (1969b). Dengue and Chikungunya virus infection in man in Thailand, 1962–1964. IV. Epidemiologic studies in the Bangkok metropolitan area. *Am. J. Trop. Med. Hyg.* **18:**997–1021.

Halstead, S. B., Shotwell, H., and Casals, J. (1973c). Studies on the pathogenesis of dengue infection in monkeys. I. Clinical laboratory responses to primary infection. *J. Infect. Dis.* **128:**7–14.

Halstead, S. B., Shotwell, H., and Casals, J. (1973d). Studies on the pathogenesis of dengue infection in monkeys. II. Clinical laboratory responses to heterologous infection. *J. Infect. Dis.* **128:**15–22.

Halstead, S. B., Udomsakdi, S., Simasathien, P., Singharaj, P., Sukhavachana, P., and Nisalak, A. (1970b). Observations related to pathogenesis of dengue hemorrhagic fever. I. Experience with classification of dengue viruses. *Yale J. Biol. Med.* **42:**261–275.

Halstead, S. B., Udomsakdi, S., Singharaj, P., and Nisalak, A. (1969c). Dengue and Chikungunya virus infection in man in Thailand, 1962–1964. III. Clinical, epidemiology, and virologic observations on disease in non-indigeneous white persons. *Am. J. Trop. Med. Hyg.* **18:**984–996.

Halstead, S. B., Venkateshan, C. N., Gentry, M. K., and Larsen, L. K. (1984). Heterogeneity of infection enhancement of dengue 2 strains by monoclonal antibodies. *J. Immunol.* **132:**1529–1532.

Hammon, W. M., Rudnick, A., and Sather, G. E. (1960). Viruses associated with epidemic hemorrhagic fevers of the Philippines and Thailand. *Science* **131:**1102–1103.

Hammon, W. M., and Sather, G. E. (1964). Virological findings in the 1960 hemorrhagic fever epidemic (dengue) in Thailand. *Am. J. Trop. Med. Hyg.* **13:**629–641.

Hawkes, R. A. (1964). Enhancement of the infectivity of arboviruses by specific antisera produced in domestic fowls. *Aust. J. Exp. Biol. Med. Sci.* **42:**465–482.
Hawkes, R. A., and Lafferty, K. J. (1967). The enhancement of virus infectivity by antibody. *Virology* **33:**250–261.
Hotta, H., and Hotta, S. (1982). Dengue virus multiplication in cultures of mouse peritoneal macrophages: Effects of macrophage activators. *Microbiol. Immunol.* **26:**665–676.
Hoxie, J. A., Fitzharris, T. P., Youngbar, P. R., Matthews, D. M., Rackowski, J. L., and Radka, S. F. (1987). Nonrandom association of cellular antigens with HTLV-III virions. *Hum. Immunol.* **18:**39–52.
Hung, S. L., Lee, P. L., Chen, H. W., Chen, L. K., Kao, C. L., and King, C. C. (1999). Analysis of the steps involved in dengue virus entry into host cells. *Virology* **257:**156–167.
Kliks, S., and Halstead, S. B. (1980). An explanation for enhanced virus plaque formation in chick embryo cells. *Nature* **285:**504–505.
Kliks, S., and Halstead, S. B. (1983). Role of antibodies and host cells in plaque enhancement of Murray Valley encephaliitis virus. *J. Virol.* **46:**394–404.
Kliks, S. C., Nimmannitya, S., Nisalak, A., and Burke, D. S. (1988). Evidence that maternal dengue antibodies are important in the development of dengue hemorrhagic fever in infants. *Am. J. Trop. Med. Hyg.* **38:**411–419.
Kliks, S. C., Nisalak, A., Brandt, W. E., Wahl, L., and Burke, D. S. (1989). Antibody-dependent enhancement of dengue virus growth in human monocytes as a risk factor for dengue hemorrhagic fever. *Am. J. Trop. Med. Hyg.* **40:**444–451.
Kobayashi, K., Blaser, M. J., and Brown, W. R. (1989). Identification of a unique IgG Fc binding site in human intestinal epithelium. *J. Immunol.* **143:**2567–2574.
Kochel, T. J., Watts, D. M., Halstead, S. B., Hayes, C. G., Espinosa, A., Felices, V., Caceda, R., Bautista, T., Montoya, Y., Douglas, S., and Russell, K. L. (2002). Neutralization of American genotype dengue 2 viral infection by dengue 1 antibodies may have prevented dengue hemorrhagic fever in Iquitos, Peru. *Lancet* **360:**310–312.
Kontny, U., Kurane, I., and Ennis, F. A. (1988). Gamma interferon augments Fc gamma receptor-mediated dengue virus infection of human monocytic cells. *J. Virol.* **62:**3928–3933.
Kouri, G. P., Guzman, M. G., Bravo, J. R., and Triana, C. (1989). Dengue haemorrhagic fever/dengue shock syndrome: Lessons from the Cuban epidemic, 1981. *Bull. World Health Org.* **67:**375–380.
Lee, E., Gubler, D. J., Weir, R. C., and Dalgarno, L. (1993). Genetic and biological differentiation of dengue-3 isolates obtained from clinical cases in Java, Indonesia, 1976–1978. *Arch. Virol.* **133:**113–125.
Leitmeyer, K. C., Vaughn, D. W., Watts, D. M., Salas, R., Villalobos, I., de, C., Ramos, C., and Rico-Hesse, R. (1999). Dengue virus structural differences that correlate with pathogenesis. *J. Virol.* **73:**4738–4747.
Loke, H., Bethell, D. B., Phuong, C. X., Dung, M., Schneider, J., White, N. J., Day, N. P., Farrar, J., and Hill, A. V. (2001). Strong HLA class I-restricted T cell responses in dengue hemorrhagic fever: A double-edged sword? *J. Infect. Dis.* **184:**1369–1373.
Mady, B. J., Erbe, D. V., Kurane, I., Fanger, M. W., and Ennis, F. A. (1991). Antibody-dependent enhancement of dengue virus infection mediated by bispecific antibodies against cell surface molecules other than Fc-gamma receptor. *J. Immunol.* **147:**3139–3144.
Mangada, M. N., and Igarashi, A. (1998). Molecular and in vitro analysis of eight dengue type 2 viruses isolated from patients exhibiting different disease severities. *Virology* **244:**458–466.

Marchette, N. J., Halstead, S. B., Falkler, W. A., Jr., Stenhouse, A., and Nash, D. (1973). Studies on the pathogenesis of dengue infection in monkeys. III. Sequential distribution of virus in primary and heterologous infections. *J. Infect. Dis.* **128:**23–30.

Marsh, M., and Helenius, A. (1989). Virus entry into animal cells. *Adv. Virus Res.* **36:**107–151.

Mascola, J. R., Matthieson, P. M., Walker, M. C., Halstead, S. B., and Burke, D. S. (1993). Summary report: Workshop on the potential risks of antibody-dependent enhancement in human HIV vaccine trials. *AIDS Res. Hum. Retrovir.* **9:**1175–1184.

McGuire, T. C., Adams, D. S., Johnson, G. C., Klevjer-Anderson, P., Barbee, P. D., and Gorham, J. R. (1990). Acute arthritis in caprine arthritis-encephalitis virus challenge exposure of vaccinated or persistently infected goats. *Am. J. Vet. Res.* **47:**537–540.

McKeating, J. A., Griffiths, P. D., and Weiss, R. A. (1990). HIV susceptibility conferred to human fibroblasts by cytomegalovirus-induced Fc receptor. *Nature* **343:**659–661.

Megret, F., Hugnot, J. P., Falconar, A., Gentry, M. K., Morens, D. M., Murray, J. M., Schlesinger, J. J., Wright, P. J., Young, P., van Regenmortel, M. H. *et al.* (1992). Use of recombinant fusion proteins and monoclonal antibodies to define linear and discontinuous antigenic sites on the dengue virus envelope glycoprotein. *Virology* **187:**480–491.

Mogensen, S. C. (1979). Role of macrophages in natural resistance to virus infections. *Microbiol. Rev.* **43:**1–26.

Morahan, P. S., Connor, J. R., and Leary, K. R. (1985). Viruses and the versatile macrophage. *Br. Med. Bull.* **41:**15–21.

Morens, D. M., and Halstead, S. B. (1987). Disease severity-related antigenic differences in dengue 2 strains detected by dengue 4 monoclonal antibodies. *J. Med. Virol.* **22:**169–174.

Morens, D. M., and Halstead, S. B. (1990). Measurement of antibody-dependent infection enhancement of four dengue virus serotypes by monoclonal and polyclonal antibodies. *J. Gen. Virol.* **71:**2909–2914.

Morens, D. M., Halstead, S. B., and Larsen, L. K. (1985a). Comparison of dengue virus plaque reduction neutralization by macro and "semi-micro" methods in LLC-MK2 cells. *Microbiol. Immunol.* **29:**1197–1205.

Morens, D. M., Halstead, S. B., and Marchette, N. J. (1987). Profiles of antibody-dependent enhancement of dengue virus type 2 infection. *Microb. Pathog.* **3:**231–237.

Morens, D. M., Halstead, S. B., Repik, P. M., Putvatana, R., and Raybourne, N. (1985b). Simplified plaque reduction neutralization assay for dengue viruses by semi-micro methods in BHK-21 cells: Comparison of the BHK suspension test with standard plaque reduction neutralization. *J. Clin. Microbiol.* **22:**250–254.

Morens, D. M., Marchette, N. J., Chu, M. C., and Halstead, S. B. (1991). Growth of dengue type 2 virus isolates in human peripheral blood leukocytes correlates with severe and mild dengue disease. *Am. J. Trop. Med. Hyg.* **45:**644–651.

Neff, J. M., Morris, L., Gonzalez-Alcover, R., Coleman, P. H., Lyss, S. B., and Negron, H. (1967). Dengue fever in a Puerto Rican community. *Am. J. Epidemiol.* **86:**162–184.

Nimmannitya, S., Halstead, S. B., Cohen, S., and Margiotta, M. R. (1969). Dengue and chikungunya virus infection in man in Thailand, 1962–1964. I. Observations on hospitalized patients with hemorrhagic fever. *Am. J. Trop. Med. Hyg.* **18:**954–971.

Okuno, Y., Harada, T., Ogawa, M., Okamoto, Y., and Maeda, K. (1989). A case of dengue hemorrhagic fever in a Japanese child. *Microbiol. Immunol.* **33:**649–655.

Paradoa Perez, M. L., Trujillo, Y., and Basanta, P. (1987). Association of dengue hemorrhagic fever with the HLA system. *Haematologia. (Budap).* **20:**83–87.

Porterfield, J. S. (1986). Antibody-dependent enhancement of viral infectivity. *Adv. Virus Res.* **31:**335–355.

Ramos-Castaneda, J., Imbert, J. L., Barron, B. L., and Ramos, C. (1997). A 65-kDa trypsin-sensible membrane cell protein as a possible receptor for dengue virus in cultured neuroblastoma cells. *J. Neurovirol.* **3:**435–440.

Rico-Hesse, R. (1990). Molecular evolution and distribution of dengue viruses type 1 and 2 in nature. *Virology* **174:**479–493.

Rico-Hesse, R., Harrison, L. M., Nisalak, A., Vaughn, D. W., Kalayanarooj, S., Green, S., Rothman, A. L., and Ennis, F. A. (1998). Molecular evolution of dengue type 2 virus in Thailand. *Am. J. Trop. Med. Hyg.* **58:**96–101.

Roehrig, J. T., Johnson, A. J., Hunt, A. R., Bolin, R. A., and Chu, M. C. (1990). Antibodies to dengue 2 virus E-glycoprotein synthetic peptides identify antigenic conformation. *Virology* **177:**668–675.

Rosen, L. (1958). Observations on the epidemiology of dengue in Panama. *Am. J. Hyg.* **68:**45–58.

Rosen, L. (1977). The emperor's new clothes revisited, or reflections on the pathogenesis of dengue hemorrhagic fever. *Am. J. Trop. Med. Hyg.* **26:**337–343.

Rosen, L. (1986). The pathogenesis of dengue haemorrhagic fever: A critical appraisal of current hypotheses. *S. Afr. Med. J.* **Suppl:** 40–42.

Rosen, L. (1989). Disease exacerbation caused by sequential dengue infections: Myth or reality? *Rev. Infect. Dis.* **11:**S840–S842.

Rosen, L. (1999). Comments on the epidemiology, pathogenesis, and control of dengue. *Med. Trop.* **59:**495–498.

Rudnick, A. (1965). Studies of the ecology of dengue in Malaysia: A preliminary report. *J. Med. Entomol.* **2:**203–208.

Rudnick, A. (1967). *Aedes aegypti* and haemorrhagic fever. *Bull. World Health Org.* **36:**528–532.

Russell, P. K., and Nisalak, A. (1967). Dengue virus identification by the plaque reduction neutralization test. *J. Immunol.* **99:**291–296.

Russell, P. K., Nisalak, A., Sukhavachana, P., and Vivona, S. (1967). A plaque reduction test for dengue virus neutralization antibodies. *J. Immunol.* **99:**285–290.

Russell, P. K., Yuill, T. M., Nisalak, A., Udomsakdi, S., Gould, D., and Winter, P. E. (1968). An insular outbreak of dengue hemorrhagic fever. II. Virologic and serologic studies. *Am. J. Trop. Med. Hyg.* **17:**600–608.

Sabin, A. B. (1952). Research on dengue during World War II. *Am. J. Trop. Med. Hyg.* **1:**30–50.

Sabin, A. B., and Schlesinger, R. W. (1945). Production of immunity to dengue with virus modified by propagation in mice. *Science* **101:**640–642.

Sangkawibha, N., Rojanasuphot, S., Ahandrik, S., Viriyapongse, S., Jatanasen, S., Salitul, V., Phanthumachinda, B., and Halstead, S. B. (1984). Risk factors in dengue shock syndrome: A prospective epidemiologic study in Rayong, Thailand. I. The 1980 outbreak. *Am. J. Epidemiol.* **120:**653–669.

Schlesinger, J. J. and Brandriss, M. W. (1981). Growth of 17D yellow fever virus in a macrophage-like cells line, U937: Role of Fc and viral receptors in antibody-mediated infection. *J. Immunol.* **127:**659–665.

Smith, T. J. (2001). Antibody interactions with rhinovirus: Lessons for mechanisms of neutralization and the role of immunity in viral evolution. *In* "Antibodies in Viral Infection" (D. R. Burton, ed.), pp. 1–28. Springer-Verlag, Berlin.

Stiasny, K., Allison, S. L., Marchler Bauer, A., Kunz, C., and Heinz, F. X. (1996). Structural requirements for low-pH-induced rearrangements in the envelope glycoprotein of tick-borne encephalitis virus. *J. Virol.* **70:**8142–8147.

Stott, J. L., Anderson, G. A., Jochim, M. M., Barber, T. L., and Osburn, B. I. (1984). Clinical expression of blue tongue disease in cattle. *Proc. Annu. Meet. U.S. Anim. Health Assoc.* **86:**126.

Sukhavachana, P., Nisalak, A., and Halstead, S. B. (1966). Tissue culture techniques for the study of dengue viruses. *Bull. World Health Org.* **35:**65–66.

Sullivan, N. J. (2001). Antibody-mediated enhancement of viral disease. *Curr. Top. Microbiol. Immunol.* **260:**145–169.

Thant, K. Z., Morita, K., and Igarashi, A. (1995). Sequences of E/NS1 gene junction from four dengue-2 viruses of northeastern Thailand and their evolutionary relationships with other dengue-2 viruses. *Microbiol. Immunol.* **39:**581–590.

Thant, K. Z., Morita, K., and Igarashi, A. (1996). Detection of the disease severity-related molecular differences among new Thai dengue-2 isolates in 1993, based on their structural proteins and major non-structural protein NS1 sequences. *Microbiol. Immunol.* **40:**205–216.

Thein, S., Aung, M. M., Shwe, T. N., Aye, M., Zaw, A., Aye, K., Aye, K. M., and Aaskov, J. (1997). Risk factors in dengue shock syndrome. *Am. J. Trop. Med. Hyg.* **56:**566–572.

Trirawatanapong, T., Chandran, B., Putnak, R., and Padmanabhan, R. (1992). Mapping of a region of dengue virus type-2 glycoprotein required for binding by a neutralizing monoclonal antibody. *Gene* **116:**139–150.

Unkeless, J. C., Scigliano, E., and Freedman, V. H. (1988). Structure and function of human and murine receptors of IgG. *Annu. Rev. Immunol.* **6:**251–281.

van der Groen, G., Vanden Berghe, D. A., and Pattyn, S. R. (1976). Interaction of mouse peritoneal macrophages with different arboviruses in vitro. *J. Gen. Virol.* **34:**353–361.

Vaughn, D. W., Green, S., Kalayanarooj, S., Innis, B. L., Nimmannitya, S., Suntayakorn, S., Endy, T. P., Raengsakulrach, B., Rothman, A. L., Ennis, F. A., and Nisalak, A. (2000). Dengue viremia titer, antibody response pattern, and virus serotype correlate with disease severity. *J. Infect. Dis.* **181:**2–9.

Wang, E., Ni, H., Xu, R., Barrett, A. D., Watowich, S. J., Gubler, D. J., and Weaver, S. C. (2000). Evolutionary relationships of endemic/epidemic and sylvatic dengue viruses. *J. Virol.* **74:**3227–3234.

Watts, D. M., Porter, K. R., Putvatana, P., Vasquez, B., Calampa, C., Hayes, C. G., and Halstead, S. B. (1999). Failure of secondary infection with American genotype dengue 2 to cause dengue haemorrhagic fever. *Lancet* **354:**1431–1434.

Wiharta, A. S., Hotta, H., Hotta, S., Matsumura, T., Sujudi, and Tsuji, M. (1985). Increased multiplication of dengue virus in mouse peritoneal macrophage cultures by treatment with extracts of *Ascaris-Parascaris* parasites. *Microbiol. Immunol.* **29:**337–348.

Winter, P. E., Nantapanich, S., Nisalak, A., Udomsakdi, S., Dewey, R. W., and Russell, P. K. (1969). Recurrence of epidemic dengue hemorrhagic fever in an insular setting. *Am. J. Trop. Med. Hyg.* **18:**573–579.

Wu, S. J., Grouard-Vogel, G., Sun, W., Mascola, J. R., Brachtel, E., Putvatana, R., Louder, M. K., Filgueira, L., Marovich, M. A., Wong, H. K., Blauvelt, A., Murphy, G. S., Robb, M. L., Innes, B. L., Birx, D. L., Hayes, C. G., and Frankel, S. S. (2000). Human skin Langerhans cells are targets of dengue virus infection. *Nature Med.* **6:**816–820.

INDEX

A

B

E

F

G

J

K

T

W

X

Y

Z

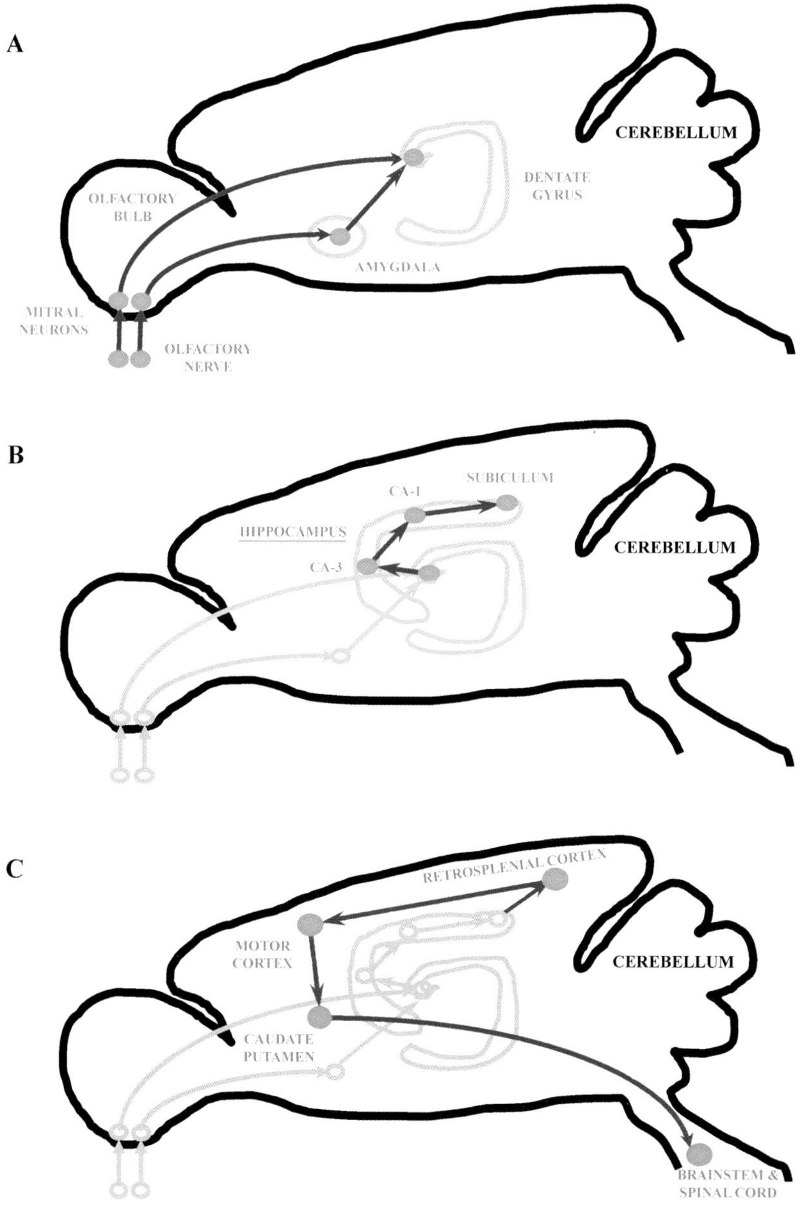

HURRELBRINK AND MCMINN, FIG 1. Diagrams of sagittal sections of mouse brain showing the mode of spread of Murray Valley encephalitis virus within the central nervous system of Swiss outbred mice after subcutaneous inoculation of 100 plaque-forming units of wild-type virus into the footpad. Specific structures within the mouse central

HURRELBRINK AND McMINN, FIG 2. The three-dimensional structure of the TBE E protein ectodomain as determined by X-ray crystallography (Rey *et al.*, 1995). The protein consists of three domains, including the central domain (domain I in red), the dimerization domain (domain II in yellow), and the immunoglobulin-like domain (domain III in blue). The fusion peptide, conserved throughout the *Flavivirus* genus and located on the distal tip of domain II, is highlighted in green.

---

nervous system were identified by reference to a stereotactic atlas of the mouse brain (Franklin and Paxinos, 1997). (A) Virus initially appears in the olfactory bulb at 4 days pi and spreads to the amygdaloid nucleus and the dentate gyrus at 5 days pi. (B) Virus spreads via neuronal connections within the hippocampus between days 5 and 6 pi, including the proximal CA-3 and distal CA-1 regions and the subiculum. (C) Virus spreads from the hippocampus to the retrosplenial cortex, the motor cortex, the caudate putamen, brainstem, and spinal cord between days 6 and 8 pi.

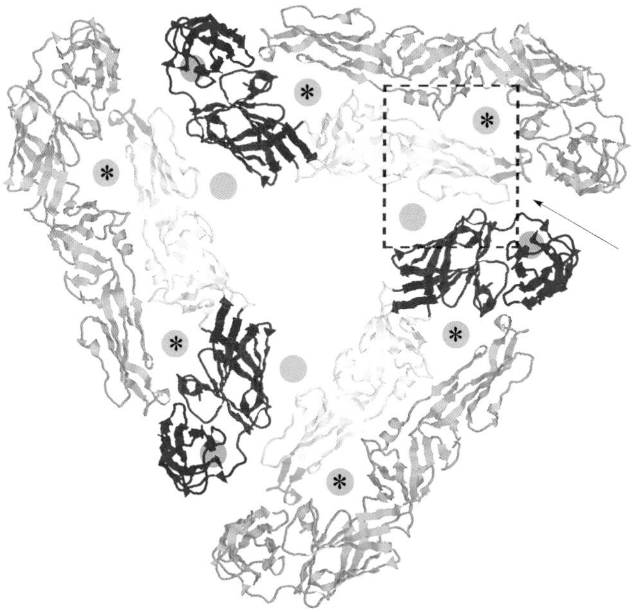

HURRELBRINK AND MCMINN, FIG 3. Proposed arrangement of structural proteins on the surface of recombinant subviral particles (RSPs) of TBE (adapted from Ferlenghi *et al.*, 2001, with permission). A similar arrangement has been proposed for DEN virions after exposure to low pH (Kuhn *et al.*, 2002). E proteins are believed to form head-to-tail dimers on the particle surface, which subsequently rearrange into trimers during acid-induced conformational change (fusion) in the endosome. Three E protein monomers (shown in color as in Fig. 2) participate in the formation of a single trimer and surround the threefold axis of symmetry. In RSPs, three copies of the prM protein are also centered on this axis (light blue circles); however, in virions, the prM protein is believed to sit inside E protein dimers rather than outside (light blue circles with asterisk). The boxed area, designated as the prM/E interface (see text), includes the stem-anchor region of E (pink circles), the prM protein, and domains I and II from adjacent E proteins.

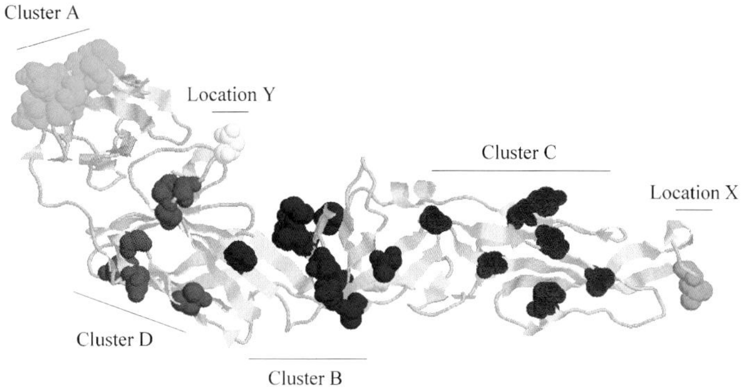

HURRELBRINK AND MCMINN, FIG 4. Superimposition of molecular determinants of virulence onto the three-dimensional structure of the TBE E protein ectodomain (Rey *et al.*, 1995). Four clusters of mutations can be seen: cluster A (green), cluster B (blue), cluster C (red), and cluster D (purple), as well as two isolated mutations located within the fusion peptide (location X in orange) and the glycosylation site (location Y in yellow), respectively.

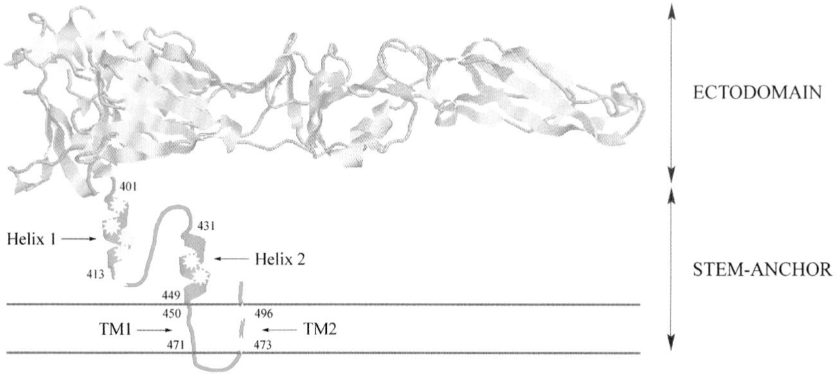

HURRELBRINK AND MCMINN, FIG 5. Schematic representation of the stem-anchor region of the TBE E protein (adapted from Allison *et al.*, 1999, with permission). Trypsin cleavage of the ectodomain (shown in gray) occurs at amino acid residue 395. Two α helices and two transmembrane domains (TM; predicted by Stiasny *et al.*, 1996) are indicated. The relative positions of mutations implicated in virulence (cluster E) are shown in blue.

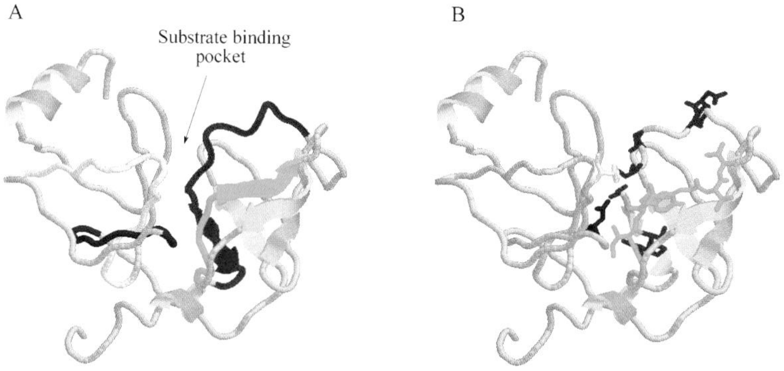

HURRELBRINK AND McMINN, FIG 6. Three-dimensional structure of the NS3 protease domain from DEN-2 (Murthy *et al.*, 1999). Conserved regions C1–C4 (A), based on amino acid sequence alignment of NS3 with flaviviral and cellular proteases, are highlighted in yellow, blue, red, and green, respectively (Bazan and Fletterick, 1989; Murthy *et al.*, 1999). Amino acid substitutions in and around the substrate binding pocket affect replication and/or virulence (B), and are shown as stick representations using the same color scheme as A.

TABLE I

MOLECULAR DETERMINANTS OF NEUROINVASIVENESS AND NEUROVIRULENCE IN THE FLAVIVIRUS E PROTEIN [HURRELBRINK AND MCMINN]

	Cluster/ location[a]	Virus[b]	Substitution	Relative position in TBE	Reference
Neuroinvasiveness	A	JE	G333D	337	Cecilia and Gould (1991)
		JE	S364F	367	Hasegawa *et al.* (1992)
		JE	N367I	368	Hasegawa *et al.* (1992)
		LGT	F333S	333	Pletnev and Men (1998)
		LGT	N389D	389	Campbell and Pletnev (2000)
		LGT	N389D	389	Pletnev and Men (1998)
		LGT	D308A	308	Campbell and Pletnev (2000)
		LIV[b]	D308N	308	Jiang *et al.* (1993)
		LIV[b]	D308N	308	Gao *et al.* (1994)
		LIV[b]	S310P	310	Jiang *et al.* (1993)
		MVE	Various at 390	387	Hurrelbrink *et al.* (2001)
		MVE	Various at 390	387	Lobigs *et al.* (1990)
		MVE	Various at 390	387	Lee and Lobigs (2000)
		TBE	Various at 308–311	308–311	Mandl *et al.* (2000)
		TBE	G368R	368	Holzmann *et al.* (1997)
		TBE	Y384H	384	Holzmann *et al.* (1990)
		WNV	K307E	309	Chambers *et al.* (1998)
		YF	Q303K	311	Jennings *et al.* (1994)
		YF	K326G	334	Chambers and Nickells (2001)
		YF	R380T	387	Chambers and Nickells (2001)
	B	LGT	G285S	285	Pletnev and Men (1998)
		JE	Q52R/K	52	Hasegawa *et al.* (1992)
		JE	I270S	272	Cecilia and Gould (1991)
		MVE	S277I	279	McMinn *et al.* (1995)
		MVE	Various at 277	279	Hurrelbrink *et al.* (2001)
		YF	R52G	52	Chambers and Nickells (2001)
	C	LGT	F119V	119	Campbell and Pletnev (2000)
		LGT	F119V	119	Pletnev and Men (1998)
		TBE	E84K	84	Labuda *et al.* (1994)
		TBE	A123K	123	Holzmann *et al.* (1997)
		WNV	L68P	68	Chambers *et al.* (1998)
	D	TBE	K171E	172	Mandl *et al.* (1989a)
		TBE	D181Y	181	Holzmann *et al.* (1997)
		YF	G155D	159	Jennings *et al.* (1994)
		YF	I173T	177	Chambers and Nickells (2001)
	E	LGT	H438Y	438	Campbell and Pletnev (2000)
		TBE	H496R	496	Gritsun *et al.* (2001)

(*continues*)

TABLE I (*continued*)

	Cluster/ location[a]	Virus[b]	Substitution	Relative position in TBE	Reference
Neurovirulence	A	DEN-1	V365I	363	dos Santos *et al.* (2000)
		DEN-2	Various at 383–385	387	Hiramatsu *et al.* (1996)
		DEN-2	H390N	391	Sanchez and Ruiz (1996)
		JE	G306E	309	Ni and Barrett (1998)
		YF	V305F	313	Schlesinger *et al.* (1996)
		YF	S305F	313	Ryman *et al.* (1998)
		YF	S325L	333	Ryman *et al.* (1998)
		YF	F380R	387	Schlesinger *et al.* (1996)
	B	DEN-1	M196V	197	dos Santos *et al.* (2000)
		DEN-2	K126E	128	Bray *et al.* (1998)
		DEN-2	K126E	128	Gualano *et al.* (1998)
		DEN-3	A54E	54	Lee *et al.* (1997)
		DEN-3	F277S	284	Lee *et al.* (1997)
		JE	E138K	140	Sumiyoshi *et al.* (1995)
		JE	E138K	140	Chen *et al.* (1996)
		JE	E138K	140	Chambers *et al.* (1999)
		JE	E138K	140	Ni *et al.* (1995)
		JE	E138K	140	Arroyo *et al.* (2001)
		JE	K279M	282	Chambers *et al.* (1999)
		JE	K279M	282	Arroyo *et al.* (2001)
		JE	K279M	282	Monath *et al.* (2002)
		YF	G52R	52	Schlesinger *et al.* (1996)
	X	JE	L107F	107	Chambers *et al.* (1999)
		JE	L107F	107	Arroyo *et al.* (2001)
	Y	TBE	N154L	154	Pletnev *et al.* (1993)
	C	DEN-2	D71E	71	Bray *et al.* (1998)
		YF	A240V	255	Ryman *et al.* (1997)
	D	DEN-3	A18S	18	Lee *et al.* (1997)
		DEN-4	T155I	157	Kawano *et al.* (1993)
		JE	I176V	172	Chambers *et al.* (1999)
		JE	I176V	172	Ni *et al.* (1995)
		YF	I173T	177	Ryman *et al.* (1997)
	E	DEN-1	T405I	406	dos Santos *et al.* (2000)
		DEN-3	E401K	406	Lee *et al.* (1997)
		DEN-3	T403I	408	Lee *et al.* (1997)
		DEN-3	K406E	409	Chen *et al.* (1995)
		DEN-4	F401L	403	Kawano *et al.* (1993)
		LGT	L416A	416	Holbrook *et al.* (2001)
		LGT	H438Y	438	Holbrook *et al.* (2001)
		LGT	V440A	440	Holbrook *et al.* (2001)
		LGT	N473K	473	Holbrook *et al.* (2001)

[a] The relative position of each mutation in the TBE E protein is shown using the same coloring scheme as that used in Fig. 4.

[b] DEN – dengue virus; JE – Japanese encephalitis virus; LGT – Langat virus; LIV – louping ill virus; MVE – Murray Valley encephalitis virus; TBE – tick-borne encephalitis virus; WNV – West Nile virus; YF – yellow fever virus.

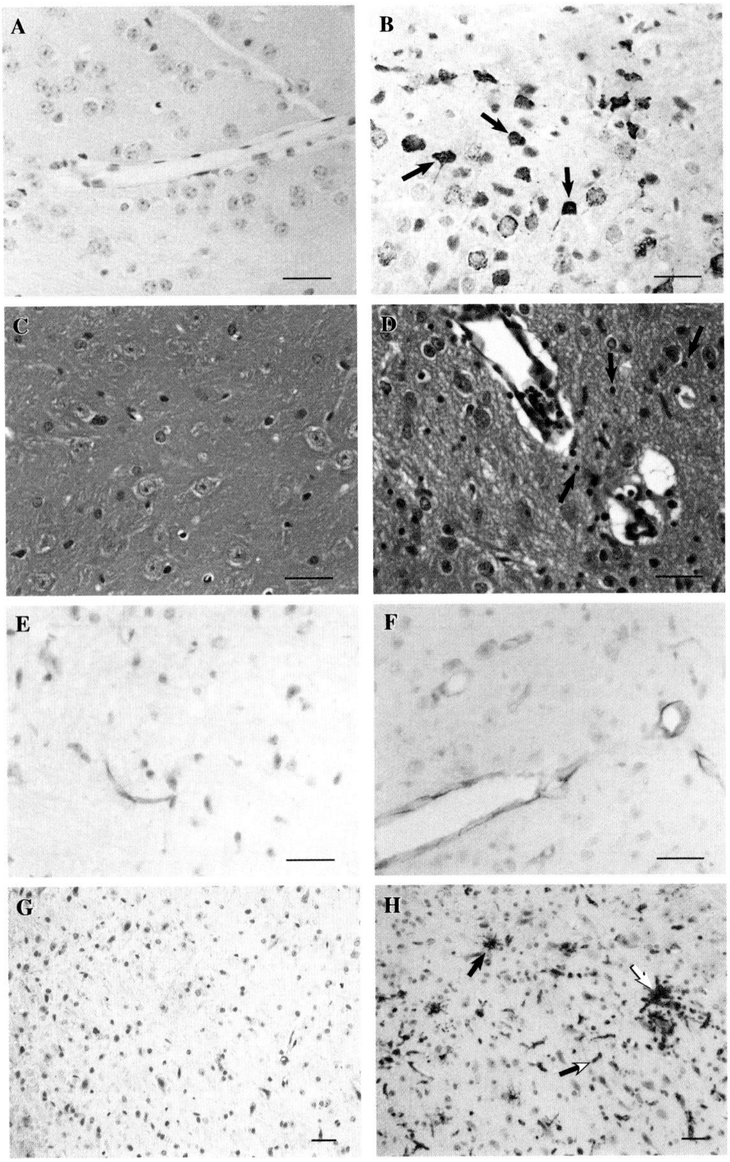
A
B
C
D
E
F
G
H

KING *ET AL.*, FIG 3. Sagittal sections of perfused, fixed brain from mock-infected (A) and day 9 WNV-infected (B) B6.IFN-$\gamma^{-/-}$ mice labeled for WNV antigen using immunoperoxidase-labeled rat anti-WNV and 3,3-diaminobenzidine tetrahydrochloride (DAB) substrate with hematoxylin as a counterstain. WNV antigen (brown) was first detected in brainstem neurons on day 6 postinfection in both strains and was found only in the cytoplasm and processes of neurons (solid arrows). Hematoxylin and eosin stained sagittal sections of perfused, fixed brains from mock-infected (C) and day 8 WNV-infected (D) B6.WT mice. Infected brain shows leukocytes infiltrating into brain parenchyma (solid arrows). Sagittal sections of perfused, fixed brains from mock-infected (E) and day 5 WNV-infected (F) B6.IFN-$\gamma^{-/-}$ mice labeled for ICAM-1 using immunoperoxidase-labeled rat anti-ICAM-1 and DAB substrate with hematoxylin as a counterstain. Increased ICAM-1 expression (brown), as measured by intensity of staining and number of vessels stained, was first detectable on day 3 in both strains. Sagittal sections of perfused, fixed brains from mock-infected (G) and day 5 WNV-infected (H) B6.WT mice labeled for activated microglia using immunoperoxidase-labeled B4 isolectin and DAB as substrate with hematoxylin as a counterstain. Activation of microglia was first detectable on day 7 and can be seen readily with dendritic (solid arrow) and migratory rod (open arrow) morphologies, which cluster to form nodules around infected neurons (white arrow). Scale bar: 30 μm throughout.

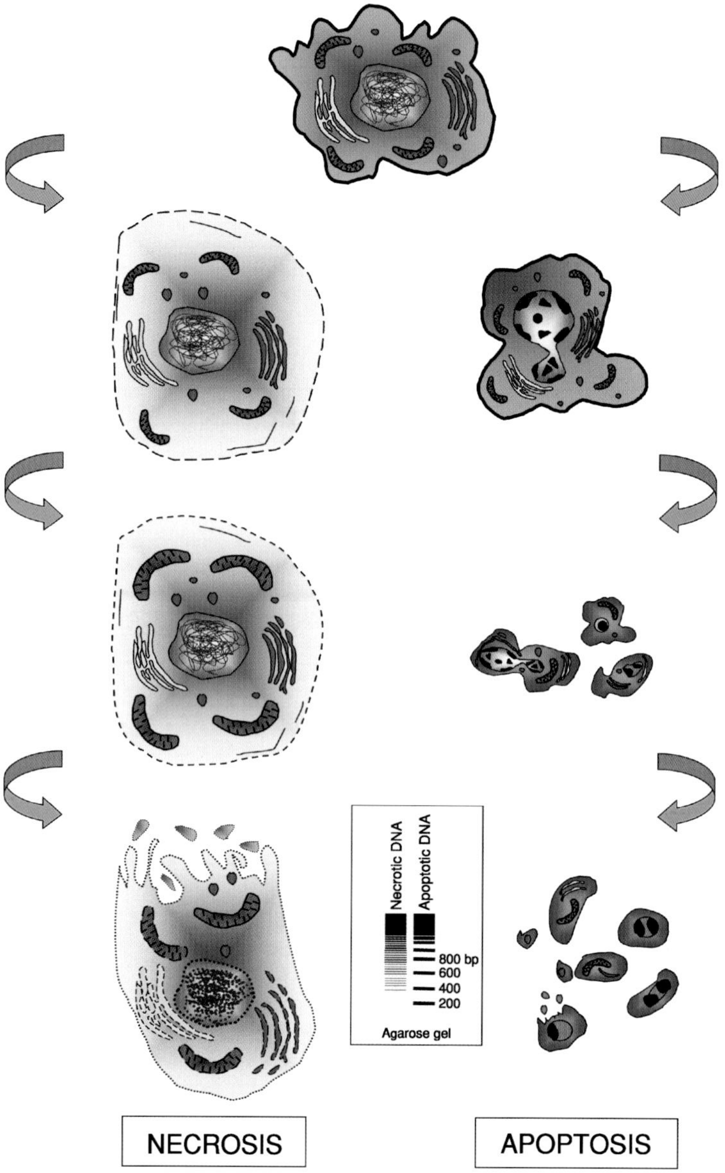

COURAGEOT *ET AL.*, FIG 1. Schematic diagram of the morphological changes that occur during necrosis and apoptosis. A normal cell is shown at the top. (Inset) Gel showing DNA degradation products in the late stages of cell death.

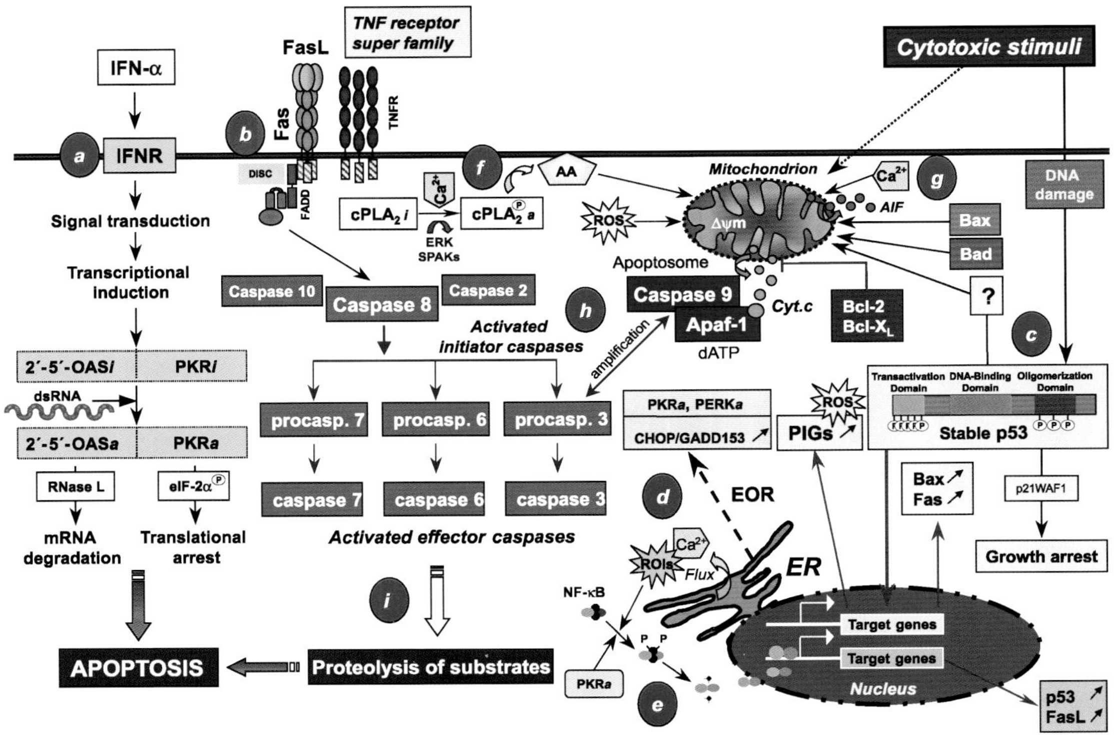

COURAGEOT *ET AL.*, FIG 2. Intracellular pathways involved in apoptosis. These pathways are mediated through the interferon system (a), the death receptor-dependent system (b), p53 (c), endoplasmic reticulum stress (d), nuclear transcription factor NF-κB (e), cytosolic phospholipase A_2 (f), death-inducing factors converging onto mitochondria (g), and caspase (h). *i*, inactive; *a*, active; AA, arachidonic acid; AIF, apoptosis-inducing factor; *Cyt.c*; cytochrome *c*; IFN, interferon; EOR, endoplasmic reticulum overload; 2′-5′OAS, 2′,5′-oligoadenylate synthetase; $cPLA_2$, cytosolic phospholipase A_2; PIGs, p53-induced genes; PKR, protein kinase RNA dependent; RNase L, ribonuclease L; ROIs, radical oxygen intermediates; ROS, reactive oxygen species; dsRNA, double-stranded RNA.

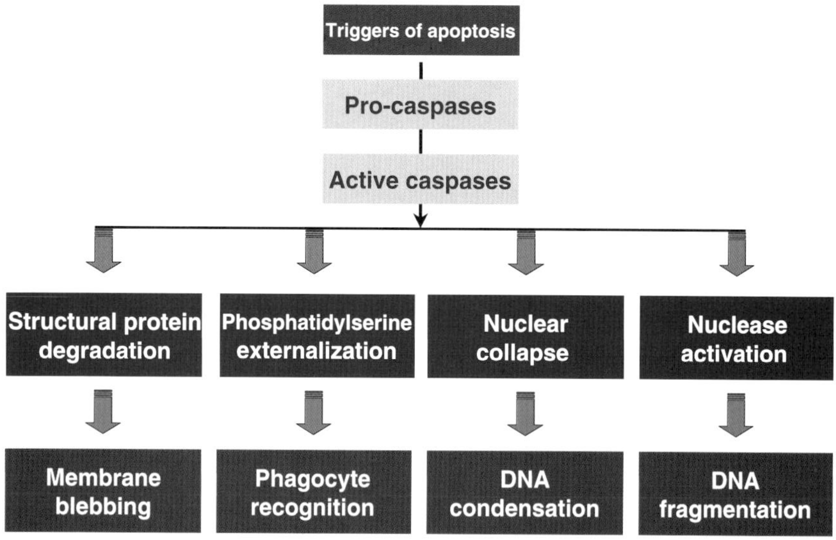

COURAGEOT *ET AL.*, FIG 3. Schematic representation showing how caspases execute the events that lead to the apoptotic death of a cell.

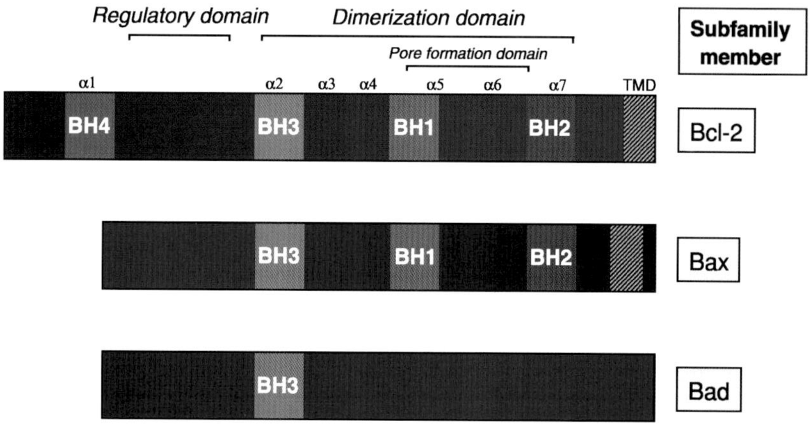

COURAGEOT *ET AL.*, FIG 4. Three members of the Bcl-2 family. Sizes of the domains are approximate and the figure is not drawn to scale. α, α helix; BH, Bcl-2 homology; TMD, transmembrane domain.

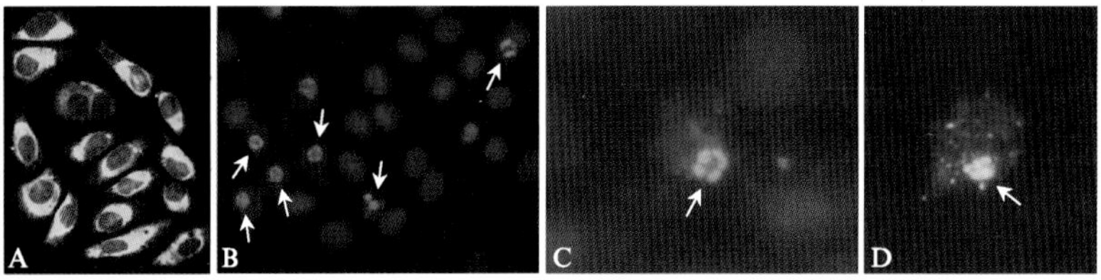

COURAGEOT *ET AL.*, FIG 5. Apoptotic cell death in DEN virus-infected Neuro 2a cells as observed by fluorescence microscopy. Paraformaldehyde-fixed cells were permeabilized with Triton X-100 and assayed for the presence of DEN antigens (A, C) at 20 h post-infection and for apoptosis (B to D) after 30 h. Intracellular DEN antigens were detected by immunofluorescence assay (A and C). The condensation of chromatin into several dense masses was visualized with the DNA intercalator propidium iodide (B and C). Apoptotic DNA degradation was demonstrated in infected cells by the TUNEL method (D). Magnification: ×200.

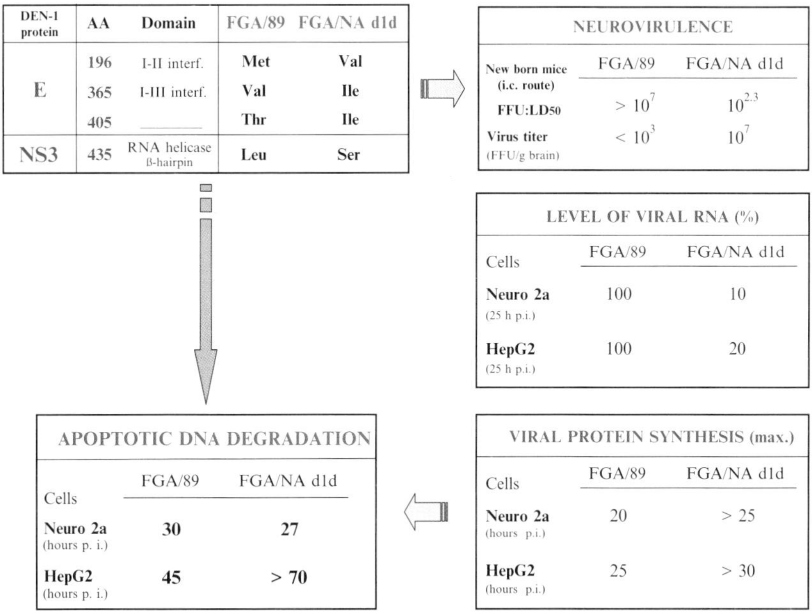

DEN-1 protein
AA
Domain
FGA/89
FGA/NA d1d
E
196 I-II interf. Met Val
365 I-III interf. Val Ile
405 Thr Ile
NS3
435 RNA helicase ß-hairpin Leu Ser
NEUROVIRULENCE
New born mice (i.c. route)
FGA/89
FGA/NA d1d
FFU:LD50 > 10^7 10^2.3
Virus titer (FFU/g brain) < 10^3 10^7
LEVEL OF VIRAL RNA (%)
Cells
FGA/89
FGA/NA d1d
Neuro 2a (25 h p.i.) 100 10
HepG2 (25 h p.i.) 100 20
APOPTOTIC DNA DEGRADATION
Cells
FGA/89
FGA/NA d1d
Neuro 2a (hours p. i.) 30 27
HepG2 (hours p. i.) 45 > 70
VIRAL PROTEIN SYNTHESIS (max.)
Cells
FGA/89
FGA/NA d1d
Neuro 2a (hours p.i.) 20 > 25
HepG2 (hours p.i.) 25 > 30

COURAGEOT *ET AL.*, FIG 6. Comparative analysis between DEN-1 virus strain FGA/89 and its neurovirulent variant FGA/NA d1d (Duarte dos Santos *et al.*, 2000). interf., interface; FFU, focus-forming unit; LD_{50}, 50% lethal dose; p.i., postinfection; i.c., intracerebral.

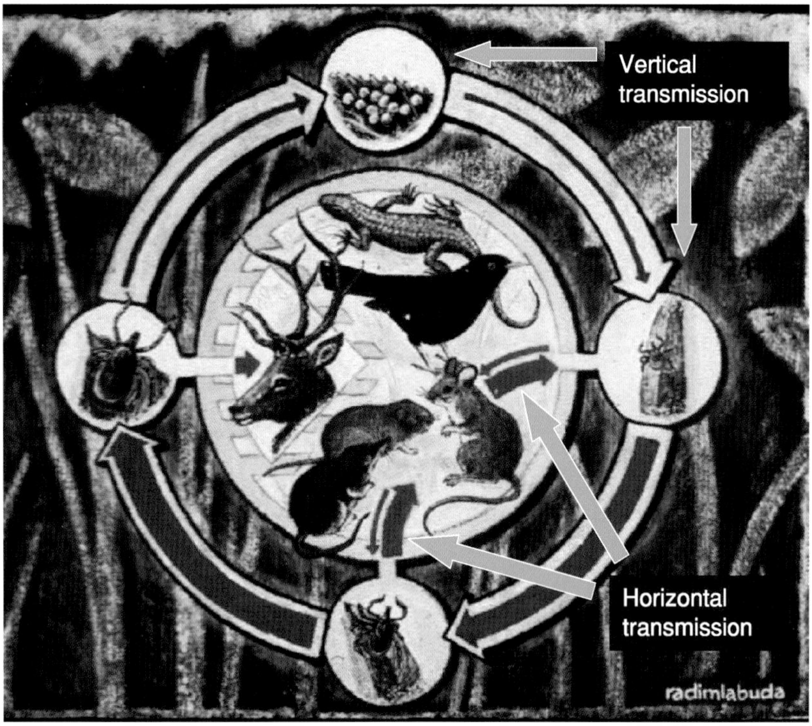

NUTTALL AND LABUDA, FIG 1. Schematic transmission cycle of TBE virus in Europe. Vertical transmission indicates virus transfers from one tick generation to the next by passage from the infected female to the eggs and the emergent larvae. Horizontal transmission, the principal transmission mechanism, indicates transmission from infected tick to uninfected host and vice versa. Width of the arrows within the transmission cycle indicates force of transmission.

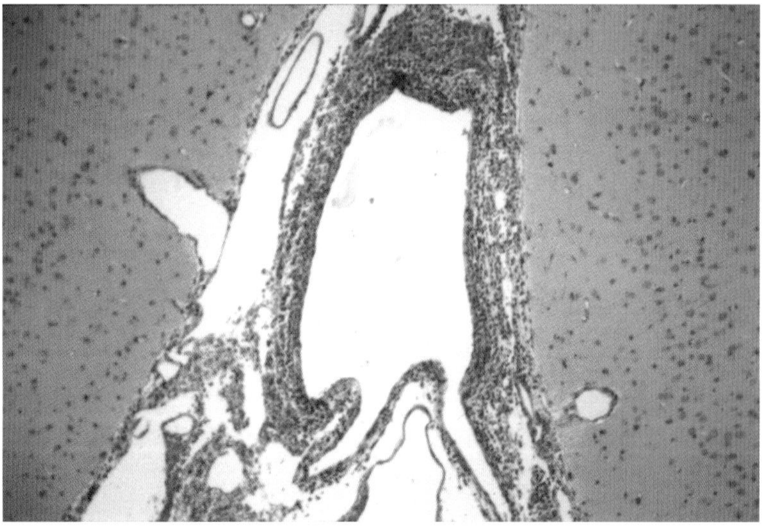

CHAMBERS AND DIAMOND, FIG 3. Yellow fever virus meningoencephalitis in the rhesus monkey showing leptomeningeal accumulation of acute inflammatory cells. Courtesy of the United States Army Medical Research Institute of Infectious Diseases (USAMRIID).

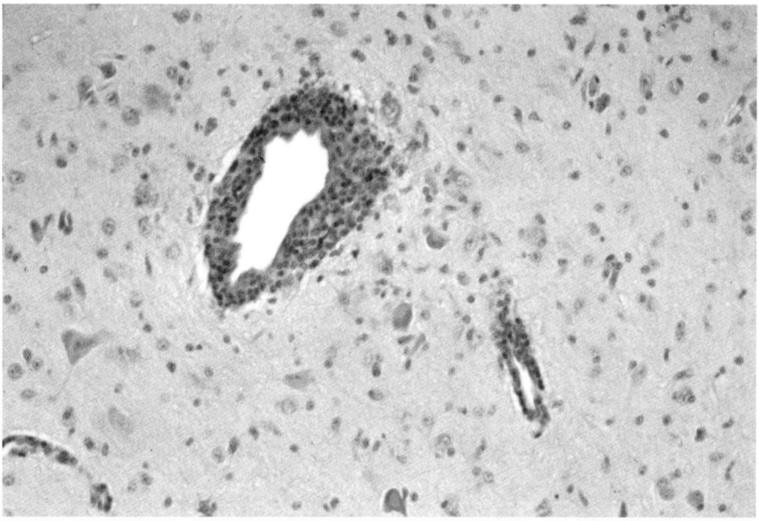

CHAMBERS AND DIAMOND, FIG 4. Yellow fever encephalitis in the rhesus monkey showing focus of perivascular infiltrate with mononuclear cells in the cerebral cortex. Courtesy of USAMRIID.

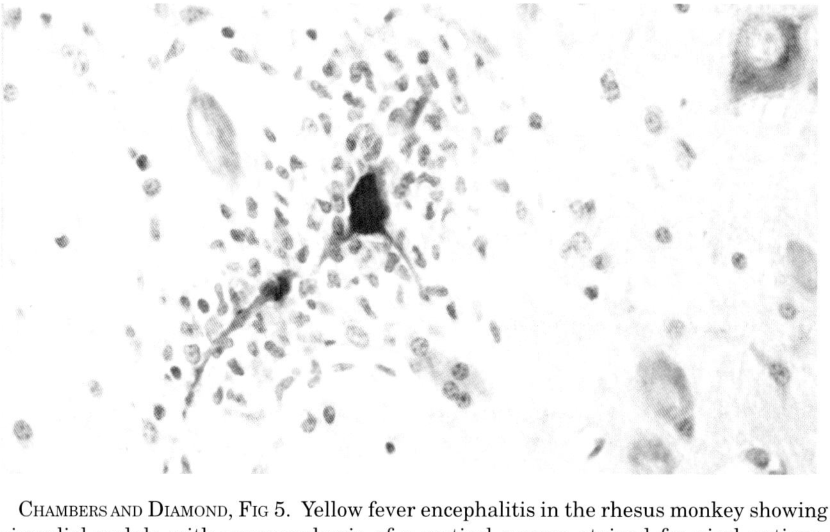

CHAMBERS AND DIAMOND, FIG 5. Yellow fever encephalitis in the rhesus monkey showing microglial nodule with neuronophagia of a cortical neuron stained for viral antigen. Courtesy of USAMRIID.

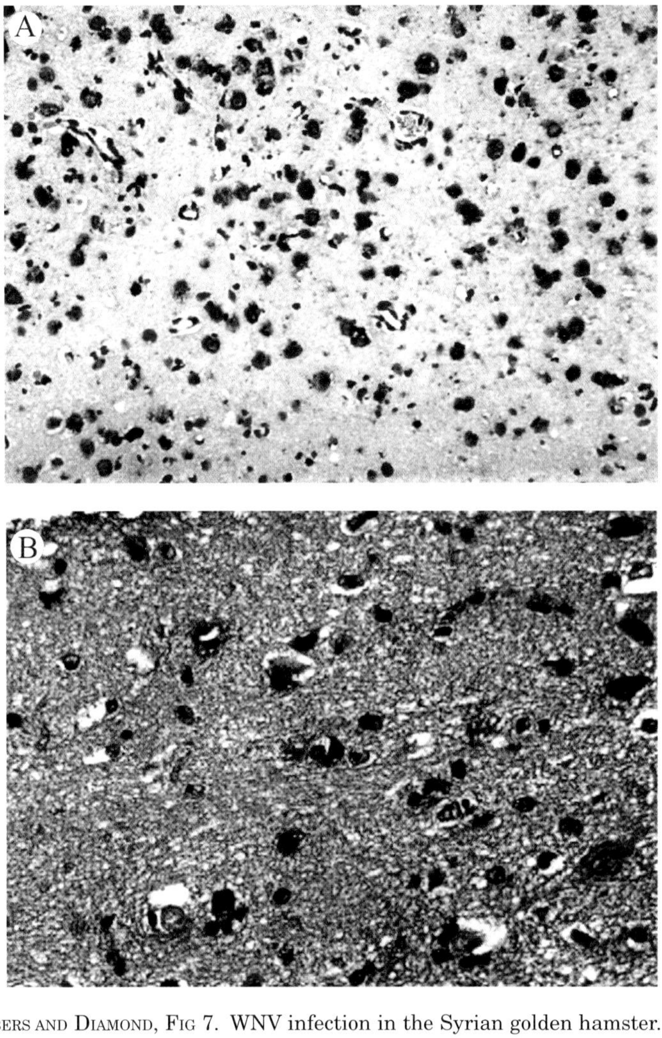

Chambers and Diamond, Fig 7. WNV infection in the Syrian golden hamster. (A) Viral antigen-positive neurons in the cerebral cortex. (B) TUNEL-positive apoptotic neurons in the cortex. Courtesy of Dr. Shu-Yan Xiao. From Xiao *et al.*, 2001.

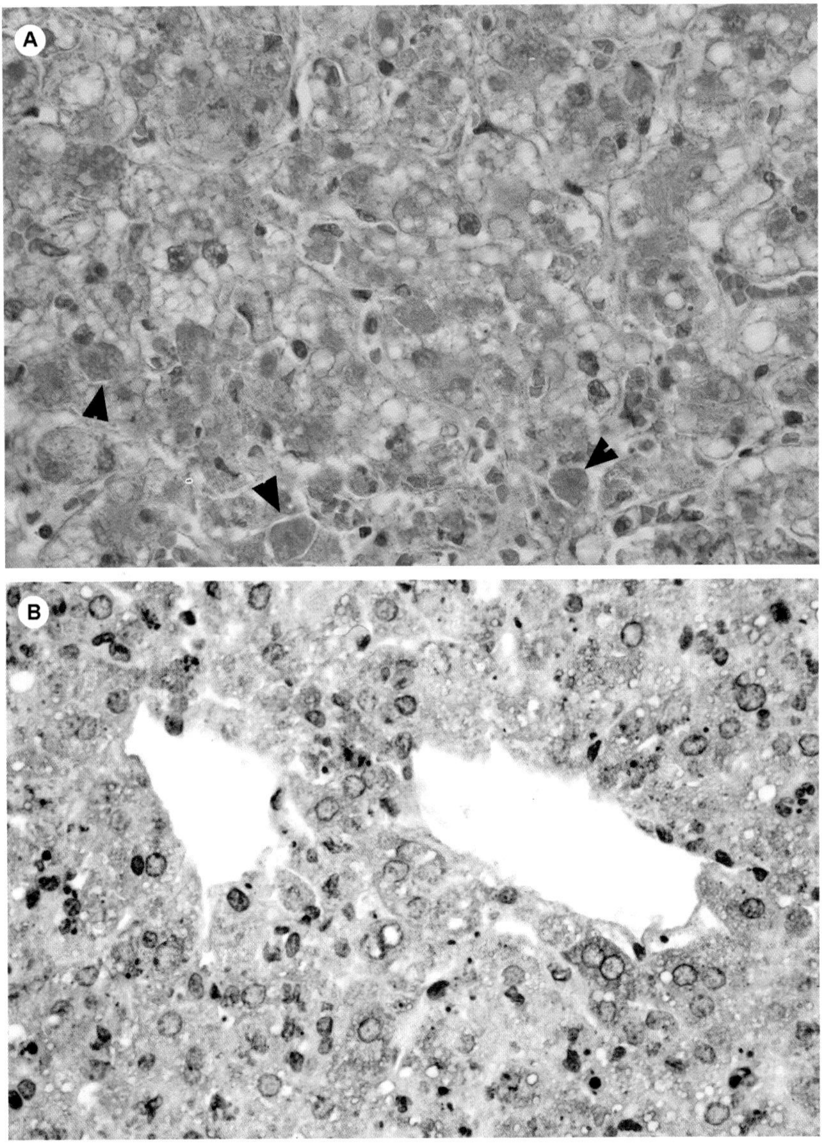

MONATH AND BARRETT, FIG 2. Histopathology of liver from a YF patient. (A) Hepatocytes exhibit cytoplasmic condensation and degeneration, many of which still retain a nucleus. Numerous councilman bodies are present in this field (arrowheads). Residual microvesicular steatosis is evident. Magnification 100×. (B) Immunohistochemical stain demonstrating the YFV antigen in the cytoplasm of many hepatocytes. The midzonal distribution of cell damage and antigen staining is less evident in this patient than in most cases; the figure is meant to illustrate the similarity of pathological effects to that seen in the hamster model (Fig. 3) Magnification 100x. Contributed by Shu-Yuan Xiao, University of Texas Medical Branch, Galveston, TX.

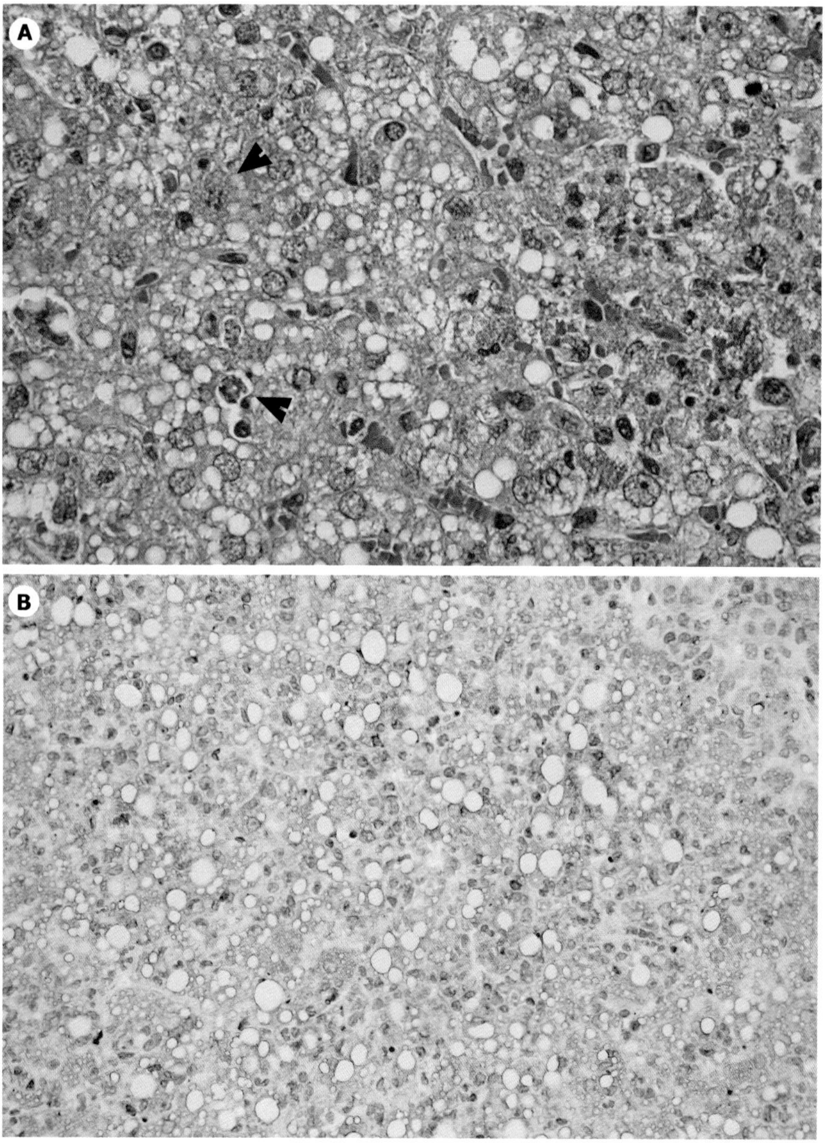

MONATH AND BARRETT, FIG 3. Histopathology of liver from a hamster infected by the Jimenez strain of yellow fever virus (postinfection day 6). (A) Hepatocytes contain small and large lipid droplets in the cytoplasm (vacuoles). Many hepatocytes are undergoing necrapoptosis (arrowhead). Hematoxylin and eosin stain. Magnification 100×. (B) Immunohistochemical staining for YFV antigen using the anti-YFV antibody. Magnification 50×. Contributed by Shu-Yuan Xiao, University of Texas Medical Branch, Galveston, TX.

ISBN: 0-12-039860-5
90000
9 780120 398607